T0295329

Non-Equilibrium Statistical Mechanics

Statistical mechanics provides a framework for relating the properties of macroscopic systems (large collections of atoms, such as in a solid) to the microscopic properties of its parts. However, what happens when macroscopic systems are not in thermal equilibrium, where time is not only a relevant variable, but also essential?

That is the province of nonequilibrium statistical mechanics – there are many ways for systems to be out of equilibrium! The subject is governed by fewer general principles than equilibrium statistical mechanics and consists of a number of different approaches for describing nonequilibrium systems.

Financial markets are analyzed using methods of nonequilibrium statistical physics, such as the Fokker-Planck equation. Any system of sufficient complexity can be analyzed using the methods of nonequilibrium statistical mechanics. The Boltzmann equation is used frequently in the analysis of systems out of thermal equilibrium, from electron transport in semiconductors to modeling the early Universe following the Big Bang.

This book provides an accessible yet very thorough introduction to nonequilibrium statistical mechanics, building on the author's years of teaching experience. Covering a broad range of advanced, extension topics, it can be used to support advanced courses on statistical mechanics, or as a supplementary text for core courses in this field.

Key Features:

- Features a clear, accessible writing style which enables the author to take a sophisticated approach to the subject, but in a way that is suitable for advanced undergraduate students and above.
- Presents foundations of probability theory and stochastic processes and treats principles and basic methods of kinetic theory and time correlation functions.
- Accompanied by separate volumes on thermodynamics and equilibrium statistical mechanics, which can be used in conjunction with this book

James H. Luscombe received a PhD in Physics from the University of Chicago in 1983. After post-doctoral positions at the University of Toronto and Iowa State University, he joined the Research Laboratory of Texas Instruments, where he worked on the development of nanoelectronic devices. In 1994, he joined the Naval Postgraduate School in Monterey, CA, where he is a professor of Physics. He was Chair of the Department of Physics from 2003 – 2009. He teaches a wide variety of topics, including general relativity, statistical mechanics, mathematical methods, and quantum computation. He has published more than 60 research articles, and has given more than 100 conference presentations; he holds 2 patents.

Non-Equilibrium Statistical Mechanics

James H. Luscombe

CRC Press is an imprint of the
Taylor & Francis Group, an **informa** business

First edition published 2025
by CRC Press
2385 NW Executive Center Drive, Suite 320, Boca Raton FL 33431

and by CRC Press
4 Park Square, Milton Park, Abingdon, Oxon, OX14 4RN

CRC Press is an imprint of Taylor & Francis Group, LLC

ISBN: 978-1-138-54295-2 (hbk)
ISBN: 978-1-032-84342-1 (pbk)
ISBN: 978-1-003-51229-5 (ebk)

DOI: 10.1201/9781003512295

Typeset in Nimbus Roman
by KnowledgeWorks Global Ltd.

Publisher's note: This book has been prepared from camera-ready copy provided by the authors.

Contents

Preface

STATISTICAL physics—a core component of university curricula—is taught at three levels: thermodynamics, equilibrium statistical mechanics, and nonequilibrium statistical mechanics.[1]

• Thermodynamics is a science of matter that remarkably presupposes no knowledge of the microscopic constituents of matter—the strength and the weakness of that theory. It introduces absolute temperature T and entropy S as properties of the equilibrium state characterized by the values of *state variables*,[2,3] measurable quantities such as pressure P and volume V. A precise definition of equilibrium is elusive; basically, a system is in equilibrium when none of its macroscopic properties appear to be changing. No system is truly quiescent, however; it's a matter of time scales. Feynman[3, p1] defined equilibrium in temporal terms: "...the state where all the fast things have happened, but the slow things have not," a working definition we'll see exemplified in this book. I have written a companion volume on thermodynamics, [2], which I refer to when specific results from that subject are required. What we can say about macroscopic systems is what we can measure. For any system there are only a handful of state variables, yet there are an enormous number of microscopic components. What we can say therefore represents a projection of huge numbers of microscopic degrees of freedom onto a small number of macro variables. It could almost be said that thermodynamics is true by definition—a phenomenological codification of our experience of the macroscopic world. That the laws of thermodynamics apply to disparate physical systems and can be expressed mathematically compels the search for a unifying set of microscopic principles.

• Statistical mechanics relates the equilibrium properties of macroscopic systems to the nature of their microscopic components. All information about a system's components and their interactions is contained in the system Hamiltonian, H. A central construct is the partition function Z, which for N identical particles is[4]

$$Z(\beta, X) \equiv \frac{1}{N!}\int \mathrm{d}\Gamma\, \mathrm{e}^{-\beta H(p,q;X)} = \frac{1}{N!}\int \mathrm{d}E\, \Omega(E)\mathrm{e}^{-\beta E}\,, \tag{1}$$

where: $\beta \equiv (kT)^{-1}$, with k Boltzmann's constant;[5] $\mathrm{d}\Gamma \equiv \mathrm{d}^N\boldsymbol{p}\mathrm{d}^N\boldsymbol{q}/h^{3N}$ is a dimensionless volume element in N-particle phase space,[6] with h Planck's constant; and X denotes external

[1] Continually distinguishing equilibrium from nonequilibrium statistical mechanics becomes tedious. Henceforth, the unadorned term statistical mechanics will be used to mean the equilibrium theory.

[2] *Italic font* is used in this book for three reasons: for emphasis, for words and phrases not from the English language, and when an important term is first used, engendering an index entry.

[3] State variables (the term is due to Gibbs[1, p2]) represent measurable quantities on macroscopic systems in thermal equilibrium, the state of which is specified by a list of the values of its state variables. The generic term *system* indicates that little (yet macroscopic) part of the universe we consider to be of interest[2, p7]. A correlative of system, *not-system*, is known as the *environment*. *Macroscopic* is another term that eludes precise definition. Even systems indiscernible to human senses (not ordinarily considered macroscopic) contain enormous numbers of atoms, and it's in that sense we use the term macroscopic: when the number of microscopic constituents is sufficiently large that we're unable to track the dynamics of individual particles.

[4] Equation (1) is classical. Appendix A reviews the foundations of statistical mechanics, classical and quantum.

[5] In this book, Boltzmann's constant is denoted k and not k_B.

[6] Familiarity with phase space is assumed, of which we make frequent use. Nolte[4] is a historical review of phase space.

parameters in the Hamiltonian.[7] The quantity $\Omega(E)$ is the density of states function, such that $\Omega(E)\mathrm{d}E$ is the number of energy levels in the range $[E, E + \mathrm{d}E]$. The Boltzmann factor $\mathrm{e}^{-\beta E}$ is proportional to the probability that the system has energy E. The partition function therefore specifies the total number of energy states available to the system at temperature T. The internal energy of a system interacting with its environment is not fixed, but fluctuates about a mean value $\langle H \rangle$ caused by energy transfers with the environment. Thermal equilibrium is not the quiescent place of thermodynamics: Measurable quantities fluctuate in time, and fluctuations motivate the use of probability and ensembles. I have written a companion volume on statistical mechanics, [5], to which I refer when specific results from that subject are required.

• Nonequilibrium—*other than equilibrium*—describes many physical systems because there are many ways *not* to be in equilibrium.[8] Statistical mechanics is a general theory of a special (yet ubiquitous) state of matter, thermal equilibrium; it's built on analytical mechanics (applies to any mechanical system) and the theory of probability (large numbers of microscopic components of macroscopic systems). It's a powerful theory: All thermodynamic quantities can be calculated from a single formula, Eq. (1). In contrast, *there is no comparably general theory of nonequilibrium systems*, because there is no one way to be out of equilibrium. Nonequilibrium statistical mechanics, the goal of which is to predict the spatiotemporal behavior of macroscopic systems given the nature of their microscopic components, consists of a collection of different theoretical approaches. What disparate nonequilibrium systems have in common is an irreversible evolution to thermal equilibrium, the state of maximum entropy consistent with macroscopic constraints,[9] and entropy is created in irreversible processes (those that can't be reversed in time). Time and irreversibility therefore play central roles in nonequilibrium theories. Strictly speaking, time plays no role in thermodynamics and statistical mechanics, although it sneaks in implicitly. Thermodynamics describes the timeless state of thermal equilibrium, yet the *direction* of time appears implicitly through the existence of irreversible processes. Thermodynamics, normally taught as a theory of transitions between equilibrium states through reversible processes,[10] contains the seeds of an extension to nonequilibrium systems through the study of irreversible processes—see Chapter 1. Time appears implicitly in statistical mechanics through the *existence* of fluctuations (forcing us to interpret measurements as average quantities), yet time is obviated (as it must be in an equilibrium theory) when time averages are replaced with ensemble averages. Time-dependent properties of fluctuations—the cornerstone of nonequilibrium statistical mechanics—are covered starting in Chapter 2.

Theories of macroscopic systems can be classified by the extent to which they include time and fluctuations; see Table 0.1. Running through statistical physics is a four-way division: microscopic and macroscopic levels of description of equilibrium and nonequilibrium systems. (One can add subdivisions into classical and quantum treatments.) Thermodynamics is a macroscopic theory of equilibrium that largely ignores fluctuations.[11] Statistical mechanics is a microscopic theory based on the existence of fluctuations *but not on their dynamical properties*. With nonequilibrium statistical mechanics we have a theory in which time and fluctuations play central roles; it supersedes thermodynamics and statistical mechanics.[12]

---

[7]External parameters are set by the environment (or by the experimenter), such as magnetic field strength. Thermodynamics distinguishes *intensive* from *extensive* variables; see [2, p5] and Section 1.2. Equilibrium is achieved when a system's intensive variables associated with conserved quantities become equal to their environmental counterparts[2, p45].

[8]Defining a subject in terms of a negative—what it's *not*—leaves open the question of what it *is*.

[9]Gibbs showed that thermal equilibrium is characterized by two equivalent criteria: entropy is a maximum for fixed energy, and energy is a minimum for fixed entropy[2, p53].

[10]The terms reversible and irreversible are defined in footnote 2 of Chapter 1.

[11]Fluctuations can be included in thermodynamics through the device of *virtual variations*, conceivable variations of state brought about by conceptually relaxing the isolation of systems from their environment[2, p45][5, p24]. Treating fluctuations as virtual processes is similar to the use of virtual displacements in classical mechanics, "mathematical experiments" consistent with existing constraints, but occurring at a fixed time.

[12]If you'll permit the digression, an analogy with relativity theory presents itself. General relativity is a theory of spacetime that supersedes special relativity and Newtonian gravitation. Special relativity is in essence a "boundary condition" on

Table 0.1 Time and fluctuations in theories of macroscopic systems

	Thermodynamics	Equilibrium statistical mechanics	Nonequilibrium statistical mechanics
Role of time	Implicit. Equilibrium theory based on time-invariant state variables. Direction of time established by irreversible processes associated with entropy creation.	Implicit. Equilibrium theory based on instantaneous fluctuations in macroscopic quantities, motivating the use of ensembles and probability. Time averages replaced with ensemble averages.	Explicit. Goal of the theory is to determine the time evolution of macroscopic systems toward thermal equilibrium.
Treatment of fluctuations	Implicit. Modeled as virtual variations of state[2, p45].	Explicit. Equal-time correlations of spatially separated fluctuations, but no dynamical properties.	Explicit. Time-dependent fluctuations modeled as stochastic processes.
Level of description	Macroscopic. Based on a handful of state variables.	Microscopic. Information about system constituents and their interactions enters the theory through the Hamiltonian. Statistical mechanics reproduces the laws of thermodynamics.	Microscopic and macroscopic. Macroscopic models of transport involving hydrodynamics and local thermodynamics. Microscopic models involving kinetic theory and time-correlation functions.

The book is organized in a way that emulates the passage from thermodynamics to statistical mechanics, from macroscopic to microscopic. We begin with the macroscopic theory of irreversibility based on local thermodynamics and hydrodynamic conservation laws obtained from phenomenological balance equations. Later (in Chapter 4) we show that balance equations and hydrodynamics follow from first principles (the Liouville equation). A balance equation for entropy is derived, allowing us to identify the fluxes and thermodynamic forces contributing to entropy creation in irreversible processes. Linear force-flux constitutive relations introduce proportionality factors—kinetic coefficients—which satisfy an experimentally verified symmetry, the Onsager reciprocity conditions. Chapter 2, on the statistical theory of fluctuations, shows that reciprocity is a consequence of the time-reversal invariance of the microscopic dynamics underlying fluctuations and the stationarity of equilibrium averages. We can treat fluctuations as dynamical variables, and classify them by their symmetries under time reversal, yet we have no first-principles way of modeling their time dependence. We use stochastic processes for that purpose which bring time into probabilistic descriptions. Appendix B is a review of probability as needed in statistical mechanics;

general relativity,[6, p24] the theory of spacetime that applies for vanishing gravity. Only in that sense does general relativity need special relativity, to handle a limiting case; general relativity is the more comprehensive theory. In the same way, the more comprehensive theory of nonequilibrium statistical mechanics relies on statistical mechanics to handle the special case of thermal equilibrium. In this sense, statistical mechanics is a boundary condition on nonequilibrium statistical mechanics.

Chapter 2 introduces stochastic processes, mathematics required in nonequilibrium statistical mechanics. Stationary and Markov processes are defined along with random walks, the master equation, and Gaussian processes. Also covered are thermal noise and Nyquist's theorem which provide a segue into the Langevin equation developed in Chapter 3 on Brownian motion and stochastic dynamics. The connection we find (with the Langevin equation) between dissipation and the strength of fluctuations is an instance of a more general result derived in Chapter 6, the fluctuation-dissipation theorem. The Fokker-Planck equation is derived. Whereas the Langevin equation supplies expressions for nonequilibrium averages of observable quantities, the Fokker-Planck equation is a differential equation for nonequilibrium probability distributions; the two are closely related. Chapter 4 is devoted to kinetic theory, a more microscopic approach to irreversibility than stochastic methods. No book on the subject is complete without an extensive treatment of the Boltzmann equation and the H-theorem. There isn't a well-defined boundary between kinetic theory and nonequilibrium statistical mechanics—kinetic theory is more concerned with the time-dependent distribution function and nonequilibrium statistical mechanics tends to emphasize the relation between irreversible processes and spontaneous fluctuations; practitioners of both subjects need to be conversant in the ways of the other. Chapter 5, which touches on the kinetic theory of plasmas, could logically have been included in Chapter 4, yet that chapter is long enough. Plasma kinetic theory is an active field of research that we make no attempt to review; it's basic principles, however, the Landau-Vlasov equations, are readily developed with material already covered in this book. We treat Landau damping, a fascinating, experimentally verified, example of irreversibility without dissipation. Chapter 6 concludes with more modern developments in nonequilibrium statistical mechanics that emphasize time correlation functions—linear response theory, fluctuation-dissipation theorems, Green-Kubo theory of transport coefficients, and the generalized Langevin equation. Nonequilibrium statistical mechanics rests on a three-legged stool of stochastic methods, kinetic theory, and time correlation functions.

Six appendices provide foundations for material used in the book. Appendix A is a review of statistical mechanics, classical and quantum (the subject *can* be succinctly summarized), as well as the topics of von Neumann entropy, the quantum generalization of Gibbs entropy, and the Weyl correspondence principle. Appendix B reviews elementary probability theory. Appendix C reviews two-body elastic collisions, essential in formulating the Boltzmann equation, and Appendix D reviews integral equations, crucial in finding so-called normal solutions of the Boltzmann equation. Appendix E is a review of the dynamical "pictures" of quantum mechanics (Schrödinger, Heisenberg, and interaction representations). Appendix F reviews the mathematics of causality: Kramers-Kronig relations, Titchmarsh's theorem, and Fourier transforms of causal functions.

Physics curricula include courses on thermodynamics at the undergraduate level. Statistical mechanics is taught at the advanced undergraduate/beginning graduate level and is a requirement of graduate physics programs. Nonequilibrium statistical mechanics is an advanced topic. A nodding acquaintance with thermodynamics and statistical mechanics is thereby presumed (which *inter alia* presumes a knowledge of elementary probability theory) as well as the standard topics of physics curricula: analytical mechanics, quantum mechanics, electrodynamics, and mathematical methods. Some exposure to tensor analysis and dyadic notation would be useful. SI units are used throughout. To reduce the complexity of equations, multiple integrals (encountered frequently) are denoted with a single integral sign, e.g., $\int \mathrm{d}\boldsymbol{v}\mathrm{d}\boldsymbol{r} \equiv \int \cdots \int \mathrm{d}v_x \mathrm{d}v_y \mathrm{d}v_z \mathrm{d}x \mathrm{d}y \mathrm{d}z$. One integral sign to rule them all.

I thank the editorial staff at CRC Press, in particular Dr. Danny Kielty and Rebecca Davies, for their forbearance of numerous blown deadlines. I thank my family, for they have seen me too often buried in a computer. My wife Lisa I thank for her encouragement and countless conversations on how not to mangle the English language. Finally, to students, try to remember that science is a "work in progress"; more is unknown than known.

James Luscombe

Monterey, California

CHAPTER 1

Irreversibility, entropy, and fluctuations

We start with the macroscopic theory of irreversibility based on thermodynamics and hydrodynamics; in this way the issues to be addressed by microscopic theories are established. Thermodynamics—the science of thermal equilibrium—contains the seeds of an extension to nonequilibrium systems (as we show), *nonequilibrium* or *irreversible thermodynamics*.[1]

1.1 ENTROPY AND IRREVERSIBILITY, CLAUSIUS INEQUALITY

The Clausius inequality for closed systems[2, p35], $\mathrm{d}S \geq \mathrm{d}Q/T$ (equality for reversible heat transfers),[2] shows that[3] $\mathrm{d}Q/T$ does not account for all contributions to $\mathrm{d}S$. Clausius[8, p363] termed the difference

$$(\mathrm{d}S - \mathrm{d}Q/T) \geq 0 \tag{1.1}$$

the *noncompensated transformation* (a term we avoid), entropy changes not "compensated" by heat transfers. The message is that *entropy is created in irreversible processes*.[4] Thermodynamics as traditionally taught treats transitions between equilibrium states through reversible processes, with (1.1) satisfied as an equality. Nonequilibrium thermodynamics studies entropy creation associated

[1]Either name has issues. Nonequilibrium thermodynamics is an oxymoron (if, as is traditional, thermodynamics is considered a theory of the equilibrium state) and irreversible thermodynamics is redundant—irreversibility is the central message of thermodynamics[2, p138]. The name notwithstanding, de Groot and Mazur[7] is the standard text on the subject.

[2]Thermodynamics distinguishes *closed* from *open* systems and *reversible* from *irreversible* processes. Closed systems allow exchanges of energy with the environment but have a fixed mass; open systems allow exchanges of matter[2, p9]. *Isolated* systems exchange neither energy nor matter. The usual expression $\mathrm{d}S = \mathrm{d}Q/T$ pertains to reversible transformations of closed systems; there is an additional contribution to the entropy of open systems associated with transfers of matter [see Eq. (1.33) and Exercise 1.7]. A process is reversible if it can be exactly reversed through infinitesimal changes in the environment; system *and* environment are restored to their original conditions[2, p14]. In irreversible processes, the system can't be restored to its original state without producing changes in the environment. The concept of reversibility is problematic because it envisions the reversal of processes in *time*. The description of equilibrium states doesn't involve time, yet reversible processes are conceived as proceeding through a succession of equilibrium states linked by infinitesimal reversible transformations. Real changes of state occur at finite rates and thus intermediate states can't rigorously be considered equilibrium states. Strictly speaking, reversible processes are idealizations that don't exist; practically speaking, there is some small yet finite rate at which processes occur such that *disequilibrium* has no observable consequences within experimental uncertainties. All processes are irreversible, *some* processes can be idealized as reversible.

[3]We dispense with the customary practice of distinguishing exact from inexact differentials; see [2, p6].

[4]I. Prigogine (1977 Nobel Prize in Chemistry), referred to the second law of thermodynamics as the *Principle of the Creation of Entropy*[9, p32]. The Clausius inequality derives from Carnot's theorem, that heat-engine efficiencies are bounded by those of reversible engines[2, p28]. A violation of Carnot's theorem is a violation of the Clausius version of the second law[2, p29]. The Clausius inequality is, therefore, another version of the second law, of which there are many equivalent ways of stating[2, p139]. Entropy is conserved in reversible processes; it's created in irreversible processes.

DOI: 10.1201/9781003512295-1

with irreversible processes.[5] The state variables of thermodynamics such as temperature T are, in irreversible thermodynamics, replaced with *fields* that vary in space and time,[6] such as temperature, $T(\boldsymbol{r}, t)$. Nonuniformities in these fields engender irreversible processes; we seek to relate irreversibility to inhomogeneities, including and especially *those created by fluctuations*.

We want irreversible thermodynamics to be a local theory.[7] We seek partial differential equations for the spatiotemporal variations of its fields, akin to Maxwell's equations or fluid dynamics. For that purpose, we rely on *balance equations* that describe the changes in time in the amount of a quantity in a region of space; see Section 1.3. A central achievement of the theory is the derivation of a balance equation for entropy (see Section 1.5), showing it can change for two reasons and two reasons only: flows, described by an entropy flux vector $\boldsymbol{J}_S$, Eq. (1.33), and entropy sources, $\sigma_S \geq 0$, Eq. (1.34), the rate per volume at which entropy is produced in irreversible processes.

1.2 EQUILIBRIUM, GLOBAL AND LOCAL

A fundamental distinction in thermodynamics is between intensive and extensive quantities. Intensive quantities have the same values throughout a system—pressure P, temperature T, or chemical potential[8] μ—and are independent of the size of the system. Extensive quantities are proportional to the number of particles in the system N: entropy[9] S, internal energy U, or magnetization[10] $\boldsymbol{M}$.

Denote the extensive variables of a system *other than entropy* $\{X_i\}_{i=1}^m$; S is a function of these quantities,[11] $S = S(X_1, \ldots, X_m)$. The first law of thermodynamics for a system of n chemical species,[12]

$$\mathrm{d}S = \frac{1}{T}\mathrm{d}U + \frac{P}{T}\mathrm{d}V - \frac{1}{T}\sum_{k=1}^{n} \mu_k \mathrm{d}N_k, \tag{1.2}$$

shows that $S = S(U, V, N_1, \ldots, N_n)$, where N_k is the number of particles of species k with μ_k the chemical potential of the kth-species. An important point is that intensive variables occur in the theory as partial derivatives of extensive variables,[13] e.g., $T^{-1} = (\partial S/\partial U)_{V,\{N_k\}}$.

---

[5] It's said that thermodynamics should be called thermostatics because of the timeless nature of equilibrium. There would be merit to that idea if thermodynamics consisted solely of the study of reversible processes. The second law, which recognizes the existence of irreversibility, prescribes a time order to the states of naturally occurring processes, namely those in which entropy increases. Time as a variable doesn't occur in thermodynamics, yet the *direction* of time does! The term thermostatics obscures the past-and-future relation of states in irreversible processes.

[6] The mathematical "arena" of thermodynamics is *state space*, a mathematical space of the values of state variables, which by definition are their equilibrium values[2, p5]. A state of equilibrium is represented by a point in this space. Nonequilibrium thermodynamics is a field theory with its variables defined on space and time.

[7] Nonlocal, nonequilibrium thermodynamics is an active field of research beyond the intended scope of this book.

[8] Chemical potential is the energy per particle required to add particles to the system under prescribed conditions (constant S, V or constant T, P). See [5, p14].

[9] Entropy is extensive; it scales with the size of the system. Nonextensive entropies lead to a puzzle known as the *Gibbs paradox*, the resolution of which is that entropy must be extensive[2, p113]. In axiomatic formulations of thermodynamics, the extensivity of entropy is taken as a postulate[10, p25].

[10] In thermodynamics, $\boldsymbol{M}$ is the total dipole moment of the system. In electromagnetic theory, $\boldsymbol{M}$ is the magnetization density, magnetic moment per volume. In thermodynamics, $\boldsymbol{M}$ is an extensive quantity; in electromagnetism $\boldsymbol{M}$ is intensive.

[11] The *zeroth law of thermodynamics* implies the existence of functional relations among state variables, *equations of state*[2, p12], of which the entropy function can be considered, $S = S(X_1, \ldots, X_m)$. In axiomatic formulations, entropy as a function of extensive variables is taken as a postulate[10, p24].

[12] Equation (1.2) is variously referred to as the *entropy form* of the first law, or the *combined first and second laws* of thermodynamics. I prefer simply the first law, the first law written in terms of exact differentials. Equation (1.2), with the $\mu\mathrm{d}N$ terms, is sometimes referred to as the Gibbs equation. The first law in the form of Eq. (1.2) is referred to as the *entropy representation*,[2, p53] where $S = S(U, V, \{N_k\})$ is the dependent variable. Writing it in the equivalent form $\mathrm{d}U = T\mathrm{d}S - P\mathrm{d}V + \sum_k \mu_k \mathrm{d}N_k$, with $U = U(S, V, \{N_k\})$ dependent, is the *energy representation*. Thermodynamics can be developed in either way; nonequilibrium thermodynamics mainly uses the entropy representation.

[13] For this reason it's sometimes said that extensive variables are the primary variables of thermodynamics. Through the use of Legendre transformations, intensive variables can be taken as independent variables in thermodynamics. See Chapter 4 of [2], *Thermodynamic potentials: The four ways to say energy*.

Extensive quantities are additive over subsystems.[14] Entropy is the sum of subsystem entropies $S^{(\alpha)}$, $S = \sum_\alpha S^{(\alpha)}$. Subsystem entropies are functions of subsystem extensive variables, $S^{(\alpha)} = S^{(\alpha)}(U^{(\alpha)}, V^{(\alpha)}, N_1^{(\alpha)}, \ldots, N_n^{(\alpha)})$, with $\sum_\alpha U^{(\alpha)} = U$, $\sum_\alpha V^{(\alpha)} = V$, and $\sum_\alpha N_k^{(\alpha)} = N_k$, $k = 1, \ldots, n$. The first law for a subsystem is [same form as Eq. (1.2)]

$$\mathrm{d}S^{(\alpha)} = \frac{1}{T^{(\alpha)}}\mathrm{d}U^{(\alpha)} + \frac{P^{(\alpha)}}{T^{(\alpha)}}\mathrm{d}V^{(\alpha)} - \frac{1}{T^{(\alpha)}}\sum_{k=1}^{n}\mu_k^{(\alpha)}\mathrm{d}N_k^{(\alpha)}, \tag{1.3}$$

where $(T^{(\alpha)})^{-1} = \left(\partial S^{(\alpha)}/\partial U^{(\alpha)}\right)_{V^{(\alpha)},\{N_k^{(\alpha)}\}}$, etc. The subsystems of a system in *global* thermodynamic equilibrium are mutually in equilibrium (zeroth law of thermodynamics), with equality of subsystem intensive variables,[15] $T^{(\alpha)} = T$, $P^{(\alpha)} = P$, and $\mu_k^{(\alpha)} = \mu_k$, $k = 1, \ldots, n$. In global equilibrium, there is no need for superscripts on the local intensive variables in Eq. (1.3).

For systems not rigorously in global equilibrium, we assume that sufficiently small volumes $\mathrm{d}V^{(\alpha)}$ (subsystems) can be found over which variations in state variables are negligible, with Eq. (1.3) holding for each subsystem, the assumption of *local thermodynamic equilibrium.*[16] For systems divided into local-equilibrium subsystems, local entropy functions $S^{(\alpha)} = S(U^{(\alpha)}, V^{(\alpha)}, N_k^{(\alpha)})$ (or local equations of state) *have the same functional form of its variables as the global entropy function, $S = S(U, V, N_k)$, or equation of state.*[17]

Thermodynamics establishes that entropy is created in irreversible processes (Clausius inequality), but nothing in that theory can tell us the *rate* of entropy creation. Nonequilibrium thermodynamics specifies a rate by dividing Eq. (1.3) by $\mathrm{d}t$:

$$T\frac{\partial S}{\partial t} \equiv \frac{\partial U}{\partial t} + P\frac{\partial V}{\partial t} - \sum_{k=1}^{n}\mu_k\frac{\partial N_k}{\partial t}, \tag{1.4}$$

where we've erased subsystem labels with the understanding that *all quantities are local.* Equation (1.4) *defines* the rate of change of entropy in terms of the variations of the quantities on the right side; it does not emerge as a result of the underlying microscopic dynamics. Such an approach can be justified only *a posteriori*, by the validity of the conclusions derived from it. Where do we get the time derivatives on the right of Eq. (1.4)? We can't invoke thermodynamics again. Fortunately for us, the extensive quantities (U, V, N_k) are associated with microscopic properties of matter. *Their rates of change are described by hydrodynamic conservation laws* (see Section 1.4).

Equation (1.4) can be rewritten in terms of *specific*, per-mass quantities. In that way, the $P\mathrm{d}V$ term is expressed as a variation in density ρ, a local system property. Let $s \equiv S/M$, $u \equiv U/M$, and $\rho^{-1} \equiv V/M$ denote specific entropy, internal energy, and volume, with M the total system mass. Thus, ρs and ρu are *per-volume* densities of entropy and internal energy. The mass of species k, M_k, has $N_k = M_k/m_k = Mc_k/m_k$ particles, with m_k the molecular mass and $c_k \equiv M_k/M$ the *mass fraction.* The specific *chemical work*[18] of the kth-species, $\mu_k\mathrm{d}N_k/M = (\mu_k/m_k)\mathrm{d}c_k \equiv \widetilde{\mu}_k\mathrm{d}c_k$, where $\widetilde{\mu}_k = \mu_k/m_k$ denotes the kth specific chemical potential. Thus, we can write Eq. (1.4)

$$T\frac{\partial s}{\partial t} = \frac{\partial u}{\partial t} + P\frac{\partial}{\partial t}\rho^{-1} - \sum_{k=1}^{n}\widetilde{\mu}_k\frac{\partial c_k}{\partial t}. \tag{1.5}$$

[14] Additivity over subsystems requires that the *thermodynamic limit* exists of $N \to \infty$, $V \to \infty$ with N/V fixed[5, p103]. The concepts of extensive and intensive don't apply to systems so small that surface effects dominate[11, p10].

[15] The general theory of thermodynamic equilibrium requires equality between subsystem-intensive variables associated with (*conjugate to*) conserved quantities—energy, volume, and particle number[2, p44].

[16] As noted in the Preface, nonequilibrium statistical physics is a collection of approaches to modeling nonequilibrium phenomena. With the picture of local equilibrium, we've specified a class of nonequilibrium systems to which macroscopic reasoning can be applied.

[17] This follows from the use of continuous functions in nonequilibrium thermodynamics to represent state variables.

[18] The small amount of chemical work (if it's not heat, it's work[2, p10]) $\mu_k\mathrm{d}N_k$ is the energy to change the number of particles of species k by $\mathrm{d}N_k$, holding S, V, and all other mole numbers $\overline{N_k}$ fixed, with $\mu_k = (\partial U/\partial N_k)_{S,V,\overline{N_k}}$[2, p39].

1.3 BALANCE EQUATIONS: FLUXES AND SOURCES

A balance equation for quantity ψ is a relation among integrals,

$$\frac{\mathrm{d}}{\mathrm{d}t}\int_V \rho_\psi \mathrm{d}\boldsymbol{r} = -\oint_S \boldsymbol{J}_\psi \cdot \mathrm{d}\boldsymbol{S} + \int_V \sigma_\psi \mathrm{d}\boldsymbol{r} \tag{1.6}$$

where ρ_ψ is the volume density of ψ, V is a volume bounded by surface S having outward-pointing surface element $\mathrm{d}\boldsymbol{S}$, $\boldsymbol{J}_\psi$ is the flux of ψ through S, and σ_ψ is a *source*, the rate per volume at which ψ is created or destroyed in V. Equation (1.6) indicates that changes in the amount of ψ in V are accounted for by 1) flows through S and 2) the creation or destruction of ψ in V by means other than flow; *there are no other possibilities.*[19] Balance equations do not presume thermal equilibrium.

By applying the divergence theorem, we have the *local form* of balance equations,

$$\frac{\partial}{\partial t}\rho_\psi + \boldsymbol{\nabla}\cdot\boldsymbol{J}_\psi = \sigma_\psi. \tag{1.7}$$

If $\sigma_\psi = 0$, ψ is said to be *conserved*; in that case, the only way ψ in V can change is to be transported through S. When ψ is carried by a fluid, *convective transport*, $\boldsymbol{J}_\psi = \rho_\psi \boldsymbol{v}$, where $\boldsymbol{v}$ is the fluid velocity. In *diffusive transport*, fluxes are proportional to gradients in system parameters Q_ψ, $\boldsymbol{J}_\psi \propto \boldsymbol{\nabla} Q_\psi$; see examples in Section 1.6. We now derive balance equations for mass, momentum, and energy;[20] readers uninterested in the details could skip to Section 1.5.

1.4 HYDRODYNAMIC CONSERVATION LAWS

1.4.1 Mass conservation, convective derivative

The balance equation for a mass of chemical species k convectively transported is, from Eq. (1.6),

$$\frac{\mathrm{d}}{\mathrm{d}t}\int_V \rho_k \mathrm{d}\boldsymbol{r} = -\oint_S \rho_k \boldsymbol{v}_k \cdot \mathrm{d}\boldsymbol{S} + \sum_{j=1}^{r}\int_V \nu_{kj} J_j \mathrm{d}\boldsymbol{r}, \tag{1.8}$$

where ρ_k is the mass density of species k, $\boldsymbol{v}_k$ is its velocity, J_j is the rate at which the jth-chemical reaction occurs,[21] and $\nu_{kj} J_j$ is the rate at which species k is produced in reaction j. The quantity ν_{kj} is the *stoichiometric coefficient* measured in units of mass density of species k in reaction j.

Example: Consider the chemical reaction $2\mathrm{H}_2 + \mathrm{O}_2 \longrightarrow 2\mathrm{H}_2\mathrm{O}$. This is a balanced chemical equation: The number of atoms of each element among the *reactants* (substances undergoing the reaction) is the same as the number of atoms among the *products* (substances created by the reaction). Chemical reactions can be written $\sum_i \nu_i^0 X_i = 0$, where X_i is the chemical symbol of the ith species and ν_i^0 is the bare stoichiometric coefficient, counted as positive for products and negative for reactants: $\nu_{\mathrm{H}_2}^0 = -2$, $\nu_{\mathrm{O}_2}^0 = -1$, $\nu_{\mathrm{H}_2\mathrm{O}}^0 = 2$. The quantities ν_{kj} in Eq. (1.8) are expressed in units of the *density* of species k appearing in reaction j; thus $\nu_{kj} \propto m_k \nu_{kj}^0$, with m_k the molecular mass. In the reaction, two H_2 molecules combine with an O_2 molecule to produce two $\mathrm{H}_2\mathrm{O}$ molecules. In terms of mass,[22] $2m_{\mathrm{H}_2} + m_{\mathrm{O}_2} = 2m_{\mathrm{H}_2\mathrm{O}}$. Whatever is the unit of density adopted in Eq. (1.8) [grams/cc or kg/(cubic mile)], we can assign (using the molecular mass of the species involved) $\nu_{\mathrm{H}_2} = -\frac{1}{9}$, $\nu_{\mathrm{O}_2} = -\frac{8}{9}$, and $\nu_{\mathrm{H}_2\mathrm{O}} = 1$. For every density unit of $\mathrm{H}_2\mathrm{O}$ produced, $\frac{1}{9}$ of a density unit of H_2 disappears, along with $\frac{8}{9}$ of a density unit of O_2.

[19]We've introduced balance equations phenomenologically, per our experience of the macroscopic world. We show in Chapter 4 that balance equations follow from first principles of physics.

[20]Balance equations for mass and energy-momentum comprise five scalar equations. In relativity theory, mass, energy, and momentum are interrelated such that there are four equations expressing conservation of energy-momentum[6, p181].

[21]No confusion should arise between J for the rate of a chemical reaction and the symbol $\boldsymbol{J}$ for current density.

[22]Mass changes due to energies released in chemical reactions ($E = mc^2$) are negligibly small.

Equation (1.8) implies the local balance equation for each species $k = 1, \ldots, n$,

$$\frac{\partial \rho_k}{\partial t} + \boldsymbol{\nabla} \cdot \rho_k \boldsymbol{v}_k = \sum_{j=1}^{r} \nu_{kj} J_j. \tag{1.9}$$

Because mass is conserved in chemical reactions, for each $j = 1, \ldots, r$,

$$\sum_{k=1}^{n} \nu_{kj} = 0 \tag{1.10}$$

(in the aforementioned example, $1 - \frac{1}{9} - \frac{8}{9} = 0$). Summing Eq. (1.9) over k and using Eq. (1.10), we have the *continuity equation* for total mass conservation:

$$\frac{\partial \rho}{\partial t} + \boldsymbol{\nabla} \cdot \rho \boldsymbol{v} = 0, \tag{1.11}$$

where $\rho \equiv \sum_k \rho_k$ is the total density and $\boldsymbol{v} \equiv \sum_k \rho_k \boldsymbol{v}_k / \rho$ is the center-of-mass velocity.

Formulas involving convected quantities can be compactly expressed when the total time derivative is written as the *convective derivative* (see [12, p73]),

$$D_w \equiv \frac{\partial}{\partial t} + \boldsymbol{w} \cdot \boldsymbol{\nabla}. \tag{1.12}$$

Consider the spatiotemporal change of a vector field $\boldsymbol{A}(\boldsymbol{r}, t)$. For small $(\mathrm{d}t, \mathrm{d}\boldsymbol{r})$, $\boldsymbol{A}(\boldsymbol{r}+\mathrm{d}\boldsymbol{r}, t+\mathrm{d}t) \approx \boldsymbol{A}(\boldsymbol{r}, t) + \mathrm{d}t\,(\partial \boldsymbol{A}/\partial t) + (\mathrm{d}\boldsymbol{r} \cdot \boldsymbol{\nabla})\,\boldsymbol{A}$; by forming the difference quotient, we have in the limit $\mathrm{d}t \to 0$, $\partial \boldsymbol{A}/\partial t + (\boldsymbol{w} \cdot \boldsymbol{\nabla})\,\boldsymbol{A} \equiv D_w \boldsymbol{A}$, with $\boldsymbol{w} \equiv \mathrm{d}\boldsymbol{r}/\mathrm{d}t$. In terms of this operator, the continuity equation has the form

$$D_v \rho = -\rho \boldsymbol{\nabla} \cdot \boldsymbol{v}. \tag{1.13}$$

If the local divergence $\boldsymbol{\nabla} \cdot \boldsymbol{v} < 0$ (> 0), fluid is being compressed (expanded) as it flows; if $\boldsymbol{\nabla} \cdot \boldsymbol{v} = 0$, the fluid is *incompressible*. Sometimes it's easier to write the continuity equation as

$$D_v\,(1/\rho) = \frac{1}{\rho} \boldsymbol{\nabla} \cdot \boldsymbol{v}. \tag{1.14}$$

An important concept is *diffusion flow*,[23] $\boldsymbol{J}_k^{\mathrm{d}} \equiv \rho_k\,(\boldsymbol{v}_k - \boldsymbol{v})$, the flux of a species relative to the center of mass velocity. Diffusion flows contribute to the entropy of open systems; see Eq. (1.33). Note that $\sum_k \boldsymbol{J}_k^{\mathrm{d}} = 0$; the total mass flux in the center of mass frame is zero (implying that only $(n-1)$ of the diffusion flows are independent). For $k = 1, \ldots, n$,

$$D_v \rho_k = -\rho_k \boldsymbol{\nabla} \cdot \boldsymbol{v} - \boldsymbol{\nabla} \cdot \boldsymbol{J}_k^{\mathrm{d}} + \sum_{j=1}^{r} \nu_{kj} J_j. \tag{1.15}$$

Summing Eq. (1.15) over k results in Eq. (1.13). Equation (1.15) simplifies with the mass fractions $c_k \equiv \rho_k / \rho$:

$$\rho D_v c_k = -\boldsymbol{\nabla} \cdot \boldsymbol{J}_k^{\mathrm{d}} + \sum_{j=1}^{r} \nu_{kj} J_j. \tag{1.16}$$

The mass fractions sum to unity, $\sum_k c_k = 1$; the sum of Eq. (1.16) over k vanishes. Only $(n-1)$ of the mass fractions are independent.

---

[23] Not the same as the gradient-driven, diffusive transport defined on page 4, $\boldsymbol{J} \propto \boldsymbol{\nabla} Q$.

1.4.2 Momentum balance, stress tensor

The balance equation for momentum is

$$\underbrace{\frac{\mathrm{d}}{\mathrm{d}t}\int_V \rho\boldsymbol{v}\mathrm{d}\boldsymbol{r}}_{\text{rate of change of momentum in } V} = \underbrace{-\oint_S \rho\boldsymbol{v}\boldsymbol{v}\cdot\mathrm{d}\boldsymbol{S}}_{\text{momentum convected through } S} + \underbrace{\oint_S \mathbf{T}\cdot\mathrm{d}\boldsymbol{S}}_{\text{momentum produced by internal forces at } S} + \underbrace{\sum_{k=1}^{n}\int_V \rho_k\boldsymbol{F}_k\mathrm{d}\boldsymbol{r}}_{\text{momentum produced by external forces}} . \tag{1.17}$$

The first term on the right accounts for the flow of momentum through S;[24] the other terms represent momentum sources—forces—those coupling to S and those coupling to V ($\boldsymbol{F}_k$ is an external force per mass coupling to species k). The *stress tensor* $\mathbf{T}$ in Eq. (1.17) represents the stresses[25,26] associated with forces acting at surfaces, $T_{ij} \equiv \lim_{\delta S_j \to 0}(\delta F_i/\delta S_j)$, the ith-component of force acting on the jth-component of surface area when surface area is represented as an outwardly oriented vector. Equation (1.17) written in terms of vector components is:

$$\int_V \frac{\partial \rho v_i}{\partial t}\mathrm{d}\boldsymbol{r} = -\oint_S \rho v_i \boldsymbol{v}\cdot\mathrm{d}\boldsymbol{S} + \sum_{j=1}^{3}\oint_S T_{ij}\mathrm{d}S_j + \sum_{k=1}^{n}\int_V \rho_k F_{k,i}\mathrm{d}\boldsymbol{r}. \qquad (i=1,2,3)$$

The form of $\mathbf{T}$ depends on the nature of the internal forces of the system. For systems with no internal torques (assumed here), $\mathbf{T}$ is symmetric,[27] $T_{ij} = T_{ji}$. A pressure P acting normally at surfaces, regardless of the orientation of the surface, is termed the *hydrostatic pressure*, with $T_{ij} = -P\delta_{ij}$. Shear stresses are modeled with the *viscous stress tensor* $\mathbf{\Pi}$, with $T_{ij} = -P\delta_{ij} + \Pi_{ij}$. The most general second-rank tensor such that in uniform rotation there is no friction due to viscosity has the form[15, p48] (we denote components of the gradient vector, $\nabla_i \equiv \partial/\partial x_i$)

$$\Pi_{ij} = \eta\big(\nabla_j v_i + \nabla_i v_j - \tfrac{2}{3}\delta_{ij}\boldsymbol{\nabla}\cdot\boldsymbol{v}\big) + \zeta\delta_{ij}\boldsymbol{\nabla}\cdot\boldsymbol{v}, \tag{1.18}$$

where $\eta > 0, \zeta > 0$ are phenomenological viscosity coefficients. The total stress tensor is written

$$\mathbf{T} = -P\mathbf{I} + \mathbf{\Pi}, \tag{1.19}$$

where $\mathbf{I}$ is the unit tensor with elements $I_{ij} = \delta_{ij}$. We return to $\mathbf{\Pi}$ in Section 4.4.4.

From Eq. (1.17) we have the equation of local momentum balance,

$$\frac{\partial}{\partial t}(\rho\boldsymbol{v}) + \boldsymbol{\nabla}\cdot(\rho\boldsymbol{v}\boldsymbol{v}) \equiv \rho D_v\boldsymbol{v} = \boldsymbol{\nabla}\cdot\mathbf{T} + \sum_{k=1}^{n}\rho_k\boldsymbol{F}_k = -\boldsymbol{\nabla}P + \boldsymbol{\nabla}\cdot\mathbf{\Pi} + \sum_{k=1}^{n}\rho_k\boldsymbol{F}_k, \tag{1.20}$$

where the divergence of a second-rank tensor—a vector—has components,

$$[\boldsymbol{\nabla}\cdot\mathbf{T}]_i \equiv \sum_j \nabla_j T_{ij}. \tag{1.21}$$

Using Eq. (1.18), Eq. (1.20) implies Newton's second law for fluids, the *Navier-Stokes equation*,

$$\rho D_v\boldsymbol{v} = -\boldsymbol{\nabla}P + \eta\nabla^2\boldsymbol{v} + \big(\zeta + \tfrac{1}{3}\eta\big)\,\boldsymbol{\nabla}(\boldsymbol{\nabla}\cdot\boldsymbol{v}) + \sum_{k=1}^{n}\rho_k\boldsymbol{F}_k. \tag{1.22}$$

---

[24] We've used *dyadic notation* (see [13, Chapter 11] or Morse and Feshbach[14, pp54–92]), a notation for tensors due to Gibbs involving the juxtaposition of vectors with $(\boldsymbol{v}\boldsymbol{v})_{ij} \equiv v_i v_j$.

[25] Stress has the dimension of force per area with SI units of Pascals (Pa).

[26] Sometimes $\rho\boldsymbol{v}\boldsymbol{v}$ is included in the stress tensor[15, p47]. The stress tensor is used in fluid mechanics and the theory of elasticity. The *negative* of the stress tensor, the *pressure tensor* $\mathbf{P} \equiv -\mathbf{T}$ is used in kinetic theory (see Chapter 4). Bold Roman notation $\mathbf{T}$ refers to tensors as basis-independent objects. Just as we distinguish vectors $\boldsymbol{A}$ (bold italic font) from their basis-dependent components A_i, we must distinguish tensors $\mathbf{T}$ from their components in a particular basis, T_{ij}.

[27] Internal angular momentum is conserved in systems with symmetric stress tensors (see Exercise 4.19) and hydrodynamics normally makes use of symmetric stress tensors. There are materials, however, requiring an antisymmetric part of the stress tensor, such as nematic liquid crystals[16, p155]. Second-rank tensors can always be decomposed into symmetric and antisymmetric parts, $T_{ij} = \frac{1}{2}(T_{ij} + T_{ji}) + \frac{1}{2}(T_{ij} - T_{ji})$.

1.4.3 Energy conservation, heat flux

We develop balance equations for the center-of-mass kinetic energy and the potential energy associated with external forces, energies we refer to as mechanical. We then consider internal energy, the energies of thermal agitation and molecular interactions.[28]

Mechanical energy

A work-energy theorem follows by projecting the momentum balance equation (1.20) onto the center-of-mass velocity $\boldsymbol{v}$ and using the results of Exercise 1.1:

$$\boldsymbol{v}\cdot\left[\frac{\partial}{\partial t}(\rho\boldsymbol{v})+\boldsymbol{\nabla}\cdot(\rho\boldsymbol{v}\boldsymbol{v})\right]=\frac{\partial}{\partial t}\left(\tfrac{1}{2}\rho v^2\right)+\boldsymbol{\nabla}\cdot\left(\tfrac{1}{2}\rho v^2\boldsymbol{v}\right)=\boldsymbol{v}\cdot\left[\boldsymbol{\nabla}\cdot\mathbf{T}+\sum_{k=1}^{n}\rho_k\boldsymbol{F}_k\right]. \quad (1.23)$$

Kinetic energy is not conserved: The source term on the right of Eq. (1.23) is the rate of work done by internal and external forces.[29] For future reference, the scalar $\boldsymbol{v}\cdot\boldsymbol{\nabla}\cdot\mathbf{T}$ can be written

$$\boldsymbol{v}\cdot[\boldsymbol{\nabla}\cdot\mathbf{T}]\equiv\sum_{ij=1}^{3}v_i\nabla_jT_{ij}=\sum_{ij=1}^{3}\nabla_j\left(v_iT_{ij}\right)-\sum_{ij=1}^{3}T_{ij}\nabla_jv_i\equiv\boldsymbol{\nabla}\cdot[\boldsymbol{v}\cdot\mathbf{T}]-\mathbf{T}{:}\boldsymbol{\nabla}\boldsymbol{v}, \quad (1.24)$$

where the double-dot notation $\mathbf{A}{:}\mathbf{B}\equiv\sum_{ij}A_{ij}B_{ij}$ indicates a contraction of second-rank tensors to produce a scalar, a generalization of the dot-product notation of vectors $\boldsymbol{A}\cdot\boldsymbol{B}$ (first-rank tensors).

Assume conservative external forces (per mass) derivable from potential functions, $\boldsymbol{F}_k = -\boldsymbol{\nabla}\Phi_k$ (so that $\rho_k\Phi_k$ is an energy density), with $\partial\Phi_k/\partial t = 0$. The total potential energy associated with external forces $\rho\Phi\equiv\sum_{k=1}^{n}\rho_k\Phi_k$ satisfies the equation of motion[30] [use Eq. (1.9)],

$$\frac{\partial}{\partial t}(\rho\Phi)=\sum_{k=1}^{n}\Phi_k\frac{\partial\rho_k}{\partial t}=-\sum_{k=1}^{n}\Phi_k\boldsymbol{\nabla}\cdot(\rho_k\boldsymbol{v}_k)+\sum_{j=1}^{r}J_j\underbrace{\sum_{k=1}^{n}\Phi_k\nu_{kj}}_{\to 0}. \quad (1.25)$$

The last term vanishes if potential energy is conserved in chemical reactions, $\sum_k\Phi_k\nu_{kj}=0$ (assumed). Using $\boldsymbol{\nabla}\cdot(\Phi_k\rho_k\boldsymbol{v}_k)=\Phi_k\boldsymbol{\nabla}\cdot(\rho_k\boldsymbol{v}_k)+\rho_k\boldsymbol{v}_k\cdot\boldsymbol{\nabla}\Phi_k$, we have from Eq. (1.25)

$$\frac{\partial}{\partial t}(\rho\Phi)+\boldsymbol{\nabla}\cdot\left(\sum_{k=1}^{n}\Phi_k\rho_k\boldsymbol{v}_k\right)=-\sum_{k=1}^{n}\rho_k\boldsymbol{v}_k\cdot\boldsymbol{F}_k=-\sum_{k=1}^{n}\boldsymbol{J}_k^{\mathrm{d}}\cdot\boldsymbol{F}_k-\boldsymbol{v}\cdot\sum_{k=1}^{n}\rho_k\boldsymbol{F}_k. \quad (1.26)$$

The loss of potential energy associated with the work done by diffusion flows against external forces, $-\sum_k\boldsymbol{J}_k^{\mathrm{d}}\cdot\boldsymbol{F}_k$, appears as a gain in the internal energy of the system; see Eq. (1.31). The loss of potential energy associated with the power expended by external forces, $-\boldsymbol{v}\cdot\sum_{k=1}^{n}\rho_k\boldsymbol{F}_k$, appears in Eq. (1.23) as a kinetic energy gain. The source terms in Eq. (1.26) represent conversions of potential energy into internal energy and center-of-mass kinetic energy.

Adding Eqs. (1.23) and (1.26), we have a balance equation for mechanical energy,

$$\begin{aligned}\frac{\partial}{\partial t}\left(\tfrac{1}{2}\rho v^2+\rho\Phi\right)&=-\boldsymbol{\nabla}\cdot\left[\tfrac{1}{2}\rho v^2\boldsymbol{v}+\sum_{k=1}^{n}\Phi_k\rho_k\boldsymbol{v}_k-\boldsymbol{v}\cdot\mathbf{T}\right]-\sum_{k=1}^{n}\boldsymbol{J}_k^{\mathrm{d}}\cdot\boldsymbol{F}_k-\mathbf{T}{:}\boldsymbol{\nabla}\boldsymbol{v}\\&\equiv-\boldsymbol{\nabla}\cdot\boldsymbol{J}_{\text{mech}}+\sigma_{\text{mech}},\end{aligned}$$

[28]Don't be misled—all energies here are mechanical in origin; the distinction with internal energy is traditional in thermodynamics.

[29]The divergence of $\mathbf{T}$ is a force density.

[30]The term *equation of motion* is used generically to indicate the time derivative of a quantity (a common practice in theoretical physics); it's not necessarily intended as its meaning in classical mechanics.

with

$$\boldsymbol{J}_{\text{mech}} = \tfrac{1}{2}\rho v^2 \boldsymbol{v} + \sum_{k=1}^{n} \Phi_k \rho_k \boldsymbol{v}_k - \boldsymbol{v}\cdot\mathbf{T}$$

$$\sigma_{\text{mech}} = -\sum_{k=1}^{n} \boldsymbol{J}_k^{\text{d}} \cdot \boldsymbol{F}_k - \mathbf{T}{:}\nabla\boldsymbol{v}. \tag{1.27}$$

Internal energy

Mechanical energy is not conserved:[31] $\sigma_{\text{mech}} \neq 0$. We have to account for internal energy, the kinetic energy of velocities relative to the center-of-mass velocity (known as *thermal motions*) and the potential energy of intermolecular forces. Working with specific quantities, define the total energy density ρe as the sum of mechanical and internal energy densities,[32] $\rho e \equiv \rho\left(\frac{1}{2}v^2 + \Phi + u\right)$. We "want" energy conservation, $\partial(\rho e)/\partial t + \nabla\cdot\boldsymbol{J}_E = 0$, where $\boldsymbol{J}_E$ is to be determined. If mechanical energy isn't conserved, neither is internal energy, implying a balance equation,

$$\frac{\partial \rho u}{\partial t} + \nabla\cdot\boldsymbol{J}_U = \sigma_U, \tag{1.28}$$

where $\boldsymbol{J}_U$ and σ_U are to be determined. Combine these definitions,

$$\frac{\partial \rho e}{\partial t} + \nabla\cdot\boldsymbol{J}_E = \frac{\partial \rho u}{\partial t} + \frac{\partial}{\partial t}\rho\left(\tfrac{1}{2}v^2 + \Phi\right) + \nabla\cdot\boldsymbol{J}_E = \sigma_U + \sigma_{\text{mech}} + \nabla\cdot\left(\boldsymbol{J}_E - \boldsymbol{J}_{\text{mech}} - \boldsymbol{J}_U\right).$$

Total energy conservation is achieved theoretically by choosing

$$\sigma_U = -\sigma_{\text{mech}} \qquad \boldsymbol{J}_E = \boldsymbol{J}_{\text{mech}} + \boldsymbol{J}_U. \tag{1.29}$$

The rate at which mechanical energy is lost is balanced by the rate at which internal energy is gained. That leaves $\boldsymbol{J}_U$ undetermined. Allowing for convection of internal energy, take

$$\boldsymbol{J}_U \equiv \rho u \boldsymbol{v} + \boldsymbol{J}_Q, \tag{1.30}$$

which passes the buck to $\boldsymbol{J}_Q$, the *heat flux.*[33] We then have a balance equation for internal energy

$$\frac{\partial \rho u}{\partial t} + \nabla\cdot(\rho u\boldsymbol{v}) \equiv \rho D_v u = -\nabla\cdot\boldsymbol{J}_Q - P\nabla\cdot\boldsymbol{v} + \mathbf{\Pi}{:}\nabla\boldsymbol{v} + \sum_{k=1}^{n}\boldsymbol{J}_k^{\text{d}}\cdot\boldsymbol{F}_k. \tag{1.31}$$

1.5 ENTROPY SOURCES: FLUXES AND THERMO-FORCES

We're now in a position to assemble a balance equation for entropy. Using Eqs. (P1.1) and (P1.2),

$$\frac{\partial(\rho s)}{\partial t} + \nabla\cdot(\rho s\boldsymbol{v}) = \rho D_v s = \frac{\rho}{T}\left(T D_v s\right) = \frac{1}{T}\left[\rho D_v u + \rho P D_v \rho^{-1} - \rho\sum_{k=1}^{n}\widetilde{\mu}_k D_v c_k\right] \tag{1.32}$$

$$= \frac{1}{T}\left[-\nabla\cdot\boldsymbol{J}_Q + \mathbf{\Pi}{:}\nabla\boldsymbol{v} + \sum_{k=1}^{n}\boldsymbol{J}_k^{\text{d}}\cdot\boldsymbol{F}_k + \sum_{k=1}^{n}\widetilde{\mu}_k\nabla\cdot\boldsymbol{J}_k^{\text{d}} + \sum_{j=1}^{r} J_j A_j\right],$$

where we've used Eqs. (1.31), (1.14), and (1.16), and where $A_j \equiv -\sum_{k=1}^{n}\widetilde{\mu}_k\nu_{kj}$ denotes the *chemical affinity.*[34] Affinities play the role of driving forces in chemical reactions[2, p89]. In equilibrium, $A_j = 0$. Note the absence of pressure P in the final equality of Eq. (1.32): Hydrostatic pressure (which is isotropic) *makes no contribution to entropy creation.*[35]

---

[31] Thermodynamics pertains to mechanically nonconservative systems[2, p9].

[32] No confusion should arise between e as the total energy per mass and the common notation for electrical charge.

[33] Equation (1.30) is phenomenological; convection has been put in "by hand" with no way of specifying $\boldsymbol{J}_Q$. We show in Section 4.4.2 that internal energy convection occurs from first principles and we derive a microscopic expression for $\boldsymbol{J}_Q$.

[34] Chemical affinities have the dimension energy density. The term *affinity* is used in different ways in physics and chemistry—its use is not unambiguous. We use the term as introduced by De Donder and Van Rysselberghe[17] to represent Clausius's noncompensated heat associated with irreversible chemical reactions.

[35] Cosmology students take note. The Friedmann equations predict a reversible, isentropic expansion of the universe[6, p351]. Pressure contributes to gravitation through the equivalent mass density P/c^2, but not to entropy creation.

Using the identities

$$\nabla \cdot \left(\frac{\boldsymbol{J}_Q}{T}\right) = \frac{1}{T}\nabla \cdot \boldsymbol{J}_Q + \boldsymbol{J}_Q \cdot \nabla\left(\frac{1}{T}\right)$$

$$\nabla \cdot \left(\widetilde{\mu}_k \frac{\boldsymbol{J}_k^{\mathrm{d}}}{T}\right) = \frac{\widetilde{\mu}_k}{T}\nabla \cdot \boldsymbol{J}_k^{\mathrm{d}} + \boldsymbol{J}_k^{\mathrm{d}} \cdot \nabla\left(\frac{\widetilde{\mu}_k}{T}\right),$$

Eq. (1.32) can be put into the form of a balance equation, $\partial(\rho s)/\partial t + \nabla \cdot \boldsymbol{J}_S = \sigma_S$, with

$$\boldsymbol{J}_S \equiv \rho s \boldsymbol{v} + \frac{1}{T}\boldsymbol{J}_Q - \frac{1}{T}\sum_{k=1}^{n} \widetilde{\mu}_k \boldsymbol{J}_k^{\mathrm{d}} \tag{1.33}$$

$$\sigma_S \equiv \frac{1}{T}\sum_{j=1}^{r} J_j A_j + \boldsymbol{J}_Q \cdot \nabla\left(\frac{1}{T}\right) + \sum_{k=1}^{n} \boldsymbol{J}_k^{\mathrm{d}} \cdot \left[\frac{\boldsymbol{F}_k}{T} - \nabla\left(\frac{\widetilde{\mu}_k}{T}\right)\right] + \frac{1}{T}\nabla \boldsymbol{v} : \boldsymbol{\Pi}. \tag{1.34}$$

Equation (1.34), *an explicit formula for entropy sources*, is a significant achievement. In thermodynamics, heat is the difference between changes in internal energy and the work done on or by the system,[2, p8] $Q \equiv \Delta U - W$. The formulas for $\boldsymbol{J}_S$ and σ_S are manifestations of that reasoning—energy and work are associated with hydrodynamic conservation laws.

Of the terms in the entropy flux $\boldsymbol{J}_S$, we have entropy convection $\rho s \boldsymbol{v}$, entropy transport from heat flows $\boldsymbol{J}_Q/T$, and entropy transport from diffusion flows $\widetilde{\mu}_k \boldsymbol{J}_k^{\mathrm{d}}/T$. The source σ_S is a sum of four terms, each of which is a product of two factors: a *rate* at which processes occur, generally referred to as *fluxes*, and a term associated with inhomogeneities, known as *thermodynamic forces*.[36] In each of the four terms in σ_S, there is a coupling of forces and fluxes of equal tensor rank.[37] We have in Eq. (1.34): 1) products of scalars (zeroth-rank tensors), $\sum_{j=1}^{r} J_j A_j$; 2) scalar products of vectors (first-rank tensors); and 3) a contraction of second-rank tensors $\nabla \boldsymbol{v} : \boldsymbol{\Pi}$ (the gradient of a vector is a second-rank tensor). The entropy source is bilinear[38] in fluxes $\mathbb{J}$ (whether scalar, vector, or tensor) and forces $\mathbb{F}$ (whether scalar, vector, or tensor), the form of which we can write

$$\sigma_S = \sum_{\alpha} \mathbb{J}_\alpha \circ \mathbb{F}_\alpha, \tag{1.35}$$

where α indicates scalar, vector, tensor, and $\circ$ denotes the operation required to produce a scalar from the composition $\mathbb{J} \circ \mathbb{F}$, whatever the tensor character of $\mathbb{J}$ and $\mathbb{F}$.

1.6 LINEAR FORCE-FLUX RELATIONS, KINETIC COEFFICIENTS

The form of Eq. (1.35) seemingly implies a connection between *causes* of nonequilibrium phenomena, forces $\mathbb{F}$, and their associated *effects*, fluxes $\mathbb{J}$, but that's misleading because *they're not independent physical effects* (as we show in Section 1.7). Temperature gradients are produced by placing different system boundaries in contact with heat reservoirs at different temperatures; gradients in chemical potentials are maintained through concentration differences; shear flows are maintained through suitable boundary conditions. Such influences set up the system's response in the form of fluxes. The distinction between cause and effect disappears, however, when it's realized

[36] This description applies to the chemical affinities A_j, which are gradients of thermodynamic potentials (U, H, F, G, see [2, p55]) with respect to reaction coordinates ξ_j; for example, $A_j = -(\partial G/\partial \xi_j)_{T,P}$[2, p89]. Nonzero chemical affinities are a measure of the extent to which reactions are not in chemical equilibrium. Thermodynamic forces have nothing to do with Newtonian forces. Thermodynamic forces are also referred to as affinities—another use of that term.

[37] By *Curie's theorem* (which according to one author, Curie neither stated nor proved[18, p35]), forces of a given tensor rank do not give rise to fluxes of another tensor rank in isotropic media. See de Groot and Mazur[7, p58].

[38] A bilinear form is a function $f(\boldsymbol{u}, \boldsymbol{v})$ linear in both arguments: $f(\boldsymbol{u}+\boldsymbol{v}, \boldsymbol{w}) = f(\boldsymbol{u}, \boldsymbol{w}) + f(\boldsymbol{v}, \boldsymbol{w})$ and $f(\boldsymbol{u}, \boldsymbol{v}+\boldsymbol{w}) = f(\boldsymbol{u}, \boldsymbol{v}) + f(\boldsymbol{u}, \boldsymbol{w})$, and, for scalar λ, $f(\lambda \boldsymbol{u}, \boldsymbol{v}) = \lambda f(\boldsymbol{u}, \boldsymbol{v}) = f(\boldsymbol{u}, \lambda \boldsymbol{v})$. See [6, p89] or [13, p342].

that temperature differences (for example) in the interior of a system, away from boundaries, are maintained by heat currents. Forces set up fluxes and fluxes induce forces.[39] For the theory to have predictive power, *it must be augmented with constitutive relations between fluxes and forces.*

Equation (1.34) has limitations apart from the need for constitutive relations to obtain closure. It was derived by assembling balance equations for the quantities on the right of Eq. (1.5) (first law of thermodynamics), where each brings with it a flux for the associated quantity. Balance equations for mass, momentum, and energy [Eqs. (1.11), (1.20), and (1.31)], were derived assuming convection to be the mechanism responsible for establishing fluxes. Equation (1.34) therefore applies to isotropic multi-component fluids, but not to solids or liquid crystals which are said to have *broken symmetries.*[40] Additional slow modes emerge in such systems that must be taken into account, e.g., long-wavelength spin waves in magnets or phonons in crystalline solids. All slow processes must be included in models of nonequilibrium phenomena.[41] Balance equations for mechanical quantities are not limited to systems in equilibrium. The balance equation for entropy, however, is valid only for macroscopic systems *slightly* out of equilibrium; Eq. (1.5) holds for changes slow compared with the time scale over which local equilibrium is established.

Diffusive transport is caused by the spontaneous movement in some quantity from regions of higher concentration to regions of lower concentration. Under a wide range of experimental conditions, diffusive flows are described by linear relations between fluxes and gradients in system properties Q_ψ,

$$\boldsymbol{J}_\psi = -\text{ (transport coefficient) } \boldsymbol{\nabla} Q_\psi,$$

with the *transport coefficient* as the proportionality factor. Examples include: *Fourier's law*, a relation between heat flux $\boldsymbol{J}_Q$ and temperature gradient, $\boldsymbol{J}_Q = -\kappa \boldsymbol{\nabla} T$, where κ is the thermal conductivity; *Fick's law*, a relation between particle flux $\boldsymbol{J}_n$ and concentration gradient, $\boldsymbol{J}_n = -D\boldsymbol{\nabla} n$, where D is the *diffusion coefficient* and n is the *number density*, number of particles per volume; and *Ohm's law*, a relation between charge flux, the current density $\boldsymbol{J}_q$, and electric field, the negative gradient of the electrostatic potential ϕ, $\boldsymbol{J}_q = -\sigma\boldsymbol{\nabla}\phi$, where σ is the electrical conductivity. Gradient relations extend to tensor quantities. In fluids, elements of the viscous stress tensor Π_{ij}, the flux of the ith-component of momentum through the jth-component of surface area, are proportional to gradients of the velocity field, $\Pi_{ij} = -\eta \partial v_i/\partial x_j$, with η the viscosity coefficient, *Newton's law of viscosity*.[42] These relations are summarized in Table 1.1.

Table 1.1 Basic gradient-driven transport processes

Flux	Force	Transport relation	Name
Heat, $\boldsymbol{J}_Q$	$\boldsymbol{\nabla}T$	$\boldsymbol{J}_Q = -\kappa\boldsymbol{\nabla}T$	Fourier's law
Particle, $\boldsymbol{J}_n$	$\boldsymbol{\nabla}n$	$\boldsymbol{J}_n = -D\boldsymbol{\nabla}n$	Fick's law
Charge, $\boldsymbol{J}_q$	$\boldsymbol{\nabla}\phi$	$\boldsymbol{J}_q = -\sigma\boldsymbol{\nabla}\phi$	Ohm's law
(Momentum)$_x$ through $(\delta\boldsymbol{S})_y$, Π_{xy}	$\dfrac{\partial v_x}{\partial y}$	$\Pi_{xy} = -\eta\dfrac{\partial v_x}{\partial y}$	Newton's law of viscosity

[39] Ohm's law provides an example where an electric current (flux) $I = \Delta V/R$ flows between the nodes of a circuit when a voltage difference (force) ΔV is maintained between them, where R is the resistance. A current I, however, maintained between the same nodes establishes a voltage difference $\Delta V = IR$. Which description is "right"?

[40] We're referring to *continuous symmetries*, such as translational or rotational, where an infinitesimal parameter—rotation angle, for example—can be continuously varied, as opposed to *discrete symmetries* such as parity or time reversal.

[41] Recall Feynman's definition: equilibrium is the state where the fast things have happened but the slow things not.

[42] A *Newtonian fluid* is one in which viscous stresses are linearly correlated with the local strain rate[12, p146].

A complication can and does occur where a process characterized by the gradient in one quantity causes the occurrence of other gradient-driven processes. For example, when a temperature gradient is established, not only is a flow of heat set up but also an electric field (*thermoelectricity*). The strength of the field is proportional to the temperature gradient, $\boldsymbol{E} = P\boldsymbol{\nabla}T$, where P is the *thermopower* of the material.[43] In electrically neutral systems, a region depleted of charge carriers sets up an electric field to oppose the motion of charge.[44] The voltage difference ΔV between two points held at a temperature difference ΔT is $\Delta V = P\Delta T$, the *Seebeck effect*. Charges arriving from a region of higher temperature carry "excess" energy relative to the environment at the colder location. The rate of energy deposition, $\dot{Q} = \Pi I$, is proportional to the current I (the *Peltier effect*), with Π the *Peltier coefficient*. The Peltier effect is said to be *conjugate* to the Seebeck effect. In the Seebeck effect, charges in motion (because of ΔT) create a voltage difference ΔV; in the Peltier effect, charges in motion (because of ΔV) create heat. Another conjugate pair is the coupling of particle diffusion and heat conduction. In the *Soret effect* (*thermophoresis*), a concentration gradient is established as a result of a temperature gradient. In the conjugate *Dufour effect*, a temperature gradient arises from a concentration gradient.

Conjugate effects can be modeled by adding appropriate terms to the transport relations in Table 1.1. One could add a temperature gradient to Fick's law, $\boldsymbol{J}_n = -D_1\boldsymbol{\nabla}n - D_2\boldsymbol{\nabla}T$, to produce a model in which particle diffusion occurs because of a concentration gradient and a temperature gradient. One could add a concentration gradient to Fourier's law, $\boldsymbol{J}_Q = -\kappa_1\boldsymbol{\nabla}T - \kappa_2\boldsymbol{\nabla}n$, for a model in which heat flow occurs because of a temperature gradient and a concentration gradient. The point is, *the same forces ($\boldsymbol{\nabla}T$ and $\boldsymbol{\nabla}n$) can give rise to different fluxes.*

To enable the treatment of irreversible phenomena of arbitrary complexity, we allow that *any force can contribute to any flux*, which we express as a system of linear equations,[45]

$$J_i = \sum_{k=1}^{n} L_{ik}F_k, \qquad (i = 1, \ldots, n) \tag{1.36}$$

where J_i, F_k are the Cartesian components of fluxes and forces, with the phenomenological terms L_{ik} the *kinetic coefficients*. The diagonal terms L_{ii} are related to basic transport coefficients and the off-diagonal terms L_{ik}, $k \neq i$, are connected to conjugate phenomena. Combining Eq. (1.36) with the "component" form of Eq. (1.35),

$$\sigma_S = \sum_i J_iF_i = \sum_i\sum_k L_{ik}F_iF_k \geq 0. \tag{1.37}$$

The square matrix of kinetic coefficients $[L_{ik}]$ must therefore be positive definite (from the physics), which constrains the diagonal terms to be non-negative, $L_{ii} \geq 0$, and the off-diagonal terms to satisfy the inequality $L_{ii}L_{kk} \geq \frac{1}{4}\left(L_{ik} + L_{ki}\right)^2$. There is no constraint on the sign of the off-diagonal terms. We'll see that the matrix $[L_{ik}]$ is symmetric.

1.7 FORCES AND FLUXES LINKED THOUGH FLUCTUATIONS

We now come arguably to the most important section of this chapter. Onsager developed a general theory of linear irreversible processes based on what they have in common, entropy creation[19][20]. A fundamental result of that theory is, for appropriately defined forces and fluxes,

[43]The thermopower is often denoted with the symbol S. To avoid confusion with entropy, we've used the symbol P.

[44]Charge carriers are much less massive than the ions they leave behind.

[45]The restriction to linear relations between fluxes and forces specifies a class of phenomena, those described by linear irreversible thermodynamics. Extensions to nonlinear processes is an active field of research.

the kinetic coefficients are symmetric, the *Onsager reciprocal relations*,[46]

$$L_{ik} = L_{ki}. \qquad (i,k = 1,\ldots,n) \tag{1.38}$$

These relations are a kind of Newton's third law, that if flux J_i is influenced by force F_k, then flux J_k is influenced by force F_i with equal strength, $L_{ik} = L_{ki}$. This simplifies the analysis of coupled irreversible processes.[47] The reciprocal relations have been verified experimentally[22][23] and could be considered another law of thermodynamics at the macroscopic level.[48] We show in Chapter 2 that Eq. (1.38) follows from basic ideas of statistical mechanics.

To explain "appropriate definition," we must examine fluctuations because *forces and fluxes are linked through fluctuations*.[49] Denote equilibrium values of extensive quantities X_i^0, and let

$$\alpha_i \equiv X_i - X_i^0 \qquad (i = 1,\ldots,n) \tag{1.39}$$

represent the instantaneous deviation[50] of X_i from X_i^0. Entropy is maximized in equilibrium: $S^0 \equiv S(X_1^0,\ldots,X_n^0)$ is the maximum value S can have subject to macroscopic constraints[2, p53]. Associated with a prescribed set of fluctuations $\{\alpha_1,\ldots,\alpha_n\}$ is an entropy fluctuation $\Delta S \equiv S - S^0$ which, to lowest order in small quantities, is found from the expression[51]

$$\Delta S(\alpha_1,\ldots,\alpha_n) \approx \frac{1}{2}\sum_{jk=1}^{n} \frac{\partial^2 S^0}{\partial X_j \partial X_k}\alpha_j\alpha_k \equiv -\frac{1}{2}\sum_{jk=1}^{n} g_{jk}\alpha_j\alpha_k, \tag{1.40}$$

where the "metric" $g_{jk} = -\partial^2 S^0/\partial X_j \partial X_k$ is symmetric, $g_{jk} = g_{kj}$. Fluctuations about equilibrium result in momentary *decreases* in entropy (the stability condition of thermodynamics[5, p26]), and thus the matrix $[g_{jk}]$ is positive definite.[52] The maximum value S^0 is associated with the state of greatest disorder consistent with macroscopic constraints; fluctuations are momentary creations of more ordered (less disordered) system configurations, implying $\Delta S < 0$. Fluctuations create nonequilibrium configurations from which the system returns to the state of maximum entropy.[53] Entropy creation is associated with irreversibility; irreversibility is implied by entropy creation.

A rate of entropy change associated with fluctuations can be found by dividing Eq. (1.40) by $\mathrm{d}t$,

$$\frac{\mathrm{d}}{\mathrm{d}t}\Delta S(\alpha_1,\ldots,\alpha_n) = \sum_k \frac{\partial \Delta S}{\partial \alpha_k}\frac{\mathrm{d}}{\mathrm{d}t}\alpha_k = -\sum_{kj} g_{kj}\alpha_j \frac{\mathrm{d}}{\mathrm{d}t}\alpha_k, \tag{1.41}$$

where the first equality follows from the chain rule with fluctuations treated as differentiable dynamical variables; the second equality follows from Eq. (1.40) and the symmetry of g_{jk}. The middle expression in Eq. (1.41) for $\mathrm{d}\Delta S/\mathrm{d}t$ has the *form* of the middle expression in Eq. (1.37) for σ_S. That invites us, after we get the dimensions right,[54] to identify the rate J_k as a time derivative of α_k and the force F_k as the partial derivative of ΔS with respect to α_k:

[46]Onsager received the 1968 Nobel Prize in Chemistry for discovery of the reciprocal relations. Equation (1.38) applies in the absence of external magnetic fields. We show in Section 2.2 that the reciprocal relations are modified for systems in a magnetic field $\boldsymbol{B}$, $L_{ik}(\boldsymbol{B}) = L_{ki}(-\boldsymbol{B})$, a result due to Casimir[21].

[47]The number of independent kinetic coefficients is reduced from n^2 to $\frac{1}{2}n(n+1)$.

[48]The reciprocal relations are sometimes referred to as the "fourth law of thermodynamics."

[49]Kinetic coefficients connect forces and fluxes phenomenologically in Eq. (1.36); at the microscopic level, forces and fluxes are linked through fluctuations. The connection with fluctuations is the link with nonequilibrium statistical mechanics.

[50]The state variables of thermodynamics are by definition their time-independent equilibrium values[2, p5]. Here we're allowing dynamic deviations from equilibrium. The n fluctuations in Eq. (1.39) are all measured at the same time. It's traditional to denote fluctuations as in Eq. (1.39) with α_i. The notation $\delta X_i \equiv X_i - X_i^0$ is also prevalent.

[51]First-order terms in the Taylor series of Eq. (1.40) vanish; see [2, p44]. Fluctuations in conserved extensive quantities (energy, volume, particle number) occur with no change in entropy to first order in small quantities.

[52]Entropy is a concave function[5, p321]. Because of the minus sign in Eq. (1.40), the matrix $[g_{jk}]$ is positive definite.

[53]We return to this idea in Chapter 6 with the fluctuation-dissipation theorem.

[54]The quantity σ_S is the rate of entropy creation per volume, $\sigma_S = (1/V)\mathrm{d}\Delta S/\mathrm{d}t$.

$$J_k \equiv \frac{1}{V}\frac{\mathrm{d}\alpha_k}{\mathrm{d}t} \tag{1.42}$$

$$F_k \equiv \frac{\partial \Delta S}{\partial \alpha_k} = -\sum_j g_{kj}\alpha_j. \tag{1.43}$$

We take Eqs. (1.42) and (1.43) as *provisional* definitions of J_k, F_k—there is wiggle room involving multiplicative factors so that the product $J_k F_k$ has the correct dimension.[55] Equation (1.42) is the *Onsager regression hypothesis*, that the rate of irreversible processes is set by the time rate of change of fluctuations.

The entropy source σ_S is a locally intensive quantity, the rate of entropy creation per volume. The term $\mathrm{d}\alpha_k/\mathrm{d}t$ in Eq. (1.41) specifies an extensive quantity (fluctuations as defined here are extensive).[56] We must divide by V in Eq. (1.42) so that the product $J_k F_k$ has the dimension of σ_S. Consider the dimensions of the terms in Eq. (1.34). In the rate of entropy production associated with viscous stresses, $\boldsymbol{\nabla v}{:}\boldsymbol{\Pi}$, $\boldsymbol{\nabla v}$ has the dimension of rate, $(\text{time})^{-1}$, and $\boldsymbol{\Pi}$ has the dimension of pressure, or energy density.[57] The same is true of chemical reactions: J_i is a rate and the chemical affinity A_i is an energy density. The use of the term *flux* is therefore misleading (if it's actually a rate). Only for vector processes does the "flux" have the traditional dimension of flux. Through dimensional analysis (square brackets denote dimension) we have, using Eqs. (1.42) and (1.43),

$$[J_k F_k] = \left[\frac{1}{V}\frac{\mathrm{d}X_k}{\mathrm{d}t}\frac{\partial S}{\partial X_k}\right] = \text{Rate} \times \text{entropy density.}$$

One might wonder how thermodynamic forces associated with gradients arise, as in Eq. (1.34). Consider that if we "borrow" a length factor from $V = L^3$,

$$[J_k F_k] = \left[\frac{1}{L^2}\frac{\mathrm{d}X_k}{\mathrm{d}t} \times \nabla\left(\frac{\partial S}{\partial X_k}\right)\right] = \text{Rate} \times \text{entropy density,}$$

except now the flux term is a flux in the traditional sense.

Consider an entropy current defined in terms of traditional fluxes,

$$\boldsymbol{J}_S \equiv \sum_k F_k \boldsymbol{J}_k, \tag{1.44}$$

where $F_k = \partial S/\partial X_k$ is an intensive parameter (start with $\mathrm{d}S = \sum_k F_k \mathrm{d}X_k$), and $\boldsymbol{J}_k$ is a current density associated with extensive parameter X_k. (One should check that $\boldsymbol{J}_S$ has the dimensions of entropy per area per time.) From Eq. (1.44),

$$\boldsymbol{\nabla}\cdot\boldsymbol{J}_S = \sum_k \boldsymbol{J}_k\cdot\boldsymbol{\nabla}F_k + \sum_k F_k\boldsymbol{\nabla}\cdot\boldsymbol{J}_k. \tag{1.45}$$

Extensive variables (other than entropy) are conserved and satisfy continuity equations,

$$\frac{\partial x_k}{\partial t} + \boldsymbol{\nabla}\cdot\boldsymbol{J}_k = 0, \tag{1.46}$$

where $x_k \equiv X_k/V$ is a per-volume density (as opposed to per mass). Combining Eqs. (1.46) and (1.45),

$$\sum_k F_k\frac{\partial x_k}{\partial t} + \boldsymbol{\nabla}\cdot\boldsymbol{J}_S = \sum_k \boldsymbol{J}_k\cdot\boldsymbol{\nabla}F_k. \tag{1.47}$$

Equation (1.47) is in the form of an entropy balance equation, implying the right side is an entropy source expressed in terms of gradients of intensive thermodynamic parameters.

---

[55] To quote C. Kittel[24, p163]: "It is rarely a trivial problem to find the correct choice of forces and fluxes applicable to the Onsager relation."

[56] The other equation (1.43) for the thermodynamic force F_k specifies an intensive quantity.

[57] When divided by T in Eq. (1.34), $\boldsymbol{\Pi}/T$ has the dimension of entropy density.

1.8 THERMOELECTRICITY, KELVIN RELATION

The simplest application of nonequilibrium thermodynamics is the thermoelectric effect, where a temperature difference ΔT produces a heat current in addition to an electric current, which in turn establishes a potential difference ΔV. Numerous other applications could be covered (see de Groot and Mazur[7] or Kreuzer[25]). Our purpose is not to provide an exhaustive coverage of irreversible thermodynamics, but rather to establish a general framework for macroscopic descriptions of nonequilibrium phenomena.

The first question to ask is: What are the forces and the fluxes? Ignoring viscous effects and chemical reactions, we have from Eq. (1.34),

$$\sigma_S = \boldsymbol{J}_Q \cdot \boldsymbol{\nabla}\left(\frac{1}{T}\right) + \boldsymbol{J}_E \cdot \left(\frac{\boldsymbol{F}}{T} - \boldsymbol{\nabla}\left(\frac{\widetilde{\mu}}{T}\right)\right) \equiv \boldsymbol{J}_Q \cdot \boldsymbol{X}_Q + \boldsymbol{J}_E \cdot \boldsymbol{X}_E, \tag{1.48}$$

where $\boldsymbol{J}_E$ is a "provisional" electric current density. We need to be mindful of units here. The units of $\boldsymbol{J}_Q$ are W m^{-2}. The units of what we've called the "electric" current, $\boldsymbol{J}_E$, are those of mass flux, kg m^{-2} s^{-1}. The units of $\boldsymbol{X}_E$ in Eq. (1.48) are force per mass per Kelvin, N kg^{-1} K^{-1} ($\boldsymbol{F}$ is the Coulomb force per mass). Thus, $\boldsymbol{J}_E \cdot \boldsymbol{X}_E$ has units of entropy per volume per time, which we want. To remove confusion over units, redefine $\boldsymbol{J}_E$ and $\boldsymbol{X}_E$ such that $\boldsymbol{J}_E \cdot \boldsymbol{X}_E = \widetilde{\boldsymbol{J}}_E \cdot \widetilde{\boldsymbol{X}}_E$, where we multiply and divide by the unit of charge so that $\widetilde{\boldsymbol{J}}_E$ has the usual units of A m^{-2} and

$$\widetilde{\boldsymbol{X}}_E = \frac{\boldsymbol{E}}{T} - \frac{1}{e}\boldsymbol{\nabla}\left(\frac{\mu}{T}\right) \equiv \frac{1}{T}\boldsymbol{\mathcal{E}} \tag{1.49}$$

has units of V m^{-1} K^{-1}, where $\boldsymbol{E}$ is the electric field, e is the magnitude of the electron charge, and μ is the chemical potential in Joules. The quantity $\boldsymbol{\mathcal{E}}$ in Eq. (1.49) is an effective electric field.

From Eq. (1.36), we write, grouping terms into vectors,

$$\widetilde{\boldsymbol{J}}_E = L_{11}\widetilde{\boldsymbol{X}}_E + L_{12}\boldsymbol{X}_Q = \frac{L_{11}}{T}\boldsymbol{\mathcal{E}} - \frac{L_{12}}{T^2}\boldsymbol{\nabla}T \tag{1.50}$$

$$\boldsymbol{J}_Q = L_{21}\widetilde{\boldsymbol{X}}_E + L_{22}\boldsymbol{X}_Q = \frac{L_{21}}{T}\boldsymbol{\mathcal{E}} - \frac{L_{22}}{T^2}\boldsymbol{\nabla}T, \tag{1.51}$$

where the kinetic coefficients are such that $L_{12} = L_{21}$ (Onsager reciprocity).[58] The quantity L_{11} has units of S K m^{-1}, L_{12} and L_{21} have units of A K m^{-1}, and L_{22} has units of W K m^{-1}. How to "dig out" the transport coefficients from these relations?

For $\boldsymbol{\nabla}T = 0$ in Eq. (1.50), we identify $L_{11} = T\sigma$, where $\sigma > 0$ is the electrical conductivity. Also for $\boldsymbol{\nabla}T = 0$, we have the ratio of the magnitudes of the currents,

$$\frac{J_Q}{\widetilde{J}_E} \equiv \Pi = \frac{L_{21}}{L_{11}}, \tag{1.52}$$

where Π is the Peltier coefficient, which can be of either sign.[59]

For $\boldsymbol{\nabla}T \neq 0$, no current flows in an open circuit. Setting $\widetilde{\boldsymbol{J}}_E = 0$ in Eq. (1.50), we have $\boldsymbol{\mathcal{E}} = (L_{12}/(TL_{11}))\,\boldsymbol{\nabla}T$. Combining with Eq. (1.51),

$$\boldsymbol{J}_Q = -\frac{1}{T^2}\left[L_{22} - \frac{L_{21}L_{12}}{L_{11}}\right]\boldsymbol{\nabla}T \equiv -\kappa\boldsymbol{\nabla}T.$$

We therefore have another connection between kinetic coefficients and a transport coefficient,

$$\kappa = \frac{1}{T^2}\left[L_{22} - \frac{L_{21}L_{12}}{L_{11}}\right].$$

[58] We see from Eqs. (1.50) and (1.51) that the same forces give rise to different fluxes, as noted in Section 1.6.

[59] Charge carriers can be of either sign.

We have three transport coefficients Π, σ, κ involving the three independent kinetic coefficients, L_{11}, L_{12}, L_{22}. The voltage difference ΔV is obtained by integrating $\boldsymbol{\mathcal{E}}$, with the result that

$$\Delta V = \int_1^2 \boldsymbol{\mathcal{E}} \cdot \mathrm{d}\boldsymbol{r} = \frac{L_{12}}{TL_{11}} \int_1^2 \boldsymbol{\nabla} T \cdot \mathrm{d}\boldsymbol{r} = \frac{L_{12}}{TL_{11}} \Delta T,$$

where we've used Eq. (1.50) with $\widetilde{\boldsymbol{J}}_E = 0$. The linear relation between ΔV and ΔT is the Seebeck effect, $P \equiv \Delta V / \Delta T$. Thus,

$$P = \frac{1}{T} \frac{L_{12}}{L_{11}} = \frac{1}{T} \Pi, \tag{1.53}$$

where we've used Eq. (1.52). Equation (1.53) is the *Kelvin relation*,[60] which has been verified experimentally[22]. Note how the phenomenological kinetic coefficients have been used as a "bootstrap" to provide a testable prediction, independent of the L_{ij}.

1.9 STEADY STATES AND MINIMUM ENTROPY PRODUCTION

Equilibrium is a time-invariant state of maximum entropy. In formulating statistical mechanics, we recognize that measurements of equilibrium systems fluctuate from the random motions of microscopic components. That leads to the use of ensembles classified by the types of interactions they permit with the environment[5, Chapter 4]. The canonical ensemble is composed of systems allowing energy exchanges with the environment; the grand canonical ensemble allows exchanges of matter and energy. Equilibrium can be characterized as *the state in which there is no net flow of matter or energy between system and environment.* A system without flows is a state of zero entropy production, $\sigma_S = 0$; thus, zero entropy production is another way to characterize equilibrium (a state of time-invariant entropy).

With nonequilibrium systems, we encounter a new type of time-invariant state, *steady* or *stationary* states, in which flows of heat, particles, or electricity occur at constant rates across system boundaries. Under such conditions, the state of the system at any point is unchanging in time. The system is time invariant, yet not in equilibrium because of the dissipative processes associated with $\sigma_S \neq 0$. Nonequilibrium thermodynamics is thus more comprehensive than traditional thermodynamics; equilibrium is the steady state in which fluxes from the environment approach zero.

Steady states are characterized by an extremum principle, that (in steady state) the rate of entropy production has the minimum value it can have consistent with prescribed boundary conditions, a result established by Prigogine[26, p76] and de Groot and Mazur[7, Chapter 5]; see the review of Jaynes[27]. Starting from Eq. (1.37), the general expression for entropy production, $\sigma_S = \sum_i J_i F_i = \sum_{ik} L_{ik} F_i F_k$, and form the derivative with respect to a given force (holding the others fixed),

$$\left(\frac{\partial \sigma_S}{\partial F_l}\right)_{F_j} \overset{L_{ik}\text{ constant}}{\overset{\downarrow}{=}} \sum_{ik} L_{ik} \left(F_i \delta_{k,l} + F_k \delta_{i,l}\right) \overset{\text{indices}}{\overset{\downarrow}{=}} \sum_i \left(L_{il} + L_{li}\right) F_i \overset{\text{reciprocity}}{\overset{\downarrow}{=}} 2 \sum_i L_{li} F_i \overset{\text{Eq. (1.36)}}{\overset{\downarrow}{=}} 2J_l.$$

Thus, $J_l = 0$ (a stationary state) is equivalent to $(\partial \sigma_S / \partial F_l)_{F_j} = 0$, implying σ_S has a minimum value[61] with respect to variations in F_l (with the remaining forces held fixed). If another flux J_m is zero, as well as J_l, the value of σ_S is at a *smaller* minimum. As the number of flows which are zero is increased (for fixed forces), the minimum value of σ_S becomes progressively smaller. The limiting

[60]Kelvin derived Eq. (1.53) in 1854 using a now-discredited theory. He got the right answer using incorrect reasoning. Boltzmann attempted without success to justify Kelvin's approach. The correct derivation had to wait for Onsager.

[61]Because σ_S is a positive definite quadratic form, the extremum condition refers to a minimum, not a maximum.

case of all flows zero corresponds to thermodynamic equilibrium where σ_S has its smallest possible value, namely $\sigma_S = 0$. This theorem applies only if the kinetic coefficients are constant (implying gradients in thermodynamic parameters are sufficiently small) and that Onsager reciprocity holds.

Consider a system featuring two flow processes, a flow of matter (which could be charged species) J_{m}, and a flow of energy, J_{th}. From Eq. (1.36), let

$$\begin{aligned} J_{\text{m}} &= L_{11}F_{\text{m}} + L_{12}F_{\text{th}} \\ J_{\text{th}} &= L_{21}F_{\text{m}} + L_{22}F_{\text{th}}, \end{aligned} \tag{1.54}$$

where F_{m} and F_{th} are the thermodynamic forces associated with fluxes J_{m} and J_{th}. Using Eq. (1.37), we find, invoking $L_{12} = L_{21}$,

$$\sigma_S = L_{11}F_{\text{m}}^2 + 2L_{12}F_{\text{m}}F_{\text{th}} + L_{22}F_{\text{th}}^2, \tag{1.55}$$

and thus

$$\frac{\partial}{\partial F_{\text{m}}}\sigma_S = 2\left(L_{11}F_{\text{m}} + L_{12}F_{\text{th}}\right) = 2J_{\text{m}}, \tag{1.56}$$

where we've used Eq. (1.54) in the final equality. Minimum entropy production, $\partial\sigma_S/\partial F_{\text{m}} = 0$, is associated with steady state, $J_{\text{m}} = 0$, and vice versa. The state of no particle current is, for a fixed temperature gradient, the state of minimum entropy production.

SUMMARY

Systems out of thermodynamic equilibrium evolve irreversibly toward equilibrium and entropy is created in irreversible processes. We considered the macroscopic theory of irreversibility based on thermodynamic and hydrodynamic principles, which, in spite of the many equations in this chapter, is based on four fundamental equations: (1.1), (1.36), (1.37), and (1.38). The macroscopic theory reveals the issues to be explained by microscopic approaches to nonequilibrium systems.

- Entropy is created but not destroyed in irreversible processes [the Clausius inequality, (1.1)], implying the existence of a non-negative entropy source (rate of entropy creation per volume), $\sigma_S \geq 0$, with equality for reversible processes. Entropy production is minimized in steady-state processes.

- The assumption of local equilibrium, which, together with macroscopic conservation laws, leads to an expression for σ_S in the form $\sigma_S = \sum_k J_k F_k$, Eq. (1.37), where the terms J_k are referred to as fluxes (but are often simply rates) and the terms F_k are known as thermodynamic forces. Establishing the form of σ_S constitutes the heavy lifting in this chapter.

- Linear relations between forces and fluxes, $J_i = \sum_j L_{ij}F_j$, Eq. (1.36), with the phenomenological terms L_{ij} known as kinetic coefficients. Equation (1.36) recognizes that any force can contribute to any flux. Said differently, the same forces can give rise to different fluxes.

- The Onsager reciprocal relations, the experimentally verified symmetry of kinetic coefficients $L_{ij} = L_{ji}$, Eq. (1.38), follow from basic ideas of statistical mechanics (see Chapter 2) and form a gateway into the study of time-dependent fluctuations. We identified fluxes as the time rate of change of fluctuations, the Onsager regression hypothesis. That forces and fluxes are linked through fluctuations, Eqs. (1.42), (1.43), implies the fundamental conclusion that *dissipation is associated with fluctuations*, of which we'll see in examples in Chapters 2, 3 and we'll study systematically in Chapter 6.

EXERCISES

1.1 a. Show that the convective derivative, Eq. (1.12), satisfies the product rule of calculus (for any differentiable functions ϕ, ψ, and for any velocity $\boldsymbol{w}$)

$$D_w(\phi\psi) = \phi D_w \psi + \psi D_w \phi.$$

b. Show for the special case of the center-of-mass velocity $\boldsymbol{v}$, for any function ϕ, that

$$\rho D_v \phi = \frac{\partial}{\partial t}(\rho\phi) + \boldsymbol{\nabla} \cdot (\phi\rho\boldsymbol{v}). \tag{P1.1}$$

Use the continuity equation; ϕ may be a function or the component of a vector or tensor.

1.2 Derive Eq. (1.15).

1.3 Show that $[\boldsymbol{\nabla} \cdot (\rho\boldsymbol{v}\boldsymbol{v})]_i = \rho(\boldsymbol{v} \cdot \boldsymbol{\nabla})v_i + v_i \boldsymbol{\nabla} \cdot \rho\boldsymbol{v}$. See page 6 for the meaning of dyadic notation and the divergence of a second-rank tensor.

1.4 Fill in the steps leading to the Navier-Stokes equation, (1.22).

1.5 Derive Eq. (1.31).

1.6 We defined a time rate of change of entropy by dividing the first law of thermodynamics by dt, which we expressed in Eq. (1.5) as a relation among partial time derivatives. Show that the same form holds as a relation among convective derivatives,

$$TD_v s = D_v u + PD_v \rho^{-1} - \sum_{k=1}^{n} \widetilde{\mu_k} D_v c_k, \tag{P1.2}$$

where all quantities are specific quantities. Thus, Eq. (1.4) remains valid for a mass element followed with the center of mass motion. The specific entropy is a function of the variables describing the macroscopic state of the system, $s = s(u, \rho^{-1}, c_k)$. Show using standard thermodynamic identities (see [2, p42]) that

$$T\boldsymbol{\nabla} s = \boldsymbol{\nabla} u + P\boldsymbol{\nabla}\rho^{-1} - \sum_{k=1}^{n} \widetilde{\mu}_k \boldsymbol{\nabla} c_k.$$

Note that for any function, $\mathrm{d}f = \boldsymbol{\nabla} f \cdot \mathrm{d}\boldsymbol{r}$.

1.7 Using the balance equation for entropy $\partial(\rho s)/\partial t + \boldsymbol{\nabla} \cdot \boldsymbol{J}_S = \sigma_S$, show that we have a generalization of the Clausius inequality for open systems, $\mathrm{d}S/\mathrm{d}t \geq -\oint \boldsymbol{J}_S \cdot \mathrm{d}\boldsymbol{S}$.

1.8 Referring to Section 1.8, derive expressions for the Onsager coefficients in terms of the transport coefficients. Show that $\kappa > 0$. A: $L_{11} = T\sigma$, $L_{12} = T\sigma\Pi$, $L_{22} = T^2\kappa + T\sigma\Pi$.

1.9 *Isentropic flow* is one for which entropy is conserved, $\sigma_S = 0$.

a. Use Eq. (1.34) to argue that isentropic flow implies no heat flow, no diffusion flows, no chemical reactions, and neglect of viscous forces. There is only convection of entropy, as per Eq. (1.33).

b. For isentropic flow show that

$$D_v \rho s = -\rho s \boldsymbol{\nabla} \cdot \boldsymbol{v}.$$

Argue that isentropic flow (the entropy of a small volume of fluid doesn't change as it flows) is also incompressible flow.

c. Enthalpy, $H \equiv U + PV$, is the heat added at constant pressure[2, p57]. Define a specific enthalpy $h \equiv H/M$, so that $h = u + P/\rho$.

   i. Show that the first law of thermodynamics can be written $\mathrm{d}h = T\mathrm{d}s + (1/\rho)\mathrm{d}P$. For isentropic flow, $\boldsymbol{\nabla} h = \boldsymbol{\nabla}(P/\rho)$.

   ii. For isentropic flows, show, assuming $\boldsymbol{F}_k = -\boldsymbol{\nabla}\Phi_k$, that

$$D_v \boldsymbol{v} = -\boldsymbol{\nabla}\left(h + (1/\rho)\sum_k \rho_k \Phi_k\right). \tag{P1.3}$$

d. Derive, or otherwise verify, the vector identity

$$\boldsymbol{v} \times \boldsymbol{\nabla} \times \boldsymbol{v} = \frac{1}{2}\boldsymbol{\nabla} v^2 - (\boldsymbol{v} \cdot \boldsymbol{\nabla})\,\boldsymbol{v}.$$

e. Show in the case of isentropic flow that

$$\frac{\partial \boldsymbol{v}}{\partial t} = -\boldsymbol{\nabla}\left(h + (1/\rho)\sum_k \rho_k \Phi_k + \frac{1}{2}v^2\right) + \boldsymbol{v} \times \boldsymbol{\nabla} \times \boldsymbol{v}. \tag{P1.4}$$

f. Finally, from Eq. (P1.4), show that

$$\frac{\partial}{\partial t}(\boldsymbol{\nabla} \times \boldsymbol{v}) = \boldsymbol{\nabla} \times \left[\boldsymbol{v} \times (\boldsymbol{\nabla} \times \boldsymbol{v})\right].$$

Thus, if $\boldsymbol{\nabla} \times \boldsymbol{v} = 0$ at an instant of time, the velocity field remains curl-free for all times.

CHAPTER 2

Fluctuations as stochastic processes

THE Onsager reciprocal relations are a consequence of the time-dependent properties of fluctuations. Although the ensemble-based framework of statistical mechanics is predicated on the existence of fluctuations, nowhere are their dynamical properties used in establishing that theory. We know that spatial correlations exist among fluctuations (from elastic scattering experiments[5, Section 6.6]). To prove the experimentally verified Onsager relations, and to secure a foothold in nonequilibrium statistical mechanics, we must consider temporal correlations. Fluctuations are modeled as stochastic processes—time sequences of random events—a topic not typically included in science curricula. In this chapter, we develop stochastic processes as a mathematical tool (Sections 2.3, 2.4, and 2.5)—a necessary foundation for nonequilibrium statistical physics[1]—as well as associated physical ideas. We begin with the simplest theory of fluctuations, that due to Einstein.

2.1 EINSTEIN FLUCTUATION THEORY

The transition from thermodynamics to statistical mechanics requires a generalization of our concepts of equilibrium and measurement. In thermodynamics, state variables have fixed values that persist unchanged in time. Microscopically, however, the local density and other quantities fluctuate due to the erratic, random motions of a system's myriad components.[2] Let $\hat{X}$ denote a measurable quantity. In thermodynamics, the equilibrium value X^0 is obtained in a single measurement. In statistical mechanics, we interpret the equilibrium value as the time-invariant mean of a large number of measurements[3] of $\hat{X}$. One can envision a series of measurements made at many times on the same system, from which we calculate time averages,[4] $\overline{X}$. Or, one could envision measurements made *at the same time* on a collection of *macroscopically identical* systems termed an ensemble, so as to sample the effects of varied microscopic initial conditions, in which case we speak of ensemble averages, $\langle X \rangle$. Statistical mechanics is concerned with ensemble averages.[5]

[1] Statistical mechanics requires probability theory: combinatorics, random variables, probability distributions, limit theorems, characteristic functions, and cumulants. Nonequilibrium statistical mechanics requires in addition stochastic processes.

[2] Motion in classical mechanics is a deterministic unfolding of initial conditions, quantities that we have no ability to control or measure in macroscopic systems. We have no alternative to treating fluctuations as random processes.

[3] The *law of large numbers* (a theorem in the theory of probability) guarantees the existence of the mean[5, p74].

[4] The existence of time averages is guaranteed by *Birkhoff's theorem*,[28][5, p48] which pertains to time averages calculated from the microscopic dynamics of the system. It's one of two theorems on which the construction of statistical mechanics relies (the other is Liouville's theorem).

[5] Calculating time averages is impossible, which would require us to calculate the motion of large numbers of interacting particles, a task beyond our ken (for practical as well as fundamental reasons). The *ergodic hypothesis* equates time averages with ensemble averages. Statistical mechanics based on this assumption is highly successful, i.e., it works.

DOI: 10.1201/9781003512295-2

A measurement of $\hat{X}$ on a randomly selected member of the ensemble will in general return a value different from the mean. Can we deduce the probability $P(X)\mathrm{d}X$ that the result of measurement lies in the range $[X, X+\mathrm{d}X]$? That sounds impossibly general, yet we're not dealing with an arbitrary ensemble, but one composed of equilibrium systems. The Boltzmann entropy formula is a bridge between microscopic and macroscopic,

$$S = k \ln W, \tag{2.1}$$

allowing us to calculate S given W, the number of ways a macrostate (specified by state variables) can be realized from the microstates of a system (specified by initial conditions). Inverting Eq. (2.1), $W = \exp(S/k)$, we have the number of ways by which a state of entropy S can be achieved. With S^0 the equilibrium entropy, by writing[6] $S = S^0 + \Delta S$, $W = \exp(S^0/k)\exp(\Delta S/k)$. Dividing by an appropriate constant, we have the probability[7] of an entropy fluctuation ΔS,

$$P(\Delta S) = K\mathrm{e}^{\Delta S/k}, \tag{2.2}$$

where K is a normalization factor ($\Delta S < 0$ in fluctuations; see Section 1.7).

How are entropy fluctuations related to fluctuations in observable properties? From Eq. (1.40) (reproduced here), fluctuations $\alpha_i \equiv X_i - X_i^0$ in a system's extensive variables are associated with entropy fluctuations ΔS at lowest order in small quantities through

$$\Delta S(\alpha_1, \ldots, \alpha_n) \approx -\frac{1}{2}\sum_{j,k=1}^{n} g_{jk}\alpha_j\alpha_k, \tag{1.40}$$

where $g_{jk} = -\left(\partial^2 S/\partial X_j \partial X_k\right)^0$ and where the number of variables n is left unspecified.

By combining Eq. (1.40) with Eq. (2.2), we have the joint probability density (see Appendix B), the probability that the value of each fluctuation lies in the range $[\alpha_i, \alpha_i + \mathrm{d}\alpha_i]$,

$$P(\alpha_1, \ldots, \alpha_n) = K \exp\left(-\frac{1}{2k}\boldsymbol{\alpha}^T \boldsymbol{G}\boldsymbol{\alpha}\right), \tag{2.3}$$

where we've "packaged" fluctuations into a vector[8] $\boldsymbol{\alpha}^T \equiv (\alpha_1, \ldots, \alpha_n)$ with T transpose and the $n \times n$ symmetric matrix $\boldsymbol{G} \equiv [g_{ij}]$. We require that the probability be normalized:[9]

$$\int_{-\infty}^{\infty} \cdots \int_{-\infty}^{\infty} P(\alpha_1, \ldots, \alpha_n)\mathrm{d}\alpha_1 \cdots \mathrm{d}\alpha_n = 1. \tag{2.4}$$

Combining Eq. (2.3) with Eq. (2.4), we find $K = \sqrt{\det \boldsymbol{G}}/\,(2\pi k)^{n/2}$ (see Exercise 2.1). We'll refer to Eq. (2.3) as the *Einstein probability distribution.* Mathematically it has the form of a multivariate *Gaussian* or *normal* distribution (see Section 2.9).[10] Keep in mind that the Einstein distribution is approximate; it's based on the second-order Taylor expansion of $\Delta S(\alpha_1, \ldots, \alpha_n)$.

[6] We're "pushing" thermodynamics to tell us something about nonequilibrium systems. Entropy is a property of the equilibrium state; we're extending it to small deviations from equilibrium, with $|\Delta S|/S^0 \ll 1$.

[7] The probability of an event is proportional to the number of ways it can occur; see Appendix B. Equation (2.2) was first applied to the study of fluctuations by Einstein in 1910. See Pais[29, Chapter 4] for a review of Einstein's contributions to statistical physics beyond his 1905 investigations of Brownian motion.

[8] Representing fluctuations as a vector is for notational convenience and shouldn't be taken literally. The argument of the exponential in Eq. (2.3) must be dimensionless, which is made so by the metric $\boldsymbol{G}$.

[9] Equation (2.3) assumes small fluctuations (so that truncating the Taylor expansion in Eq. (1.40) is accurate), yet we're allowing $-\infty < \alpha_i < \infty$ in Eq. (2.4). $P(\alpha_1, \ldots, \alpha_n)$ is presumed sharply peaked about $P(0, \ldots, 0)$ so that extending the limits of integration introduces negligible error. It's not strictly necessary to normalize probability distributions, although highly convenient. At a minimum we require that probabilities be normalizable, whether or not we actually normalize.

[10] The misleading adjective "normal" was introduced by Karl Pearson, one of the founders of mathematical statistics, who later regretted it[30].

Example. Consider a system in thermal contact with a heat reservoir such that U and V fluctuate. The matrix $\boldsymbol{G}$ for this system is

$$\boldsymbol{G} = -\begin{pmatrix} \dfrac{\partial^2 S}{\partial U^2} & \dfrac{\partial^2 S}{\partial U \partial V} \\ \dfrac{\partial^2 S}{\partial V \partial U} & \dfrac{\partial^2 S}{\partial V^2} \end{pmatrix} = \begin{pmatrix} \dfrac{1}{T^2 C_V} & -\dfrac{1}{C_V T}\left(\dfrac{\alpha}{\beta_T} - \dfrac{P}{T}\right) \\ -\dfrac{1}{C_V T}\left(\dfrac{\alpha}{\beta_T} - \dfrac{P}{T}\right) & \dfrac{1}{\beta_T T V} + \dfrac{1}{C_V}\left(\dfrac{\alpha}{\beta_T} - \dfrac{P}{T}\right)^2 \end{pmatrix}, \tag{2.5}$$

where C_V is the constant-volume heat capacity, and α, β_T are the thermal expansivity and isothermal compressibility[2, p19]. You are asked in Exercise 2.2 to evaluate the derivatives in Eq. (2.5). Note that $\det \boldsymbol{G} = 1/(\beta_T T^3 V C_V)$. From thermodynamics, β_T and C_V are always positive, the conditions of mechanical and thermal stability[2, p48].

2.1.1 Characteristic functions and moments

It turns out that Fourier transforms of probability distributions, known as *characteristic functions*, often have a more direct relation to physical quantities of interest than the distributions themselves.[11] The characteristic function of an n-variate probability density $P(\alpha_1, \ldots, \alpha_n)$ is defined

$$\Phi(\omega_1, \ldots, \omega_n) \equiv \int_{-\infty}^{\infty} \cdots \int_{-\infty}^{\infty} \exp(\mathrm{i}\boldsymbol{\omega}^T \boldsymbol{\alpha}) P(\alpha_1, \ldots, \alpha_n) \mathrm{d}^n \alpha \equiv \langle \exp(\mathrm{i}\boldsymbol{\omega}^T \boldsymbol{\alpha}) \rangle, \tag{2.6}$$

where $\boldsymbol{\omega} \equiv (\omega_1, \ldots, \omega_n)^T$ is a vector of transform variables ω_i (one for each α_i), $\mathrm{d}^n\alpha \equiv \mathrm{d}\alpha_1 \cdots \mathrm{d}\alpha_n$, and the angular brackets denote an average. It's often easier to find $\Phi(\boldsymbol{\omega})$ than $P(\boldsymbol{\alpha})$, in which case $P(\boldsymbol{\alpha})$ is obtained through inverse Fourier transformation. Approximations are also easier to develop for $\Phi(\boldsymbol{\omega})$ than for $P(\boldsymbol{\alpha})$ (see Section 2.9.1). Note that $\Phi(\boldsymbol{\omega} = 0) = 1$ is the normalization integral on $P(\boldsymbol{\alpha})$. General theorems on characteristic functions are covered in Cramér[31, Chapter 10] or in [5, p75].

One use of characteristic functions is to find the *moments* of probability distributions (see Appendix B).[12] The moment $\langle \alpha_1^{k_1} \cdots \alpha_n^{k_n} \rangle$ is obtained from derivatives of $\Phi(\omega_1, \ldots, \omega_n)$,

$$\langle \alpha_1^{k_1} \cdots \alpha_n^{k_n} \rangle \equiv \int_{-\infty}^{\infty} \cdots \int_{-\infty}^{\infty} \alpha_1^{k_1} \cdots \alpha_n^{k_n} P(\alpha_1, \ldots, \alpha_n) \mathrm{d}^n \alpha = (-\mathrm{i})^k \left. \frac{\partial^k \Phi(\omega_1, \ldots, \omega_n)}{\partial^{k_1}\omega_1 \cdots \partial^{k_n}\omega_n} \right|_{\boldsymbol{\omega}=0}, \tag{2.7}$$

where $k \equiv k_1 + \cdots + k_n$. Thus Φ must be differentiable at $\boldsymbol{\omega} = 0$ for moments to exist.[13] If one knew all the moments, but not $P(\boldsymbol{\alpha})$ itself, one could in principle reconstruct $\Phi(\boldsymbol{\omega})$,[14] from which one could infer $P(\boldsymbol{\alpha})$. For $k = 1 + 1 = 2$, Eq. (2.7) generates the *covariance matrix*

$$\langle \alpha_r \alpha_m \rangle = -\left. \frac{\partial^2}{\partial \omega_r \partial \omega_m} \Phi(\omega_1, \ldots, \omega_n) \right|_{\boldsymbol{\omega}=0}. \tag{2.8}$$

---

[11] The term characteristic function is typically unfamiliar to science students. Such students are, however, quite familiar with the role of integral transforms in simplifying the analysis of physical problems; probability theory is no different.

[12] Characteristic functions are also called moment generating functions, and moments are also called correlation functions. We presume familiarity with bracket notation $\langle\rangle$ as indicating averages.

[13] The Cauchy distribution $P(\alpha) = (1/\pi)1/(1+\alpha^2)$ is an example of a normalized probability distribution for which $\Phi(\omega) = \mathrm{e}^{-|\omega|}$ exists, but is not differentiable at $\omega = 0$[5, p73]. The Cauchy distribution has no moments higher than the zeroth. Consider that for the second moment, $\int_{-\infty}^{\infty} \frac{x^2}{1+x^2} \mathrm{d}x = \int_{-\infty}^{\infty} \left[1 - \frac{1}{1+x^2}\right] \mathrm{d}x$.

[14] One must check that certain relations among the moments hold; see [32].

The characteristic function associated with the Einstein distribution is, combining Eq. (2.3) with Eq. (2.6),

$$\Phi(\boldsymbol{\omega}) = K\int_{-\infty}^{\infty}\cdots\int_{-\infty}^{\infty}\exp\left(-\frac{1}{2k}\boldsymbol{\alpha}^T\boldsymbol{G}\boldsymbol{\alpha} + \mathrm{i}\boldsymbol{\omega}^T\boldsymbol{\alpha}\right)\mathrm{d}^n\alpha = \exp\left(-\frac{k}{2}\boldsymbol{\omega}^T\boldsymbol{G}^{-1}\boldsymbol{\omega}\right), \tag{2.9}$$

where $\boldsymbol{G}^{-1}$ is the matrix inverse of $\boldsymbol{G}$; see Exercise 2.5. Combining Eqs. (2.9) and (2.7), we find

$$\begin{aligned}\langle\alpha_r\rangle &= 0\\ \langle\alpha_r\alpha_m\rangle &= k\left(\boldsymbol{G}^{-1}\right)_{rm}.\end{aligned} \tag{2.10}$$

Positive fluctuations are as equally likely as negative, hence $\langle\alpha_r\rangle = 0$ (more generally there can be symmetry-breaking mechanisms so that $\langle\alpha_r\rangle \neq 0$). The second result is highly useful.

Example. From the matrix $\boldsymbol{G}$ in Eq. (2.5),

$$\boldsymbol{G}^{-1} = \begin{pmatrix} C_VT^2 + \beta_TVT^3\left(\frac{\alpha}{\beta_T} - \frac{P}{T}\right)^2 & \beta_TVT^2\left(\frac{\alpha}{\beta_T} - \frac{P}{T}\right) \\ \beta_TVT^2\left(\frac{\alpha}{\beta_T} - \frac{P}{T}\right) & \beta_TVT \end{pmatrix} = \frac{1}{k}\begin{pmatrix}\langle(\Delta U)^2\rangle & \langle\Delta U\Delta V\rangle \\ \langle\Delta V\Delta U\rangle & \langle(\Delta V)^2\rangle\end{pmatrix} \tag{2.11}$$

where we've used Eq. (2.10) in the second equality. Note for the ideal gas[15] that $\alpha/\beta_T = P/T$.

The Einstein distribution follows from representing ΔS with the second term in its Taylor series in fluctuations of extensive quantities. It can be shown that Eq. (2.10) is rigorously obtained even when the truncation of ΔS implied by Eq. (1.40) is not made[33]. The higher-order moments calculated from the Einstein distribution are incorrect, however.[16] Fortunately the first two moments are most frequently used in physical applications.

2.1.2 Force-fluctuation correlations

The thermodynamic force F_k conjugate to fluctuation α_k is, from Eq. (1.43), $F_k = \partial\Delta S/\partial\alpha_k = -\sum_j g_{kj}\alpha_j$, or in vector form, $\boldsymbol{F} = -\boldsymbol{G}\boldsymbol{\alpha}$, where $\boldsymbol{F} = (F_1,\ldots,F_n)^T$. Thus, $\boldsymbol{\alpha} = -\boldsymbol{G}^{-1}\boldsymbol{F}$, or in component form,

$$\alpha_i = -\sum_k\left(\boldsymbol{G}^{-1}\right)_{ik}F_k, \tag{2.12}$$

implying

$$\frac{\partial\alpha_i}{\partial F_l} = -\left(\boldsymbol{G}^{-1}\right)_{il} = -\frac{1}{k}\langle\alpha_i\alpha_l\rangle,$$

where we've used Eq. (2.10). For the correlation of fluctuations and forces,

$$\langle\alpha_jF_l\rangle = -\sum_m g_{lm}\langle\alpha_j\alpha_m\rangle = -k\sum_m \boldsymbol{G}_{lm}\left(\boldsymbol{G}^{-1}\right)_{mj} = -k\delta_{lj}, \tag{2.13}$$

where we've used Eqs. (1.43) and (2.10). Even though each fluctuation α_i has instantaneous projections along thermodynamic forces [see Eq. (2.12)], on average there is a nonzero correlation[17] only among conjugate force-fluctuation pairs, Eq. (2.13), and they're anticorrelated (negative correlation). Equation (2.13) is important in the Onsager theory of irreversible processes; it can be derived directly, without the use of $\boldsymbol{G}$; see Exercise 2.10.

---

[15] For the ideal gas, internal energy is independent of volume (Joule's law[2, p21]); U and V are uncorrelated.

[16] To quote H.B. Callen[10, p280]: "The moments of fluctuating extensive parameters were first computed by Einstein in 1910. Einstein's method was an approximate one, which happens to give precisely the correct results for the second moments but which gives inexact results for the higher moments."

[17] Quantities having zero covariance are said to be uncorrelated.

2.2 MICROSCOPIC REVERSIBILITY AND ONSAGER RECIPROCITY

Onsager reciprocity follows from the time-reversal properties of fluctuations[18] and the stationarity of equilibrium averages. Remarkably, kinetic coefficients—associated with irreversibility—have a fundamental symmetry as a result of microscopic reversibility. Why microscopic theories feature time reversibility yet macroscopically there is a unique direction of time is a perennial question.[19] The governing theory of irreversibility is at root the second law of thermodynamics, which cannot be reduced to microdynamics.[20]

Consider fluctuation α_i at time t and fluctuation α_j at time $t+\tau$ with $\tau > 0$ and form their product $\alpha_i(t)\alpha_j(t+\tau)$. Define the time average of the product,

$$\overline{\alpha_i(t)\alpha_j(t+\tau)} \equiv \lim_{T\to\infty} \frac{1}{T}\int_0^T \alpha_i(t)\alpha_j(t+\tau)\mathrm{d}t. \tag{2.14}$$

By Birkhoff's theorem[28] the limit exists when the time dependence of the integrand is generated by Hamilton's equations of motion;[21] $\lim_{T\to\infty}(1/T)\int_{-T}^{0}$ also exists. The limit is independent of the origin of time: $\lim_{T\to\infty}(1/T)\int_{t_0}^{T+t_0}\alpha_i(t)\alpha_j(t+\tau)\mathrm{d}t$ is independent[22] of t_0. A proof of Onsager reciprocity requires us to distinguish fluctuations that are even and odd in the velocities of particles. Following Casimir[21], we refer to fluctuations even in velocities as α-variables and those odd in velocities as β-variables. We discuss α-variables first.

Fluctuations even under time inversion ($t \to -t$) have the property

$$\overline{\alpha_i(t)\alpha_j(t+\tau)} = \overline{\alpha_i(t)\alpha_j(t-\tau)}. \tag{2.15}$$

The correlation of $\alpha_i(t)$ and $\alpha_j(t+\tau)$ is the same as that of $\alpha_i(t)$ and $\alpha_j(t-\tau)$.[23] It's immaterial whether or not α_j occurs before α_i; *only the relative time between events is significant*. By shifting the origin of time $t \to t+\tau$ on the right side of Eq. (2.15),

$$\overline{\alpha_i(t)\alpha_j(t+\tau)} = \overline{\alpha_i(t+\tau)\alpha_j(t)}. \tag{2.16}$$

Equation (2.16) is Onsager reciprocity in disguised form, a "time-influence" symmetry of fluctuations, that if $\alpha_i(t)$ is correlated with (influences) $\alpha_j(t+\tau)$ then $\alpha_j(t)$ influences $\alpha_i(t+\tau)$ in the same way. It's a consequence of the time-reversal invariance of the microscopic dynamics and the stationarity of equilibrium averages (no unique origin in time).

At this point subtract $\overline{\alpha_i(t)\alpha_j(t)}$ from both sides of Eq. (2.16) and divide by τ:

$$\frac{1}{\tau}\overline{\alpha_i(t)\left[\alpha_j(t+\tau)-\alpha_j(t)\right]} = \frac{1}{\tau}\overline{\left[\alpha_i(t+\tau)-\alpha_i(t)\right]\alpha_j(t)}.$$

---

[18] Under $t \to t' = -t$, $\boldsymbol{p} \to \boldsymbol{p}' = -\boldsymbol{p}$, $\boldsymbol{q} \to \boldsymbol{q}' = \boldsymbol{q}$, and $H(\boldsymbol{p}',\boldsymbol{q}') = H(-\boldsymbol{p},\boldsymbol{q}) = H(\boldsymbol{p},\boldsymbol{q})$; Hamilton's equations of motion are therefore time-reversal invariant. The same is true at the quantum level. For every solution of the Schrödinger equation $\psi(t)$ (having a self-adjoint Hamiltonian), there is a solution $\psi^*(-t)$[5, p34]; the system is dynamically reversible.

[19] The discovery in 1964 of CP violations in weak decays (1980 Nobel Prize in Physics) implies that time reversal symmetry *is* broken for sub-nuclear processes involving weak interactions. At the *super* macroscopic level, we have an expanding universe, which provides a sense for the direction of time.

[20] The second law is the explicit recognition in the pantheon of physical laws of the existence of irreversibility. Unless we seek to derive the second law from something more fundamental, we don't have to account for irreversibility; it simply is. We tend to think microscopic laws are more "real" than macroscopic; we're accustomed to breaking things apart in the attempt to say that the whole can explained by the sum of its parts, and sometimes it works that way, but not always. Unless and until a more fundamental understanding emerges, laws have their domains of validity. The second law is not inherent in microscopic laws of motion; it's a separate principle that stands on its own. "More is different," to quote P.W. Anderson[34]. That makes nonequilibrium statistical mechanics interesting but challenging.

[21] See [5, p48]. Birkhoff's theorem also applies to fluctuations treated as random variables; see Section 2.5.

[22] If $\lim_{T\to\infty}(1/2T)\int_{-T}^{T}|f(t)|^2\mathrm{d}t$ exists, then so does $\lim_{T\to\infty}(1/2T)\int_{-T+t_0}^{T+t_0}|f(t)|^2\mathrm{d}t$; see Wiener[35, p38].

[23] Consider a movie of a kicker kicking a ball and a blocker jumping up to grab the ball. The blocker grabs the ball after it's been kicked. Now run the movie backwards. In the reversed sense of time, the ball leaves the blocker's hands before it arrives at the kicker's foot.

Take the limit $\tau \to 0$ (see Section 2.3.2 for the conditions under which that's allowed),

$$\overline{\alpha_i(t)\dot{\alpha}_j(t)} = \overline{\dot{\alpha}_i(t)\alpha_j(t)}. \tag{2.17}$$

In Eq. (2.17), use Eq. (1.42) which equates fluxes with time derivatives of fluctuations (Onsager regression hypothesis), and combine with Eq. (1.36), the linear force-flux relation:

$$\sum_m L_{jm}\overline{\alpha_i F_m} = \sum_n L_{in}\overline{\alpha_j F_n}.$$

Equate time averages with ensemble averages (ergodic hypothesis), $\overline{\alpha_i F_m} = \langle \alpha_i F_m \rangle = -k\delta_{im}$ [see Eq. (2.13)] and we arrive at the reciprocal relations, $L_{ji} = L_{ij}$.

In systems subject to external magnetic fields, under time reversal[24] $\boldsymbol{B} \to -\boldsymbol{B}$. The Lorentz force $q\boldsymbol{v} \times \boldsymbol{B}$ is therefore invariant under time reversal. The Onsager relations must be modified to indicate that fluctuations on the left side of Eq. (2.16) are functions of $\boldsymbol{B}$ but those on the (time-reversed) right side are functions of $-\boldsymbol{B}$. Equation (1.38) is therefore written $L_{ij}(\boldsymbol{B}) = L_{ji}(-\boldsymbol{B})$. Are there *other* velocity-dependent forces? In a reference frame rotating uniformly with angular velocity $\boldsymbol{\Omega}$, particles experience the Coriolis force $2m\boldsymbol{v} \times \boldsymbol{\Omega}$. Under time reversal, $\boldsymbol{\Omega} \to -\boldsymbol{\Omega}$. Thus, fluctuations on the left side of Eq. (2.16) are functions of $\boldsymbol{\Omega}$ and those on the right are functions of $-\boldsymbol{\Omega}$. The Onsager relations are therefore $L_{ij}(\boldsymbol{B}, \boldsymbol{\Omega}) = L_{ji}(-\boldsymbol{B}, -\boldsymbol{\Omega})$.

Equation (2.15), and everything subsequent to it, holds for fluctuations even under time reversal. It's traditional to denote fluctuations odd under time reversal as β_i. The analogs of Eq. (2.15) for these variables are $\overline{\alpha_i(t)\beta_j(t+\tau)} = -\overline{\alpha_i(t)\beta_j(t-\tau)}$ and $\overline{\beta_i(t)\beta_j(t+\tau)} = \overline{\beta_i(t)\beta_j(t-\tau)}$. The Onsager relations in their most general form can be written

$$L_{ij}(\boldsymbol{B}, \boldsymbol{\Omega}) = \epsilon_i \epsilon_j L_{ji}(-\boldsymbol{B}, -\boldsymbol{\Omega}),$$

with ϵ_i, ϵ_j the *signatures* of fluctuations under time reversal, $\epsilon = 1$ if even and $\epsilon = -1$ if odd.[25]

2.3 STOCHASTIC PROCESSES: ADDING TIME TO PROBABILITY

We've classified fluctuations according to their time-reversal properties, yet we have no way of calculating them from first principles. We turn to *stochastic processes*, the branch of probability theory that adds time to probabilistic descriptions.[26] In what follows, we consider the case where a single quantity can fluctuate,[27] such as internal energy, where $\alpha \equiv U - \langle U \rangle$ is a random variable.[28]

Definition. *A stochastic process is a family of random variables* $\{\alpha(t){:}t \in T\}$ *where* T *is a set of real numbers, usually interpreted as time. For every* $t \in T$*, a random number* $\alpha(t)$ *is observed. If* T *is denumerable, it's a discrete-time process; if* T *is not countable, it's a continuous-time process.*

Stochastic processes are families of random variables indexed by time, a definition too general to provide a way of constructing them. Instead, they're specified by a *hierarchy* of joint probability densities obeying prescribed rules. A stochastic process *is* its hierarchy of joint probabilities.

[24] Under time reversal, currents that establish $\boldsymbol{B}$ are reversed, implying the reversal of $\boldsymbol{B}$.

[25] Equation (2.16) should therefore be written $\overline{\alpha_i(t)\alpha_j(t+\tau)} = \epsilon_i\epsilon_j\overline{\alpha_i(t+\tau)\alpha_j(t)}$.

[26] There is no clear-cut reason to use the term stochastic over probabilistic. Stochastic is often followed by the word *process*; stochastic processes emulate data that's collected sequentially in time. To quote J.L. Doob[36], "A stochastic process is the mathematical abstraction of an empirical process whose development is governed by probabilistic laws."

[27] A stochastic process involves a single random variable sampled at various times. Such processes can be defined for all other fluctuating quantities in a system, and correlations between different stochastic processes can be investigated.

[28] Random variables assign a number to every outcome of an experiment; see Appendix B.

2.3.1 The consistency conditions

We found in Section 2.1 the joint probability density for n random variables $(\alpha_1, \ldots, \alpha_n)$ representing fluctuations in n different physical quantities, all measured at the same time.[29] Here we consider fluctuations in the *same* quantity observed at n different times, $(t_1, \ldots, t_n)$. The joint probability density for an n-time stochastic process is a function of $2n$ variables, the n times $(t_1, \ldots, t_n)$ and the n random variables[30] $(\alpha(t_1), \ldots, \alpha(t_n))$, indicated notationally as a function of the times t_i and the values of random variables $y_i = \alpha(t_i)$,

$$P_n(y_1, t_1; y_2, t_2; \ldots; y_n, t_n) = \text{Prob}\left(\alpha(t_1) = y_1; \alpha(t_2) = y_2; \ldots; \alpha(t_n) = y_n\right).$$

The subscript n on P_n indexes a collection of joint probabilities which, to comprise a stochastic process, must possess certain properties known as the *consistency conditions*:[31]

(a) $P_n \geq 0$ (they *are* probabilities);

(b) The *symmetry condition*, that P_n is invariant under the interchange[32] of pairs (y_i, t_i) and (y_j, t_j) for $i, j \in 1, \ldots, n$. There is no loss of generality therefore in ordering the times $t_1 < \cdots < t_n$;

(c) $\int \mathrm{d}y_n P_n(y_1, t_1; y_2, t_2; \ldots; y_n, t_n) = P_{n-1}(y_1, t_1; y_2, t_2; \ldots; y_{n-1}, t_{n-1})$, the condition that higher-order distributions imply the lower-order ones;

(d) $\int \mathrm{d}y_1 P_1(y_1, t_1) = 1$.

Note that integrations are over the y-variables only. The quantity P_n is defined for $t_1 \neq t_2 \neq \cdots \neq t_n$. If two times are identical, such as $t_n = t_{n-1}$, we have the stochastic process $P_{n-1}(y_1, t_1; \ldots; y_{n-1}, t_{n-1})$. No contradiction occurs if we define

$$P_n(y_1, t_1; \ldots; y_{n-1}, t_{n-1}; y_n, t_n = t_{n-1}) = \delta(y_n - y_{n-1}) P_{n-1}(y_1, t_1; \ldots; y_{n-1}, t_{n-1}). \quad (2.18)$$

Functions $\{P_n\}$ satisfying the consistency conditions completely specify stochastic processes.[33]

We require conditional probabilities. Let $P_{1|1}(y_2, t_2 | y_1, t_1)$ denote the probability that $\alpha(t_2) = y_2$, given that $\alpha(t_1) = y_1$. Conditional probabilities specify *subensembles*, that of the realizations of $\alpha(t)$ having a value between $[y_1, y_1 + \mathrm{d}y_1]$ at time t_1, the fraction $P_{1|1}(y_2, t_2 | y_1, t_1)$ has the value between $[y_2, y_2 + \mathrm{d}y_2]$ at t_2. More generally,

$$\begin{aligned} &P_{l|k}(y_{k+1}, t_{k+1}; \ldots; y_{k+l}, t_{k+l} | y_1, t_1; \ldots; y_k, t_k) \\ &\qquad \equiv \frac{P_{k+l}(y_1, t_1; \ldots; y_k, t_k; y_{k+1}, t_{k+1}; \ldots; y_{k+l}, t_{k+l})}{P_k(y_1, t_1; \ldots; y_k, t_k)}, \end{aligned} \quad (2.19)$$

i.e., $P_{l|k}$ is the probability that l samplings of $\alpha(t)$ have the values indicated, given that k samplings are known to have produced the values indicated. Conditional probabilities satisfy the requirements:

(a) $P_{n|v} \geq 0$;

(b) $\int \mathrm{d}y_1 \cdots \mathrm{d}y_n P_{n|v} = 1$;

---

[29] The "prescribed set" of fluctuations introduced in Section 1.7 were all observed at the same time, "instantaneous."

[30] The values $\alpha(t_j)$ would be known if $\alpha(t)$ were an ordinary function of t; as it is, however, $\alpha(t)$ is a random variable.

[31] Compare with similar requirements on probability in Appendix B. We're being cavalier in not always distinguishing probabilities from probability densities (ditto with joint probabilities and conditional probabilities). It should be clear from the context what is meant. We sum over discrete probabilities and integrate over probability densities.

[32] Joint probabilities are in the form of a series of "and" statements, the probability that $\alpha(t_1) = y_1$, and $\alpha(t_2) = y_2$, and so on. The order in which one makes these statements is immaterial.

[33] A theorem due to Kolmogorov[37, p37] states that any functions satisfying the consistency conditions (and only the consistency conditions) define a stochastic process. Specifying the joint probabilities P_n (that satisfy the consistency conditions) is a way of specifying stochastic processes that bears a direct relation to the physics of the problem.

(c)

$$P_n(y_1,t_1;\ldots;y_n,t_n) = \int\cdots\int \mathrm{d}y_{n+1}\cdots\mathrm{d}y_{n+v}\Big[P_{n|v}(y_1,t_1;\ldots;y_n,t_n|y_{n+1},t_{n+1};\ldots;y_{n+v},t_{n+v}) \\ \times P_v(y_{n+1},t_{n+1};\ldots;y_{n+v},t_{n+v})\Big].$$

2.3.2 Stochastic calculus

The Onsager regression hypothesis relies on the ability to differentiate fluctuations, and the reciprocal conditions rely on the ability to integrate fluctuations. Can calculus be applied to stochastic processes? The definite integral of a stochastic process $y(t)$ over $a \le t \le b$ is defined in the usual way as the limit of approximating sums:

$$\int_a^b y(t)\mathrm{d}t \equiv \lim \sum_{k=1}^{n} y(t'_k)\,(t_k - t_{k-1}), \tag{2.20}$$

where the limit is over subdivisions of $[a,b]$, $a = t_0 < t_1 < \cdots < t_n = b$, with $t_{k-1} \le t'_k \le t_k$, as the maximum length of the sub-intervals $(t_k - t_{k-1})$ tends to zero. The integral $\int_{-\infty}^{\infty} y(t)dt$ is the limit of Eq. (2.20) as $a \to -\infty$ and $b \to \infty$. Although Eq. (2.20) is the customary definition of Riemann integral, it presents us with a question specific to this case: Does an infinite sum of random variables possess a limit? A finite sum of random variables is itself a random variable. The question becomes, can we speak of the *convergence* of a sequence of random variables?[34] There are several ways by which convergence of random variables is defined; see Parzen[38, Chapter 10]. We adopt the following: A sequence of random variables $Z_1,\ldots,Z_n$ is said to *converge in mean square* to the random variable Z if[35] $\lim_{n\to\infty}\langle |Z - Z_n|^2\rangle = 0$. A necessary and sufficient condition (Loève[39, p472]) for the approximating sums in Eq. (2.20) to have a limit (in mean square) is that the correlation function $\langle y(s)y(t)\rangle$ be integrable over $(a \le s \le b, a \le t \le b)$. Assuming the integrability of $\langle y(s)y(t)\rangle$, the linear operations of integration and calculating expectation values commute,

$$\left\langle \int_a^b y(t)\mathrm{d}t \right\rangle = \int_a^b \langle y(t)\rangle \mathrm{d}t.$$

The derivative of a random process $y(t)$ having a finite second moment is defined as the limit of the difference quotient,

$$y'(t) \equiv \lim_{h\to 0}\frac{1}{h}\left[y(t+h) - y(t)\right],$$

where the limit is taken as convergence in mean square. The limit exists (Loève[39, p470]) if the correlation function $K(s,t) \equiv \big\langle \left(y(s) - \langle y(s)\rangle\right)\left(y(t) - \langle y(t)\rangle\right) \big\rangle$ has continuous mixed second partial derivatives $\partial^2 K(s,t)/\partial s\partial t$. A process $y(t)$ is said to be *differentiable in mean square* if $y'(t)$ exists as a limit in mean square. Under these conditions, the operations of differentiation and expectation values commute,

$$\left\langle y'(t) \right\rangle = \frac{\mathrm{d}}{\mathrm{d}t}\langle y(t)\rangle. \tag{2.21}$$

Thus, the question of whether the rules of calculus can be applied to stochastic processes reduces to the integrability and differentiability of the correlation function $\langle y(t)y(s)\rangle$ as a function of (t,s). For future reference, a *stochastic differential equation*[40] (as applied to dynamical systems) is one subject to stochastic driving forces. The solution of a stochastic differential equation is itself

[34] It helps not to get hung up on the word *random*. One might think that a truly random variable would occasionally have large, ostensibly infinite, values. Random variables refer to unpredictable outcomes of experiments (see Appendix B), and experiments on physical systems will not produce indefinitely large values of measured quantities.

[35] The convergence criterion implies that $\langle |Z - Z_n|^2\rangle < \epsilon$ for $n > N(\epsilon)$, for every positive number ϵ.

a stochastic process. If an input function $I(t)$ is a stochastic process, then the solution $X(t)$ of a linear differential equation $a_0(t)X^{(n)}(t)+a_1(t)X^{(n-1)}(t)+\cdots+a_n(t)X(t)=I(t)$ is a stochastic process, where the coefficients $a_k(t)$ are non-random functions of time.

2.4 STATIONARY, INDEPENDENT, AND MARKOV PROCESSES

The definitions given in Section 2.3 pertain to any stochastic process and comprise a level of generality at which little else can be accomplished. To make progress, ideas motivated by physical conditions must be brought to bear. In this section, three classes of stochastic processes are defined. An additional two will be introduced in later sections (Gaussian and differential).

2.4.1 Stationary processes: No unique origin of time

Definition. *A stochastic process is stationary in the strict sense if, for all n and τ,*

$$P_n(y_1,t_1;y_2,t_2;\ldots;y_n,t_n)=P_n(y_1,t_1+\tau;y_2,t_2+\tau;\ldots;y_n,t_n+\tau). \tag{2.22}$$

Stationary processes have no unique origin in time and are therefore a *proxy for the dynamics of fluctuations*; their statistical properties remain the same if $t_1,t_2,\ldots,t_n$ are each shifted by τ. Thus, P_1 is independent of time, $P_1(y_1,t_1)=P_1(y_1)$, and $P_2(y_1,t_1;y_2,t_2)$ is a function of the single time $t_1-t_2=\tau$, $P_2(y_1,t+\tau;y_2,t)$. The mean of a stationary process is therefore a constant

$$\langle y(t)\rangle=\int \mathrm{d}y y P_1(y,t)=c=\text{constant}, \tag{2.23}$$

and the *autocorrelation function* is a function only of the relative time between events

$$R(\tau)\equiv\langle y(t+\tau)y(t)\rangle=\int\int \mathrm{d}y_1\mathrm{d}y_2 y_1 y_2 P_2(y_1,t+\tau;y_2,t). \tag{2.24}$$

Clearly $R(0)>0$. From Exercise 2.12, $R(\tau)=R(-\tau)$ and $|R(\tau)|\le R(0)$ for all τ. For every system there is a time τ_c, the *correlation time*, such that $R(\tau)\overset{\tau\gg\tau_c}{\longrightarrow}\langle y(t)\rangle\langle y(t+\tau)\rangle$, i.e., $y(t)$ and $y(t+\tau)$ become uncorrelated for $\tau\gg\tau_c$. Do stationary processes have time derivatives? The function $R(\tau)$ has a continuous second derivative for any τ if it has a continuous second derivative at $\tau=0$. It may happen that a process satisfies Eqs. (2.23) and (2.24) but Eq. (2.22) cannot be checked for all n. Processes are termed *stationary in the wide sense* if, for all τ, Eq. (2.22) is satisfied for $n=1,2$. Theorems that might be easy to show for wide-sense stationarity often become difficult for strict-sense stationarity.[36]

For a process to be stationary, it must have lasted forever and it must last forever, $-\infty<\tau<\infty$. Physical processes can't truly be stationary, therefore, which are of limited duration. Stationarity can be invoked to a high degree of accuracy for systems that last longer than dynamical phenomena of interest, such as in thermal equilibrium.[37]

2.4.2 Independent processes: Present is independent of the past

Suppose in a coin-flipping experiment one has obtained a long string of heads, $HHH\cdots H$. What are the odds of obtaining H in the next flip? It might seem that the chances of tails would go up—one just "has" to obtain tails at some point. Unless the coins are correlated in an unusual way, the odds of obtaining heads or tails on the next flip are the same; the outcome of the next flip is *independent of* (has no memory of) the history of the process. Many problems in the theory of probability are based on independent events. If only physics were so simple.

[36] Strict-sense stationarity implies wide-sense stationarity but not conversely.

[37] Equilibrium is the state where "all the 'fast' things have happened and all the 'slow' things not"[3, p1].

Definition. *An independent stochastic process is one for which either statement holds (for all k):*

$$P_k(y_1, t_1; \ldots ; y_k, t_k) = \prod_{l=1}^{k} P_1(y_l, t_l) \tag{2.25}$$
$$P_{1|k-1}(y_k, t_k | y_1, t_1; \ldots ; y_{k-1}, t_{k-1}) = P_1(y_k, t_k).$$

For independent processes, a knowledge of P_1 is a complete solution of the problem. Because $P_{1|k-1}$ is replaced with P_1, the outcome of the next trial is independent of the system's past. Independent processes are sometimes referred to as *purely random processes.*

2.4.3 Markov processes: Present depends on the immediate past

Flipping coins is a discrete-time process for which there is obviously a time interval between flips. One could flip a coin and wait a week before flipping the next; *independent events are uncorrelated and are not affected by the time between them.* Processes exist, however, where $y(t_1 + \tau)$ is not independent of $y(t_1)$ for all $\tau \geq 0$, where $y(t_2)$ and $y(t_1)$ are *correlated* if $|t_2 - t_1|$ is small enough. Time correlations do not exist in independent processes. We need a richer framework; we must be able to incorporate *memory* of past events.

The simplest way to do that is with a *Markov process,* in which $y(t_n)$ depends only on the most recent past observation $y(t_{n-1})$ and not on observations made at earlier times.[38]

Definition. *In a Markov process, $P_{n|v>1}$ is replaced with $P_{n|1}$,*

$$P_{n|v}(y_1, t_1; \ldots ; y_n, t_n | y_{n+1}, t_{n+1}; \ldots ; y_{n+v}, t_{n+v}) \overset{\substack{\text{Markov} \\ \text{approximation}}}{\longrightarrow} P_{n|1}(y_1, t_1; \ldots ; y_n, t_n | y_{n+1}, t_{n+1}),$$

for $t_1 > \cdots > t_n > t_{n+1} > \cdots > t_{n+v}$.

The history recorded in $P_{n|v>1}$ is therefore not fully retained. If the times $(t_1, \ldots, t_n)$ occur in discrete multiples of a time unit, $t_n = n\Delta$, a Markov process is referred to as a *Markov chain.*

Markov processes are determined by P_2. One can show, by iterating the definition of conditional probability, that for $t_1 > \cdots > t_n$,

$$P_n(y_1, t_1; \ldots ; y_n, t_n) = P_{1|n-1}(y_1, t_1 | y_2, t_2; \ldots ; y_n, t_n) P_{1|n-2}(y_2, t_2 | y_3, t_3; \ldots ; y_n, t_n) \cdots$$
$$P_{1|1}(y_{n-1}, t_{n-1} | y_n, t_n) P_1(y_n, t_n). \tag{2.26}$$

Apply the Markov approximation to the conditional probabilities in Eq. (2.26) (compare with Eq. (2.25)):

$$P_n(y_1, t_1; \ldots ; y_n, t_n) \overset{\substack{\text{Markov} \\ \downarrow}}{=} \prod_{i=1}^{n-1} P_{1|1}(y_i, t_i | y_{i+1}, t_{i+1}) P_1(y_n, t_n). \tag{2.27}$$

All members of the hierarchy therefore follow from knowledge of $P_{1|1}$ and P_1. But, because

$$P_{1|1}(y_i, t_i | y_{i+1}, t_{i+1}) = \frac{P_2(y_i, t_i; y_{i+1}, t_{i+1})}{P_1(y_{i+1}, t_{i+1})} = \frac{P_2(y_i, t_i; y_{i+1}, t_{i+1})}{\int \mathrm{d}y_i P_2(y_i, t_i; y_{i+1}, t_{i+1})},$$

one could derive the hierarchy from knowledge of just P_2.

One can't choose $P_{1|1}$ arbitrarily, however; it's constrained by a consistency relation we now derive. From Eq. (2.27), we have for $n = 3$, where $t_1 > t_2 > t_3$,

$$P_3(y_1, t_1; y_2, t_2; y_3, t_3) = P_{1|1}(y_1, t_1 | y_2, t_2) P_{1|1}(y_2, t_2 | y_3, t_3) P_1(y_3, t_3). \tag{2.28}$$

[38] The mathematician A.A. Markov studied (in 1907) the alternation between vowels and consonants in a piece of Russian literature. He found that the occurrence of a vowel or a consonant depended strongly on whether the immediately preceding letter was a vowel or a consonant, but only weakly on the character of earlier letters.

Integrating Eq. (2.28) over y_2, we find

$$P_{1|1}(y_1,t_1|y_3,t_3) = \int \mathrm{d}y_2 P_{1|1}(y_1,t_1|y_2,t_2)P_{1,1}(y_2,t_2|y_3,t_3). \tag{2.29}$$

Equation (2.29) is the *Smoluchowski-Chapman-Kolmogorov equation* (SCK),[39] a basic equation of Markov processes.[40] It has an interpretation reminiscent of quantum physics: For $t_1 > t_2 > t_3$, the probability of $\alpha(t_1) = y_1$ given that $\alpha(t_3) = y_3$ can be found by considering the process as a sum over compound processes consisting of the probability of $\alpha(t_2) = y_2$ given $\alpha(t_3) = y_3$ multiplied by the probability of $\alpha(t_1) = y_1$ given $\alpha(t_2) = y_2$. Given this (appealing) interpretation, it's easy to forget that the SCK equation has been derived within the Markov approximation.[41,42] The conditional probability $P_{1|1}(y_1,t_1|y_2,t_2)$ is referred to as a *transition probability*: the probability the system will be in state y_1 at time t_1 when it was in state y_2 at time t_2. Said differently, the system makes a transition from y_2 to y_1 in the time interval $(t_1 - t_2)$ with probability $P_{1|1}$.

Independent processes are specified by P_1 and Markov processes are specified by P_2. One could go on and consider processes specified by P_3. Markov processes, however, find such prevalent use that higher-order processes are simply referred to as *non-Markovian*. Wide-sense stationary processes are also specified by P_2. Processes that are stationary *and* Markov are of great interest in modeling fluctuations.

2.5 ERGODIC AND SPECTRAL PROPERTIES

2.5.1 Ergodicity

In proving Onsager reciprocity we considered time correlation functions defined as time averages, $\overline{y(t)y(t+\tau)} \equiv \lim_{T\to\infty}(1/T)\int_0^T y(t)y(t+\tau)\mathrm{d}t$. In completing the derivation, we equated time averages with ensemble averages (see page 24); we invoked ergodicity.[43] Fluctuations were treated as dynamical variables in applying Birkhoff's theorem;[44] we did not treat fluctuations as stochastic processes. Birkhoff's theorem, however, applies to stationary stochastic processes.[45] *Stationary processes must exhibit ergodicity to serve as models of fluctuations.* For $\{y(t_i), i = 1,2,\ldots\}$ a strictly stationary stochastic process, Birkhoff and others have shown that for a function g for which $\langle g(y(t))\rangle$ exists (and is independent of t; see Eq. (2.23)), the time average $(1/T)\sum_{i=1}^{T} g(y(t_i))$ converges as $T \to \infty$ and is equal to the ensemble average when certain supplementary conditions are met.[46] Ergodicity for strictly stationary processes is beyond the scope of this book.[47] Processes stationary in the wide sense, however, obey an ergodic theorem that's easier to digest.

---

[39] Equation (2.29) was derived by Smoluchowski in 1906, Chapman in 1916, and Kolmogorov in 1931. It's frequently referred to as the Chapman-Kolmogorov equation, and infrequently as the Smoluchowski equation.

[40] One could start with the SCK equation as the definition of Markov process.

[41] The Markovian character of the SCK equation is exhibited by the fact that the probability of the transition $y_2 \to y_1$ is not affected by the previous transition $y_3 \to y_2$.

[42] We note there are non-Markovian processes satisfying the SCK equation[41, p203].

[43] Ergodic is an adjective, ergodicity is its noun form. The word *ergodic* was coined by Boltzmann as a combination of the Greek words *εργον* (*ergon*—"work") and *οδός* (*odos*—"path").

[44] Birkhoff's theorem has two parts. It's primarily concerned with the existence of time averages; a corollary provides the conditions under which time averages equal phase averages.

[45] What stationary processes and Hamiltonian dynamics have in common are *measure-preserving* transformations. The natural motion in phase space (see [5, p33]) imposed by Hamilton's equations of motion is a one-to-one measure-preserving mapping of phase space onto itself at different times, basically the content of Liouville's theorem. (The time evolution in phase space is equivalent to a mapping of the canonical variables at time t into canonical variables at time $t+\mathrm{d}t$—a canonical transformation—and the Jacobian of canonical transformations is unity[5, p331].) Stationary processes are invariant under time translations, another type of measure-preserving transformation.

[46] Such as "metric transitivity." Although Birkhoff's theorem provides necessary and sufficient conditions for the equality of time and phase averages, namely metric transitivity, proving whether that condition holds is difficult. To quote M. Kac: "As is well known, metric transitivity plays an important role in ergodic theory; however, it is almost impossible to decide what Hamiltonians give rise to metrically transitive transformations"[42, p67].

[47] See Doob[36, Chapters 10, 11] and Loève[39, Chapters 30–32].

Consider the time average of a process $y(t)$ over the interval $(-T, T)$,

$$\overline{y}_T \equiv \frac{1}{2T}\int_{-T}^{T} y(t)\mathrm{d}t. \tag{2.30}$$

A sequence of time averages is ergodic[48] if $\lim_{T\to\infty}\left(\langle(\overline{y}_T)^2\rangle - \langle\overline{y}_T\rangle^2\right) = 0$, if the *variances of time averages tend to zero as* $T \to \infty$. From Eq. (2.30),

$$\langle(\overline{y}_T)^2\rangle = \frac{1}{4T^2}\int_{-T}^{T}\mathrm{d}t'\int_{-T}^{T}\mathrm{d}t\langle y(t)y(t')\rangle. \tag{2.31}$$

For stationary processes, $\langle y(t)y(t')\rangle \equiv R(|\tau|)$ is a function only of $\tau \equiv t' - t$ (see Exercise 2.12). Change variables: Let $\theta \equiv \frac{1}{2}(t' + t)$ and $\tau = t' - t$, an area-preserving transformation. Thus,

$$\langle(\overline{y}_T)^2\rangle = \frac{1}{4T^2}\int_{-2T}^{2T} R(|\tau|)\left(2T - |\tau|\right)\mathrm{d}\tau = \frac{1}{T}\int_0^{2T} R(\tau)\left(1 - \frac{\tau}{2T}\right)\mathrm{d}\tau. \tag{2.32}$$

A wide-sense stationary process is ergodic if and only if (*Slutsky's theorem*[43, Chapter 13]),

$$\lim_{T\to\infty}\frac{1}{T}\int_0^{2T}\left(1 - \frac{\tau}{2T}\right)\left(R(\tau) - c^2\right)\mathrm{d}\tau = 0, \tag{2.33}$$

where $\langle\overline{y}_T\rangle = c$; see Eq. (2.23). Wide-sense stationary processes are ergodic if the autocorrelation function $R(\tau)$ decays sufficiently rapidly that Eq. (2.33) is satisfied.

2.5.2 Spectral analysis, Wiener-Khinchin theorem

Periodic functions $f(x)$ of period p (f invariant under $x \to x+p$) have Fourier series representations $f(x) = \sum_{n=-\infty}^{\infty} c_n \mathrm{e}^{2\mathrm{i}n\pi x/p}$, where it's known how to find the coefficients c_n. By allowing p to vary, and by letting $p \to \infty$, square-integrable functions[49] $f(x)$ have integral representations $f(x) = (1/2\pi)\int_{-\infty}^{\infty} F(\alpha)\mathrm{e}^{\mathrm{i}\alpha x}\mathrm{d}\alpha$, where $F(\alpha) = \int_{-\infty}^{\infty} f(x')\mathrm{e}^{-\mathrm{i}\alpha x'}\mathrm{d}x'$ is the Fourier transform of $f(x)$. These considerations suggest that stationary stochastic processes have Fourier representations (statistical properties invariant under $t \to t + \tau$ for $\tau \to \infty$).

A problem with that idea is that stochastic processes do not have well-defined limits as $t \to \pm\infty$. A rigorous mathematical theory that avoids the problem, *harmonic analysis*, is beyond the level of this book.[50] To ensure the convergence of Fourier integrals, introduce a truncated stochastic process:

$$y_T(t) \equiv \begin{cases} y(t) & |t| < T \\ 0 & |t| \geq T, \end{cases}$$

where ultimately we'll let $T \to \infty$. The Fourier transform of $y_T(t)$ is defined

$$\widetilde{y}_T(\omega) = \int_{-\infty}^{\infty} y_T(t)\mathrm{e}^{\mathrm{i}\omega t}\mathrm{d}t = \int_{-T}^{T} y(t)\mathrm{e}^{\mathrm{i}\omega t}\mathrm{d}t,$$

with the inverse relation

$$y_T(t) = \frac{1}{2\pi}\int_{-\infty}^{\infty}\widetilde{y}_T(\omega)\mathrm{e}^{-\mathrm{i}\omega t}\mathrm{d}\omega. \tag{2.34}$$

Equation (2.34) shows that $\widetilde{y}_T(\omega)$ is the *spectral content* of $y_T(t)$ at frequency ω.

---

[48] Ergodicity for wide-sense stationary processes is referred to as ergodic in the mean or mean-ergodic.

[49] That is, $\int_{-\infty}^{\infty}|f(x)|^2\mathrm{d}x < \infty$, implying that $f(x) \to 0$ as $x \to \pm\infty$.

[50] See Wiener[35] and the article by S.O. Rice, *Mathematical Analysis of Random Noise*, in Wax[44, pp133–294].

Let's look at the spectral content of the second moment. Start with the time average

$$\overline{y^2} = \lim_{T\to\infty} \frac{1}{2T}\int_{-T}^{T} y^2(t)\mathrm{d}t = \lim_{T\to\infty} \frac{1}{2T}\int_{-T}^{T} y_T^2(t)\mathrm{d}t. \tag{2.35}$$

By substituting two copies of Eq. (2.34),

$$\begin{aligned}\int_{-T}^{T} y_T^2(t)\mathrm{d}t &= \frac{1}{4\pi^2}\int_{-\infty}^{\infty} \mathrm{d}\omega \widetilde{y}_T(\omega)\int_{-\infty}^{\infty} \mathrm{d}\omega' \widetilde{y}_T(\omega')\int_{-T}^{T} \mathrm{d}t\mathrm{e}^{-\mathrm{i}t(\omega+\omega')} \\ &= \frac{1}{4\pi^2}\int_{-\infty}^{\infty}\int_{-\infty}^{\infty} \mathrm{d}\omega\mathrm{d}\omega' \widetilde{y}_T(\omega)\widetilde{y}_T(\omega')\frac{2}{\omega+\omega'}\sin[(\omega+\omega')T].\end{aligned}$$

Because we want the limit $T \to \infty$ in Eq. (2.35), we can, by the product law of limits, use the representation of the Dirac delta function[13, p105]

$$\lim_{T\to\infty} \frac{\sin[(\omega+\omega')T]}{\omega+\omega'} = \pi\delta(\omega+\omega')$$

to infer that

$$\overline{y^2} = \frac{1}{2\pi}\lim_{T\to\infty}\frac{1}{2T}\int_{-\infty}^{\infty} \mathrm{d}\omega \left|\widetilde{y}_T(\omega)\right|^2 = \frac{1}{2\pi}\int_{-\infty}^{\infty} S(\omega)\mathrm{d}\omega, \tag{2.36}$$

where the *spectral density* or *power spectrum* is defined

$$S(\omega) \equiv \lim_{T\to\infty}\frac{1}{2T}\left|\widetilde{y}_T(\omega)\right|^2. \tag{2.37}$$

We see from Eq. (2.36) that $S(\omega)\mathrm{d}\omega$ is the spectral contribution to $\overline{y^2}$ between ω and $\omega + \mathrm{d}\omega$. Note that $S(\omega) = S(-\omega)$ and $S(\omega) \geq 0$. If $y(t)$ is a fluctuating current, $S(\omega)$ has SI units A^2/Hz; if $y(t)$ is a fluctuating voltage, $S(\omega)$ has SI units V^2/Hz.

Equation (2.36) can be generalized to the autocorrelation function of a stationary process:

$$R(\tau) = \lim_{T\to\infty}\frac{1}{2T}\int_{-T}^{T} y(t+\tau)y(t)\mathrm{d}t = \lim_{T\to\infty}\frac{1}{2T}\int_{-T}^{T} y_T(t+\tau)y_T(t)\mathrm{d}t. \tag{2.38}$$

Substituting two copies of Eq. (2.34) and repeating the steps above, we find

$$R(\tau) = \frac{1}{2\pi}\int_{-\infty}^{\infty} S(\omega)\mathrm{e}^{-\mathrm{i}\omega\tau}\mathrm{d}\omega = \frac{1}{2\pi}\int_{-\infty}^{\infty} S(\omega)\cos\omega\tau\mathrm{d}\omega = \frac{1}{\pi}\int_{0}^{\infty} S(\omega)\cos\omega\tau\mathrm{d}\omega. \tag{2.39}$$

Equation (2.39) is the *Wiener-Khinchin theorem*, that the autocorrelation function of a wide-sense stationary process has a spectral decomposition given by the power spectrum of that process. The inverse Fourier transform (valid when all questions of convergence make sense),

$$S(\omega) = \int_{-\infty}^{\infty} R(\tau)\mathrm{e}^{\mathrm{i}\omega\tau}\mathrm{d}\tau = \int_{-\infty}^{\infty} R(\tau)\cos\omega\tau\mathrm{d}\tau = 2\int_{0}^{\infty} R(\tau)\cos\omega\tau\mathrm{d}\tau, \tag{2.40}$$

is also called the Wiener-Khinchin theorem, in which case the spectral density of a wide-sense stationary process is the Fourier transform of its autocorrelation function.

Suppose $R(\tau) = c\delta(\tau)$, implying $S(\omega) = c$ for all ω. A spectrum having equal intensity at all frequencies (an infinite amount of energy) is termed *white*. A white spectrum can arise only if the process $y(t)$ is so random as to be uncorrelated with itself over any time, no matter how small. Such an example is pathological because it implies that $\langle y^2(t)\rangle$ is unbounded [see Eq. (2.36)], contradicting the assumption of stationarity. In real situations, $R(\tau)$ will not be so sharply peaked as a delta function. At the other extreme, suppose $R(\tau) = R(0)\cos\omega_0\tau$, implying that the spectral density is sharply peaked, $S(\omega) = \pi R(0)\left[\delta(\omega+\omega_0) + \delta(\omega-\omega_0)\right]$. This example is also unphysical: For ergodic stationary processes, $R(\tau)$ must decay sufficiently rapidly that Eq. (2.33) is satisfied.

2.6 THERMAL NOISE, NYQUIST THEOREM

Random thermal motions of charge carriers in electrical circuits generate *thermal noise* or *Johnson noise*.[51] Consider the RLC circuit depicted in Fig. 2.1, where the system is in thermal equilibrium

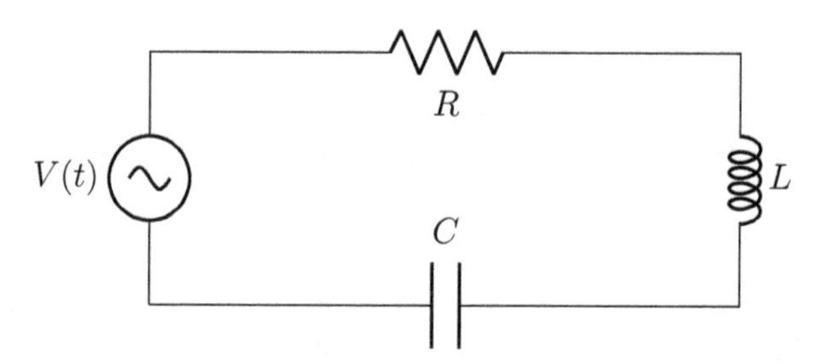

Figure 2.1 Thermal agitations of charge carriers produce a fluctuating voltage $V(t)$.

at temperature T. Thermal agitations of charge carriers produce fluctuating currents $I(t)$, which, through the usual circuit law, are associated with fluctuating voltages $V(t)$,

$$V(t) = L\frac{\mathrm{d}I}{\mathrm{d}t} + RI + \frac{1}{C}\int I(t)\mathrm{d}t. \tag{2.41}$$

Fluctuations are stationary, implying we can bring to bear the tools of spectral analysis. Introduce the Fourier transforms, $V(t) = (1/2\pi)\int_{-\infty}^{\infty} \mathrm{e}^{\mathrm{i}\omega t}\widetilde{V}(\omega)\mathrm{d}\omega$ and $I(t) = (1/2\pi)\int_{-\infty}^{\infty} \mathrm{e}^{\mathrm{i}\omega t}\widetilde{I}(\omega)\mathrm{d}\omega$. Using Eq. (2.41), we find the relation between $\widetilde{V}(\omega)$, $\widetilde{I}(\omega)$:

$$\widetilde{V}(\omega) = (R + \mathrm{i}\,[\omega L - 1/(\omega C)])\,\widetilde{I}(\omega) \equiv Z(\omega)\widetilde{I}(\omega), \tag{2.42}$$

where $Z(\omega)$ is the circuit impedance at frequency ω. The mean energy stored in the inductor associated with the fluctuating current is, from the equipartition theorem,[52]

$$\left\langle \frac{1}{2}LI^2 \right\rangle = \frac{1}{2}kT. \tag{2.43}$$

Equation (2.43) must be consistent with the circuit equation (2.42). Let's see what that implies. From Eq. (2.36), we have the time average[53]

$$\begin{aligned}\overline{I^2} &= \frac{1}{2\pi}\lim_{T\to\infty}\frac{1}{2T}\int_{-\infty}^{\infty}\mathrm{d}\omega\left|\widetilde{I}_T(\omega)\right|^2 = \frac{1}{2\pi}\lim_{T\to\infty}\frac{1}{2T}\int_{-\infty}^{\infty}\mathrm{d}\omega\frac{\left|\widetilde{V}_T(\omega)\right|^2}{|Z(\omega)|^2} \\ &= \frac{1}{2\pi}\int_{-\infty}^{\infty}\mathrm{d}\omega\frac{S(\omega)}{R^2 + [\omega L - 1/(\omega C)]^2},\end{aligned} \tag{2.44}$$

where $S(\omega) \equiv \lim_{T\to\infty}(1/2T)|\widetilde{V}_T(\omega)|^2$ is the spectral density [see Eq. (2.37)]. Invoking ergodicity, equate $\langle I^2\rangle$ from Eq. (2.43) with $\overline{I^2}$ from Eq. (2.44). We find

$$\frac{kT}{L} = \frac{1}{2\pi R^2}\int_{-\infty}^{\infty}\mathrm{d}\omega\frac{S(\omega)}{1 + (L/(\omega R))^2\,[\omega^2 - \omega_0^2]^2},$$

[51] Shot noise, another kind of noise, originates from the discreteness of electrical charge, from fluctuations in the number of charge carriers. Thermal noise arises from voltage fluctuations produced by randomness in charge motions.

[52] The equipartition theorem of classical statistical mechanics[5, p97] holds that energy equipartition (see [5, p91]) applies not just to kinetic energy but to any *quadratic degree of freedom*, any degree of freedom that adds a quadratic term to the Hamiltonian or energy function, e.g., $p^2/(2m)$ or $\frac{1}{2}m\omega^2x^2$. To apply the theorem in this example, the energy of the charge carriers must consist of the term $\frac{1}{2}LI^2$ plus any other terms involving generalized coordinates and momenta of these particles *not involving* I. The equipartition theorem breaks down at low temperatures where quantum effects manifest.

[53] T in Eq. (2.44) is the observation time, not the temperature.

where $\omega_0 \equiv 1/\sqrt{LC}$ is the resonant frequency. Assume that L is so large[54] that the integrand is sharply peaked about $\omega = \omega_0$, i.e., we have a *tuned circuit*. Under these conditions, we can approximate $S(\omega) \approx S(\omega_0)$ so that, with $y = (2L/R)(\omega - \omega_0)$,

$$\frac{kT}{L} \approx \frac{S(\omega_0)}{2\pi R^2}\left(\frac{R}{2L}\right)\int_{-\infty}^{\infty}\frac{\mathrm{d}y}{1+y^2} = \frac{S(\omega_0)}{4LR},$$

presenting us with a *formula* for the spectral density:

$$S(\omega) = 4RkT. \tag{2.45}$$

We've erased the subscript from ω_0 in Eq. (2.45) because the resonant frequency can be freely varied by varying circuit parameters. Equation (2.45) is *Nyquist's theorem*[55][45], published in 1928 concurrently with the discovery and measurement of thermal noise by Johnson[46]. It tells us the *noise power*, the second moment of the fluctuating voltage $\langle V^2\rangle = 4RkT\Delta f$ over the frequency bandwidth Δf that measurements are made.[56] Nyquist's theorem can be tested by using it to infer the value of Boltzmann's constant from noise measurements (found to differ by less than 1% from its value obtained by other means[47]). *The strength of thermal noise is proportional to the resistance and the absolute temperature.* Sensitive electronic equipment, such as radio telescopes, are cooled to cryogenic temperatures to reduce thermal noise.

Our derivation of Eq. (2.45) differs from Nyquist's, which involved an ingenious analysis of a transmission line terminated at both ends with resistance R (schematically shown in Fig. 2.2), with the entire system in equilibrium at temperature T. The transmission line is chosen so that its

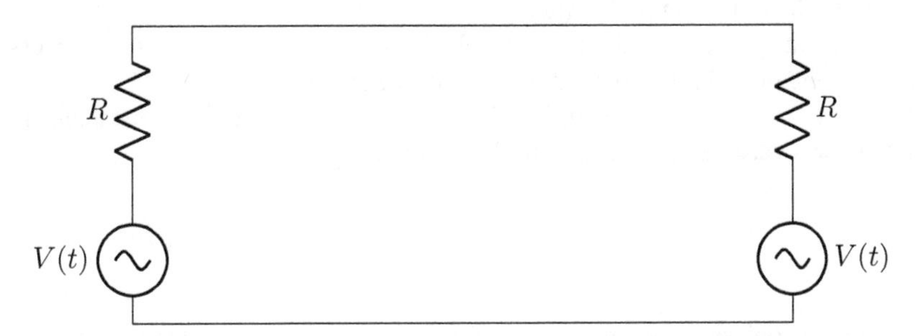

Figure 2.2 Idealized transmission line impedance matched to R, which act as black bodies.

characteristic impedance matches that of the resistor. As a consequence, electrodynamic modes on the transmission line are absorbed by the terminating resistors without reflection. The resistors then act as *black bodies*, objects that reflect nothing[2, p68] and emit energy modes distributed with the Planck spectrum. The mean energy of a mode of frequency ω is $\hbar\omega/(\mathrm{e}^{\beta\hbar\omega}-1)$, where $\beta = (kT)^{-1}$. There's more to the argument, but basically we can replace $kT \longrightarrow \hbar\omega/\left(\mathrm{e}^{\beta\hbar\omega}-1\right)$, and thus

$$S(\omega) = 4R\frac{\hbar\omega}{\mathrm{e}^{\beta\hbar\omega}-1} \tag{2.46}$$

is the Nyquist formula corrected for quantum effects. Thermal noise is therefore not white noise, as one would infer from Eq. (2.45), which we now see applies for $\omega \ll kT/\hbar$ (terahertz at room temperature). The breakdown of equipartition averts the ultraviolet catastrophe.

[54] We want $\omega_0(L/R) \gg 1$ (or $L \gg CR^2$) so that many oscillations of the circuit occur within the time L/R.
[55] Not to be confused with the Nyquist-Shannon sampling theorem for digital signal processing, also from 1928.
[56] For $\Delta f = N$ Hz, at room temperature $kT\Delta f \approx 4.1N \times 10^{-21}$ W.

2.7 THE RANDOM WALK IN ONE DIMENSION, DIFFUSION

We consider the *random walk* of a mythical drunkard who emerges from a drinking establishment sufficiently inebriated so as to take steps with equal probability to the right or left on a level street (a *symmetric* random walk). This problem is a stripped-down model of Brownian motion (see Chapter 3), where a large particle (larger than molecular sizes), suspended in a fluid, undergoes erratic motions from random collisions with fluid molecules. The random walk can be treated with more traditional methods of probability theory (the *binomial distribution*, see Exercise 2.22); here we analyze it as a Markov stochastic process.

Consider a one-dimensional lattice of points at positions $x = ma$, $m = 0, \pm 1, \pm 2, \cdots$. At any time the "walker" is at one of these sites. At times $t = n\tau$, $n = 0, 1, 2, \ldots$, the particle jumps to a neighboring site to the right or left with equal probability. With the process started at $x = 0$ at $t = 0$, we seek the conditional probabilities $P_{1|1}(m, n|0, 0)$. The form of the SCK equation appropriate to a discrete state space[57] is

$$P_{1|1}(m, n|0, 0) = \sum_k P_{1|1}(m, n|k, n-1) P_{1|1}(k, n-1|0, 0), \tag{2.47}$$

where the sum on k is over the intermediate states accessible to the system between time steps $n-1$ and n, in this case $k = m \pm 1$. To save writing, let $P_{1|1}(m, n|0, 0) \equiv p(m, n)$, in which case Eq. (2.47) implies

$$p(m, n) = \frac{1}{2} p(m-1, n-1) + \frac{1}{2} p(m+1, n-1), \tag{2.48}$$

where we've used the transition probability $P_{1|1}(m, n|k, n-1) = \frac{1}{2}\delta_{k,m-1} + \frac{1}{2}\delta_{k,m+1}$. Note there is no time derivative on the left of Eq. (2.48); we have a stochastic process, a succession in time (labeled by n) of the probabilities of finding the particle at site m.

Equation (2.48) can be handled with Fourier methods. Define the Fourier transform of $p(m, n)$, where the transform variable is conjugate to the spatial index m,

$$\widetilde{p}(q, n) \equiv \sum_{m=-\infty}^{\infty} p(m, n) e^{-iqm}, \qquad (-\pi < q < \pi) \tag{2.49}$$

where q is a dimensionless parameter restricted to $(-\pi, \pi)$, and where we've extended the spatial size of the system to $\pm\infty$. The inverse of Eq. (2.49) is

$$p(m, n) = \frac{1}{2\pi} \int_{-\pi}^{\pi} e^{iqm} \widetilde{p}(q, n) dq. \tag{2.50}$$

By applying Eq. (2.49) to Eq. (2.48), we find

$$\widetilde{p}(q, n) = \widetilde{p}(q, n-1) \cos q, \qquad (n = 1, 2, \ldots)$$

an equation which is easily iterated:

$$\widetilde{p}(q, n) = \cos^n(q) \widetilde{p}(q, 0). \qquad (n = 0, 1, \ldots) \tag{2.51}$$

Combining Eq. (2.51) with Eq. (2.50), the solution of Eq. (2.48) has the integral representation

$$p(m, n) = \frac{1}{2\pi} \int_{-\pi}^{\pi} e^{iqm} (\cos q)^n \, \widetilde{p}(q, 0) dq.$$

[57] State space is the set of values of a random variable, in this case the probability of the walker at site m.

With $p(m,0) = \delta_{m,0}$, $\widetilde{p}(q,0) = 1$. The solution of Eq. (2.48) associated with the initial condition $p(m,0) = \delta_{m,0}$ has the representation

$$p(m,n) = \frac{1}{2\pi}\int_{-\pi}^{\pi} \mathrm{e}^{\mathrm{i}qm} (\cos q)^n \,\mathrm{d}q. \tag{2.52}$$

For $n = 1$, we find that $p(m,1) = \frac{1}{2}(\delta_{m,-1} + \delta_{m,1})$, as it must for the given initial condition. Equation (2.52) can be evaluated for any n, but developing a formula for general n isn't illuminating. We need to look at its large-n behavior.[58]

Probabilities are real. If $\mathrm{e}^{\mathrm{i}qm}$ is replaced with its real part, $\cos qm$, Eq. (2.52) would be a Fourier cosine transform, which applies to even functions. It behooves us to examine the parity properties of the integrand. With a bit of analysis (see Exercise 2.28) one can show that $p(m,n) = 0$ for $n+m$ an odd number, whereas for $n+m$ an even number $p(m,n) \neq 0$, with

$$p(m,n) = \begin{cases} (1/\pi)\int_{-\pi/2}^{\pi/2} (\cos q)^n \cos(mq)\mathrm{d}q & (n+m = \text{even}) \\ 0. & (n+m = \text{odd}) \end{cases} \tag{2.53}$$

That $p(m,n) = 0$ for $m+n$ an odd number implies that events with $m+n$ odd are impossible to achieve starting from the prescribed initial condition; see Exercise 2.29.

Note that $p(m,n)$ is even in m, $p(m,n) = p(-m,n)$ (symmetric random walk). The net displacement is thus zero:

$$\langle m \rangle \equiv \sum_{m=-\infty}^{\infty} m p(m,n) = 0.$$

The variance is more difficult. Reach for the trick that should be familiar from statistical mechanics: Generate the sum you want by taking derivatives, $m^2 \cos(qm) = -\partial^2 \cos(qm)/\partial q^2$,

$$\langle m^2 \rangle = \sum_{m=-\infty}^{\infty} m^2 p(m,n) = -\frac{1}{\pi}\sum_{m=-\infty}^{\infty} \int_{-\pi/2}^{\pi/2} \mathrm{d}q\,(\cos q)^n \frac{\partial^2}{\partial q^2}\cos(mq). \tag{2.54}$$

Now reach for the even older trick of integrating by parts, twice:

$$\begin{aligned} -\frac{1}{\pi}\int_{-\pi/2}^{\pi/2} \cos^n q \frac{\partial^2}{\partial q^2}\cos(mq)\mathrm{d}q &= \frac{n}{\pi}\int_{-\pi/2}^{\pi/2} \left[n\cos^n q - (n-1)\cos^{n-2} q\right]\cos(mq)\mathrm{d}q \\ &= n^2 p(m,n) - n(n-1)p(m,n-2). \end{aligned} \tag{2.55}$$

Combining Eqs. (2.55) and (2.54) (and using the normalization on $p(m,n)$, see Exercise 2.27), we have the characteristic result of a random walk,

$$\langle m^2 \rangle = n. \tag{2.56}$$

The standard deviation of the process grows like $\sqrt{n}$.

2.7.1 Asymptotic form

Instead of attempting to evaluate Eq. (2.53) as $n \to \infty$ for all m, let's restrict ourselves to cases of physical interest, $|m| \lesssim \sqrt{n}$. Scale the "wavenumber" q in units of $1/\sqrt{n}$; let $q \equiv y/\sqrt{n}$. With that substitution, Eq. (2.53) is equivalent to

$$p(m,n) = \frac{1}{\pi\sqrt{n}}\int_{-\pi\sqrt{n}/2}^{\pi\sqrt{n}/2} \cos^n\left(\frac{y}{\sqrt{n}}\right)\cos(my/\sqrt{n})\mathrm{d}y.$$

[58]The time τ represents an average time between collisions of the Brownian-particle with fluid molecules, a time short relative to the times over which macroscopic observations are made.

For large n, the major contribution to the integral occurs for $|y| \ll \sqrt{n}$, for which

$$\cos^n\left(\frac{y}{\sqrt{n}}\right) \approx \left(1 - \frac{y^2}{2n}\right)^n \overset{n\to\infty}{\longrightarrow} \exp(-y^2/2),$$

where we've used the Euler definition of the exponential function. Thus, we have the asymptotic form for $n \to \infty$, $m/\sqrt{n} = \text{finite}$ ($\int_{-\infty}^{\infty} \mathrm{e}^{-\beta x^2}\cos(bx)\mathrm{d}x = \sqrt{\pi/\beta}\exp[-b^2/(4\beta)]$)

$$p(m,n) \sim \frac{1}{\pi\sqrt{n}}\int_{-\infty}^{\infty} \mathrm{e}^{-y^2/2}\cos(my/\sqrt{n})\mathrm{d}y = \sqrt{\frac{2}{\pi n}}\exp(-m^2/(2n)). \tag{2.57}$$

2.7.2 Passing to the continuous limit, Einstein diffusion equation

We can use Eq. (2.57) to pass from a description involving the probability distribution $p(m,n)$ on lattice points $x = ma$ at discrete times $t = n\tau$ to a continuous description involving positions x and times t. Under the substitutions $m = x/a$ and $n = t/\tau$, we seek the probability density $p(x,t)$ such that $p(x,t)\Delta x$ is the probability the walker is in an interval Δx around x at time t, where $\Delta x = a\Delta m$, with $1 \ll \Delta m \ll \sqrt{n}$, so that

$$p(x,t)\Delta x \approx \sum_{m\le m_1\le m+\Delta m} p(m_1, t/\tau) \approx \sqrt{\frac{2\tau}{\pi t}}\frac{1}{a}\exp\left(-\frac{x^2\tau}{2a^2t}\right)a\Delta m, \tag{2.58}$$

because $p(m,n)$ varies slowly over the interval $\Delta m \ll \sqrt{n}$. Define the diffusion coefficient[59]

$$D \equiv a^2/(2\tau), \tag{2.59}$$

so that, from Eq. (2.58), $p(x,t) \approx (1/\sqrt{\pi Dt})\mathrm{e}^{-x^2/(4Dt)}$. Let's tentatively adopt that form of $p(x,t)$ to apply for all x and t, *such that it's properly normalized*. As can be shown (see Exercise 2.31),

$$p(x,t) = \frac{1}{\sqrt{4\pi Dt}}\exp\left(-\frac{x^2}{4Dt}\right) \tag{2.60}$$

is a Gaussian such that $\int_{-\infty}^{\infty} p(x,t)\mathrm{d}x = 1$ for all times and moreover that $p(x,0) = \delta(x)$.

Equation (2.60), which is strictly valid for $t \to \infty$ and $x/\sqrt{Dt} = \text{finite}$, is commonly taken as a model to apply for all x and t for systems having the initial condition $p(x,0) = \delta(x)$. As one can show, $p(x,t)$ in Eq. (2.60) is a solution of the *diffusion equation,*[60]

$$\frac{\partial}{\partial t}p(x,t) = D\frac{\partial^2}{\partial x^2}p(x,t) \tag{2.61}$$

(which one would find from the continuity equation together with Fick's law). Thus $p(x,t)$, derived from the asymptotic form of the random-walk discrete distribution, Eq. (2.57), *satisfies a partial differential equation*, a result we'll see again in Chapter 3 (Fokker-Planck equation). We could have developed Eq. (2.61) directly from the stochastic process. Subtract $p(m,n-1)$ from Eq. (2.48),

$$\frac{1}{\tau}\left(p(m,n) - p(m,n-1)\right) = \frac{a^2}{2\tau}\frac{1}{a^2}\left[p(m-1,n-1) + p(m+1,n-1) - 2p(m,n-1)\right],$$

in which we recognize the finite-difference approximations of the derivatives in Eq. (2.61). In this way we see how naturally $D = a^2/(2\tau)$ arises as the diffusion coefficient. If this analysis were repeated for a random walk on a d-dimensional integer lattice, one would find $D = a^2/(2d\tau)$.

[59]The dimensions of D are length^2 per time. Committing that to memory will hold you in good stead.

[60]The diffusion equation is the prototype parabolic partial differential equation. That we've arrived at a time-irreversible differential equation is a consequence of starting from a Markov process which assume one-way flows of time.

From Eq. (2.60) we find the second moment of the displacement, the *Einstein diffusion equation*,

$$\langle x^2 \rangle_t \equiv \int_{-\infty}^{\infty} x^2 p(x,t) \mathrm{d}x = 2Dt, \tag{2.62}$$

the continuum analog of Eq. (2.56), a result obtained in 1905.[61] Equation (2.59) could be considered microscopic in character—we've related the phenomenological transport coefficient D to model parameters, yet we haven't related a and τ to mechanical properties of the system. We'll do that in Chapter 3. Einstein's famous result (another!), Eq. (2.62), is valid at long times but requires modification at short times; see Section 3.2. Equation (2.62) is an instance of a more general result derived in Chapter 6, the Green-Kubo theory of transport coefficients.

2.7.3 Wiener-Levy process

The *Wiener-Levy process* is a continuous-time Markov process for diffusion (the limit of the random walk for continuous time and infinitesimally small step size) specified by the probabilities[62]

$$\begin{aligned} P_1(y,t) &= \frac{1}{\sqrt{4\pi Dt}} \exp\left[-y^2/(4Dt)\right] & (t \geq 0) \\ P_{1|1}(y_2,t_2|y_1,t_1) &= \frac{1}{\sqrt{4\pi D(t_2-t_1)}} \exp\left[-(y_2-y_1)^2/(4D(t_2-t_1))\right]. & (t_2 > t_1 > 0) \end{aligned} \tag{2.63}$$

One can verify that $\langle y^2 \rangle = 2Dt$ using P_1 and that the SCK equation is satisfied by the Wiener-Levy form of $P_{1|1}(y_2,t_2|y_1,t_1)$. We return to the Wiener-Levy process in Section 3.2.

2.8 THE MASTER EQUATION

The *master equation* [see Eq. (2.66)] is a form of the SCK equation that finds wide use.[63] Depending on the nature of the state space, it's either a set of differential-difference equations or an integro-differential equation[64] for time-dependent probabilities.

2.8.1 Derivation

Consider a small time interval between events $t_1 - t_2 \equiv \Delta t > 0$. Write $P_{1|1}(y_1,t_1|y_2,t_2) = P_{1|1}(y_1,t_2+\Delta t|y_2,t_2)$ in the form

$$P_{1|1}(y_1,t_2+\Delta t|y_2,t_2) = (1 - a_0(y_2)\Delta t)\,\delta(y_1-y_2) + \Delta t W_{t_2}(y_1|y_2), \tag{2.64}$$

where $W_{t_2}(y_1|y_2)$ (a positive quantity) is a *transition rate*, the probability per unit time of a transition from state y_2 at t_2 to state y_1 at $t_2+\Delta t$ and $(1-a_0(y_2)\Delta t)$ is the probability of *no* transition in the time Δt. In this form, $P_{1,1}$ smoothly approaches a Dirac delta function as $\Delta t \to 0^+$. Integrating Eq. (2.64) over y_1 (using the rules on page 25), we find $a_0(y_2) = \int W_{t_2}(y'|y_2)\mathrm{d}y'$, the rate of all transitions from state y_2 at time t_2 (y' is a dummy variable). Combining Eq. (2.64) with the SCK equation (see Exercise 2.35), we find in the limit $\Delta t \to 0$ ($y \equiv y_1, t \equiv t_1$)

$$\frac{\partial}{\partial t} P_{1|1}(y,t|y_3,t_3) = \int \mathrm{d}y_2 W_t(y|y_2) P_{1|1}(y_2,t|y_3,t_3) - P_{1|1}(y,t|y_3,t_3) \int \mathrm{d}y' W_t(y'|y). \tag{2.65}$$

[61] Fürth[48] is a valuable collection of Einstein's articles on Brownian motion translated into English.

[62] Markov processes are specified by P_1 and $P_{1|1}$; see Section 2.4.3.

[63] The term *master equation* first appeared in 1940 in a statistical analysis of cosmic-ray showers[49]. The name arose from the role of a general equation from which other results are derived.

[64] In an *integro-differential equation*, the unknown function appears in integrals and derivatives of the function[50].

Equation (2.65) is the *differential form* of the SCK equation. It can be simplified by multiplying by $P_1(y_3, t_3)$ and integrating over y_3. We find (dropping the subscript on W_t for simplicity[65])

$$\frac{\partial}{\partial t} P_1(y, t) = \int \big[\underbrace{W(y|y')P_1(y', t)}_{\text{gain}} - \underbrace{W(y'|y)P_1(y, t)}_{\text{loss}}\big] \mathrm{d}y'. \tag{2.66}$$

The rate at which $P_1(y, t)$ changes is a balance between rates of *gain* processes, in which transitions occur *into* the state y from all other states y', and *loss* processes in which transitions to all other states y' occur from y. The master equation finds wide use as it's often a simple matter to come up with approximate expressions for the transition rates $W(y|y')$. For a discrete state space labeled by index n, the master equation has the form

$$\frac{\partial}{\partial t} P(n, t) = \sum_{n' \neq n} \left[W(n|n')P(n', t) - W(n'|n)P(n, t)\right]. \tag{2.67}$$

The sum is restricted to $n' \neq n$; the contribution of $n' = n$ formally drops out of the summation.

Example. The transition rate for the symmetric random walk is (α is a rate)

$$W(n|n') = \alpha \left(\delta_{n', n+1} + \delta_{n', n-1}\right), \tag{2.68}$$

and thus

$$\frac{\partial}{\partial t} P(n, t) = \alpha \left[P(n+1, t) + P(n-1, t) - 2P(n, t)\right]. \tag{2.69}$$

The right side of Eq. (2.69) is the finite-difference approximation to the diffusion equation, (2.61).

Example. A sample of radioactive material has n_0 active, identical nuclei at time $t = 0$. Let $n'(t)$ denote the number of active nuclei at time $t > 0$, those that have not decayed. Assume a Markov process, that the probability distribution at time t depends on the state of the system at time $t' < t$, but not on its entire past history. Let α denote the probability per unit time that a surviving nucleus decays, with the transition probabilities for $n' \neq n$ given by

$$P_{1|1}(n|n') = \begin{cases} 0 & n > n' \\ n'\alpha\Delta t & n = n' - 1 \\ O(\Delta t)^2. & n < n' - 1 \end{cases}$$

The first requirement is that no new radioactive nuclei are created in the process (for example, $n = 5$ and $n' = 3$), and by the second requirement Δt must be sufficiently small that the number of radioactive nuclei changes by one (effectively enforced by the third requirement). The transition rates are thus summarized by the formula $W(n|n') = \alpha n' \delta_{n, n'-1}$, implying from Eq. (2.67) that

$$\frac{\partial}{\partial t} P(n, t) = \alpha \left[(n+1)P(n+1, t) - nP(n, t)\right] \qquad (n = 0, 1, 2, \ldots) \tag{2.70}$$

subject to the initial condition $P(n, 0) = \delta_{n, n_0}$. The solution to this system of equations is developed in Exercises 2.38 and 2.39.

---

[65] Time-independent transition rates are referred to as *temporally homogeneous*.

2.8.2 Detailed balance

From Eq. (2.67), the master equation has a stationary solution $P_{\text{eq}}(n)$ when the following holds[66]

$$\sum_{n'} W(n|n')P_{\text{eq}}(n') = \Big(\sum_{n'} W(n'|n)\Big)P_{\text{eq}}(n). \tag{2.71}$$

Equation (2.71) is a relation among the transition rates $W(n|n')$; the coefficients $P_{\text{eq}}(n)$ are known from statistical mechanics. Equation (2.71) indicates that in equilibrium the sum of all transitions (per unit time) $n' \to n$ is balanced by the sum of transitions $n \to n'$ to all other states. A stronger (yet simpler) condition is the requirement that *detailed balance* holds *for each pair* (n, n'),

$$W(n|n')P_{\text{eq}}(n') = W(n'|n)P_{\text{eq}}(n), \tag{2.72}$$

that in equilibrium the rate of the transition $n' \to n$ is the same as that for the reverse transition $n \to n'$, a *dynamic* characterization of equilibrium. If Eq. (2.72) holds, so does Eq. (2.71).

Detailed balance does not have to be imposed as a separate requirement on transition rates, it's built in from the time-reversal symmetry of microscopic motions (just as with the Onsager reciprocal relations). A key point is that the states n, n' in Eq. (2.72) represent macroscopic quantities, *functions* of the microscopic coordinates of the system point in Γ-space (see Appendix A).[67] Let Γ_n ($\Gamma_{n'}$) denote the region of Γ-space associated with state n (n'). A *part* (subset) of Γ_n, denoted $\Gamma_{n\to n'}$, flows in time t to a subset of $\Gamma_{n'}$, effecting the transition $n \to n'$. The transition probability is the ratio of two phase-space volumes,

$$\begin{aligned} P_{1|1}(n'|n) &= \frac{\text{Vol}(\Gamma_{n\to n'})}{\text{Vol}(\Gamma_n)} = \frac{\text{Vol}(\Gamma_{n\to n'})}{P_{\text{eq}}(n)} \\ P_{1|1}(n|n') &= \frac{\text{Vol}(\Gamma_{n'\to n})}{\text{Vol}(\Gamma_{n'})} = \frac{\text{Vol}(\Gamma_{n'\to n})}{P_{\text{eq}}(n')}, \end{aligned} \tag{2.73}$$

where Vol denotes the $6N$-dimensional volume of a region in Γ-space, and where we've equated $P_{\text{eq}}(n) = \text{Vol}(\Gamma_n)$, a step that fixes a measure on phase space but does not affect the conclusion we reach. The natural motion in phase space generated by Hamiltonian dynamics is a one-to-one, measure-preserving mapping of Γ-space onto itself at different instants of time. A phase point at time t_1 uniquely determines its position in Γ-space at time t_2, regardless of whether $t_2 > t_1$ or $t_2 < t_1$. Because of these properties, $\text{Vol}(\Gamma_{n\to n'}) = \text{Vol}(\Gamma_{n'\to n})$; the transition $n' \to n$ is the time-reversal of the transition $n \to n'$. From the two relations in Eq. (2.73), we infer that $P_{1|1}(n'|n)P_{\text{eq}}(n) = P_{1|1}(n|n')P_{\text{eq}}(n')$, and then, with $P_{1|1}(n|n') = W(n|n')\Delta t$, we have detailed balance, Eq. (2.72). This argument is due to Wigner[52].[68] A fuller derivation is given by de Groot and Mazur.[7, pp92–100] We've relied on classical mechanics; the quantum version is discussed in van Kampen[51, pp452–458].

2.8.3 Matrix form

The master equation can be written in matrix form by introducing the *transition matrix* W having elements,[69]

$$W_{n,n'} \equiv W(n|n') - \delta_{n,n'} \sum_{n''} W(n''|n). \tag{2.74}$$

---

[66] We assume only one stationary solution, the equilibrium probability distribution. Systems with multiple stationary solutions exist, a topic beyond the intended level of this book. See van Kampen[51, p101].

[67] Many micro-states are associated with macroscopically specified states, the basic message of Boltzmann entropy.

[68] E.P. Wigner received the 1963 Nobel Prize in Physics for the use of symmetry principles in fundamental physics.

[69] The transition rates $W(n|n')$ are positive, as are the off-diagonal elements of the transition matrix $W_{n,n'}$, $n \neq n'$, but the diagonal elements of the transition matrix $W_{n,n}$ are negative. Note that $W_{n,n'}$ need not be symmetric.

In that way, the master equation (2.67) is equivalent to

$$\frac{\partial}{\partial t}P(n,t) = \sum_{n'} W_{n,n'}P(n',t). \tag{2.75}$$

Example. Using the transition rates associated with the random walk, Eq. (2.68), we have from Eq. (2.74),

$$W_{n,n'} = \alpha\left(\delta_{n',n-1} + \delta_{n',n+1} - 2\delta_{n,n'}\right).$$

For the decay process with $W(n|n') = \alpha n'\delta_{n,n'-1}$, Eq. (2.70), we have

$$W_{n,n'} = \alpha\left[(n+1)\delta_{n',n+1} - n\delta_{n',n}\right].$$

2.8.4 Eigenfunction expansion

By considering $P(n,t)$ $(n = 1,2,\ldots)$ as the elements of a column vector, $P(t)$, the master equation in (2.75) can be written $\partial P/\partial t = WP(t)$, which has the formal solution,

$$P(t) = \mathrm{e}^{tW}P(0). \tag{2.76}$$

The matrix exponential function is defined by the infinite series,[70] $\exp(tW) \equiv \sum_{n=0}^{\infty}(t^n/n!)W^n$, where W^n is the nth power of W. From Eq. (2.74), $W_{n,n'} \geq 0$ for $n \neq n'$, which is a necessary and sufficient condition for e^{Wt} to be non-negative[53, p172]. Thus, $P(t)$ in Eq. (2.76) is positive and bounded[71] for all $t \geq 0$.

Equation (2.75) is analogous to the Schrödinger equation $\partial\psi/\partial t = -(\mathrm{i}/\hbar)H\psi$, where H is a Hermitian operator. For $H \neq H(t)$, $\psi(t) = \exp\left(-(\mathrm{i}t/\hbar)H\right)\psi(0)$. To handle this expression one considers the associated eigenproblem, $H\phi_n = \lambda_n\phi_n$ $(n = 0,1,\ldots)$. The eigenfunctions of a Hermitian operator form a complete orthonormal set of functions (see [13, p66]). Thus, one can represent the initial condition as an infinite linear combination of eigenfunctions, $\psi(0) = \sum_{n=0}^{\infty} c_n\phi_n$, where the expansion coefficients are found from the inner product $c_n = \langle\phi_n|\psi(0)\rangle$. The solution has the form $\psi(t) = \sum_{n=0}^{\infty} c_n \mathrm{e}^{-(\mathrm{i}t/\hbar)\lambda_n}\phi_n$ (see Exercise 2.41).

We can't directly apply this familiar reasoning to the master equation because the transition matrix W need not be symmetric. When combined with detailed balance, however, there is a way forward. Note that the sum of the elements in any column of W vanishes (use Eq. (2.74)),

$$\sum_n W_{n,n'} = 0. \qquad \text{(for each } n'\text{)} \tag{2.77}$$

Equation (2.77) expresses *conversation of probability*; by combining Eq. (2.77) with Eq. (2.75), the total probability $\sum_n P(n,t)$ is fixed in time.[72] Equation (2.77) informs us that W has a left eigenvector $\psi \equiv (1,1,\ldots,1)$ with zero eigenvalue, $\psi \cdot W = 0$, implying that W has a right eigenvector with zero eigenvalue (in principle there could be more than one; see Exercise 2.42). The detailed balance condition, Eq. (2.72), expressed in terms of transition matrices is

$$W_{n,n'}P_{\text{eq}}(n') = W_{n',n}P_{\text{eq}}(n). \tag{2.78}$$

[70] Equation (2.76) is a formal solution because, even though e^{tW} formally solves Eq. (2.75), one can't sum the infinite series (that e^{tW} represents) to obtain a useful closed-form expression (other than denoting it the exponential). In Eq. (2.76), W is independent of time. There are specialized methods (for obtaining formal solutions) when $W = W(t)$.

[71] The infinite series for e^{Wt} exists for all matrices W for fixed values of t, and for all t for any fixed W[53, p166].

[72] The hallmark of conserved quantities is that associated with a change in time is a flow in space (see Section 1.3). We'll see in Section 3.3 that probability *flows* in the state space of random variables.

From Eq. (2.75), and using Eqs. (2.78), (2.77),

$$\frac{\partial}{\partial t}P_{\text{eq}}(n) = \sum_{n'} W_{n,n'}P_{\text{eq}}(n') = \sum_{n'} W_{n',n}P_{\text{eq}}(n) = 0.$$

Thus, W has a right eigenvector of zero eigenvalue, the equilibrium probability distribution.

The transition matrix W is *similar* to a symmetric matrix, V. Define the matrix elements

$$V_{n,n'} \equiv \frac{1}{\sqrt{P_{\text{eq}}(n)}}W_{n,n'}\sqrt{P_{\text{eq}}(n')} \tag{2.79}$$

for which $V_{n,n'} = V_{n',n}$ by the detailed balance condition, Eq. (2.78). Equation (2.79) is a similarity transformation,[13, p27] $V = S^{-1}WS$, with $S_{ij} = \sqrt{P_{\text{eq}}(i)}\delta_{ij}$. The eigenvalues of W are therefore real and, as it turns out, nonpositive (see Exercise 2.45), an expression of the irreversibility of the evolution of the nonequilibrium probability distribution into its equilibrium form.[73]

We can express the eigenproblem for W in the form

$$\sum_{n'} W_{n,n'}\phi_m(n') = -\lambda_m\phi_m(n), \tag{2.80}$$

where the λ_m are positive with ϕ_m the associated eigenfunction. Combining Eqs. (2.79), (2.80),

$$\sum_{n'} \underbrace{\frac{1}{\sqrt{P_{\text{eq}}(n)}}W_{n,n'}\sqrt{P_{\text{eq}}(n')}}_{V_{n,n'}} \underbrace{\frac{1}{\sqrt{P_{\text{eq}}(n')}}\phi_m(n')}_{\widetilde{\phi}_m(n')} = -\lambda_m \underbrace{\frac{1}{\sqrt{P_{\text{eq}}(n)}}\phi_m(n)}_{\widetilde{\phi}_m(n)}. \tag{2.81}$$

Equation (2.80) is thus equivalent to $V\widetilde{\phi}_m = -\lambda_m\widetilde{\phi}_m$, where the eigenfunctions form a complete orthonormal set (V is real and symmetric). We therefore have an orthonormality relation for the eigenfunctions of W with respect to a weighting function,

$$\langle\phi_m|\phi_{m'}\rangle \equiv \sum_n \frac{1}{P_{\text{eq}}(n)}\left(\phi_m(n)\right)^T\phi_{m'}(n) = \delta_{m,m'}. \tag{2.82}$$

With Eq. (2.82) established, the initial condition in Eq. (2.76) can be represented as an eigenfunction expansion, $P(0) = \sum_m c_m\phi_m = \sum_m\langle\phi_m|P(0)\rangle\phi_m$. Note that $c_0 = 1$ (where $\phi_0 = P_{\text{eq}}$). We therefore have the solution of the master equation,

$$P(n,t) = P_{\text{eq}}(n) + \sum_{m>0} c_m\phi_m(n)\text{e}^{-\lambda_m t}. \tag{2.83}$$

We see that the nonequilibrium probability distribution approaches the equilibrium distribution as $t \to \infty$. If we order the eigenvalues $\lambda_1 < \lambda_2 < \cdots$, λ_1 controls the approach to equilibrium.

2.9 GAUSSIAN PROCESSES

A *Gaussian process* is a type of stochastic process occurring with sufficient regularity in applications as to warrant separate treatment. A Gaussian process is an extension of the properties of Gaussian distributions to stochastic processes.[74] We found in Section 2.1 a multivariate Gaussian distribution and its characteristic function, Eqs. (2.3) and (2.9). Here we develop properties of Gaussians not covered in Section 2.1 and then we introduce Gaussian processes.

[73] The master equation presumes a one-way flow of time between transitions, $\Delta t > 0$.

[74] Gaussian distributions occur generically in statistical physics. As an example, the probability $P_N(m)$ that, in a one-dimensional random walk of N steps, the walker is found m steps to the right of the starting point is given by the binomial distribution; see Exercise 2.22. The limit $N \to \infty$ of the binomial distribution is precisely the Gaussian, the *DeMoivre-*

2.9.1 Gaussian distributions and cumulants

A Gaussian function is the exponential of a concave quadratic function, $P(x) \equiv K\exp(-ax^2+bx)$, $a > 0$ and $-\infty < x < \infty$. Using $\int_{-\infty}^{\infty} e^{-ax^2+bx}dx = \sqrt{\pi/a}e^{b^2/(4a)}$ $(a > 0)$, choose $K = \sqrt{a/\pi}e^{-b^2/(4a)}$ so that $\int_{-\infty}^{\infty} P(x)dx = 1$. The constants a, b are conventionally parameterized in terms of the mean $\mu \equiv b/(2a)$ and the variance $\sigma^2 \equiv 1/(2a)$, implying

$$P(x) = \frac{1}{\sqrt{2\pi\sigma^2}}\exp\left(-\frac{(x-\mu)^2}{2\sigma^2}\right). \tag{2.84}$$

One can show that $\langle x\rangle \equiv \int_{-\infty}^{\infty} xP(x)dx = \mu$ is the mean and $\langle x^2\rangle \equiv \int_{-\infty}^{\infty} x^2P(x)dx = \sigma^2 + \mu^2$, and thus $\sigma^2 = \langle x^2\rangle - \langle x\rangle^2$ is the variance. Gaussians have the property that their Fourier transforms are Gaussian.[75] As one can show,

$$\Phi(\omega) \equiv \int_{-\infty}^{\infty} P(x)e^{i\omega x}dx = \frac{1}{\sqrt{2\pi\sigma^2}}\int_{-\infty}^{\infty} e^{i\omega x}e^{-(x-\mu)^2/(2\sigma^2)}dx = \exp\left(i\omega\mu - \omega^2\sigma^2/2\right). \tag{2.85}$$

As discussed in Section 2.1, one role of the characteristic function Φ is to provide a generating function for the moments of the distribution; moments are found from derivatives of Φ. For a single-variable probability density $P(\alpha)$, $\Phi(\omega) = \langle e^{i\omega\alpha}\rangle = \sum_{n=0}^{\infty}(i\omega)^n\langle\alpha^n\rangle/n!$, and thus $\langle\alpha^n\rangle = (-i)^n\partial^n\Phi(\omega)/\partial\omega^n|_{\omega=0}$. Another role of Φ (not treated in Section 2.1.1) is to provide the *cumulant generating function* $C(\omega)$, defined as

$$C(\omega) \equiv \ln\Phi(\omega) = \ln\left(\sum_{n=0}^{\infty}\frac{(i\omega)^n}{n!}\langle\alpha^n\rangle\right) \equiv \sum_{n=1}^{\infty}\frac{C_n}{n!}(i\omega)^n, \tag{2.86}$$

where the expansion coefficient C_n, the nth-order *cumulant*, is found by taking derivatives,

$$C_n = (-i)^n\frac{\partial^n}{\partial\omega^n}C(\omega)\bigg|_{\omega=0} = (-i)^n\frac{\partial^n}{\partial\omega^n}\ln\Phi(\omega)\bigg|_{\omega=0}. \tag{2.87}$$

Note that $C_0 = 0$; $\Phi(0) = 1$ for a normalized probability distribution. The cumulant expansion therefore starts at $n = 1$ in Eq. (2.86). Cumulants are functions of moments. Using the power series $\ln(1+z) = z - \frac{1}{2}z^2 + \frac{1}{3}z^3 - \cdots$, one can show[76] for the first four cumulants,

$$\begin{aligned} C_1 &= \langle\alpha\rangle \\ C_2 &= \langle\alpha^2\rangle - \langle\alpha\rangle^2 \\ C_3 &= \langle\alpha^3\rangle - 3\langle\alpha^2\rangle\langle\alpha\rangle + 2\langle\alpha\rangle^3 \\ C_4 &= \langle\alpha^4\rangle - 4\langle\alpha^3\rangle\langle\alpha\rangle - 3\langle\alpha^2\rangle^2 + 12\langle\alpha^2\rangle\langle\alpha\rangle^2 - 6\langle\alpha\rangle^4. \end{aligned} \tag{2.88}$$

Laplace theorem; see Cramér[31, pp198–203], Sinai[54, p30], or [5, p72]. Gaussians also occur in the *central limit theorem*, a general theorem of probability theory; see [5, p75] or Papoulis[43, p218]. Consider $n \gg 1$ independent random variables $\Delta x_1, \ldots, \Delta x_n$ representing the step lengths in a random walk having zero expectation values and variances $\sigma_i^2 = \langle(\Delta x_i)^2\rangle$. The step lengths are uncorrelated (statistically independent), $\langle\Delta x_i\Delta x_j\rangle = \langle\Delta x_i\rangle\langle\Delta x_j\rangle = 0$. Let $x_n \equiv \sum_{i=1}^{n}\Delta x_i$ be the sum of the step lengths, and let $s_n^2 \equiv \sum_{i=1}^{n}\sigma_i^2$ be the sum of the individual variances. Define a new random variable $y_n \equiv x_n/s_n$. The central limit theorem states that, as $n \to \infty$, the probability distribution for y_n approaches a Gaussian with unity variance, $P_n(y_n) \overset{n\to\infty}{\longrightarrow} P(y) = (1/\sqrt{2\pi})\exp(-y^2/2)$.

[75] It's sometimes said, erroneously, that Gaussians are the only function having the property of being its own Fourier transform. A Gaussian multiplied by a Hermite polynomial, for example, is its own Fourier transform, up to a multiplicative constant. Considering the Fourier transform as a linear transformation in a Hilbert space of square integrable functions, one is seeking the eigenfunctions of the Fourier transform operator, the number of which is unlimited.

[76] Put aside concerns over the existence of moments or the convergence of series. You're formally matching powers of $(i\omega)$ between the two expansions in Eq. (2.86), a procedure that relies on the uniqueness of power series. See [13, p124].

Explicit expressions for the first 10 cumulants are listed in [55]. The nth cumulant[77] is a function of moments of order $k \leq n$ (conversely, the nth moment is a function of cumulants of order $k \leq n$; see Exercise 2.48). The first two cumulants C_1, C_2 are the mean and the variance. Cumulants beyond second order are related to higher-order fluctuation moments. As one can show,

$$\begin{aligned} C_3 &= \langle (\alpha - \langle \alpha \rangle)^3 \rangle \\ C_4 &= \langle (\alpha - \langle \alpha \rangle)^4 \rangle - 3\langle (\alpha - \langle \alpha \rangle)^2 \rangle. \end{aligned} \tag{2.89}$$

Because fluctuations are typically small in macroscopic systems (exception: critical phenomena), an approximation scheme is to set cumulants beyond a certain order to zero.[78]

The connection $\Phi(\omega) = \mathrm{e}^{C(\omega)}$ between the cumulant function $C(\omega)$ and the characteristic function $\Phi(\omega)$ generalizes the relation $Z(\beta) = \mathrm{e}^{-\beta F}$ in statistical mechanics between the free energy and the partition function; see Eq. (A.15).[79] A key property of cumulants is that they can be found for *any* probability distribution: Whether equilibrium or nonequilibrium, the meaning of the average symbols $\langle\rangle$ in Eq. (2.88) has not been specified. In the statistical mechanics of interacting particles, one develops a cumulant expansion for the free energy using a probability function based on noninteracting particles[5, Section 6.3]. Cumulants also play a role in theories of real-space renormalization[5, p280].

Suppose that $P(\alpha) = \delta(\alpha - \alpha_0)$, i.e., the event $\alpha = \alpha_0$ occurs with certainty and with no variance. The associated cumulants all vanish except for C_1. From $\Phi(\omega) = \int_{-\infty}^{\infty} \mathrm{e}^{\mathrm{i}\omega\alpha}\delta(\alpha - \alpha_0)\mathrm{d}\alpha = \mathrm{e}^{\mathrm{i}\omega\alpha_0}$, $C(\omega) = \ln \Phi(\omega) = \mathrm{i}\omega\alpha_0$, implying from Eq. (2.86) that $C_1 = \alpha_0$ and $C_{n>1} = 0$. The moments in this case factorize trivially, $\langle \alpha^n \rangle = \langle \alpha \rangle^n$, and thus, from Eq. (2.88), $C_2 = C_3 = C_4 = 0$.

We've made this foray into cumulants to show that cumulants associated with Gaussian distributions vanish beyond second order, $C_{n>2} = 0$. From Eq. (2.85), $C(\omega) = \ln \Phi(\omega) = \mathrm{i}\omega\mu - \omega^2\sigma^2/2$, implying from Eq. (2.86) that $C_1 = \mu$, $C_2 = \sigma^2$, and $C_{n>2} = 0$. Gaussians are unique in this regard;[80] one could *define* a Gaussian by the requirement that $C_{n>2} = 0$. *Gaussian processes are the simplest generalization of deterministic processes* (wherein the same outcome occurs with no variance). Gaussians have one more parameter (than deterministic processes) describing the width of the distribution such that $C_{n>2} = 0$. The vanishing of higher-order cumulants implies that the moments factorize in a characteristic way. From Eq. (2.88), we have for Gaussian statistics

$$\begin{aligned} C_3 = 0 &\implies \langle \alpha^3 \rangle = 3\langle \alpha^2 \rangle \langle \alpha \rangle - 2\langle \alpha \rangle^3 = 3\sigma^2\mu + \mu^3 \\ C_4 = 0 &\implies \langle \alpha^4 \rangle = 3\langle \alpha^2 \rangle^2 - 2\langle \alpha \rangle^4 = 3\sigma^4 + 6\sigma^2\mu^2 + \mu^4. \end{aligned} \tag{2.90}$$

2.9.2 Multivariate Gaussian distributions

Multivariate Gaussians are generalizations of the single-variable form to n random variables:

$$P(\alpha_1, \ldots, \alpha_n) \equiv K \exp\left(-\frac{1}{2} \sum_{ij=1}^{n} A_{ij}\alpha_i\alpha_j + \sum_{i=1}^{n} B_i\alpha_i \right), \tag{2.91}$$

[77] Cumulants are sometimes written $\langle \alpha^n \rangle_c$ to signify the nth *cumulant average*. The average of an exponential, $\Phi(\omega) = \langle \mathrm{e}^{\mathrm{i}\omega\alpha} \rangle$, is the exponential of a particular kind of average, $\Phi(\omega) = \exp C(\omega) = \exp\left[\sum_{n=1}^{\infty} (\mathrm{i}\omega)^n \langle \alpha^n \rangle_c / n!\right] = \exp\left[\mathrm{i}\omega\langle\alpha\rangle + \frac{1}{2}(\mathrm{i}\omega)^2(\langle \alpha^2 \rangle - \langle \alpha \rangle^2) + \frac{1}{3!}(\mathrm{i}\omega)^3(\langle \alpha^3 \rangle - 3\langle \alpha^2 \rangle\langle \alpha \rangle + 2\langle \alpha \rangle^3) + \cdots\right]$.

[78] Cumulants provide a systematic way to introduce approximations to $\Phi(\omega)$, with the associated approximate probability distribution $P(\alpha)$ found from inverse Fourier transformation[5, p76].

[79] The partition function is the Laplace transform of the density of states function,[5, p86] $Z(\beta) = \int_0^{\infty} \mathrm{e}^{-\beta E}\Omega(E)\mathrm{d}E$. First derivatives of $\ln Z(\beta)$ generate thermodynamic quantities, e.g., $P = kT\partial \ln Z/\partial V$, and second derivatives generate measures of fluctuations such as the heat capacity, $\partial^2 \ln Z/\partial\beta^2 = kT^2C_V$. The characteristic function is the Fourier transform of a probability distribution, equilibrium or not, $\Phi(\omega) = \int_{-\infty}^{\infty} \mathrm{e}^{\mathrm{i}\omega\alpha}P(\alpha)\mathrm{d}\alpha$. Derivatives of $\ln \Phi(\omega)$ (of any order) generate cumulants (of that order) [see Eq. (2.87)], which are functions of the moments of the distribution.

[80] The only smooth distribution having finitely many nonzero cumulants is the Gaussian[56].

where A_{ij} are elements of a real symmetric matrix and the quantities B_i are symmetry-breaking terms. Equation (2.91) can be written in vector notation[81]

$$P(\boldsymbol{\alpha}) = K \exp\Big(-\frac{1}{2}\boldsymbol{\alpha}^T \boldsymbol{A}\boldsymbol{\alpha} + \boldsymbol{B}^T\boldsymbol{\alpha}\Big). \tag{2.92}$$

If the random variables $\alpha_1, \ldots, \alpha_n$ are statistically independent, $\boldsymbol{A}$ is diagonal (Exercise 2.50). As one can show, using the results of Exercise 2.5,[82]

$$\int \exp\Big(-\frac{1}{2}\boldsymbol{\alpha}^T \boldsymbol{A}\boldsymbol{\alpha} + \boldsymbol{B}^T\boldsymbol{\alpha}\Big)\mathrm{d}^n\alpha = \frac{(2\pi)^{n/2}}{\sqrt{\det \boldsymbol{A}}}\exp\Big(\frac{1}{2}\boldsymbol{B}^T\boldsymbol{A}^{-1}\boldsymbol{B}\Big).$$

Thus, take $K = \sqrt{\det \boldsymbol{A}}\exp\left(-\frac{1}{2}\boldsymbol{B}^T\boldsymbol{A}^{-1}\boldsymbol{B}\right)/(2\pi)^{n/2}$. An equivalent way of writing Eq. (2.91) (analogous to Eq. (2.84)) is, if we identify $B_i = \sum_j A_{ij}\mu_j$ where $\mu_i \equiv \langle \alpha_i \rangle$ (see Exercise 2.51),

$$P(\alpha_1, \ldots, \alpha_n) = \frac{\sqrt{\det \boldsymbol{A}}}{(2\pi)^{n/2}}\exp\Big(-\frac{1}{2}\sum_{ij=1}^{n}(\alpha_i - \mu_i)A_{ij}(\alpha_j - \mu_j)\Big). \tag{2.93}$$

Combining Eq. (2.92) with Eq. (2.6), we have the characteristic function (a generalization of Eq. (2.9))

$$\Phi(\omega_1, \ldots, \omega_n) = \exp\Big(\mathrm{i}\boldsymbol{\omega}^T\boldsymbol{\mu} - \frac{1}{2}\boldsymbol{\omega}^T\boldsymbol{A}^{-1}\boldsymbol{\omega}\Big). \tag{2.94}$$

The characteristic function of a multivariate Gaussian is therefore a multivariate Gaussian, implying that cumulants of multivariate Gaussians vanish beyond second order (multivariate cumulants are defined in Eq. (2.98)). All Gaussian distributions, multivariate or not, have the same property of their cumulants. By combining Eq. (2.94) with Eq. (2.8) we find the generalization of Eq. (2.10)

$$\langle (\alpha_i - \mu_i)(\alpha_j - \mu_j)\rangle = -\frac{\partial^2 \Phi(\omega_1, \ldots, \omega_n)}{\partial\omega_i\partial\omega_j}\bigg|_{\boldsymbol{\omega}=0} = (\boldsymbol{A}^{-1})_{ij}. \tag{2.95}$$

2.9.3 Many-time cumulants

We defined in Eq. (2.6) the characteristic function $\Phi(\omega_1, \ldots, \omega_n)$ of a multivariate probability distribution $P(\alpha_1, \ldots, \alpha_n)$, where n random variables $\alpha_1, \ldots, \alpha_n$ are measured at the same time. In Section 2.3 we defined stochastic processes characterized by a joint probability density $P_n(\alpha_1, t_1; \cdots; \alpha_n, t_n)$ that a single random variable α has the values $\alpha_i \equiv \alpha(t_i)$ at the n times $t_1, \ldots, t_n$. We now define the *many-time characteristic function* of an n-time stochastic process

$$\Phi_n(\omega_1, t_1; \ldots; \omega_n, t_n) \equiv \int P_n(\alpha_1, t_1; \ldots; \alpha_n, t_n)\mathrm{e}^{\mathrm{i}(\omega_1\alpha_1 + \cdots + \omega_n\alpha_n)}\mathrm{d}^n\alpha, \tag{2.96}$$

i.e., there's a transform variable ω_i associated with each $\alpha_i = \alpha(t_i)$. The general n-time correlation function is found by differentiating Eq. (2.96) (compare with Eq. (2.7)), where $k \equiv k_1 + \cdots + k_n$,

$$\begin{aligned}\langle \alpha^{k_1}(t_1)\cdots\alpha^{k_n}(t_n)\rangle &= \int \alpha_1^{k_1}\cdots\alpha_n^{k_n}P_n(\alpha_1, t_1; \ldots; \alpha_n, t_n)\mathrm{d}^n\alpha \\ &= (-\mathrm{i})^k\frac{\partial^k}{\partial^{k_1}\omega_1\cdots\partial^{k_n}\omega_n}\Phi_n(\omega_1, t_1; \ldots; \omega_n t_n)\bigg|_{\boldsymbol{\omega}=0}. \end{aligned} \tag{2.97}$$

[81]Multivariate Gaussians are often defined in the form $P(\boldsymbol{\alpha}) \propto \exp(-\frac{1}{2}\boldsymbol{\alpha}^T\boldsymbol{\Sigma}^{-1}\boldsymbol{\alpha})$ with a matrix inverse, $\boldsymbol{\Sigma}^{-1}$.

[82]We're using a single integral sign to indicate a multiple integration (a common practice in advanced work)—integrate over the variables specified by the differential volume element using whatever limits of integration are appropriate.

We also define the *many-time cumulant* (compare with Eq. (2.87))

$$C_n(t_1, \ldots, t_n) \equiv (-\mathrm{i})^n \frac{\partial^n}{\partial\omega_1 \cdots \partial\omega_n} \ln \Phi_n(\omega_1, t_1; \cdots; \omega_n, t_n)\bigg|_{\omega=0} . \tag{2.98}$$

Using Eq. (2.98), we can write down the first few many-time cumulants:

$$\begin{aligned} C_1(t_1) &= \langle\alpha(t_1)\rangle \\ C_2(t_1, t_2) &= \langle\alpha(t_1)\alpha(t_2)\rangle - \langle\alpha(t_1)\rangle\langle\alpha(t_2)\rangle \\ C_3(t_1, t_2, t_3) &= \langle\alpha(t_1)\alpha(t_2)\alpha(t_3)\rangle - \langle\alpha(t_1)\rangle\langle\alpha(t_2)\alpha(t_3)\rangle \\ &\quad - \langle\alpha(t_2)\rangle\langle\alpha(t_3)\alpha(t_1)\rangle - \langle\alpha(t_3)\rangle\langle\alpha(t_1)\alpha(t_2)\rangle \\ &\quad + 2\langle\alpha(t_1)\rangle\langle\alpha(t_2)\rangle\langle\alpha(t_3)\rangle . \end{aligned} \tag{2.99}$$

Cumulants have the property of vanishing (for $C_{n>1}$) when the random variables involved are statistically independent; see Kubo[57] or [5, p77]. Nonzero cumulants probe many-particle correlations. The vanishing of cumulants *in time* probes the time over which random variables remain correlated.

Example. Consider the construction of the third cumulant. Start by subtracting from the average $\langle\alpha_1\alpha_2\alpha_3\rangle$ the contribution when $\alpha_1, \alpha_2, \alpha_3$ are each statistically independent of the others, $\langle\alpha_1\rangle\langle\alpha_2\rangle\langle\alpha_3\rangle$. Next subtract the value obtained when α_1 is independent of α_2 and α_3, $\langle\alpha_1\rangle\left(\langle\alpha_2\alpha_3\rangle - \langle\alpha_2\rangle\langle\alpha_3\rangle\right)$. Repeat for α_2 independent of α_3 and α_1, and α_3 independent of α_1 and α_2. The final result is the expression for C_3 shown in Eq. (2.99).

2.9.4 Gaussian stochastic processes, Doob's theorem

A stochastic process is Gaussian if its hierarchy of joint probabilities consists of multivariate Gaussians, i.e., for $n = 1, 2, \ldots,$

$$P_n(y_1, t_1; \cdots; y_n, t_n) = \frac{\sqrt{\det \boldsymbol{A}}}{(2\pi)^{n/2}} \exp\left[-\frac{1}{2}\sum_{ij=1}^{n}(y_i - \mu_i)A_{ij}(y_j - \mu_j)\right], \tag{2.100}$$

where $y_i \equiv y(t_i)$ and $\mu_i = \langle y(t_i)\rangle$ is the expectation value of $y(t)$ at time t_i, and $\boldsymbol{A}$ is a positive-definite $n \times n$ matrix. The many-time characteristic function follows by combining Eqs. (2.100) and (2.96),

$$\Phi_n(\omega_1, t_1; \cdots; \omega_n, t_n) = \exp\left[\mathrm{i}\sum_{j=1}^{n}\mu_j\omega_j - \frac{1}{2}\sum_{jk=1}^{n}(\boldsymbol{A}^{-1})_{jk}\omega_j\omega_k\right]. \tag{2.101}$$

From Eq. (2.97) combined with Eq. (2.101), we find the generalization of Eq. (2.95) to include time,

$$(\boldsymbol{A}^{-1})_{ij} = \Big\langle(y_i - \mu_i)(y_j - \mu_j)\Big\rangle \equiv \Big\langle(y(t_i) - \langle y(t_i)\rangle)(y(t_j) - \langle y(t_j)\rangle)\Big\rangle. \tag{2.102}$$

We should check that the requirements on hierarchies listed on page 25 are satisfied by the sequence in Eq. (2.100). The first two requirements are manifestly satisfied ($P_n \geq 0$ and the symmetry condition) and the last is satisfied by construction (normalization on P_1). The consistency condition $\int \mathrm{d}y_n P_n(y_1, \ldots, y_n) = P_{n-1}(y_1, \ldots, y_{n-1})$ requires a few steps, however. Use the fact that $P_n(y_1, \ldots, y_n)$ is the inverse Fourier transform of the associated characteristic function, $P_n(y_1,\ldots,y_n) = 1/(2\pi)^n \int \mathrm{d}\omega_1 \cdots \mathrm{d}\omega_n \mathrm{e}^{-\mathrm{i}(\omega_1 y_1 + \cdots \omega_n y_n)} \Phi_n(\omega_1,\ldots,\omega_n)$. Then,

$$\begin{aligned}
\int \mathrm{d}y_n P_n(y_1,\ldots,y_n) &= \frac{1}{(2\pi)^n} \int \mathrm{d}\omega_1 \cdots \mathrm{d}\omega_n \mathrm{e}^{-\mathrm{i}(\omega_1 y_1 + \cdots + \omega_{n-1} y_{n-1})} \Phi_n(\omega_1,\ldots,\omega_n) \\
&\quad \times \int \mathrm{d}y_n \mathrm{e}^{-\mathrm{i}\omega_n y_n} \\
&= \frac{1}{(2\pi)^{n-1}} \int \mathrm{d}\omega_1 \cdots \mathrm{d}\omega_n \mathrm{e}^{-\mathrm{i}(\omega_1 y_1 + \cdots + \omega_{n-1} y_{n-1})} \delta(\omega_n) \Phi_n(\omega_1,\ldots,\omega_n) \\
&= \frac{1}{(2\pi)^{n-1}} \int \mathrm{d}\omega_1 \cdots \mathrm{d}\omega_{n-1} \mathrm{e}^{-\mathrm{i}(\omega_1 y_1 + \cdots + \omega_{n-1} y_{n-1})} \Phi_{n-1}(\omega_1,\ldots,\omega_{n-1}) \\
&= P_{n-1}(y_1,\ldots,y_{n-1}).
\end{aligned}$$

In the first line we interchanged the order of integration and in the second we used the representation of the Dirac delta function, $\delta(x) = 1/(2\pi) \int_{-\infty}^{\infty} \mathrm{d}k \mathrm{e}^{-\mathrm{i}kx}$. One might wonder where properties specific to Gaussians were used in this derivation. We used the minimal property that $\Phi_n(\omega_1,\ldots,\omega_{n-1},\omega_n = 0) = \Phi_{n-1}(\omega_1,\ldots,\omega_{n-1})$, as can be seen from Eq. (2.101).[83]

2.9.4.1 Stationary Gaussian processes

Gaussian processes are specified by the covariance matrix $\boldsymbol{A}^{-1}$, which we illustrate with the simplest example of finding the form of bivariate stationary Gaussian processes, $P_2(y_1,t_1;y_2,t_2)$, where for convenience we consider a process $y(t)$ with zero expectation value $\langle y(t_i)\rangle = 0$ (otherwise define a new variable $\alpha_i \equiv y_i - \mu_i$.) In that case, from Eq. (2.102),

$$\boldsymbol{A}^{-1} = \begin{pmatrix} \langle y^2(t_1)\rangle & \langle y(t_1)y(t_2)\rangle \\ \langle y(t_1)y(t_2)\rangle & \langle y^2(t_2)\rangle \end{pmatrix} \equiv \begin{pmatrix} C & C_{12} \\ C_{12} & C \end{pmatrix}, \tag{2.103}$$

where we've used that for stationary processes $\langle y^2(t_1)\rangle = \langle y(t_2)^2\rangle = \langle y^2\rangle = C$, a constant (Eq. (2.23)); the autocorrelation function C_{12} is a function of the time difference $t_1 - t_2$, Eq. (2.24). From the form of $\boldsymbol{A}^{-1}$ in Eq. (2.103), we find for $\boldsymbol{A}$

$$\begin{aligned}
\boldsymbol{A} &= \frac{1}{C^2 - C_{12}^2} \begin{pmatrix} C & -C_{12} \\ -C_{12} & C \end{pmatrix} = \frac{1}{C(1-(C_{12}/C)^2)} \begin{pmatrix} 1 & -C_{12}/C \\ -C_{12}/C & 1 \end{pmatrix} \\
&\equiv \frac{1}{C(1-\Gamma_{12}^2)} \begin{pmatrix} 1 & -\Gamma_{12} \\ -\Gamma_{12} & 1 \end{pmatrix},
\end{aligned} \tag{2.104}$$

where

$$\Gamma_{12} \equiv \frac{\langle y(t_1)y(t_2)\rangle}{\langle y^2\rangle} \tag{2.105}$$

is the *normalized autocorrelation function*. With the matrix $\boldsymbol{A}$ determined, we have from Eq. (2.100)

$$P_2(y_1,t_1;y_2,t_2) = \frac{1}{2\pi C\sqrt{1-\Gamma_{12}^2}} \exp\left[-\frac{1}{2C(1-\Gamma_{12}^2)}\left(y_1^2 + y_2^2 - 2\Gamma_{12}y_1y_2\right)\right]. \tag{2.106}$$

P_2 is stationary through the stationarity of the correlation function $\langle y(t_1)y(t_2)\rangle = \Gamma(t_1 - t_2)$. If $\Gamma_{12} = 0$, i.e., if $y(t_1)$, $y(t_2)$ are uncorrelated, P_2 factors into a product of independent Gaussians.

With P_2 established, we can find P_1 by integrating $\int P_2(y_1,t_1;y_2,t_2)\mathrm{d}y_2 = P_1(y_1,t_1)$,

$$P_1(y,t) = \frac{1}{\sqrt{2\pi C}} \exp\left(-\frac{y^2}{2C}\right) \tag{2.107}$$

[83] One might also wonder why we've taken the diversion through the characteristic function; why not simply integrate over P_n directly? Such an approach is more difficult and requires a certain dexterity with determinants, but the end result is the same. The characteristic function doesn't have a multiplicative factor involving determinants, a great simplification.

(the same as we'd find from Eq. (2.100) for $n = 1$). We see that, indeed, $P_1(y,t) = P_1(y)$ for stationary processes (see Section 2.4.1). We can also find the transition probability associated with a stationary Gaussian process. From Eq. (2.19),

$$P_{1|1}(y_1,t_1|y_2,t_2) = \frac{P_2(y_1,t_1;y_2,t_2)}{P_1(y_2,t_2)} = \frac{1}{\sqrt{2\pi C(1-\Gamma_{12}^2)}}\exp\left[-\frac{1}{2C(1-\Gamma_{12}^2)}(y_1-\Gamma_{12}y_2)^2\right]. \tag{2.108}$$

We use this expression in Chapter 3.

2.9.4.2 *Doob's theorem*

Does the transition probability in Eq. (2.108) represent a Markov process? It would appear so—the probability of the transition $y_2 \to y_1$ does not depend on the past history of the system. Markov processes satisfy the SCK equation (Section 2.4.3). Let's check if, using Eq. (2.108), whether $P_{1|1}(y_1,t_1|y_3,t_3) = \int \mathrm{d}y_2 P_{1|1}(y_1,t_1|y_2,t_2)P_{1|1}(y_2,t_2|y_3,t_3)$. Setting $C = 1$ for convenience, we find

$$\begin{aligned} &\frac{1}{2\pi\sqrt{(1-\Gamma_{12}^2)(1-\Gamma_{23}^2)}}\int \mathrm{d}y_2 \exp\left[-\frac{1}{2(1-\Gamma_{12}^2)}(y_1-\Gamma_{12}y_2)^2\right] \\ &\qquad\qquad\times \exp\left[-\frac{1}{2(1-\Gamma_{23}^2)}(y_2-\Gamma_{23}y_3)^2\right] \\ &= \frac{1}{\sqrt{2\pi(1-\Gamma_{12}^2\Gamma_{23}^2)}}\exp\left[-\frac{1}{2(1-\Gamma_{12}^2\Gamma_{23}^2)}(y_1-\Gamma_{12}\Gamma_{23}y_3)^2\right] \stackrel{?}{=} P_{1|1}(y_1,t_1|y_3,t_3). \end{aligned} \tag{2.109}$$

Comparing the result of Eq. (2.109) with the form on the right side of Eq. (2.108) (with C=1), the SCK equation is satisfied if the autocorrelation functions have the composition property

$$\Gamma_{13} = \Gamma_{12}\Gamma_{23}. \tag{2.110}$$

Equation (2.110) is a *functional equation*, an equation where the unknowns are functions.[84] Based on the stationarity of $\Gamma_{ij} \equiv \langle y(t_i)y(t_j)\rangle \equiv \phi(|t_i - t_j|)$, and the ordering of the times $t_1 > t_2 > t_3$, Eq. (2.110) is equivalent to the functional equation $\phi(x+y) = \phi(x)\phi(y)$, the solution of which is $\phi(x) = a^x = \mathrm{e}^{x\ln a}$ for some base a. We require that, for a positive constant c,

$$\Gamma(\tau) = \exp(-c|\tau|) \qquad (c \geq 0) \tag{2.111}$$

because $\Gamma(\tau) \leq \Gamma(0) = 1$ and $\Gamma(\tau) = \Gamma(-\tau)$ (see Exercise 2.12). Equation (2.111) is known as *Doob's theorem*,[85] that the autocorrelation function of a stationary Markov Gaussian process has the exponential form $\Gamma(\tau) = \exp(-c|\tau|)$.

SUMMARY

An introduction to the statistical theory of fluctuations was presented, an approach that follows from the classic works of Einstein and Onsager and comprises a necessary foundation for the study of nonequilibrium systems.

- The Einstein theory of fluctuations (Section 2.1) draws on the Boltzmann entropy formula to establish the probability $P(\Delta S) \propto \mathrm{e}^{\Delta S/k}$ that fluctuations about equilibrium are associated with entropy fluctuations ΔS, where $\Delta S < 0$. Entropy fluctuations, in turn, are related,

[84] A useful reference on functional equations is Aczel[58].

[85] There are many Doob theorems in the literature. We're referring to Theorem 1.1 in [59], reprinted in Wax[44, pp319–337]. Wax is an invaluable resource to students of nonequilibrium statistical physics.

through the first law of thermodynamics, to fluctuations $\alpha_i \equiv X_i - X_i^0$ in the extensive quantities of the system X_i, where X_i^0 denotes the equilibrium value. With the Einstein theory, one can calculate *static* properties of fluctuations, such as their average magnitudes and covariances. It provides, in the form of a multivariate Gaussian function, the joint probability density $P(\alpha_1, \ldots, \alpha_n)$ that n fluctuating extensive quantities have the instantaneous values $\alpha_1, \ldots, \alpha_n$. We found a nonzero correlation between fluctuations and thermodynamic forces, $\langle \alpha_j F_l \rangle = -k\delta_{jl}$, a key result in the Onsager theory of irreversible processes. Through scattering experiments we know that long-range spatial correlations can develop among fluctuations (such as in critical opalescence). The time-dependent properties of fluctuations are not used in setting up statistical mechanics but are crucial in nonequilibrium statistical mechanics.

- Fluctuations are dynamical processes, and by recognizing that we're led to consider the correlation of fluctuations in time. We defined time correlation functions in terms of time averages, $\overline{\alpha_i(t)\alpha_j(t+\tau)} \equiv \lim_{T\to\infty}(1/T)\int_0^T \alpha_i(t)\alpha_j(t+\tau)\mathrm{d}t$, Eq. (2.14). In statistical mechanics, time averages are equated with ensemble averages (ergodic hypothesis) to such an extent that only infrequently does one encounter time averages. The ergodic hypothesis does not apply to systems out of equilibrium and consequently one often encounters quantities defined as time averages in nonequilibrium statistical mechanics. In our initial look at time correlation functions (Section 2.2), we treated fluctuations as dynamical quantities subject to Hamilton's equations of motion. This was done so we could invoke Birkhoff's theorem on the properties of time averages and so that we could invoke the time-reversal symmetry of the microscopic equations of motion. As a result, we arrived at Eq. (2.16), $\overline{\alpha_i(t)\alpha_j(t+\tau)} = \overline{\alpha_i(t+\tau)\alpha_j(t)}$, a kind of "time-influence" symmetry of fluctuations, that if $\alpha_i(t)$ is correlated with (influences) $\alpha_j(t+\tau)$ with $\tau > 0$, then $\alpha_j(t)$ influences $\alpha_i(t+\tau)$ in the same way. Equation (2.16) is a consequence of the time-reversal invariance of the microscopic dynamics underlying fluctuations and the time-translational invariance of equilibrium averages. Even though we can't calculate the time dependence of $\alpha_i(t)$ from first principles, Eq. (2.16) is a fundamental property of fluctuations about the equilibrium state. And, as discussed in Section 2.2, Eq. (2.16) is Onsager reciprocity in disguised form. Remarkably, kinetic coefficients, introduced as phenomenological parameters in the description of irreversible processes, obey the experimentally verified reciprocal relations from the time-reversibility of the microscopic motions of system components.

- Although we treated fluctuations as dynamical quantities in Section 2.2 and classified them by their time-reversal properties, we have no way of modeling their time dependence from first principles. For that purpose we turned to stochastic processes (Section 2.3), the branch of probability theory devoted to time-dependent random processes. A stochastic process is a family of random variables $\{\alpha(t_i)\}$ indexed by time. Stochastic processes are specified by a hierarchy of joint probability densities meeting prescribed rules. Whereas in Section 2.1 we found the probability density for n variables $\alpha_1, \ldots, \alpha_n$ representing fluctuations in n different physical quantities, all measured at the same time, in Section 2.3 we considered a random variable α representing the same quantity observed at n different times, $(t_1, \ldots, t_n)$. The joint probability density for an n-time stochastic process is a function of $2n$ variables, the n times $(t_1, \ldots, t_n)$ and the n random variables $(\alpha(t_1), \ldots, \alpha(t_n))$, indicated notationally $P_n(y_1, t_1; y_2, t_2; \ldots; y_n, t_n)$ as a function of the times t_i and the values of random variables $y_i = \alpha(t_i)$. Note the subscript on P_n; the set of functions P_n for $n = 1, 2, \ldots$ comprises a hierarchy of joint probabilities describing the stochastic process in successively more detail. Using joint probability densities, one can calculate various correlation functions; e.g., $\langle y(t_1)y(t_2)\rangle = \int\int y_1 y_2 P_2(y_1, t_1; y_2, t_2)\mathrm{d}y_1\mathrm{d}y_2$. There are three main classes of stochastic processes used in nonequilibrium statistical physics: independent, stationary, and Markov.

- A stochastic process is stationary if, for all n and τ,

$$P_n(y_1,t_1;y_2,t_2;\ldots;y_n,t_n) = P_n(y_1,t_1+\tau;y_2,t_2+\tau;\ldots;y_n,t_n+\tau), \qquad (2.22)$$

i.e., stationary processes have no unique origin in time. Stationarity implies that P_1 is independent of time, $P_1(y_1,t_1) = P_1(y_1)$, and that $P_2(y_1,t_1;y_2,t_2)$ is a function of the single time $\tau \equiv t_1 - t_2$, $P_2(y_1,t+\tau;y_2,t)$. As a consequence, the mean of stationary processes is a constant, $\langle y(t)\rangle = \int \mathrm{d}y y P_1(y,t) =$ constant, and the autocorrelation function $R(\tau) \equiv \langle y(t+\tau)y(t)\rangle = \int\int \mathrm{d}y_1\mathrm{d}y_2 y_1 y_2 P_2(y_1,t+\tau;y_2,t)$ has the properties $R(\tau) = R(-\tau)$ and $|R(\tau)| \le R(0)$. Moreover, $R(\tau)$ factorizes for $\tau \gg \tau_c$, where τ_c is a system-dependent characteristic time, $R(\tau) \overset{\tau\gg\tau_c}{\longrightarrow} \langle y(t)\rangle\langle y(t+\tau)\rangle$, i.e., the process becomes uncorrelated for $\tau \gg \tau_c$. To be useful as dynamical models of fluctuations, stationary processes must be ergodic (Section 2.5), that time averages $\overline{y}_T \equiv (1/2T)\int_{-T}^{T} y(t)\mathrm{d}t$ approach the ensemble mean as $T \to \infty$, $\overline{y}_T \longrightarrow \langle \overline{y}_T\rangle$. This requirement is met when $R(\tau)$ decays sufficiently rapidly in time; see Eq. (2.33). We derived the Wiener-Khinchin theorem, that for stationary processes the spectral density (defined in Eq. (2.37)) is the Fourier transform of the autocorrelation function. As an illustration of an ergodic stationary process, we analyzed in Section 2.6 the noise generated in conductors by the random thermal agitations of charge carriers and derived the experimentally verified Nyquist theorem relating the noise power spectral density to the resistance and the absolute temperature.

- The conditional probability $P_{n|v}(y_1,t_1;\ldots;y_n,t_n|y_{n+1},t_{n+1};\ldots;y_{n+v}t_{n+v})$ is the probability that n samplings of $\alpha(t)$ have the values indicated, given that v samplings are known to have produced the values indicated for $t_1 > \cdots > t_n > t_{n+1} > \cdots > t_{n+v}$; see Eq. (2.19). The Markov approximation consists of replacing $P_{n|v}$ with $P_{n|1}$,

$$P_{n|v}(y_1,t_1;\ldots;y_n,t_n|y_{n+1},t_{n+1};\ldots;y_{n+v},t_{n+v}) \overset{\substack{\text{Markov}\\ \text{approximation}}}{\longrightarrow} P_{n|1}(y_1,t_1;\ldots;y_n,t_n|y_{n+1},t_{n+1}).$$

In a Markov process the probability of the current state depends only on the most recent past observation, and not on observations made at earlier times; the "present" step is conditioned by memory only of the most recent past. The probability $P_{1|1}$ is well suited to physical descriptions in a way similar to quantum mechanics: $P_{1|1}(y_1,t_1|y_2,t_2)$ is the probability that $\alpha(t_1) = y_1$ given that $\alpha(t_2 < t_1) = y_2$ is known to have occurred. $P_{1|1}(y_1,t_1|y_2,t_2)$ is referred to as a transition probability; the system makes a transition from y_2 to y_1 in the time interval $(t_1 - t_2)$ with probability $P_{1|1}$. The basic equation of Markov processes is the SCK equation

$$P_{1|1}(y_1,t_1|y_3,t_3) = \int \mathrm{d}y_2 P_{1|1}(y_1,t_1|y_2,t_2)P_{1,1}(y_2,t_2|y_3,t_3). \qquad (2.29)$$

For $t_1 > t_2 > t_3$, the probability of $\alpha(t_1) = y_1$ given that $\alpha(t_3) = y_3$ can be found by considering the process as a sum over compound processes consisting of the probability of $\alpha(t_2) = y_2$ given $\alpha(t_3) = y_3$ multiplied by the probability of $\alpha(t_1) = y_1$ given $\alpha(t_2) = y_2$.

- We considered the symmetric random walk problem in one dimension as an illustration of a Markov process (Section 2.7). A particle situated at one of the discrete positions $x = ma$, $m = 0, \pm1, \pm2, \cdots$, makes a transition (at a discrete time $t = n\tau$, $n = 0,1,2,\ldots$) to one of the neighboring sites $m \pm 1$ with equal probability. The goal is to calculate the conditional probability $p(m,n)$ that a "walker" started at location $m = 0$ at time $n = 0$ is found at $x = ma$ at time $t = n\tau$. This problem can be treated with traditional methods of probability and combinatorics (see Exercise 2.22). We treated this problem as a stochastic process using the discrete form of the SCK equation; see Eq. (2.47). There are many connections between random walks and diffusion. We derived the Einstein result that in diffusive motion the mean-square displacement $\langle x^2\rangle \propto t$.

- The master equation is an equivalent form of the SCK equation (and hence applies to Markov processes) that finds wide use in physics and chemistry. For continuous-time processes, the master equation is an integro-differential equation for the nonequilibrium probability distribution,

$$\frac{\partial}{\partial t}P(y,t) = \int \left[W(y|y')P(y',t) - W(y'|y)P(y,t)\right]\mathrm{d}y', \tag{2.66}$$

  where $W(y|y') \geq 0$ is a rate, the transition probability per unit time from state y' to state y. The characteristic feature of the master equation is its "gain-loss form," that the rate at which $P(y,t)$ changes is a balance between rates of gain processes in which transitions occur into the state y from all other states y' and loss processes in which transitions to all other states y' occur from y.

  For a discrete state space labeled by index n, the master equation has the form

$$\frac{\partial}{\partial t}P(n,t) = \sum_{n'\neq n}\left[W(n|n')P(n',t) - W(n'|n)P(n,t)\right] \equiv \sum_{n'} W_{n,n'}P(n',t),$$

  where $W_{n,n'}$ is an element of the transition matrix (see Eq. (2.74)). The detailed balance condition, $W(n|n')P_{\text{eq}}(n') = W(n'|n)P_{\text{eq}}(n)$ (or $W_{n,n'}P_{\text{eq}}(n') = W_{n',n}P_{\text{eq}}(n)$), is a dynamic characterization of equilibrium, where for any pair of states n, n', the rate of the transition $n' \to n$ is the same as that of the reverse transition $n \to n'$. Detailed balance does not have to be imposed as a requirement on transition rates; it follows from fundamental physics, the time-reversal invariance of microscopic motions (just as with the Onsager relations). The transition matrix elements have the property $\sum_n W_{n,n'} = 0$ for each n', implying that the total probability $\sum_n P(n,t)$ is conserved in time. It also implies, when combined with detailed balance, that the equilibrium probability distribution $P_{\text{eq}}(n)$ is an eigenvector of W having zero eigenvalue, $\sum_{n'} W_{n,n'}P_{\text{eq}}(n') = 0$. It can be shown that the nonzero eigenvalues of the transition matrix are negative (see Exercise 2.45), implying that the nonequilibrium probability distribution evolves in time to the equilibrium distribution (see Eq. (2.83)).

- Gaussian processes (Section 2.9) are extensions of Gaussian probability distributions to include stochastic variables and time-dependent parameters. Gaussians occur generically in statistical physics through the central limit theorem, and they occur in nonequilibrium processes as well. The theory of cumulants was introduced as a way of parameterizing Fourier transforms (characteristic functions) of probability distributions primarily because Gaussians have the unique property that their associated cumulants vanish except for the first two. Gaussian processes are the simplest generalization of deterministic processes wherein the same outcome occurs with no variance; they have one more parameter (than deterministic processes) describing the width of the distribution in such a way that higher-order cumulants vanish. Gaussian processes are specified by the covariance matrix, Eq. (2.102), the $n \times n$ matrix of time correlation functions among the random variables of an n-time stochastic process. We outlined the proof of Doob's theorem that the autocorrelation function of a stationary Markov Gaussian process decays exponentially in time.

EXERCISES

2.1 In the statistical theory of fluctuations we encounter multiple integrals of the form

$$I \equiv \int_{-\infty}^{\infty}\cdots\int_{-\infty}^{\infty}\exp\Big(-\sum_{ij=1}^{n}A_{ij}x_i x_j\Big)\mathrm{d}x_1\cdots\mathrm{d}x_n, \tag{P2.1}$$

where A_{ij} are elements of an $n\times n$ symmetric, positive-definite matrix $\boldsymbol{A}$. The purpose of this exercise is to guide you through one approach to evaluating this integral. The strategy is to change variables such that the quadratic form[86] $\sum_{ij} A_{ij}x_i x_j \equiv \boldsymbol{x}^T \boldsymbol{A}\boldsymbol{x}$ is simplified, where T denotes transpose and $\boldsymbol{x}^T = (x_1, \ldots, x_n)$. Make a linear transformation $x_i = \sum_k L_{ik}y_k$, or in matrix notation $\boldsymbol{x} = \boldsymbol{L}\boldsymbol{y}$, such that $\boldsymbol{x}^T\boldsymbol{A}\boldsymbol{x} = \boldsymbol{y}^T\boldsymbol{y}$. We require $\boldsymbol{y}^T\boldsymbol{L}^T\boldsymbol{A}\boldsymbol{L}\boldsymbol{y} = \boldsymbol{y}^T\boldsymbol{y}$, or $\boldsymbol{L}^T\boldsymbol{A}\boldsymbol{L} = \boldsymbol{I}$, where $\boldsymbol{I}$ is the $n \times n$ identity matrix. From the rules of determinants ($\det \boldsymbol{L}^T = \det \boldsymbol{L}$), $(\det \boldsymbol{L})^2 \det \boldsymbol{A} = 1$, implying $\det \boldsymbol{L} = 1/\sqrt{\det \boldsymbol{A}}$ ($\det \boldsymbol{L}$ exists if $\det \boldsymbol{A} \neq 0$). The volume element in Eq. (P2.1) transforms as $\mathrm{d}x_1 \cdots \mathrm{d}x_n = \det \boldsymbol{L} \mathrm{d}y_1 \cdots \mathrm{d}y_n$ ($\det \boldsymbol{L}$ is the Jacobian). Show that

$$I = \frac{\pi^{n/2}}{\sqrt{\det \boldsymbol{A}}}. \tag{P2.2}$$

Hint: $\int_{-\infty}^{\infty} \mathrm{e}^{-y^2} \mathrm{d}y = \sqrt{\pi}$.

2.2 Evaluate the derivatives in Eq. (2.5). Refer to [2, p19] for definitions of α, β_T, C_V.

a. Show that $\left(\frac{\partial^2 S}{\partial U^2}\right)_V = \left(\frac{\partial}{\partial U}\left(\frac{1}{T}\right)\right)_V = -\frac{1}{T^2 C_V}$.

b. Show that

$$\frac{\partial^2 S}{\partial U \partial V} = \left(\frac{\partial}{\partial U}\left(\frac{P}{T}\right)\right)_V = \frac{1}{T}\left(\frac{\partial P}{\partial U}\right)_V - \frac{P}{T^2}\left(\frac{\partial T}{\partial U}\right)_V = -\frac{1}{C_V T}\left(\frac{P}{T} - \frac{\alpha}{\beta_T}\right).$$

Use the handy trick $\left(\frac{\partial P}{\partial U}\right)_V = \left(\frac{\partial P}{\partial T}\right)_V \left(\frac{\partial T}{\partial U}\right)_V$.

c. Show that $\left(\frac{\partial^2 S}{\partial V^2}\right)_U = \left(\frac{\partial}{\partial V}\left(\frac{P}{T}\right)\right)_U = \frac{1}{T}\left(\frac{\partial P}{\partial V}\right)_U - \frac{P}{T^2}\left(\frac{\partial T}{\partial V}\right)_U$. Use the cyclic relation[2, p18] or perhaps your skill with Jacobian determinants[2, p49] to show

$$\left(\frac{\partial P}{\partial V}\right)_U = \frac{\partial(P,U)}{\partial(V,U)} = \frac{\partial(P,U)}{\partial(V,T)}\frac{\partial(V,T)}{\partial(V,U)} = \frac{1}{C_V}\frac{\partial(P,U)}{\partial(V,T)} = -\left[\frac{1}{\beta_T V} + \frac{\alpha}{\beta_T C_V}\left(\frac{\partial U}{\partial V}\right)_T\right].$$

In the same way, show that

$$\left(\frac{\partial T}{\partial V}\right)_U = \frac{\partial(T,U)}{\partial(V,U)} = \frac{\partial(T,U)}{\partial(T,V)}\frac{\partial(T,V)}{\partial(V,U)} = -\frac{1}{C_V}\left(\frac{\partial U}{\partial V}\right)_T.$$

Show we have the intermediate result

$$\left(\frac{\partial^2 S}{\partial V^2}\right)_U = -\left[\frac{1}{\beta_T T V} + \frac{1}{C_V T}\left(\frac{\partial U}{\partial V}\right)_T \left(\frac{\alpha}{\beta_T} - \frac{P}{T}\right)\right].$$

It remains to evaluate the derivative $(\partial U/\partial V)_T$, which can be found from an analysis of the first law of thermodynamics together with a Maxwell relation,[2, p20]

$$\left(\frac{\partial U}{\partial V}\right)_T = T\frac{\alpha}{\beta_T} - P.$$

Conclude that

$$\left(\frac{\partial^2 S}{\partial V^2}\right)_U = -\left[\frac{1}{\beta_T T V} + \frac{1}{C_V}\left(\frac{\alpha}{\beta_T} - \frac{P}{T}\right)^2\right].$$

[86] A quadratic form is a homogeneous quadratic polynomial in any number of variables. In three variables, $ax^2 + by^2 + cz^2 + dxy + exz + fyz$ is a quadratic form for constants $(a, \cdots, f)$. A quadratic form in n variables can be generated by an $n \times n$ symmetric matrix. A *positive-definite* quadratic form is positive for any nonzero values of its variables.

2.3 Let $\{X_i\}_{i=1}^n$ be a set of statistically independent random variables (see Appendix B), each with its own characteristic function $\Phi_{X_i}(\omega)$, and let Y be a random variable resulting from a linear combination $Y \equiv \sum_{i=1}^n a_i X_i$, where the a_i are constants. Show that the characteristic function associated with Y is the product of characteristic functions for the X_i,

$$\Phi_Y(\omega) = \prod_{i=1}^n \Phi_{X_i}(\omega).$$

Hint: $\Phi_Y(\omega) = \langle \mathrm{e}^{\mathrm{i}\omega Y}\rangle = \langle \mathrm{e}^{\mathrm{i}\omega \sum_{i=1}^n a_i X_i}\rangle = \langle \mathrm{e}^{\mathrm{i}\omega a_1 X_1} \cdots \mathrm{e}^{\mathrm{i}\omega a_n X_n}\rangle$ and the joint probability distribution $P(X_1, \ldots, X_n)$ for independent random variables factorizes.

2.4 Verify Eq. (2.7).

2.5 Guided exercise: Consider the following integral, the form of which we have in Eq. (2.9) (a multivariate Gaussian integral, see Section 2.9.2),

$$I \equiv \int_{-\infty}^{\infty} \cdots \int_{-\infty}^{\infty} \exp\left(-a\boldsymbol{r}^T \boldsymbol{G}\boldsymbol{r} + \mathrm{i}\boldsymbol{\omega}^T \boldsymbol{r}\right) \mathrm{d}x_1 \cdots \mathrm{d}x_n, \tag{P2.3}$$

where $a > 0$, $\boldsymbol{r} \equiv (x_1, \ldots, x_n)^T$, $\boldsymbol{G}$ is a real symmetric $n \times n$ matrix, and $\boldsymbol{\omega} = (\omega_1, \ldots, \omega_n)^T$ is a constant vector. As always, the strategy is to make a suitable change in variables.[87] Because $\boldsymbol{G}$ is real and symmetric, there exists an orthogonal matrix $\boldsymbol{L}$ such that $\boldsymbol{L}^T\boldsymbol{G}\boldsymbol{L} = \boldsymbol{\Lambda}$, where $\boldsymbol{\Lambda}$ is a diagonal matrix with the eigenvalues of $\boldsymbol{G}$ as its entries[13, p41]. Introduce a new vector $\boldsymbol{y}$ such that $\boldsymbol{r} = \boldsymbol{L}\boldsymbol{y}$. The argument of the exponential in Eq. (P2.3) transforms as

$$-a\boldsymbol{r}^T\boldsymbol{G}\boldsymbol{r} + \mathrm{i}\boldsymbol{\omega}^T\boldsymbol{r} \to -a\boldsymbol{y}^T\boldsymbol{\Lambda}\boldsymbol{y} + \mathrm{i}\boldsymbol{\omega}^T\boldsymbol{L}\boldsymbol{y} = \sum_{k=1}^n \left[-a\lambda_k y_k^2 + \mathrm{i}\left(\boldsymbol{\omega}^T\boldsymbol{L}\right)_k y_k\right],$$

where λ_k are the eigenvalues of $\boldsymbol{G}$. Because $\boldsymbol{L}$ is orthogonal, $\mathrm{d}^n x = \mathrm{d}^n y$. Thus, Eq. (P2.3) is equivalent to

$$I = \prod_{k=1}^n \int_{-\infty}^{\infty} \exp\left[-a\lambda_k y_k^2 + \mathrm{i}\left(\boldsymbol{\omega}^T\boldsymbol{L}\right)_k y_k\right] \mathrm{d}y_k. \tag{P2.4}$$

Using standard methods of integration, show that Eq. (P2.4) is equivalent to

$$I = \prod_{k=1}^n \sqrt{\frac{\pi}{a\lambda_k}} \exp\left(-\frac{(\boldsymbol{\omega}^T\boldsymbol{L})_k^2}{4a\lambda_k}\right) = \frac{(\pi/a)^{n/2}}{\sqrt{\prod_{k=1}^n \lambda_k}} \exp\left(-\sum_{k=1}^n (\boldsymbol{\omega}^T\boldsymbol{L})_k^2/(4a\lambda_k)\right). \tag{P2.5}$$

Show that

$$\boldsymbol{\omega}^T\boldsymbol{L}\boldsymbol{\Lambda}^{-1}\boldsymbol{L}^T\boldsymbol{\omega} = \sum_k \frac{(\boldsymbol{\omega}^T\boldsymbol{L})_k^2}{\lambda_k}. \tag{P2.6}$$

Hint: First show that $\left(\boldsymbol{L}^T\boldsymbol{\omega}\right)_i = \left(\boldsymbol{\omega}^T\boldsymbol{L}\right)_i$. Then show that $\boldsymbol{\Lambda}^{-1} = \boldsymbol{L}^T\boldsymbol{G}^{-1}\boldsymbol{L}$, and hence

$$I = \frac{(\pi/a)^{n/2}}{\sqrt{\det \boldsymbol{G}}} \exp\left(-\frac{1}{4a}\boldsymbol{\omega}^T\boldsymbol{G}^{-1}\boldsymbol{\omega}\right). \tag{P2.7}$$

The determinant of a matrix is the product of its eigenvalues[13, p38].

2.6 Show the results in Eq. (2.10). Hint: The inverse of a symmetric matrix is symmetric.

2.7 Derive the inverse matrix $\boldsymbol{G}^{-1}$ in Eq. (2.11) starting from the matrix $\boldsymbol{G}$ in Eq. (2.5).

[87] Note the difference between the approach here and that of Exercise 2.1.

2.8 Show from the example where U and V are allowed to fluctuate (page 22), that

$$\langle(\Delta U)^2\rangle = C_V kT^2 + \langle(\Delta V)^2\rangle \left(\frac{\partial U}{\partial V}\right)_T^2$$

$$\langle\Delta U \Delta V\rangle = \langle(\Delta V)^2\rangle \left(\frac{\partial U}{\partial V}\right)_T .$$

From the first formula we see that $\langle(\Delta U)^2\rangle$ is not independent of $\langle(\Delta V)^2\rangle$, which is reflected in the nonzero correlation $\langle\Delta U \Delta V\rangle$ in the second formula.

2.9 Show that $\langle F_i F_j\rangle = k g_{ij}$. Hint: Combine two copies of Eq. (1.43) with Eq. (2.10).

2.10 Equation (2.13) (required in the proof of the Onsager relations) can be derived directly, without using the matrix $\boldsymbol{G}$.

a. Show that $F_j = k\dfrac{\partial \ln P}{\partial \alpha_j}$. Use Eq. (2.2) together with Eq. (1.43).

b. Show that $\langle \alpha_i F_j\rangle = k\int_{-\infty}^{\infty}\cdots\int_{-\infty}^{\infty} \alpha_i \dfrac{\partial P}{\partial \alpha_j} \mathrm{d}^n\alpha = -k\delta_{ij}$. Consider separately the cases $j \neq i$ and $j = i$. Integrate by parts.

2.11 Fill in the steps between Eq. (2.17) and the Onsager relations, Eq. (1.38).

2.12 For a stationary stochastic process, show that:

a. The autocorrelation function, Eq. (2.24), has the property $R(\tau) = R(-\tau)$. Hint: Use the symmetry condition under interchange of arguments on page 25.

b. $|R(\tau)| \leq R(0)$. Hint: Show that

$$0 \leq \langle(y(\tau) \pm y(0))^2\rangle = \langle y^2(\tau)\rangle + \langle y^2(0)\rangle \pm 2\langle y(\tau)y(0)\rangle.$$

2.13 Show that the joint probability densities P_k in Eq. (2.25) for independent processes satisfy the consistency conditions listed on page 25.

2.14 Derive Eq. (2.26).

2.15 Show in the Markov approximation that

$$P_3(y_1, t_1; y_2, t_2; y_3, t_3) = \frac{P_2(y_1, t_1; y_2, t_2)P_2(y_2, t_2; y_3, t_3)}{P_1(y_2, t_2)}.$$

2.16 Fill in the steps between Eqs. (2.28) and (2.29), the SCK equation.

2.17 Show by multiplying Eq. (2.29) by $P_1(y_3, t_3)$ and integrating over y_3 (using standard rules for joint and conditional probability densities), one is led to an identity, $P_1(y_1, t_1) = P_1(y_1, t_1)$. Thus one might expect the SCK equation to be a general relation because it reduces to an identity, yet it's derived using the Markov approximation. There are non-Markovian processes for which their transition probabilities satisfy the SCK equation[41, p203].

2.18 Fill in the steps between Eq. (2.31) and Eq. (2.32).

2.19 Suppose the autocorrelation function decays exponentially, $R(\tau) = R(0)\mathrm{e}^{-\gamma|\tau|}$, where $\gamma > 0$ is a constant. Use the Wiener-Khinchin theorem to derive the associated spectral density. A:

$$S(\omega) = R(0)\frac{2\gamma}{\gamma^2 + \omega^2}.$$

2.20 For a real-valued function $f(x)$, show that its Fourier transform $F(\omega)$ is such that $F^*(\omega) = F(-\omega)$, where F^* is the complex conjugate of F. Hint: Let $f(x) = \int_{-\infty}^{\infty} \mathrm{e}^{\mathrm{i}\omega x} F(\omega)\mathrm{d}\omega$.

2.21 What is the root-mean-square of voltage fluctuations at room temperature in a 100 Ω resistor with a measurement bandwidth of 1 MHz? A: 1.29 μV.

2.22 The one-dimensional random walk can be treated with the binomial distribution (see [5, p70]; see also Chapter 1 of Reif[60]). Consider a walker that takes n_r (n_l) steps to the right (left), for a total of $N = n_r + n_l$ steps. We seek the probability $P_N(m)$ that in N steps the walker is at site $m = n_r - n_l$ given that it was initially at the origin. For the symmetric random walk show that

$$P_N(m) = \frac{N!}{\left(\frac{1}{2}(N+m)\right)!\left(\frac{1}{2}(N-m)\right)!}\frac{1}{2^N}. \tag{P2.8}$$

Note that $(N+m)$ and $(N-m)$ are even integers (as can be shown). Show that Eq. (P2.8) solves Eq. (2.48) when we identify $P_N(m) \equiv p(m,n)$, i.e., N steps imply n time steps.

2.23 Show from Eq. (2.49) that the Fourier transform satisfies the relation $\widetilde{p}(q+Q,n) = \widetilde{p}(q,n)$ for $Q = 2\pi t$, $t = \pm 1, \pm 2, \ldots$, i.e., $\widetilde{p}(q,n)$ is periodic in q with period 2π. That's why q in $\widetilde{p}(q,n)$ is restricted to the interval $(-\pi,\pi)$, the familiar *Brillouin zone* of one-dimensional lattices—nothing new to be learned by extending q outside of the first Brillouin zone.

2.24 Show that Eq. (2.50) is the inverse of Eq. (2.49). First show that $\int_{-\pi}^{\pi} \mathrm{e}^{\mathrm{i}(k-m)}\mathrm{d}q = 2\pi\delta_{k,m}$ for integer (k,m).

2.25 Verify by direct substitution that the integral representation of $p(m,n)$ in Eq. (2.52) satisfies Eq. (2.48).

2.26 Use Eq. (2.52) to find $p(m,2)$. A: $p(m,2) = \frac{1}{4}\left(\delta_{m,-2} + 2\delta_{m,0} + \delta_{m,2}\right)$. Does this make sense given that $p(m,0) = \delta_{m,0}$?

2.27 Show that the normalization on $p(m,n)$, $\sum_{m=-\infty}^{\infty} p(m,n)$, is preserved by Eq. (2.52). Hint: $\sum_{m=-\infty}^{\infty} \mathrm{e}^{\mathrm{i}qm} = 2\pi\delta(q)$. See [13, p97].

2.28 Substantiate the claims made in Eq. (2.53). The following strategy may help. Define a function $f(q) \equiv (\cos q)^n \cos mq$. Show that

$$f(q) = (-1)^{n+m} f(\pi - q).$$

If $n+m$ is even (odd), then $f(q)$ is even (odd) about $q = \pi/2$.

2.29 From Eq. (2.53), $p(m,n) = 0$ for $m+n$ an odd number. Give an explanation why $p(\pm 1, 2) = 0$, i.e., why is this probability zero, given that $p(m,0) = \delta_{m,0}$.

2.30 Generalize the symmetric random walk of Section 2.7 to a *biased* random walk where at each site there is a constant probability α (β) of a step to the right (left), where $\alpha + \beta = 1$.

a. Show, using the appropriate generalization of Eq. (2.48), that Eq. (2.52) generalizes to

$$p(m,n) = \frac{1}{2\pi}\int_{-\pi}^{\pi} \mathrm{e}^{\mathrm{i}qm}\left(\alpha \mathrm{e}^{-\mathrm{i}q} + \beta \mathrm{e}^{\mathrm{i}q}\right)^n \mathrm{d}q \tag{P2.9}$$

where $p(m,n) \equiv P_{1|1}(m,n|0,0)$ with the initial condition $p(m,0) = \delta_{m,0}$. Show from Eq. (P2.9) that $\sum_{m=-\infty}^{\infty} p(m,n) = 1$, for all n. Hint: $\sum_{m=-\infty}^{\infty} \mathrm{e}^{\mathrm{i}qm} = 2\pi\delta(q)$. Show that $p(m,n) = p(-m,n)$ only if $\alpha = \beta$. Thus, $\langle m \rangle \neq 0$ when $\alpha \neq \beta$.

b. Apply the *binomial theorem*, $(x+y)^N = \sum_{k=0}^{N} \binom{N}{k} x^{N-k} y^k$,[5, p62] to Eq. (P2.9) to derive the generalization of Eq. (P2.8),

$$p(m,n) = \frac{n!}{\left(\frac{1}{2}(n-m)\right)!\left(\frac{1}{2}(n+m)\right)!}\alpha^{(n+m)/2}\beta^{(n-m)/2}.$$

c. Using Eq. (P2.9), show that

$$mp(m,n) = n\left[\alpha p(m-1,n-1) - \beta p(m+1,n-1)\right].$$

Hint: $m\mathrm{e}^{\mathrm{i}qm} = -\mathrm{i}\partial/\partial q(\mathrm{e}^{\mathrm{i}qm})$. Then show that in an n-step ensemble,

$$\langle m \rangle_n = n(\alpha - \beta), \tag{P2.10}$$

i.e., there is a constant *drift* characterized by the difference $(\alpha - \beta)$. With $m = x/a$ and $n = t/\tau$, argue that Eq. (P2.10) provides an estimate of the drift speed $(a/\tau)(\alpha - \beta)$.

d. Show that

$$m^2 p(m,n) = np(m,n) + n(n-1)\left[\alpha^2 p(m-2,n-2) - 2\alpha\beta p(m,n-2) + \beta^2 p(m+2,n-2)\right].$$

Hint: $m^2\mathrm{e}^{\mathrm{i}qm} = -\partial^2/\partial q^2(\mathrm{e}^{\mathrm{i}qm})$. Then show that

$$\langle m^2 \rangle_n = n + n(n-1)\left(\alpha - \beta\right)^2. \tag{P2.11}$$

Thus there is a drift component to the spread of the distribution normally associated with diffusion. Equation (P2.11) generalizes Eq. (2.56).

e. Show that $\langle m^2 \rangle_n - \langle m \rangle_n^2 = 4\alpha\beta n$. The variance grows linearly with n.

2.31 Show that $p(x,t)$ in Eq. (2.60) is: 1) normalized, $\int_{-\infty}^{\infty} p(x,t) = 1$ for all times; and 2) reduces to the Dirac function, $\lim_{t\to 0} p(x,t) = \delta(x)$. Hint: Show that $\delta_\epsilon(x) \equiv (1/\sqrt{\pi\epsilon})\mathrm{e}^{-x^2/\epsilon}$ is a sequence of functions such that $\lim_{\epsilon\to 0}\int \delta_\epsilon(x)f(x) = f(0)$ for any smooth function $f(x)$. See [13, p89].

2.32 Show that $p(x,t)$ in Eq. (2.60) is the solution of the diffusion equation, (2.61).

2.33 Show how Eq. (2.62) follows from Eq. (2.56) using the substitutions $x = ma$, $t = n\tau$, and $D = a^2/(2\tau)$.

2.34 The Wiener-Levy process is specified by the probability distributions in Eq. (2.63).

a. Verify that $\langle y^2 \rangle = 2Dt$ using P_1. Gaussian integrals are highly useful (remember where they are for future reference):

$$\int_{-\infty}^{\infty} x^{2n}\mathrm{e}^{-ax^2}\mathrm{d}x = \sqrt{\frac{\pi}{a}}\frac{1\cdot 3\cdot 5\cdots(2n-1)}{(2a)^n} \qquad (n \geq 1)$$

$$\int_{-\infty}^{\infty} x^{2n+1}\mathrm{e}^{-ax^2}\mathrm{d}x = 0. \qquad (n \geq 0)$$

b. Show that the SCK equation is satisfied by the Wiener-Levy form of $P_{1|1}(y_2, t_2|y_1, t_1)$.

2.35 Derive Eqs. (2.65) and (2.66). Hint: To derive Eq. (2.65), substitute just one copy of Eq. (2.64) in the right side of Eq. (2.29). The time t_1 on the left side is $t_2 + \Delta t$.

2.36 Derive the master equations (2.34) and (2.70) from the transition rates given in the examples on page 38. Hint: Use Eq. (2.67).

2.37 By introducing the Fourier transform $\widetilde{P}(q,t) \equiv \sum_{n=-\infty}^{\infty} \mathrm{e}^{-iqn} P(n,t)$, show that Eq. (2.34) is equivalent to

$$\frac{\partial}{\partial t}\widetilde{P}(q,t) = -4\alpha \sin^2(q/2)\widetilde{P}(q,t). \quad \text{(P2.12)}$$

Equation (P2.12) can be integrated to obtain a closed-form expression for $\widetilde{P}(q,t)$, from which $P(n,t)$ can be found through inverse Fourier transformation.

2.38 a. Because of a special trick, useful information can be extracted from Eq. (2.70) without having to obtain an explicit solution. Multiply Eq. (2.70) by n, and sum. Show that

$$\frac{\partial}{\partial t}\left(\sum_{n=0}^{\infty} nP(n,t)\right) = -\alpha \sum_{n=0}^{\infty} nP(n,t),$$

and thus

$$\frac{\partial}{\partial t}\langle n(t)\rangle = -\alpha\langle n(t)\rangle.$$

The average number of surviving nuclei decays exponentially, $\langle n(t)\rangle = n_0 \mathrm{e}^{-\alpha t}$, implying that the probability of a nucleus surviving to time t is $p(t) = \mathrm{e}^{-\alpha t}$.

b. The solution of Eq. (2.70) can be derived with elementary methods of probability theory. Let p denote the probability of a single nucleus surviving to time t. Argue that the probability of n nuclei surviving to time t is

$$P(n,t) = \binom{n_0}{n} p^n (1-p)^{n_0 - n},$$

where $\binom{n_0}{n}$ is the binomial coefficient (see for example [5, p62]).

c. Show that

$$P(n,t) = \binom{n_0}{n} \mathrm{e}^{-n\alpha t}\left(1 - \mathrm{e}^{-\alpha t}\right)^{n_0 - n} \qquad (n = 0, 1, \ldots, n_0) \quad \text{(P2.13)}$$

solves the master equation, (2.70), subject to the initial condition, $P(n,0) = \delta_{n,n_0}$.

2.39 The solution to Eq. (2.70) can be found using another method, that of *generating functions*, $G(z,t)$. The form of the generating function for a particular problem depends on the range of the stochastic variables and on the nature of the state space.[88] We've already used the Fourier transforms of probability distributions (a type of generating function), such as Eq. (2.6) for continuous random variables or Eq. (2.49) for discrete random variables having the range $(-\infty, \infty)$, $\widetilde{p}(q,n) = \sum_{m=-\infty}^{\infty} \mathrm{e}^{-iqm} p(m,n)$. For random variables restricted to positive discrete values (as in Eq. (2.70)), one might try a discrete Laplace transform $g(s,t) \equiv \sum_{n=0}^{\infty} \mathrm{e}^{-ns} P(n,t)$, or more generally the generating function

$$G(z,t) \equiv \sum_{n=0}^{\infty} z^n P(n,t), \quad \text{(P2.14)}$$

[88] The use of generating functions is therefore an art; there are no definite rules to guide.

where the nature of the transform variable z emerges from the problem at hand.[89] Note that $G(1,t) = 1$ (for normalized distributions) and $G(0,t) = P(0,t)$, for all t. The quantity $G(z,0) = \sum_{n=0}^{\infty} z^n P(n,0)$ is determined by the initial condition $P(n,0)$.

a. Show by differentiating Eq. (P2.14) and using the master equation (2.70) that

$$\begin{aligned}\frac{\partial}{\partial t}G(z,t) &= \alpha \sum_{n=0}^{\infty} z^n \left[(n+1)P(n+1,t) - nP(n,t)\right] \\ &= \alpha \left[\frac{\partial}{\partial z}\sum_{n=0}^{\infty} z^{n+1}P(n+1,t) - z\frac{\partial}{\partial z}\sum_{n=0}^{\infty} z^n P(n,t)\right] \\ &= \alpha \left[\frac{\partial}{\partial z}\left[G(z,t) - P(0,t)\right] - z\frac{\partial}{\partial z}G(z,t)\right] \\ &= \alpha(1-z)\frac{\partial}{\partial z}G(z,t). \end{aligned} \tag{P2.15}$$

Equation (P2.15) is a partial differential equation for the generating function associated with the master equation (2.70). Note from Eq. (P2.15) that $\partial G(1,t)/\partial t = 0$ (normalization on $P(n,t)$ remains fixed).

b. Equation (P2.15) can be put in a more convenient form through a change of variables. Let $\tau \equiv \alpha t$ and $\xi \equiv -\ln(1-z)$ (so that $z = 1 - \mathrm{e}^{-\xi}$). Show that Eq. (P2.15) can be written

$$\frac{\partial}{\partial \tau}G(z,t) = \frac{\partial}{\partial \xi}G(z,t). \tag{P2.16}$$

c. Show that Eq. (P2.16) is satisfied by any differentiable function f of a *single variable*, $G(z,t) = f(\tau + \xi)$. Show for the initial condition $P(n,0) = \delta_{n,n_0}$ that $G(z,0) = z^{n_0}$. Argue that $f(\xi) = \left(1 - \mathrm{e}^{-\xi}\right)^{n_0}$, and therefore that

$$G(z,t) = \left[1 - (1-z)\mathrm{e}^{-\alpha t}\right]^{n_0}. \tag{P2.17}$$

Verify by direct substitution that Eq. (P2.17) solves Eq. (P2.15).

d. Show that Eq. (P2.17) is equivalent to the sum

$$G(z,t) = \sum_{k=0}^{n_0} \binom{n_0}{k} \left(1 - \mathrm{e}^{-\alpha t}\right)^{n_0 - k} \left(z\mathrm{e}^{-\alpha t}\right)^k$$

and thus, comparing with Eq. (P2.14),

$$P(k,t) = \binom{n_0}{k} \left(1 - \mathrm{e}^{-\alpha t}\right)^{n_0 - k} \mathrm{e}^{-k\alpha t}, \qquad (k = 0, 1, \ldots, n_0) \tag{P2.18}$$

in agreement with Eq. (P2.13). Hint: Use the binomial theorem. The binomial coefficients have the property that $\binom{N}{k} = 0$ for $k > N$, and thus the generating function in this case is a finite series. We've relied on the uniqueness of power series in arriving at Eq. (P2.18).

[89] A generating function is a way of encoding information about the *family* of probabilities $P(n,t)$, $n = 0, 1, 2, \ldots$, by treating them as the expansion coefficients of a power series (which represents an analytic function). The generating function for Legendre polynomials, for example (see [13, p153]), $G(x,y) \equiv \sum_{n=0}^{\infty} P_n(x)y^n$, has the closed-form expression $G(x,y) = (1 - 2xy + y^2)^{-1/2}$ (but not all generating functions are known in closed form). To quote G. Polya, "A generating function is a device somewhat similar to a bag. Instead of carrying many little objects detachedly, which could be embarrassing, we put them all in a little bag, and then we have only one object to carry, the bag."[61, p101]. Mathematically, the transform variable z need not have a physical interpretation, although it often does in applications.

2.40 Show that Eq. (2.75) is equivalent to Eq. (2.67) when the general form of the transition matrix Eq. (2.74) is used.

2.41 a. Show from the infinite series representation of the matrix exponential (see page 40) that $(\mathrm{d}/\mathrm{d}t)\mathrm{e}^{Wt} = W\mathrm{e}^{Wt} = \mathrm{e}^{Wt}W$.

b. Show that Eq. (2.76) formally solves Eq. (2.75).

c. Show that if the eigenfunctions and eigenvalues of an operator A are known, $A\phi_n = \lambda_n\phi_n$, $n = 1, 2, \ldots$, then, for each n, $\mathrm{e}^{A}\phi_n = \mathrm{e}^{\lambda_n}\phi_n$.

2.42 The distinction between left and right eigenvectors is often a topic to which students of the physical sciences are not exposed (hence this exercise).

a. Show that Eq. (2.77) follows from the definition of transition matrix, Eq. (2.74). Show that Eq. (2.77) implies the existence of a left eigenvector (row vector) $\psi = (1, 1, \ldots, 1)$ having zero eigenvalue, where in multiplying from the left, $\psi \cdot W = 0$.

b. Consider a square matrix M that has a right eigenvector (column vector) ϕ associated with eigenvalue λ, $M \cdot \phi = \lambda\phi$. Show, for any right eigenvector ϕ of M, there is a left eigenvector ϕ^T of M^T having the same eigenvalue, where T denotes transpose. (The characteristic polynomial of M is the same as for M^T; the determinant of a matrix equals that of its transpose.) For symmetric matrices, therefore ($M = M^T$), for any right eigenvector ϕ of M, there is a left eigenvector ϕ^T of M having the same eigenvalue.

c. Now show, for a non-symmetric square matrix M, if it has a right eigenvector $M \cdot \phi = \lambda\phi$ and a left eigenvector $\psi \cdot M = \mu\psi$, that if $\psi \cdot \phi \neq 0$, then $\mu = \lambda$, but ϕ is not necessarily the same as ψ^T. Note that ψ and ϕ are left and right eigenvectors of the *same* matrix M (not M^T). There could be more than one right eigenvector ϕ having the same eigenvalue as a left eigenvector ψ, as long as $\psi \cdot \phi \neq 0$.

2.43 Show that Eq. (2.78) (detailed balance in terms of transition matrices) follows from Eq. (2.72) and Eq. (2.74).

2.44 Show that the matrix elements defined in Eq. (2.79) are symmetric, $V_{n,n'} = V_{n',n}$.

2.45 In this exercise we show the eigenvalues of the matrix V in Eq. (2.79) are nonpositive. We do so by constructing the quadratic form associated with V. A quadratic form Q is a polynomial of degree two in a set of real variables $(y_1, y_2, \ldots)$,

$$Q(y_1, y_2, \ldots) \equiv \sum_{n,m} V_{n,m} y_n y_m, \tag{P2.19}$$

where $V_{n,m}$ are the elements of a symmetric matrix.

a. Start by showing from Eqs. (2.79) and (2.74) that

$$V_{n,m} = \sqrt{\frac{P_{\text{eq}}(m)}{P_{\text{eq}}(n)}} \left[W(n|m) - \delta_{n,m} \sum_k W(k|n) \right],$$

where $W(n|m)$ is the probability per unit time of the transition from state $m \to n$.

b. Show that the diagonal elements are negative,

$$V_{n,n} = -\sum_{k \neq n} W(k|n),$$

and the off-diagonal elements are positive

$$V_{n,m} = \sqrt{\frac{P_{\text{eq}}(m)}{P_{\text{eq}}(n)}} W(n|m). \qquad (m \neq n)$$

Combine these results with Eq. (P2.19) to show that

$$Q = -\sum_n y_n^2 \sum_{k \neq n} W(k|n) + \sum_n \sum_{m \neq n} \sqrt{\frac{P_{\text{eq}}(m)}{P_{\text{eq}}(n)}} y_n y_m W(n|m). \tag{P2.20}$$

c. Show that Eq. (P2.20) is equivalent to

$$Q = -\frac{1}{2} \sum_{n,m} W(n|m) P_{\text{eq}}(m) \left(\frac{y_n}{\sqrt{P_{\text{eq}}(n)}} - \frac{y_m}{\sqrt{P_{\text{eq}}(m)}} \right)^2, \tag{P2.21}$$

and thus that Q is negative for any sets of variables $(y_1, y_2, \ldots)$. Note there is no contribution for $m = n$, so the sum is restricted to $m \neq n$. The simplest way to proceed is to show that Eq. (P2.21) implies Eq. (P2.20). Expand the quadratic term in Eq. (P2.21) and make use of detailed balance in the form of Eq. (2.72).

d. In a basis of eigenfunctions, V is diagonal with its eigenvalues on the diagonal. In that case, Q (from Eq. (P2.19)) has the form $Q = \sum_n \lambda_n y_n^2$. We conclude that the eigenvalues are negative semidefinite, $\lambda_n \leq 0$. (If you're wondering whether one could have some positive eigenvalues and still have the sum be negative, other methods of analysis show that the real parts of the eigenvalues are nonpositive).

2.46 Derive Eq. (2.83) for the form of the solution to the master equation.

2.47 Derive Eq. (2.85). Along the way you'll need to reckon the fact that, for real constant α,

$$\int_{-\infty}^{\infty} e^{-x^2} dx = \int_{-\infty}^{\infty} e^{-(x+i\alpha)^2} dx,$$

i.e., in this case, the integral with a complex offset is independent of offset. Show this follows from Cauchy's integral theorem, that the closed contour integral in the complex plane of an entire function is zero, e.g., $\oint e^{-z^2} dz = 0$. Consider a rectangle in the complex plane with vertices at $z = -R$, $z = R$, $z = R + i\alpha$, $z = -R + i\alpha$ and let $R \to \infty$. See [13, p246].

2.48 Show that the relation between cumulants and moments in Eq. (2.88) can be inverted:

$$\begin{aligned}
\langle \alpha \rangle &= C_1 \\
\langle \alpha^2 \rangle &= C_2 + C_1^2 \\
\langle \alpha^3 \rangle &= C_3 + 3C_2C_1 + C_1^3 \\
\langle \alpha^4 \rangle &= C_4 + 4C_3C_1 + 3C_2^2 + 6C_2C_1^2 + C_1^4.
\end{aligned}$$

Thus, the nth moment is a function of cumulants of order $k \leq n$.

2.49 Show that the expressions for C_3, C_4 in Eq. (2.89) follow from the results in Eq. (2.88).

2.50 a. Show that, if the matrix $\boldsymbol{A}$ in the multivariate Gaussian Eq. (2.91) is diagonal, the random variables are statistically independent. Hint: Consider a joint probability distribution obtained from the product of single-variable Gaussians:

$$P(\alpha_1, \ldots, \alpha_n) = \prod_{k=1}^{n} \frac{1}{\sqrt{2\pi\sigma_k^2}} \exp\left(-\frac{\alpha_k^2}{2\sigma_k^2} \right).$$

Conversely, one can always diagonalize a real symmetric matrix (spectral theorem). One can therefore find linear combinations of correlated Gaussian variables that are statistically independent.

b. Show the closely related statement that linear combinations of independent Gaussian variables are Gaussian. A *Gaussian variable* is one that is distributed with a Gaussian probability distribution. That is, let $\{X_i\}_{i=1}^{n}$ be a set of independent Gaussian variables and form a linear combination $Y \equiv \sum_{i=1}^{n} a_i X_i$, where the a_i are constants. Show that Y is a Gaussian variable. Hint: This is easily done using characteristic functions.

2.51 Show that Eq. (2.93) is equivalent to Eq. (2.91) when the choice for K is made such that the distributions are normalized to unity. Hint: You should find that $\boldsymbol{B} = \boldsymbol{A}\boldsymbol{\mu}$ ($B_i = \sum_j A_{ij}\mu_j$). Show that $\boldsymbol{\mu}^T \boldsymbol{A}\boldsymbol{\mu} = \boldsymbol{B}^T \boldsymbol{A}^{-1} \boldsymbol{B}$. For an invertible square matrix $\boldsymbol{A}$, $\left(\boldsymbol{A}^{-1}\right)^T = \left(\boldsymbol{A}^T\right)^{-1}$, and in this example $\boldsymbol{A}$ is symmetric. The inverse of a symmetric matrix is symmetric.

2.52 Derive Eq. (2.95).

2.53 Show that Eq. (2.107) follows by integrating Eq. (2.106).

2.54 Perform the integration indicated in Eq. (2.109).

CHAPTER 3

Brownian motion and stochastic dynamics

BROWNIAN motion is a central problem in statistical physics, the perpetual irregular motion exhibited by colloidal[1] or Brownian or B-particles, shown in Fig. 3.1. Incessant erratic

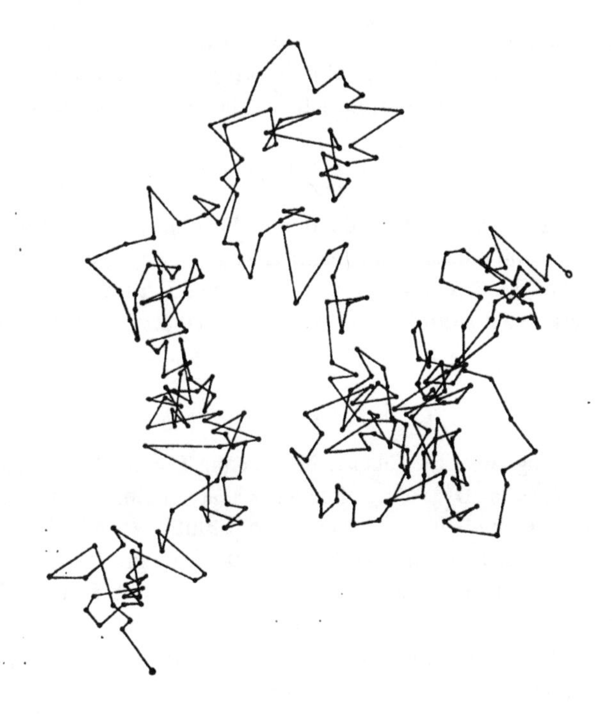

Figure 3.1 Brownian motion, the positions of a small particle suspended in water recorded at regular time intervals. From [62, p116]. The length scale is 3 μm.

motions are maintained by collisions with the molecules of the suspension medium, impacts occurring at such high rates[2] that B-particle motion can only be modeled as a stochastic process. We

[1] A *colloid* is a mixture in which dispersed particles of size $\approx$ 1μm (large relative to molecular dimensions) are suspended in a substance such as a liquid, aerosol, or gel. A colloid has a dispersed phase of suspended particles and a host phase, the suspension medium.

[2] See Exercise 3.1.

DOI: 10.1201/9781003512295-3

can't expect to follow the trajectories of B-particles as in classical mechanics; the very concept of path breaks down in Brownian motion,[3] reminiscent of quantum mechanics. Two approaches are in wide use, the *Langevin equation* and the *Fokker-Planck equation*, the subjects of this chapter.

3.1 LANGEVIN EQUATION, EINSTEIN RELATION

3.1.1 Thermal noise

We begin with the electrical analog of Brownian motion.[4] Consider the RL circuit in Fig. 3.2 where

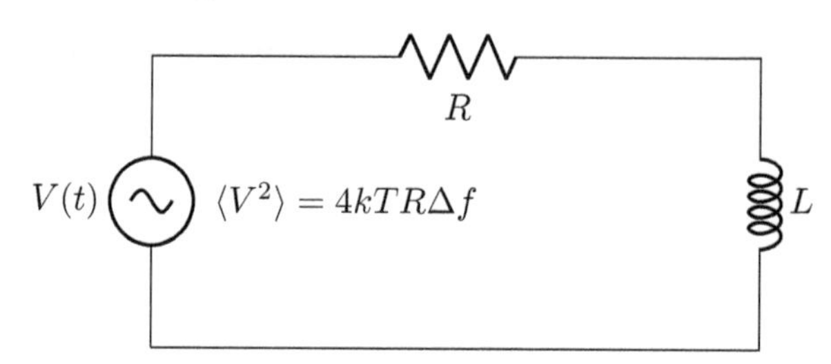

Figure 3.2 RL series circuit showing the fluctuating voltage $V(t)$ associated with R.

the indicated voltage source $V(t)$ is a fluctuating voltage with $\langle V\rangle = 0$ and second moment given by Nyquist's theorem, i.e., $V(t)$ is the thermal noise associated with resistor R. Dropping the capacitance in Eq. (2.41), the fluctuating current satisfies the differential equation

$$\frac{\mathrm{d}I}{\mathrm{d}t} + \frac{1}{\tau_c}I = \frac{1}{L}V(t), \tag{3.1}$$

where $\tau_c \equiv L/R$ is a characteristic time. Equation (3.1) is an instance of a Langevin equation, a stochastic differential equation modeling a dynamical system subject to a random driving force.

The formal solution[5] of Eq. (3.1) is developed in Exercise 3.2, from which one finds $\langle I(t)\rangle = 0$. Using Eq. (P3.1), the current autocorrelation is connected to that of the voltage through the relation,

$$\langle I(t)I(t+\tau)\rangle_{I_0} = I_0^2\mathrm{e}^{-(2t+\tau)/\tau_c} + \frac{1}{L^2}\mathrm{e}^{-(2t+\tau)/\tau_c}\int_0^t\int_0^{t+\tau}\mathrm{d}t'\mathrm{d}t''\mathrm{e}^{(t'+t'')/\tau_c}\langle V(t')V(t'')\rangle_{I_0}, \tag{3.2}$$

where $\langle\rangle_{I_0}$ denotes an average over the subensemble having $I = I_0$ at $t = 0$ and where cross terms have been dropped ($\langle V(t)\rangle_{I_0} = 0$). Voltage fluctuations are stationary and hence $\langle V(t')V(t'')\rangle_{I_0}$ is a function of the relative time, $K(t''-t')$. The substitutions $s \equiv t''-t'$ and $u \equiv t''+t'$ suggest themselves as a means of simplifying the double integral in Eq. (3.2). The change of variables is straightforward, but what are the limits of integration? Figure 3.3 shows the geometry of the coordinate transformation where a rectangle of area $t(t+\tau)$ in the t''-t' plane is rotated 45° into a rectangle in the u-s plane of area $2t(t+\tau)$ (the Jacobian equals $\frac{1}{2}$). We find, for $J \equiv \int_0^t\int_0^{t+\tau}\mathrm{d}t'\mathrm{d}t''\exp\left((t'+t'')/\tau_c\right)K(t''-t')$, that

$$2J = \int_{-t}^{0}K(s)\mathrm{d}s\int_{-s}^{2t+s}\mathrm{e}^{u/\tau_c}\mathrm{d}u + \int_0^{\tau}K(s)\mathrm{d}s\int_s^{2t+s}\mathrm{e}^{u/\tau_c}\mathrm{d}u + \int_{\tau}^{t+\tau}K(s)\mathrm{d}s\int_s^{2(t+\tau)-s}\mathrm{e}^{u/\tau_c}\mathrm{d}u. \tag{3.3}$$

[3] The small black circles in Fig. 3.1 indicate the position of the suspended particle recorded every 30 seconds, published in 1913 by Jean Perrin, who received the 1926 Nobel Prize in Physics for confirming the atomic picture of matter. The straight lines connecting the dots in Fig. 3.1 do not represent the particle's path. To quote Perrin, "If, in fact, one were to mark second by second, each of the straight line segments would be replaced by a polygon path of 30 sides." The path is jagged at all length scales accessible to observation. Even today, Perrin's *Atoms*[62] is worth reading.

[4] It's well known that electrical networks can represent mechanical problems.

[5] We say formal because we don't know the functional form of $V(t)$, only its statistical properties.

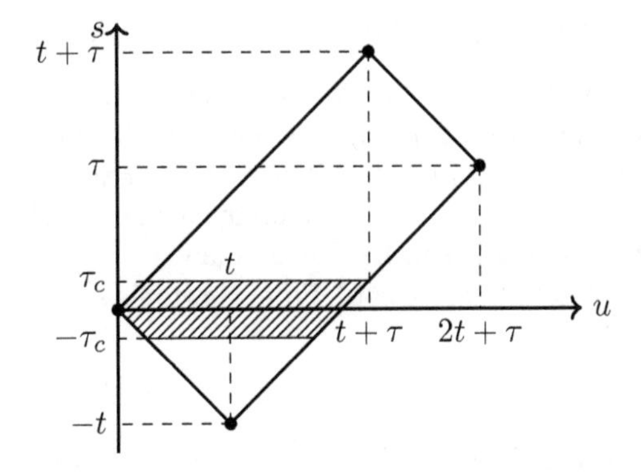

Figure 3.3 Domain of integration in the (u,s) plane corresponding to that in the (t'',t') plane associated with the double integral in Eq. (3.2). Crosshatching indicates the physically relevant region $|s| \lesssim \tau_c \ll \tau$. The rectangle has area $2t(t+\tau)$.

As a check, if $K(s)$ and e^{u/τ_c} are replaced with unity in Eq. (3.3), the net effect of the integrals is to generate the area $2t(t+\tau)$.

Equation (3.3) is as far as we can take an exact calculation without the explicit form of $K(s)$. Whatever its form, $K(s)$ is appreciable only for $|s| \lesssim \tau_c \ll \tau$, the crosshatched region in Fig. 3.3. We can therefore ignore the third double integral on the right of Eq. (3.3). Performing the remaining integrations over u and approximating $\mathrm{e}^{|s|/\tau_c} \approx 1$, we find

$$J \approx \frac{\tau_c}{2}\left(\mathrm{e}^{2t/\tau_c} - 1\right)\int_{-\infty}^{\infty} K(s)\mathrm{d}s, \tag{3.4}$$

where we've extended the limits of integration on s to $\pm\infty$ (where $K(s)$ is negligibly small). Combining Eq. (3.4) with Eq. (3.2),

$$\langle I(t)I(t+\tau)\rangle_{I_0} = I_0^2\mathrm{e}^{-(2t+\tau)/\tau_c} + \frac{\tau_c}{2L^2}\mathrm{e}^{-\tau/\tau_c}\left(1 - \mathrm{e}^{-2t/\tau_c}\right)C, \tag{3.5}$$

where $C \equiv \int_{-\infty}^{\infty} K(s)\mathrm{d}s$ is the area under the curve. Equation (3.5) is not exact; it's an asymptotic result for $t, \tau \gg \tau_c$. The fine details of the short-time behavior have been smoothed over.

For short times, fluctuations are determined by initial values, but at long times initial conditions are forgotten. With $t = 0$ in Eq. (3.5), $\langle I(0)I(\tau)\rangle_{I_0} = I_0^2\mathrm{e}^{-\tau/\tau_c}$; τ_c is the time scale over which current fluctuations become uncorrelated. For $\tau = 0$ and $t \to \infty$,

$$\langle I^2(t)\rangle \overset{t\to\infty}{\longrightarrow} \frac{\tau_c}{2L^2}C \overset{\substack{\text{equipartition}\\ \downarrow}}{=} \frac{kT}{L}, \tag{3.6}$$

where we've equated $\langle I^2(\infty)\rangle$ with its equilibrium value. At long times, current fluctuations are driven by voltage fluctuations. Estimating $C \approx 2\tau_c\langle V^2\rangle$ in Eq. (3.6), we have $\langle I^2\rangle \sim \langle V^2\rangle/R^2$, "Ohm's law." Equation (3.6) implies a connection between R and the strength of fluctuations:

$$R = \frac{1}{2kT}\int_{-\infty}^{\infty} K(s)\mathrm{d}s. \tag{3.7}$$

Equation (3.7) is an instance of a general result, the *fluctuation-dissipation theorem*, which relates dissipation[6] (associated with entropy creation) to the temporal properties of fluctuations. *Wherever there is damping there are fluctuations*, and conversely. Indeed, *dissipation is essential to maintaining thermal equilibrium*. We return to this fundamental result in Chapter 6.

[6]Dissipated energy is energy diverted by the production of entropy into a form not available for work[5, p14].

3.1.2 Brownian motion of free particles; the random force

Modeling Brownian motion requires us to give up on exact descriptions of motion in favor of stochastic treatments. With that said, consider how one might approach the problem. A particle of mass M (with M much larger than molecular masses, see Exercise 3.14), injected into a fluid with velocity[7] v_0, experiences, by the laws of hydrodynamics, a velocity-dependent *drag force* $F_d = -\alpha v$, with α a friction coefficient. By Stokes's law, $\alpha = 6\pi R\nu$ for a sphere of radius R moving slowly in a fluid of viscosity ν[15, p66]. From Newton's second law, thereore,

$$M\frac{dv}{dt} = -\alpha v, \qquad \text{(macroscopic)} \tag{3.8}$$

implying $v(t) = v_0 \exp(-\alpha t/M)$ for $t \geq 0$; $\tau_c \equiv M/\alpha$ is a characteristic time for dissipation of kinetic energy (the *relaxation time*). This naive application of classical dynamics fails to capture an essential part of the physics: The model predicts the velocity decaying to zero, yet Brownian motion is incessant, and, from statistical mechanics, $\langle v^2 \rangle = kT/M$ (equipartition theorem[5, p97]). How to do better? Clearly, take into account more of the interactions with the suspension medium. Yet we're unable to model the detailed motions of fluid molecules, whose impacts with the B-particle are frequent and of irregular strength and direction. An approach introduced in 1908 by P. Langevin obviates that problem by postulating a *random force* $F(t)$ representing the effects of collisions.[8] We append Newton's law of motion:

$$M\frac{dv}{dt} = \underbrace{-\alpha v}_{\substack{\text{systematic}\\ \text{force}}} + \underbrace{F(t)}_{\substack{\text{random}\\ \text{force}}} . \qquad \text{(microscopic)} \tag{3.9}$$

In the parlance of the Langevin equation, one speaks of *systematic forces*, those affecting the average motion and the random force.[9] B-particles not subject to systematic forces other than the drag force are known as *free Brownian particles*.[10] The motion of free B-particles is inertial for short times ($t \ll M/\alpha$); see Eq. (3.18).

In that it models *effects* of microscopic properties, we can say the Langevin equation is microscopic. We can't say anything, however, about the random force other than specifying its statistical properties. With the systematic force governing the average force experienced by B-particles, *the average of the random force must be zero* over a subensemble specified by the initial condition $v = v_0$,

$$\langle F(t) \rangle_{v_0} = 0. \tag{3.10}$$

Equation (3.10) embodies the key assumption that the *random force is independent of macroscopic initial conditions*.[11] Equation (3.10) implies, from Eq. (3.9), that the *average* velocity decays to zero,

$$\langle v(t) \rangle_{v_0} = v_0 \exp(-t/\tau_c). \tag{3.11}$$

Whatever the initial momentum of the B-particle, it's transferred to the molecules of the suspension medium, and the average velocity decays to zero.

More information is contained in the second moment, $\langle v^2(t) \rangle_{v_0}$, the average of a positive quantity. The random force is stationary, and thus $\langle F(t_1)F(t_2) \rangle = K(t_1 - t_2)$. We expect there to be

---

[7] We restrict ourselves to one-dimensional descriptions, appropriate for isotropic systems. The generalization to three dimensions is not difficult; see Exercise 3.6. We ignore the effects of gravity here. See Section 3.4.5.

[8] An English translation of Langevin's 1908 paper appears in [63]. Our treatment follows that of Uhlenbeck and Ornstein[64], reprinted in Wax[44, pp93–111].

[9] The drag force is a dissipative systematic force supplied by the suspension medium, a source of irreversibility present in any treatment of Brownian motion; it follows from hydrodynamics and is macroscopic in character.

[10] Analogous to *free fall* in gravitational physics, motion with no acceleration other than that provided by gravity[6, p188].

[11] Said differently, the properties of the surrounding medium are unaffected by the motion of the B-particle; the suspension medium is a stationary environment. A justification of this key assumption is given in Section 6.5.

almost no correlation between impacts at different times,[12] and thus $K(\tau)$ is expected to be sharply peaked about $\tau = 0$. Without doing more math, we have from Eq. (3.5) under the substitutions $I \to v$, $L \to M$, and $R \to \alpha$,

$$\langle v(t)v(t+\tau)\rangle_{v_0} = v_0^2 \mathrm{e}^{-(2t+\tau)/\tau_c} + \frac{\tau_c}{2M^2}\mathrm{e}^{-\tau/\tau_c}\left(1-\mathrm{e}^{-2t/\tau_c}\right)C, \tag{3.12}$$

with the analog of Eq. (3.7), $\alpha = 1/(2kT)\int_{-\infty}^{\infty} K(s)\mathrm{d}s \equiv C/(2kT)$. What matters in the connection between dissipation and the strength of fluctuations is the total area under the curve of the correlation function $K(s)$. One typically takes $K(t-t')$ as a delta function (*delta-correlated noise*),

$$\langle F(t)F(t')\rangle = \Gamma\delta(t-t') \tag{3.13}$$

with

$$\Gamma = 2kT\alpha. \tag{3.14}$$

Setting $\tau = 0$ in Eq. (3.12), we find

$$\langle v^2(t)\rangle_{v_0} = \langle v^2\rangle_{\text{eq}} + \left(v_0^2 - \langle v^2\rangle_{\text{eq}}\right)\mathrm{e}^{-2t/\tau_c}, \tag{3.15}$$

where $\langle v^2\rangle_{\text{eq}} = kT/M$, the *thermal speed.* Particles with v_0 less than (greater than) $\sqrt{kT/M}$ approach that value from below (above). For $v_0^2 = \langle v^2\rangle_{\text{eq}}$, $\langle v^2(t)\rangle$ is time invariant; systems in equilibrium stay in equilibrium.[13]

To find $\langle x^2(t)\rangle$, one could integrate the velocity, $x(t) \sim \int v(t)\mathrm{d}t$, square the result, and take the average. A more direct way is to develop an equation of motion for $\langle x^2(t)\rangle$. Multiply Eq. (3.9) by $x(t)$ (the instantaneous position) and use $\mathrm{d}(x\dot{x})/\mathrm{d}t = x\ddot{x} + \dot{x}^2$, $x\dot{x} = \frac{1}{2}\mathrm{d}x^2/\mathrm{d}t$, and that $\langle xF(t)\rangle = 0$. We find:

$$\frac{\mathrm{d}^2}{\mathrm{d}t^2}\langle x^2(t)\rangle + \frac{1}{\tau_c}\frac{\mathrm{d}}{\mathrm{d}t}\langle x^2(t)\rangle = 2\langle v^2(t)\rangle. \tag{3.16}$$

The solution of Eq. (3.16) when Eq. (3.15) for $\langle v^2(t)\rangle_{v_0}$ is substituted on the right side, is

$$\langle x^2(t)\rangle_{v_0} = v_0^2\tau_c^2\left(1-\mathrm{e}^{-t/\tau_c}\right)^2 + \langle v^2\rangle_{\text{eq}}\tau_c^2\left[2\frac{t}{\tau_c} - \left(1-\mathrm{e}^{-t/\tau_c}\right)\left(3-\mathrm{e}^{-t/\tau_c}\right)\right], \tag{3.17}$$

for the initial conditions $\langle x^2\rangle|_{t=0} = 0$ and $\mathrm{d}\langle x^2\rangle/\mathrm{d}t|_{t=0} = 0$. For short times, Eq. (3.17) implies

$$\langle x^2(t)\rangle_{v_0} = v_0^2 t^2 + O(t/\tau_c)^3, \qquad t \ll \tau_c \tag{3.18}$$

that B-particles behave initially as free particles moving at constant speed, and for long times

$$\langle x^2(t)\rangle_{v_0} \sim 2\langle v^2\rangle_{\text{eq}}\tau_c t \equiv 2Dt, \qquad t \gg \tau_c \tag{3.19}$$

that B-particles behave like diffusing particles executing a random walk.

The Langevin equation leads to a picture of Brownian motion as inertial at short times and diffusive at long times. In diffusion, $\langle x^2\rangle = 2Dt$ [Eq. (2.62)], where $D = a^2/(2\tau)$ [Eq. (2.59)] involves the random-walk parameters a and τ. From Eq. (3.19) we identify a diffusion coefficient for Brownian motion, $D = \langle v^2\rangle_{\text{eq}}\tau_c = kT/\alpha$. The "jump length" a (from the random walk) can be estimated from Brownian-motion parameters, $a \approx \tau_c\sqrt{2\langle v^2\rangle_{\text{eq}}}$. The formula

$$D = \frac{kT}{\alpha} \tag{3.20}$$

is known as *Einstein's relation*. Perrin combined Eq. (3.20) with his measurements of $\langle x^2(t)\rangle$ for particles executing Brownian motion to infer reasonably good values of the Boltzmann constant.[14]

---

[12] We expect, as solutions of the Langevin equation, that $v(t)$ and $v(t+\Delta t)$ should differ infinitesimally for small Δt, and hence there must be almost no correlation between $F(t)$ and $F(t+\Delta t)$.

[13] Thus, thermal equilibrium of B-particles at temperature T is achieved through numerous collisions with the particles of the suspension medium that are themselves in equilibrium at temperature T.

[14] See Exercise 3.13. One compares the measured values of $\langle x^2(t)\rangle/(2t)$ with Eq. (3.20), $\lim_{t\to\infty}\langle x^2(t)\rangle/(2t) = D = kT/\alpha$. Clearly one needs an independent measurement of the friction coefficient, α.

3.1.3 Uniform external force; drift speed

Consider a B-particle of charge q in a uniform[15] electric field E. Adding the Coulomb force as another systematic force,

$$M\frac{\mathrm{d}v}{\mathrm{d}t} = qE - \alpha v + F(t), \tag{3.21}$$

where v is the velocity in the direction selected by E. In steady state[16] $\mathrm{d}\langle v(t)\rangle/\mathrm{d}t = 0$ and thus

$$0 = qE - \alpha\langle v\rangle_{\mathrm{d}}, \tag{3.22}$$

where $\langle v\rangle_{\mathrm{d}}$, the *drift speed*, is the terminal speed attained by charged particles in a resistive medium. Experimentally, $\langle v\rangle_{\mathrm{d}}$ is linearly proportional to the electric field, $\langle v\rangle_{\mathrm{d}} = \mu E$ (for E not too large), where μ, the *electrical mobility*, is a material-specific parameter. From Eq. (3.22), $\mu = q/\alpha$; mobility is inversely related to friction. With $\alpha = q/\mu$ in Eq. (3.20), we have the Einstein relation for charged particles

$$D = (kT/q)\,\mu. \tag{3.23}$$

Equation (3.23) shows a universal connection between transport coefficients,[17] a result to be explained by nonequilibrium statistical mechanics; see Chapter 6. Equation (3.23) specifies a voltage (energy per charge), the *thermal voltage* ≈ 26 mV at room temperature.

3.2 LANGEVIN EQUATION AS A STOCHASTIC PROCESS

The Langevin equation finds wide use in applications and is a successful theory by any measure.[18] There's a point of mathematical consistency, however, that should be addressed: Solutions of the Langevin equation are *continuous but not differentiable*.[19] That creates a logical predicament. The Langevin equation, involving the derivative $\mathrm{d}v/\mathrm{d}t$, is used to find a solution $v(t)$ not having a derivative! Physical theories can often be justified only *a posteriori*, by the validity of the conclusions derived from them. Given the success of the Langevin equation, we shouldn't be surprised if another mathematical formulation were to be found that circumvents this problem.

Doob did that by interpreting the Langevin equation as a stochastic process. Start by multiplying Eq. (3.9) by $\mathrm{d}t$:

$$\mathrm{d}v(t) = -\beta v(t)\mathrm{d}t + A(t)\mathrm{d}t, \qquad (\mathrm{d}t > 0) \tag{3.24}$$

where $\beta \equiv \alpha/M$ and $A(t) \equiv F(t)/M$. Equation (3.24) indicates that a small change in velocity is produced by the frictional force acting over the next $\mathrm{d}t$ seconds (β has dimension time^{-1}) and by a random increment $A(t)\mathrm{d}t$, which occurs with a distribution of values, positive and negative. There are two times scales: time intervals over which variations in velocity $v(t)$ are small, yet which, over the same interval, $A(t)$ undergoes numerous fluctuations, i.e., *$A(t)$ varies extremely rapidly compared to variations in $v(t)$*.[20] Doob interpreted $A(t)\mathrm{d}t$ as a stochastic process, writing $A(t)\mathrm{d}t \equiv \mathrm{d}B(t)$, where the notation $\mathrm{d}B(t)$ for a *differential stochastic process* must be explained.

---

[15]Uniform force fields are a convenient idealization. We take up spatially dependent forces in Section 3.4.

[16]Steady state, where time derivatives vanish, is not necessarily the same as equilibrium. See Section 1.9.

[17]Historically, the Einstein relation provided the first clear way to experimentally verify our picture of Brownian motion as due to thermal agitations of molecules, a picture which may seem obvious today, but was not in 1905. The history of Brownian motion is recounted in Mazo[65].

[18]Coffey[66] summarizes applications of Brownian motion theory. See also Risken[67].

[19]See Doob[59], reprinted in Wax[44, pp319–337]. In the Einstein theory, the derivative $v = \mathrm{d}x/\mathrm{d}t$ doesn't exist if $\langle x^2\rangle = 2Dt$ holds for $t \to 0$, an issue known to Einstein; see Exercise 3.11. With the Langevin equation, $\langle x^2\rangle \propto t^2$ at short times, Eq. (P3.8), implying the existence of $\mathrm{d}x/\mathrm{d}t$. The variance of the velocity, however, is $\propto |t|$ for short times (Equation (1.1.9) in [59]), implying that the derivative $\mathrm{d}v/\mathrm{d}t$ doesn't exist (contrary to what was assumed at the outset).

[20]The time between successive collisions of the B-particle with molecules of the suspension medium is truly microscopic, of order 10^{-18} sec; see Exercise 3.1. See [68] for observations of the velocity of a Brownian particle at μs time scales.

Doob wrote the Langevin equation as a relation among differentials (one in which the existence of the limit $\mathrm{d}v/\mathrm{d}t$ is not assumed),

$$\mathrm{d}v(t) = -\beta v(t)\mathrm{d}t + \mathrm{d}B(t). \qquad (\mathrm{d}t > 0) \tag{3.25}$$

The formal solution of Eq. (3.25) is

$$v(t) = \mathrm{e}^{-\beta t}\left[v_0 + \int_0^t \mathrm{e}^{\beta t'}\mathrm{d}B(t')\right]. \tag{3.26}$$

The integral in Eq. (3.26), a *Stieltjes integral* (not typically included in science curricula), is the limit of approximating sums,[21]

$$\int_0^t \mathrm{e}^{\beta y}\mathrm{d}B(y) = \lim_{n\to\infty}\sum_{m=0}^{n-1}\mathrm{e}^{\beta m\Delta t}\int_{m\Delta t}^{(m+1)\Delta t}\mathrm{d}B = \lim_{n\to\infty}\sum_{m=0}^{n-1}\mathrm{e}^{\beta m\Delta t}\left[B((m+1)\Delta t) - B(m\Delta t)\right],$$

where we've divided $(0,t)$ into subintervals, with $n\Delta t = t$. Thus,

$$B((m+1)\Delta t) - B(m\Delta t) = \int_{m\Delta t}^{(m+1)\Delta t}\mathrm{d}B = \int_{m\Delta t}^{(m+1)\Delta t} A(t')\mathrm{d}t'.$$

The difference $B((m+1)\Delta t) - B(m\Delta t)$ is the change in velocity of the B-particle over the time interval $m\Delta t \to (m+1)\Delta t$; $\mathrm{d}B(t)$ indicates the change in velocity for small Δt at time t.

Doob took $B(t)$ to be a *process with independent increments* (also called a differential process), one in which, for $t_1 < \cdots < t_n$, the *differences* $\{B(t_2) - B(t_1), \ldots, B(t_n) - B(t_{n-1})\}$ are a family of mutually independent random variables[36, Chapter 8]. Such a process is Markov:[22] $B(t_{i+1}) - B(t_i)$ is independent of $B(t)$ for $t < t_i$. The process is assumed stationary in the sense that the probability distribution of the differences $B(t+s) - B(s)$ is independent of s. As such, the first moment is constant, taken to be zero,

$$\langle B(t) - B(s)\rangle = 0. \tag{3.27}$$

On physical grounds we take $\langle B(t)\rangle = 0$. The variance of processes with these properties (stationary independent increments and $\langle B(t)\rangle = 0$), is proportional to time (see Exercise 3.15),

$$\langle [B(t) - B(s)]^2\rangle = \gamma|t - s|, \tag{3.28}$$

where γ is a positive constant determined by the physics of the problem. The Wiener-Levy process (see Section 2.7.3) meets these requirements, implying a Gaussian distribution of $B(t) - B(0)$ with zero mean and variance γt.

For $f(t)$ a continuous function, the integral $\int f(t)\mathrm{d}B(t)$ is well defined even though $B(t)$ is not of bounded variation.[23] For any function $f(t)$ continuous on $[a,b]$, we have from Eq. (3.25),

$$\int_a^b f(t)\mathrm{d}v(t) = -\beta\int_a^b f(t)v(t)\mathrm{d}t + \int_a^b f(t)\mathrm{d}B(t). \tag{3.29}$$

[21] The Stieltjes integral is a generalization of the Riemann integral. For $[a,b]$ an interval of the real line with $a \equiv x_0 \le x_1 \le \cdots \le x_n \le x_{n+1} \equiv b$, and for $f(x)$ bounded on $[a,b]$, the Riemann integral is the limit of the sums $R_n \equiv f(\xi_0)(x_1 - a) + f(\xi_1)(x_2 - x_1) + \cdots + f(\xi_n)(b - x_n)$, $x_r \le \xi_r \le x_{r+1}$, denoted $\int_a^b f(x)\mathrm{d}x$. For f and g bounded functions on $[a,b]$, form the sum $S_n \equiv f(\xi_0)[g(x_1) - g(a)] + f(\xi_1)[g(x_2) - g(x_1)] + \cdots + f(\xi_n)[g(b) - g(x_n)]$, with ξ_r chosen as previously. The limit, the Stieltjes integral, is denoted $\int_{x=a}^b f(x)\mathrm{d}g(x)$, or simply $\int_a^b f\mathrm{d}g$. Nothing else in this book relies on the properties of Stieltjes integrals.

[22] The converse is not true, however. Markov processes are not necessarily independent-increment stochastic processes. Processes with independent increments form a subset of Markov processes.

[23] The existence of the Stieltjes integral $\int f\mathrm{d}g$ requires that $g(x)$ be bounded; see [69, p26]. Why that requirement can be relaxed for stochastic processes is treated in Paley and Wiener[70, pp151–157].

Taking $f(t) = e^{\beta t}$ (an integrating factor; see Exercise 3.16), Eq. (3.29) reproduces Eq. (3.26) when we set $a = 0$, $b = t$, and integrate by parts.[24] Equation (3.26) is then the complete solution of the Langevin equation,[25] a conclusion reached without positing the existence of dv/dt. Doob showed further (Theorem 3 in [59]) that only when the differential process $B(t) - B(0)$ has a Gaussian distribution of zero mean and variance $(2\beta kTt/M)$ are 1) the solutions $v(t)$ continuous and 2) only then is the Maxwell-Boltzmann distribution (see Section 4.6.2) for $v(t)$ attained as $t \to \infty$ for any initial condition $v(0)$. Through Doob's analysis we gain a deeper insight into the nature of the random force. The upshot, however, is that we can use the Langevin equation as originally stated in Eq. (3.9) knowing that a rigorous approach leads to a solution, Eq. (3.26), the form of which is also obtained through the naive assumption that dv/dt exists.

3.3 FOKKER-PLANCK EQUATION FOR ONE RANDOM VARIABLE

The Langevin equation supplies expressions for quantities such as $\langle v(t)\rangle_{v_0}$, how the average speed of a B-particle evolves in a subensemble specified by initial condition v_0. One might wonder if, instead of solving a differential equation for observables like $\langle v(t)\rangle_{v_0}$, the same time-dependent average could be calculated as an expectation value with respect to a time-dependent probability,[26]

$$\langle v(t)\rangle_{v_0,t_0} = \int_{-\infty}^{\infty} vP(v,t|v_0,t_0)dv. \tag{3.30}$$

The quantity $P(v,t|v_0,t_0)dv$ we seek is a *transition probability*,[27] the probability the velocity lies in $[v, v + dv]$ at time t, given that it had the value v_0 at time $t_0 < t$. As we show, such a function is obtained as the solution to a differential equation, the Fokker-Planck equation, (3.36).[28] Assuming no unique origin in time, $P(v,t|v_0,t_0)$ depends only on the difference $\tau \equiv t - t_0$. We erase the label t_0 in Eq. (3.30) and change notation, $P(v,t|v_0,t_0) \to P(v,\tau|v_0)$. For simplicity we treat $P(v,\tau|v_0)$ as a Markov process,[29] that the probability does not depend on the entire past history of the particle, only on $v = v_0$ at $\tau = 0$. We assume a stationary Markov process.

3.3.1 Derivation

We'd like an equation analogous to (3.30) to hold for any ensemble average, not just for the velocity. Let $P(y,\tau|y_0)$ be the conditional probability of a random variable y and let $f(y)$ be a smooth function[30] such that $f(y) \to 0$ as $y \to \pm\infty$. We want an equation like (3.30) to hold for any f,

$$\langle f(\tau)\rangle_{y_0} = \int f(y)P(y,\tau|y_0)dy. \tag{3.31}$$

Note the generality—any random variable y and any function $f(y)$. To isolate $P(y,\tau|y_0)$, one might try differentiating with respect to the parameter τ. As we show, a partial differential equation for

---

[24] Integration by parts is defined for Stieltjes integrals, $\int_a^b gdf = [fg]_a^b - \int_a^b fdg$.

[25] If Eq. (3.25) is true, it implies Eq. (3.26); conversely if $v(t)$ is given by Eq. (3.26), it satisfies Eq. (3.25).

[26] The Langevin equation is the equation of motion for observables in the presence of a random force. One solves the differential equation and then takes the ensemble average, where the average is over measurements made on an ensemble of systems having the same initial condition. With the Fokker-Planck approach, ensemble averages are calculated with respect to the transition probability $P(v,t|v_0,t_0)$ describing the same ensemble.

[27] We've erased the subscripts from $P_{1|1}(v,t|v_0,t_0)$ in Section 2.4.3.

[28] The Fokker-Planck equation was introduced by A.D. Fokker in 1913 and M. Planck in 1917. It was independently derived by Kolmogorov in 1931, where it's known as the Kolmogorov forward equation. It's also known as the Einstein-Fokker-Planck equation, which appeared without fanfare in Einstein's 1905 article on Brownian motion.

[29] Fokker-Planck equations for non-Markovian processes exist in the literature (see for example Adelman[71]), which pertain to research problems beyond the scope of this book.

[30] *Smooth function* is generally a code word for "all derivatives exist."

$P(y,\tau|y_0)$ can be developed from the construct ($x \equiv y_0$ to save writing)

$$\int \mathrm{d}y f(y)\frac{\partial}{\partial \tau}P(y,\tau|x) = \lim_{\Delta\tau\to 0}\frac{1}{\Delta\tau}\int \mathrm{d}y f(y)\left[P(y,\tau+\Delta\tau|x) - P(y,\tau|x)\right] \tag{3.32}$$
$$= \lim_{\Delta\tau\to 0}\frac{1}{\Delta\tau}\int \mathrm{d}y f(y)\left[\left(\int \mathrm{d}z P(y,\Delta\tau|z)P(z,\tau|x)\right) - P(y,\tau|x)\right],$$

where we've used the SCK equation, the hallmark of Markov processes. Expand $f(y)$ on the right side Eq. (3.32) to second order in a Taylor series, $f(y) \approx f(z) + (y-z)f'(z) + \frac{1}{2}(y-z)^2 f''(z)$ and interchange the order of integration:

$$\begin{aligned}\int \mathrm{d}y f(y)\frac{\partial}{\partial \tau}P(y,\tau|x) = \lim_{\Delta\tau\to 0}\frac{1}{\Delta\tau}\Bigg[&\int \mathrm{d}z f(z)P(z,\tau|x)\int \mathrm{d}y P(y,\Delta\tau|z)\\ &+\int \mathrm{d}z f'(z)P(z,\tau|x)\int \mathrm{d}y(y-z)P(y,\Delta\tau|z)\\ &+\frac{1}{2}\int \mathrm{d}z f''(z)P(z,\tau|x)\int \mathrm{d}y(y-z)^2 P(y,\Delta\tau|z)\\ &-\int \mathrm{d}z f(z)P(z,\tau|x)\Bigg].\end{aligned} \tag{3.33}$$

The first and last terms in Eq. (3.33) cancel (by the rules for conditional probabilities, see page 25). Equation (3.33) involves a family of integrals of the form $\int \mathrm{d}y(y-z)^n P(y,\Delta\tau|z)$ ($n = 1,2,\ldots$),[31] known as *transition moments*, which (keep in mind) we need evaluate only for small $\Delta\tau$. We assume for small $\Delta\tau$,

$$\begin{aligned}\int \mathrm{d}y(y-z)P(y,\Delta\tau|z) &\equiv M_1(z)\Delta\tau + O\left(\Delta\tau\right)^2\\ \int \mathrm{d}y(y-z)^2 P(y,\Delta\tau|z) &\equiv M_2(z)\Delta\tau + O\left(\Delta\tau\right)^2\\ \int \mathrm{d}y(y-z)^n P(y,\Delta\tau|z) &\equiv M_n(z)(\Delta\tau)^2. \qquad (n\geq 3)\end{aligned} \tag{3.34}$$

All such integrals vanish as $\Delta\tau \to 0$ (for $n \geq 1$);[32] the question is *how* they vanish for small $\Delta\tau$. It's borne out in physical theories that transition moments for $n \geq 3$ vanish at least as fast as $(\Delta\tau)^2$ (an assumption to be checked at the end of a calculation—see Exercise 3.25). Combine Eqs. (3.34) and (3.33), take the limit $\Delta\tau \to 0$, and integrate by parts,

$$\int \mathrm{d}y f(y)\frac{\partial}{\partial \tau}P(y,\tau|x) = \int \mathrm{d}z f(z)\left(-\frac{\partial}{\partial z}\left[M_1(z)P(z,\tau|x)\right] + \frac{1}{2}\frac{\partial^2}{\partial z^2}\left[M_2(z)P(z,\tau|x)\right]\right),$$

which, with a trivial change of variables, leads to

$$\int \mathrm{d}y f(y)\left(\frac{\partial P(y,\tau|x)}{\partial \tau} + \frac{\partial}{\partial y}\left[M_1(y)P(y,\tau|x)\right] - \frac{1}{2}\frac{\partial^2}{\partial y^2}\left[M_2(y)P(y,\tau|x\right]\right) = 0. \tag{3.35}$$

Because Eq. (3.35) holds for any function $f(y)$, it implies the Fokker-Planck equation[33,34]

$$\frac{\partial}{\partial \tau}P(y,\tau|x) = -\frac{\partial}{\partial y}\left[M_1(y)P(y,\tau|x)\right] + \frac{1}{2}\frac{\partial^2}{\partial y^2}\left[M_2(y)P(y,\tau|x)\right]. \tag{3.36}$$

[31] We keep transition moments for $n = 1, 2$ in Eq. (3.33) because we have limited the expansion of $f(y)$ to two term in it's Taylor series, which suffices for many physical applications. More terms could be included if need be.

[32] $\mathrm{Lim}_{\Delta\tau\to 0}P(y,\Delta\tau|z) = \delta(y-z)$; $\lim_{\Delta\tau\to 0}\int \mathrm{d}y(y-z)^n P(y,\Delta\tau|z) = \int \mathrm{d}y(y-z)^n\delta(y-z) = 0$ for $n \geq 1$.

[33] The passage from Eq. (3.35) to (3.36) is a familiar step in theoretical physics. It relies on a theorem that if $\int \eta(x)\phi(x)\mathrm{d}x = 0$ holds for all reasonable functions $\eta(x)$, then $\phi(x) = 0$ identically[72, p185].

[34] The Fokker-Planck equation has an extensive literature. Books have been written about it; see Risken[67] for example.

Equation (3.36) involves the transition probability $P(y, \tau|x)$, but the same form applies to $P_1(y,t) = \int dx P(y, t-t_0|x)P_1(x,t_0)$. Equation (3.36) is equivalent to:

$$\frac{\partial}{\partial \tau}P_1(y,\tau) = -\frac{\partial}{\partial y}\left[M_1(y)P_1(y,\tau)\right] + \frac{1}{2}\frac{\partial^2}{\partial y^2}\left[M_2(y)P_1(y,\tau)\right]. \tag{3.37}$$

Thus there are two versions of the Fokker-Planck equation: Eq. (3.36), involving the transition probability $P(y,\tau|x)$ or Eq. (3.37) involving the probability $P_1(y,t)$. Either is referred to as the Fokker-Planck equation; either can be used depending on the problem.

The Fokker-Planck equation is a partial differential equation for time-dependent probabilities. It provides for the time rate of change of probability in terms of derivatives with respect to y; *it models the flow of probability in the space of random variables*.[35] In that regard, one might wonder if there's a connection with the master equation, which also provides for a time derivative of probability. Such is indeed the case; see Exercise 3.21. Equation (3.37) has the form of a continuity equation for probability (erasing the subscript on $P_1(y,\tau)$),

$$\frac{\partial}{\partial \tau}P(y,\tau) + \frac{\partial}{\partial y}J(y,\tau) = 0, \tag{3.38}$$

where the probability current density is

$$J(y,\tau) = M_1(y)P(y,\tau) - \frac{1}{2}\frac{\partial}{\partial y}\left[M_2(y)P(y,\tau)\right]. \tag{3.39}$$

There are two (and only two) types of probability flows: drift (or convection), associated with M_1, and diffusion, associated with M_2. Whereas $M_2 > 0$ [see Eq. (3.34)], M_1 can be of either sign; M_1 probes the *asymmetry* of the transition probability; $M_1 \neq 0$ only if $P(y, \Delta\tau|z)$ is not an even function of y about z for small Δt.

3.3.2 Free Brownian particles in spatially homogeneous systems

Equation (3.37) becomes an effective differential equation for determining $P_1(y,t)$ only when the transition moments have been specified. For free Brownian motion, where the random variable is the scalar velocity of the B-particle, we find the transition moments from the associated Langevin equation. The first moment involves the average deviation of v from its initial value v_0 in the next Δt seconds (from Eq. (3.34)),[36]

$$M_1\Delta t = \int (v - v_0)P(v, \Delta t|v_0)dv = \langle v(\Delta t)\rangle_{v_0} - v_0 = -v\frac{\alpha}{M}\Delta t + O(\Delta t)^2,$$

implying for $\Delta t \to 0$,

$$M_1 = -\frac{\alpha}{M}v, \tag{3.40}$$

where we've used Eq. (3.11). $M_1 < 0$ because the average speed $\langle v(t)\rangle_{v_0} < v_0$ as a result of collisions. One can show that (see Exercise 3.22)

$$M_2 = \frac{2kT\alpha}{M^2}. \tag{3.41}$$

Thus we have the Fokker-Planck equation for free Brownian motion,

$$\frac{\partial}{\partial t}P(v,t) = \frac{\alpha}{M}\frac{\partial}{\partial v}\left[vP(v,t)\right] + \frac{kT\alpha}{M^2}\frac{\partial^2}{\partial v^2}P(v,t). \tag{3.42}$$

---

[35] It applies to processes where the jump length (see Exercise 3.21) is small enough that the methods of calculus can be applied. This is justified for Brownian motion where the mass of the B-particle is much larger than molecular masses.

[36] The first transition moment is sometimes written $M_1\Delta t = \langle \Delta v(\Delta t)\rangle$, which can be misleading: $\langle \Delta v\rangle$ vanishes in an equilibrium ensemble average. The transition moments refer to *nonequilibrium* ensemble averages at short times.

Let's find the stationary solution, which we expect to be the equilibrium probability distribution. With $\partial P/\partial t = 0$, we see from Eq. (3.38) that $J =$ constant. The "boundary condition" $J \to 0$ as $v \to \infty$ implies $J = 0$ for all v in steady state. Combining Eqs. (3.40), (3.41) with $J = 0$ in Eq. (3.39), the Fokker-Planck theory leads us to a differential equation for the stationary solution $P_{\text{st}}(v)$,

$$\left(v + \frac{kT}{M}\frac{\partial}{\partial v}\right) P_{\text{st}}(v) = 0, \tag{3.43}$$

which is solved by elementary means (K is a normalization constant),

$$P_{\text{st}}(v) = K\mathrm{e}^{-Mv^2/(2kT)} = \sqrt{\frac{M}{2\pi kT}}\mathrm{e}^{-Mv^2/(2kT)}, \tag{3.44}$$

none other than the Maxwell-Boltzmann distribution.[37] At long times, the B-particle has a distribution of speeds characteristic of thermal equilibrium at absolute temperature T.

The non-stationary solution is more difficult because we have to build in the initial condition. For that purpose work with the transition probability, for which $P(y,0|x) = \delta(y-x)$. Write the equation in terms of a dimensionless time $\tau \equiv \beta t$, where $\beta \equiv \alpha/M$ and $x \equiv v_0$:

$$\frac{\partial}{\partial \tau}P(v,\tau|x) = \frac{\partial}{\partial v}\left[vP(v,\tau|x)\right] + \frac{kT}{M}\frac{\partial^2}{\partial v^2}P(v,\tau|x). \tag{3.45}$$

There is no one way to solve differential equations. With that said, we Fourier transform with respect to v, $\widetilde{P}(q,\tau|x) \equiv \int_{-\infty}^{\infty} P(v,\tau|x)\mathrm{e}^{-iqv}\mathrm{d}v$. Multiply Eq. (3.45) by e^{-iqv} and integrate over v, which, using the results of Exercise 3.23, is equivalent to the first-order partial differential equation,[38]

$$\left(\frac{\partial}{\partial \tau} + q\frac{\partial}{\partial q}\right)\widetilde{P}(q,\tau|x) = -\frac{kT}{M}q^2\widetilde{P}(q,\tau|x), \tag{3.46}$$

which can be solved by the method of characteristics.[39] The general solution to Eq. (3.46) is of the form (as can be verified)

$$\widetilde{P}(q,\tau|x) = F(q\mathrm{e}^{-\tau})\mathrm{e}^{-kTq^2/(2M)}, \tag{3.47}$$

where F is an arbitrary differentiable function. Choose F so that the initial condition is met. With $P(v,0|x) = \delta(v-x)$, $\widetilde{P}(q,0|x) = \mathrm{e}^{-iqx}$, implying $F(q) = \mathrm{e}^{-iqx}\mathrm{e}^{kTq^2/(2M)}$. Thus,

$$\widetilde{P}(q,\tau|x) = \exp\left[-iqx\mathrm{e}^{-\tau} - (kT/2M)q^2\left(1-\mathrm{e}^{-2\tau}\right)\right], \tag{3.48}$$

a Gaussian in the variable q (see Section 2.9.1). To find $P(v,\tau|x)$, we must evaluate the inverse Fourier transform, a calculation we've already done in Exercise 2.5. We write down the final result for the transition probability associated with Brownian motion ($\tau \equiv \alpha t/M$):

$$P(v,\tau|v_0) = \sqrt{\frac{M}{2\pi kT(1-e^{-2\tau})}}\exp\left[-\frac{M}{2kT}\frac{(v-v_0\mathrm{e}^{-\tau})^2}{1-e^{-2\tau}}\right]. \tag{3.49}$$

It reproduces the correct the initial condition as $\tau \to 0$, $\lim_{\tau\to 0} P(v,\tau|v_0) = \delta(v-v_0)$ (see Exercise 3.27). As $\tau \to \infty$ initial conditions are forgotten, and we recover the Maxwell-Boltzmann distribution, Eq. (3.44). We examine the three-dimensional version of this formula in Eq. (3.90).

[37] The Maxwell-Boltzmann distribution is derived in [2, p105], in [5, p127], and in Section 4.6.2.

[38] That is, introducing an integral transform has reduced the order of the Fokker-Planck equation from two to one.

[39] Discussed in books on partial differential equations; see Sneddon[73] or Courant and Hilbert[74]. See also the last page of Wang and Uhlenbeck[75], reprinted in Wax[44, pp113–132].

3.3.3 Ornstein-Uhlenbeck process

Equation (3.49), the solution of the Fokker-Planck equation for free B-particles, has the form of the transition probability in stationary Gaussian processes; see Eq. (2.108). Doob's theorem (see Section 2.9.4.2), which applies to stationary Gaussian Markov processes,[40] on the exponential decay of time correlation functions, is satisfied. The stochastic process having Eq. (3.49) as its transition probability and P_1 in the Gaussian form of Eq. (2.107) is known as the *Ornstein-Uhlenbeck process.*[41] It's associated with a Fokker-Planck equation having a linear drift coefficient $M_1 = -\gamma v$ (with γ constant) and diffusion coefficient, $D = kT\gamma/M$. The Ornstein-Uhlenbeck process is the only process that is Markov, Gaussian, and stationary,[42] an equivalent way of stating Doob's theorem.

3.4 BROWNIAN PARTICLES IN EXTERNAL FORCE FIELDS

The Fokker-Planck equation derived in Section 3.3.1 pertains to the probability density $P_1(y,t)$ of a single random variable y, which we used in Section 3.3.2 to represent the velocity v of a free Brownian particle. More generally, a particle's position $\boldsymbol{r}$, its vector velocity $\boldsymbol{v}$, or both can be taken as random variables. Spatial variables come into play for particles in force fields and for spatially dependent initial conditions. In this section, we derive the Fokker-Planck equation for the probability density $P_1(\boldsymbol{r},\boldsymbol{v},t)$ that at time t the B-particle is "at" the point in phase space $(\boldsymbol{r},\boldsymbol{v})$.

We start by developing an integral equation[43] for $P_1(\boldsymbol{r},\boldsymbol{v},t)$. To save writing, denote the conditional probability density $P_{1|1}(\boldsymbol{r},\boldsymbol{v},t|\boldsymbol{r}-\Delta\boldsymbol{r},\boldsymbol{v}-\Delta\boldsymbol{v},t-\Delta t) \equiv \psi_{\Delta t}(\boldsymbol{r}-\Delta\boldsymbol{r},\boldsymbol{v}-\Delta\boldsymbol{v};\Delta\boldsymbol{r},\Delta\boldsymbol{v})$, the probability that the particle at $(\boldsymbol{r}-\Delta\boldsymbol{r},\boldsymbol{v}-\Delta\boldsymbol{v})$ makes the transition $(\Delta\boldsymbol{r},\Delta\boldsymbol{v})$ in the next Δt seconds. From the rules for conditional probabilities (see Section 2.3.1),

$$P_1(\boldsymbol{r},\boldsymbol{v},t) = \int \mathrm{d}(\Delta\boldsymbol{r})\mathrm{d}(\Delta\boldsymbol{v})\psi_{\Delta t}(\boldsymbol{r}-\Delta\boldsymbol{r},\boldsymbol{v}-\Delta\boldsymbol{v};\Delta\boldsymbol{r},\Delta\boldsymbol{v})P_1(\boldsymbol{r}-\Delta\boldsymbol{r},\boldsymbol{v}-\Delta\boldsymbol{v},t-\Delta t), \quad (3.50)$$

where $\mathrm{d}(\Delta\boldsymbol{v}) \equiv \mathrm{d}(\Delta v_1)\mathrm{d}(\Delta v_2)\mathrm{d}(\Delta v_3)$, etc. Assume Δt is sufficiently small that $\Delta\boldsymbol{r} = \boldsymbol{v}\Delta t$ is accurate. We can enforce that connection with a Dirac delta function,

$$\psi_{\Delta t}(\boldsymbol{r}-\Delta\boldsymbol{r},\boldsymbol{v}-\Delta\boldsymbol{v};\Delta\boldsymbol{r},\Delta\boldsymbol{v}) \equiv \delta(\Delta\boldsymbol{r}-\boldsymbol{v}\Delta t)\phi_{\Delta t}(\boldsymbol{r}-\Delta\boldsymbol{r},\boldsymbol{v}-\Delta\boldsymbol{v};\Delta\boldsymbol{v}), \quad (3.51)$$

where $\phi_{\Delta t}$ is a transition probability in velocity space. Combining Eq. (3.51) with Eq. (3.50) and integrating over $\Delta\boldsymbol{r}$, we have

$$P(\boldsymbol{r}+\boldsymbol{v}\Delta t,\boldsymbol{v},t+\Delta t) = \int \mathrm{d}(\Delta\boldsymbol{v})\phi_{\Delta t}(\boldsymbol{r},\boldsymbol{v}-\Delta\boldsymbol{v};\Delta\boldsymbol{v})P(\boldsymbol{r},\boldsymbol{v}-\Delta\boldsymbol{v},t), \quad (3.52)$$

where we've erased the subscript on P_1 and we've let $\boldsymbol{r} \to \boldsymbol{r}+\boldsymbol{v}\Delta t$ and $t \to t+\Delta t$. Assume that $\phi_{\Delta t}$ is sharply peaked about $\Delta\boldsymbol{v} = 0$ (impacts with fluid molecules produce small changes in the velocity of the more massive B-particle). Expand the quantities in Eq. (3.52) in Taylor series,

$$\begin{aligned}
P(\boldsymbol{r}+\boldsymbol{v}\Delta t,\boldsymbol{v},t+\Delta t) &= P(\boldsymbol{r},\boldsymbol{v},t) + \Delta t\left(\boldsymbol{v}\cdot\frac{\partial P}{\partial\boldsymbol{r}} + \frac{\partial P}{\partial t}\right) + O(\Delta v^2, \Delta v\Delta t, \Delta t^2) \\
\phi_{\Delta t}(\boldsymbol{r},\boldsymbol{v}-\Delta\boldsymbol{v};\Delta\boldsymbol{v}) &= \phi_{\Delta t}(\boldsymbol{r},\boldsymbol{v};\Delta\boldsymbol{v}) - \Delta\boldsymbol{v}\cdot\frac{\partial\phi_{\Delta t}}{\partial\boldsymbol{v}} + \frac{1}{2}\frac{\partial^2\phi_{\Delta t}}{\partial\boldsymbol{v}\partial\boldsymbol{v}}{:}\Delta\boldsymbol{v}\Delta\boldsymbol{v} + O(\Delta v^3) \\
P(\boldsymbol{r},\boldsymbol{v}-\Delta\boldsymbol{v},t) &= P(\boldsymbol{r},\boldsymbol{v},t) - \Delta\boldsymbol{v}\cdot\frac{\partial P}{\partial\boldsymbol{v}} + \frac{1}{2}\frac{\partial^2 P}{\partial\boldsymbol{v}\partial\boldsymbol{v}}{:}\Delta\boldsymbol{v}\Delta\boldsymbol{v} + O(\Delta v^3),
\end{aligned} \quad (3.53)$$

[40] The Fokker-Planck equation was derived under the assumption of stationary Markov processes.

[41] After [64], reprinted in Wax[44, pp93–111].

[42] There is one other, trivial, process having these properties—independent stochastic processes.

[43] In integral equations, the unknown function is inside an integral. See [13, Chapter 10] or Appendix D.

where we expand the terms inside the integral to second order and we've used dyadic notation for tensor contraction (see Section 1.4.3). Use the expansions in Eq. (3.53) in Eq. (3.52) and work consistently to second order in $\Delta \boldsymbol{v}$ on the right side and to first order in Δt on the left. We find

$$\left(\frac{\partial P}{\partial t} + \boldsymbol{v} \cdot \frac{\partial P}{\partial \boldsymbol{r}}\right) \Delta t \approx \int \mathrm{d}(\Delta \boldsymbol{v}) \left[-\Delta \boldsymbol{v} \cdot \frac{\partial (P\phi_{\Delta t})}{\partial \boldsymbol{v}} + \frac{1}{2} \frac{\partial^2 (P\phi_{\Delta t})}{\partial \boldsymbol{v} \partial \boldsymbol{v}} : \Delta \boldsymbol{v} \Delta \boldsymbol{v} \right], \tag{3.54}$$

where, for conditional probabilities, $\int \mathrm{d}(\Delta \boldsymbol{v}) \phi_{\Delta t}(\boldsymbol{v}, \boldsymbol{r}; \Delta \boldsymbol{v}) = 1$ (so that the zeroth order term cancels across the equation). Equation (3.54) implies, in the limit $\Delta t \to 0$,

$$\left(\frac{\partial}{\partial t} + \boldsymbol{v} \cdot \boldsymbol{\nabla}_r\right) P(\boldsymbol{r}, \boldsymbol{v}, t) = -\frac{\partial}{\partial \boldsymbol{v}} \cdot \left(P(\boldsymbol{r}, \boldsymbol{v}, t) \frac{\langle \Delta \boldsymbol{v} \rangle_{\Delta t}}{\Delta t} \right) + \frac{1}{2} \frac{\partial^2}{\partial \boldsymbol{v} \partial \boldsymbol{v}} : \left(P(\boldsymbol{r}, \boldsymbol{v}, t) \frac{\langle \Delta \boldsymbol{v} \Delta \boldsymbol{v} \rangle_{\Delta t}}{\Delta t} \right), \tag{3.55}$$

where $\langle \Delta \boldsymbol{v} \rangle_{\Delta t} \equiv \int \mathrm{d}(\Delta \boldsymbol{v}) \Delta \boldsymbol{v} \phi_{\Delta t}(\boldsymbol{r}, \boldsymbol{v}; \Delta \boldsymbol{v})$, etc. Equation (3.55) is the generalization of Eq. (3.37) to include a vector velocity, a spatial dependence $\boldsymbol{r}$, and transition moments having a tensor character (compare with Eq. (3.34)). It becomes an effective differential equation for determining $P(\boldsymbol{r}, \boldsymbol{v}, t)$ when the transition moments have been specified.

To do that, return to the Langevin equation in the form of Eq. (3.25),

$$\Delta \boldsymbol{v} = -(\beta \boldsymbol{v} - \boldsymbol{K}) \Delta t + \Delta \boldsymbol{B}, \tag{3.56}$$

where we include an acceleration field $\boldsymbol{K}(\boldsymbol{r})$. Inverting Eq. (3.56), $\Delta \boldsymbol{B} = \Delta \boldsymbol{v} + (\beta \boldsymbol{v} - \boldsymbol{K}) \Delta t$. By the Doob theorem mentioned on page 68, $\Delta \boldsymbol{B}$ has a Gaussian distribution of known mean and variance, implying

$$\phi_{\Delta t}(\boldsymbol{r}, \boldsymbol{v}; \Delta \boldsymbol{v}) = \frac{1}{(2\pi \sigma_0^2 \Delta t)^{3/2}} \exp\left(-[\Delta \boldsymbol{v} + (\beta \boldsymbol{v} - \boldsymbol{K}(\boldsymbol{r})) \Delta t]^2 / (2\sigma_0^2 \Delta t) \right), \tag{3.57}$$

where $\sigma_0^2 \equiv 2\beta kT/M$. It's straightforward to show that

$$\begin{aligned} \langle \Delta \boldsymbol{v} \rangle_{\Delta t} &= -\left(\beta \boldsymbol{v} - \boldsymbol{K} \right) \Delta t \\ \langle \Delta v_i \Delta v_j \rangle_{\Delta t} &= \frac{2\beta kT}{M} \delta_{ij} \Delta t + O(\Delta t)^2. \end{aligned} \tag{3.58}$$

Combining the expressions in Eq. (3.58) with Eq. (3.55) and taking the limit $\Delta t \to 0$, we have[44]

$$\begin{aligned} \frac{\mathrm{d}}{\mathrm{d}t} P(\boldsymbol{r}, \boldsymbol{v}, t) &= \left(\frac{\partial}{\partial t} + \boldsymbol{v} \cdot \boldsymbol{\nabla}_{\boldsymbol{r}} + \boldsymbol{K}(\boldsymbol{r}) \cdot \boldsymbol{\nabla}_{\boldsymbol{v}} \right) P(\boldsymbol{r}, \boldsymbol{v}, t) \\ &= \beta \boldsymbol{\nabla}_{\boldsymbol{v}} \cdot \left(\boldsymbol{v} + \frac{kT}{M} \boldsymbol{\nabla}_{\boldsymbol{v}} \right) P(\boldsymbol{r}, \boldsymbol{v}, t) \equiv \Omega P(\boldsymbol{r}, \boldsymbol{v}, t) \end{aligned} \tag{3.59}$$

where $\boldsymbol{\nabla}_{\boldsymbol{r}}$ is the gradient operator in position space with $\boldsymbol{\nabla}_{\boldsymbol{v}}$ that in velocity space, and where Ω is the *Fokker-Planck operator*,

$$\Omega \equiv \beta \boldsymbol{\nabla}_{\boldsymbol{v}} \cdot \left(\boldsymbol{v} + \frac{kT}{M} \boldsymbol{\nabla}_{\boldsymbol{v}} \right). \tag{3.60}$$

The terms on the right side of Eq. (3.59) are associated with Brownian motion and those on the left comprise the total time derivative, $\mathrm{d}P(\boldsymbol{r}, \boldsymbol{v}, t)/\mathrm{d}t$. Note that the left side changes sign under $t \to -t$ but the right side does not—the hallmark of irreversible evolution equations. Equation (3.59) cannot be solved in closed form, but it can be solved in special cases.

[44]Equation (3.59) is known as the Kramers equation or the Klein-Kramers equation; it was derived by O. Klein in 1922 and independently by Kramers[76] in 1940. The nomenclature is not standardized and can be confusing. The distinction made by some authors seems to be that the name Fokker-Planck equation is reserved for free B-particles.

3.4.1 The stationary solution

Referring to Eq. (3.59), the stationary solution depends on $\boldsymbol{r}$ and $\boldsymbol{v}$, call it $P_{\text{st}}(\boldsymbol{r},\boldsymbol{v})$. When the force field is derivable from a potential energy function, $\boldsymbol{K}(\boldsymbol{r}) = -(1/M)\boldsymbol{\nabla}\psi(\boldsymbol{r})$, $P_{\text{st}}(\boldsymbol{r},\boldsymbol{v})$ has the Boltzmann form,

$$P_{\text{st}}(\boldsymbol{r},\boldsymbol{v}) = (\text{constant})\,\mathrm{e}^{-\psi(\boldsymbol{r})/(kT)}\mathrm{e}^{-Mv^2/(2kT)}, \tag{3.61}$$

see Exercise 3.29. Note that the stationary solution is independent of the friction parameter β.

3.4.2 Passage over potential barriers: The Kramers escape problem

The stationary solution comes into play in the problem considered by Kramers[76] of the passage of particles over potential barriers as a consequence of Brownian motion (see Fig. 3.4). Particles in

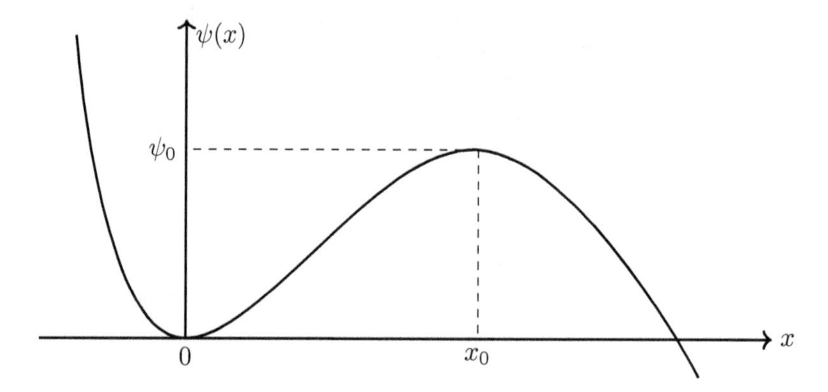

Figure 3.4 Potential energy function $\psi(x)$ with a smooth barrier ψ_0.

the potential well are subject to the force associated with the negative gradient of $\psi(x)$, but also to the friction and random force supplied by the environment. Some particles in the well escape over the barrier as a result of Brownian motion. At what rate do particles arrive at the top of the barrier? This model has applications in the study of chemical as well as nuclear reactions.

The barrier is assumed large relative to thermal energies, $\psi_0 \gg kT$, implying that *quasi-stationary* conditions prevail in which time derivatives can be neglected. Near the bottom of the well the probability distribution is given by Eq. (3.61). The equilibrium distribution, however, in that it's associated with zero current, can't be valid near the potential barrier where particles escape at a nonzero rate. We seek a steady-state solution valid in the vicinity of the barrier that supports a nonzero current. The shape of $\psi(x)$ at the top of the barrier can be fit to a downward facing parabola, $\psi(x) \approx \psi_0 - \frac{1}{2}M\omega_b^2(x-x_0)^2$ for $x \approx x_0$. Here ω_b is a parameter with dimension frequency set by the curvature at $x = x_0$, $\omega_b^2 = (-1/M)\psi''|_{x=x_0}$. No oscillations occur near the barrier.

We modify the stationary solution Eq. (3.61) by introducing an unknown multiplicative function $F(x,v)$ and where we use the form of the barrier $\psi(x) = \psi_0 - \frac{1}{2}M\omega_b^2(x-x_0)^2$,

$$P(x,v) = CF(x,v)\mathrm{e}^{-\psi_0/(kT)}\exp\left[\frac{M}{2kT}\left(\omega_b^2(x-x_0)^2 - v^2\right)\right], \tag{3.62}$$

where C is a constant. The one-dimensional version of Eq. (3.59) without the time derivative (we seek a stationary solution) and with $K(x) = \omega_b^2(x-x_0)$ is

$$\left(v\frac{\partial}{\partial x} + \omega_b^2(x-x_0)\frac{\partial}{\partial v}\right)P = \beta\frac{\partial}{\partial v}\left(v + \frac{kT}{M}\frac{\partial}{\partial v}\right)P. \tag{3.63}$$

The function F is found by substituting Eq. (3.62) in Eq. (3.63); we find it must satisfy the differential equation

$$v\frac{\partial F}{\partial x} + \left[\beta v + \omega_b^2(x-x_0)\right]\frac{\partial F}{\partial v} = \beta\frac{kT}{M}\frac{\partial^2 F}{\partial v^2}. \tag{3.64}$$

Let's attempt a solution of Eq. (3.64) by assuming F to be a function of a single variable, $F(x,v) = f[v-\alpha(x-x_0)] \equiv f(z)$, where α is to be determined. With this substitution, Eq. (3.64) is equivalent to

$$\left[(\beta-\alpha)v + \omega_b^2(x-x_0)\right] f' = \frac{\beta kT}{M} f''. \tag{3.65}$$

For this approach to work, the terms in square brackets must be a function of z only. Try something linear; let

$$(\beta-\alpha)v + \omega_b^2(x-x_0) = \lambda z = \lambda(v-\alpha(x-x_0)), \tag{3.66}$$

where λ is to be determined. Equation (3.66) is satisfied with $\lambda = \beta-\alpha$ and $\lambda\alpha = -\omega_b^2$, implying the quadratic equation $\lambda^2 - \beta\lambda - \omega_b^2 = 0$ having roots $\lambda_\pm = \frac{1}{2}(\beta \pm \sqrt{\beta^2+4\omega_b^2})$. Note that $\lambda_+ + \lambda_- = \beta$ and $\lambda_+\lambda_- = -\omega_b^2$. For either $\lambda = \lambda_\pm$, Eq. (3.65) is equivalent to

$$\lambda z f'(z) = \frac{\beta kT}{M} f''(z), \tag{3.67}$$

which is readily integrated (A and B are constants),

$$f(z) = A + B\int_0^z \mathrm{d}y \exp\left[\frac{g}{2}y^2\right], \tag{3.68}$$

where $g \equiv M\lambda/(\beta kT)$.

The root λ_+ is positive and λ_- is negative. Which should we choose, or do we work with both? What are the boundary conditions on $f(z)$? Referring to Fig. 3.4, we expect $F(x,v)$ to vanish as $x \to \infty$ for $v < 0$. With $z \equiv v - \alpha(x-x_0)$, if α is positive, we have the boundary condition $f(z) \to 0$ as $z \to -\infty$. To have $\alpha > 0$, choose the negative root, $\lambda = \lambda_- = -\omega_b^2/\lambda_+$, implying $\alpha = \lambda_+$ and $g = -M\omega_b^2/(\beta kT\lambda_+)$. Setting $A = 0$ in Eq. (3.68) and setting the lower limit to $-\infty$,

$$f(z) = B\int_{-\infty}^z \exp\left(-\frac{M\omega_b^2}{2\beta kT\lambda_+}y^2\right)\mathrm{d}y. \tag{3.69}$$

For the other boundary condition, we expect that to the left of the barrier (see Fig. 3.4) $F(x,v) \to 1$ as $x \to -\infty$; thus $f(z) \to 1$ as $z \to \infty$. This boundary condition determines B (set $z = \infty$ in Eq. (3.69)), implying $B = \sqrt{M\omega_b^2/(2\pi\beta kT\lambda_+)}$. The solution of Eq. (3.67) meeting these boundary conditions is

$$f(z) = \sqrt{\frac{M\omega_b^2}{2\pi\beta kT\lambda_+}}\int_{-\infty}^z \exp\left(-\frac{M\omega_b^2}{2\beta kT\lambda_+}y^2\right)\mathrm{d}y. \tag{3.70}$$

Combining Eq. (3.70) with Eq. (3.62), we have a stationary solution valid for $x \approx x_0$,

$$P(x,v) = C\sqrt{\frac{M\omega_b^2}{2\pi\beta kT\lambda_+}}\mathrm{e}^{-\psi_0/(kT)}\exp\left(\frac{M}{2kT}(\omega_b^2(x-x_0)^2 - v^2)\right)\int_{-\infty}^z \mathrm{d}y\exp\left(-\frac{M\omega_b^2}{2\beta kT\lambda_+}y^2\right), \tag{3.71}$$

where $z = v - \lambda_+(x-x_0)$, with $\lambda_+ = \frac{1}{2}(\beta+\sqrt{\beta^2+4\omega_b^2})$. To evaluate the constant C, consider the stationary solution Eq. (3.61) in the vicinity of the potential well at $x \approx 0$, where $\psi(x) = \frac{1}{2}M\omega_0^2x^2$, with ω_0 characterizing the curvature, $\omega_0 = (1/M)\psi''|_{x=0}$. Thus for $x \approx 0$,

$$P(x,v) = C\exp\left(-\frac{M}{2kT}(v^2+\omega_0^2x^2)\right). \tag{3.72}$$

By normalizing, with $\int_{-\infty}^{\infty}\int_{-\infty}^{\infty} P(x,v)\mathrm{d}x\mathrm{d}v = 1$, we find $C = M\omega_0/(2\pi kT)$.

The net diffusion current at the top of the barrier, $x = x_0$, is found from

$$J = \int_{-\infty}^{\infty} P(x = x_0, v) v \mathrm{d}v. \tag{3.73}$$

Using Eq. (3.71),

$$J = C\sqrt{\frac{M\omega_b^2}{2\pi\beta kT\lambda_+}} \mathrm{e}^{-\psi_0/(kT)} \int_{-\infty}^{\infty} \mathrm{d}v \, v \mathrm{e}^{-Mv^2/(2kT)} \int_{-\infty}^{v} \exp\left(-\frac{M\omega_b^2}{2\beta kT\lambda_+} y^2\right) \mathrm{d}y. \tag{3.74}$$

After integrating by parts, we have

$$\begin{aligned} J &= C\sqrt{\frac{M\omega_b^2}{2\pi\beta kT\lambda_+}} \mathrm{e}^{-\psi_0/(kT)} \left(\frac{kT}{M}\right) \int_{-\infty}^{\infty} \exp\left[-\frac{M\lambda_+}{2\beta kT} v^2\right] \\ &= \frac{\omega_0\omega_b}{\pi(\beta + \sqrt{\beta^2 + 4\omega_0^2})} \mathrm{e}^{-\psi_0/(kT)}. \end{aligned} \tag{3.75}$$

Equation (3.75) is an example of a *thermally activated process*, with activation energy ψ_0.

3.4.3 The field-free case with spatially dependent initial conditions

Free Brownian motion is an important case that we better get right. With $\boldsymbol{K} = 0$ in Eq. (3.59),

$$\left(\frac{\partial}{\partial t} + \boldsymbol{v} \cdot \boldsymbol{\nabla}_{\boldsymbol{r}}\right) P(\boldsymbol{r}, \boldsymbol{v}, t) = \beta \boldsymbol{\nabla}_{\boldsymbol{v}} \cdot \left(\boldsymbol{v} + \frac{kT}{M} \boldsymbol{\nabla}_{\boldsymbol{v}}\right) P(\boldsymbol{r}, \boldsymbol{v}, t). \tag{3.76}$$

As shown in Exercise 3.30, probability is conserved in free Brownian motion (satisfies a continuity equation) and the drag force is reproduced by the solution, an important consistency check, as it was assumed at the outset. Under the banner of free Brownian motion, we consider separately the cases of whether or not a spatial dependence is required.

3.4.3.1 *Field free, no spatial dependence*

When P is independent of position, the field-free Fokker-Planck equation has the form

$$\frac{\partial}{\partial t} P(\boldsymbol{v}, t) = \beta \boldsymbol{\nabla}_{\boldsymbol{v}} \cdot \left(\boldsymbol{v} + \frac{kT}{M} \boldsymbol{\nabla}_{\boldsymbol{v}}\right) P(\boldsymbol{v}, t) \equiv \Omega P(\boldsymbol{v}, t), \tag{3.77}$$

which as a class of problems (time dependence generated by an operator) would suggest finding the eigenvalues and eigenfunctions of Ω. The operator Ω, however, is not self-adjoint (see Exercise 3.31), which 1) complicates the analysis and 2) reflects the irreversible nature of the time evolution of nonequilibrium systems. Consider that in the matrix representation of Ω, its matrix elements $\Omega_{m,n} \equiv \langle \phi_m | \Omega \phi_n \rangle$ (in an orthonormal basis[45]) are not symmetric; we have in general $\Omega_{m,n} = \left(\Omega^\dagger_{n,m}\right)^*$ and if $\Omega^\dagger \neq \Omega$, its matrix elements cannot be symmetric or Hermitian. We see from Eq. (3.77) that Ω involves a divergence of probability currents which represent transitions in velocity space. As shown in Exercise 3.21, the Fokker-Planck equation is equivalent to the master equation, where we encounter a similar issue that the transition matrix $W_{n,n'}$ need not be symmetric (see Section 2.8); in a nonequilibrium system the rate of the transition $n \to n'$ need not be the same as for $n' \to n$. We found, however, a similarity transformation of the transition matrix utilizing the

[45] Sometimes students forget that matrix elements of operators are defined with respect to orthonormal bases. There's no loss of generality; any basis is associated with an orthonormal basis through the Gram-Schmidt process[13, p12].

detailed balance condition, Eq. (2.79). Does detailed balance pertain to the Fokker-Planck equation; is there a regime in which all transitions are balanced? Yes, when the probability current $\boldsymbol{J} = 0$, velocity transitions in one direction are balanced by those in the opposite, a type of detailed balance condition. The *Maxwellian function* $\phi(\boldsymbol{v}) \equiv \exp(-Mv^2/(2kT))$ is associated with zero current:

$$\left(\boldsymbol{v} + \frac{kT}{M}\boldsymbol{\nabla}_{\boldsymbol{v}}\right)\phi(\boldsymbol{v}) = 0.$$

These considerations suggest that to put Ω in self-adjoint form, we apply the same transformation as in Eq. (2.81) using $\phi(\boldsymbol{v})$ as the stationary distribution.

Thus, starting with the eigenvalue problem associated with Ω,

$$\Omega F(\boldsymbol{v}) = \lambda F(\boldsymbol{v}), \tag{3.78}$$

introduce the similarity transformation

$$\left(\frac{1}{\sqrt{\phi(\boldsymbol{v})}}\Omega\sqrt{\phi(\boldsymbol{v})}\right)\left(\frac{F(\boldsymbol{v})}{\sqrt{\phi(\boldsymbol{v})}}\right) = \lambda\left(\frac{F(\boldsymbol{v})}{\sqrt{\phi(\boldsymbol{v})}}\right), \tag{3.79}$$

which we symbolize

$$\widetilde{\Omega}\widetilde{F}(\boldsymbol{v}) = \lambda\widetilde{F}(\boldsymbol{v}). \tag{3.80}$$

It's straightforward to show that

$$\widetilde{\Omega} = \beta\left(\frac{kT}{M}\nabla_{\boldsymbol{v}}^2 - \frac{M}{4kT}v^2 + \frac{3}{2}\right), \tag{3.81}$$

which is self-adjoint.

With $\widetilde{\Omega}$ as in Eq. (3.81), Eq. (3.80) has the form of the time-independent Schrödinger equation for the three-dimensional isotropic harmonic oscillator (see Exercise 3.33), the solutions of which are well known. Using Eq. (P3.20) for $\widetilde{F}(\boldsymbol{v})$, with $F(\boldsymbol{v}) = \sqrt{\phi(\boldsymbol{v})}\widetilde{F}(\boldsymbol{v})$, the eigenfunctions of Ω are labeled by three non-negative integers n_1, n_2, n_3,

$$F_{n_1,n_2,n_3}(\boldsymbol{v}) = A_{n_1,n_2,n_3}\exp\left(-Mv^2/(2kT)\right)H_{n_1}(v_1/\alpha)H_{n_2}(v_2/\alpha)H_{n_3}(v_3/\alpha), \tag{3.82}$$

where $\alpha = \sqrt{2kT/M}$, A_{n_1,n_2,n_3} is a constant, and the functions H_n are Hermite polynomials. The associated eigenvalues $\lambda_{n_1,n_2,n_3} = -\beta(n_1 + n_2 + n_3)$. For simplicity we refer to the integers n_1, n_2, n_3 collectively as $\boldsymbol{n}$, and we write Eq. (3.82) in compressed form

$$F_{\boldsymbol{n}}(\boldsymbol{v}) = A_{\boldsymbol{n}}\exp\left(-Mv^2/(2kT)\right)H_{\boldsymbol{n}}\left(\boldsymbol{v}/\sqrt{2kT/M}\right). \tag{3.83}$$

Equation (3.83) is the same as Eq. (3.82). Hermite polynomials have the orthogonality property[46]

$$\int_{-\infty}^{\infty} \mathrm{e}^{-x^2}H_n(x)H_m(x)\mathrm{d}x = \sqrt{\pi}2^n n!\delta_{n,m}. \tag{3.84}$$

When we choose

$$A_{n_1,n_2,n_3} = \left[2^{n_1+n_2+n_3}n_1!n_2!n_3!\left(2\pi kT/M\right)^{3/2}\right]^{-1/2},$$

the eigenfunctions $F_{\boldsymbol{n}}$ have the orthonormality relation,[47]

$$\int \mathrm{d}^3v\frac{1}{\phi(\boldsymbol{v})}F_{\boldsymbol{n}}(\boldsymbol{v})F_{\boldsymbol{m}}(\boldsymbol{v}) = \delta_{\boldsymbol{n},\boldsymbol{m}}. \tag{3.85}$$

Compare Eq. (3.85) with Eq. (2.82).

[46] Dennery and Krzywicki[77] is a good source on orthogonal polynomials, including the Hermite polynomials. We're using the "physicist" Hermite polynomials $H_n(x)$, and not that of the "probabilists," $He_n(x)$.

[47] The functions $F_{\boldsymbol{n}}$ already have a Maxwellian built into them. To match up to the orthogonality condition Eq. (3.84), we have to divide by $\phi(\boldsymbol{v})$.

Armed with the eigenvalues and eigenfunctions of Ω, we have the solution of the Fokker-Planck equation (3.77):

$$P(\boldsymbol{v},t) = \sum_n C_n F_n(\boldsymbol{v}) \exp(\lambda_n t), \tag{3.86}$$

where the expansion coefficients are obtained from the initial condition

$$C_n = \int \mathrm{d}^3 v \frac{1}{\phi(\boldsymbol{v})} F_n(\boldsymbol{v}) P(\boldsymbol{v},0). \tag{3.87}$$

For a particle known to have velocity $\boldsymbol{v}_0$ at time $t = 0$, $P(\boldsymbol{v},0) = \delta(\boldsymbol{v}-\boldsymbol{v}_0)$, implying that $C_n = F_n(\boldsymbol{v}_0)/\phi(\boldsymbol{v}_0)$. In that case, we can write Eq. (3.86) as

$$P(\boldsymbol{v},t|\boldsymbol{v}_0) = \frac{1}{\phi(\boldsymbol{v}_0)} \sum_n F_n(\boldsymbol{v}_0) F_n(\boldsymbol{v}) \exp(\lambda_n t). \tag{3.88}$$

The series in Eq. (3.88) can be summed using a property of Hermite polynomials, *Mehler's formula*, that for $|z| < 1$,[48]

$$\sum_{n=0}^{\infty} \left(\frac{z}{2}\right)^n \frac{1}{n!} H_n(x) H_n(y) = \frac{1}{\sqrt{1-z^2}} \exp\left(\frac{2xyz - z^2(x^2+y^2)}{1-z^2}\right). \tag{3.89}$$

Using Eq. (3.89) in Eq. (3.88), we find

$$P(\boldsymbol{v},t|\boldsymbol{v}_0) = \left[\frac{M}{2\pi kT(1-\mathrm{e}^{-2\beta t})}\right]^{3/2} \exp\left[-\frac{M}{2kT(1-\mathrm{e}^{-2\beta t})}\left(\boldsymbol{v}-\boldsymbol{v}_0 e^{-\beta t}\right)^2\right], \tag{3.90}$$

a three-dimensional version of the Ornstein-Uhlenbeck process (compare with Eq. (3.49)). We previously found the Ornstein-Uhlenbeck transition probability using the method of characteristics (see Section 3.3.2), a way of solving first-order partial differential equations. Here we've solved the problem by finding the eigenvalues and eigenfunctions of the Fokker-Planck operator.

3.4.3.2 *Field free, spatially dependent*

A spatial dependence of the nonequilibrium probability distribution can occur when a particle's initial position is specified as well as its initial velocity. The theoretical treatment of such systems builds on our previous analysis. Start with a Fourier transform with respect to spatial variables,

$$\widetilde{P}(\boldsymbol{k},\boldsymbol{v},t) \equiv \int \mathrm{d}^3 x \exp(\mathrm{i}\boldsymbol{k}\cdot\boldsymbol{r}) P(\boldsymbol{r},\boldsymbol{v},t). \tag{3.91}$$

Using Eq. (3.76) (the relevant Fokker-Planck equation), it's straightforward to show that

$$\frac{\partial}{\partial t}\widetilde{P}(\boldsymbol{k},\boldsymbol{v},t) = (\Omega + \mathrm{i}\boldsymbol{k}\cdot\boldsymbol{v})\,\widetilde{P}(\boldsymbol{k},\boldsymbol{v},t). \tag{3.92}$$

Thus we have an equation like Eq. (3.77) but with a modified operator, $\Omega \to \Omega' = \Omega + \mathrm{i}\boldsymbol{k}\cdot\boldsymbol{v}$. We write the associated eigenvalue problem

$$(\Omega + \mathrm{i}\boldsymbol{k}\cdot\boldsymbol{v})\, F' = \lambda' F', \tag{3.93}$$

[48] Mehler's formula is not a standard topic, even though it was published in 1866 (Szego[78] cites the original reference). Rainville[79, p197] supplies a derivation. It's listed without attribution as Theorem 53 in Titchmarsh[80, p77]. Mehler's formula works in summing Eq. (3.88) because the eigenvalue spectrum of Ω is equally spaced, like the harmonic oscillator. Indeed, it's used to find the quantum-mechanical propagator for the harmonic oscillator (see Sakurai[81, p119]).

where the primes distinguish the eigenfunctions and eigenvalues from those in Eq. (3.78). Perform the same transformation on Eq. (3.93) as with Eq. (3.79). We find

$$\left(\widetilde{\Omega} + \mathrm{i}\boldsymbol{k} \cdot \boldsymbol{v}\right) f = \lambda' f, \tag{3.94}$$

where $f(\boldsymbol{v}) \equiv F'(\boldsymbol{v})/\sqrt{\phi(\boldsymbol{v})}$ and $\widetilde{\Omega}$ is given in Eq. (3.81). Equation (3.94) is equivalent to

$$\left[\frac{kT}{M}\nabla_v^2 - \frac{M}{4kT}v^2 + \frac{3}{2} + \mathrm{i}\beta^{-1}\boldsymbol{k}\cdot\boldsymbol{v}\right] f = \beta^{-1}\lambda' f. \tag{3.95}$$

Let $v^2 \to v^2 - 4\mathrm{i}D\boldsymbol{k}\cdot\boldsymbol{v} - 4D^2k^2 + 4\mathrm{i}D\boldsymbol{k}\cdot\boldsymbol{v} + 4D^2k^2$. Using Eq. (3.20), we have the identity

$$-\frac{M}{4kT}v^2 = -\frac{M}{4kT}\left(\boldsymbol{v} - 2\mathrm{i}D\boldsymbol{k}\right)^2 - \mathrm{i}\beta^{-1}\boldsymbol{k}\cdot\boldsymbol{v} - \beta^{-1}Dk^2.$$

Equation (3.95) is equivalent to

$$\left[\frac{kT}{M}\nabla_v^2 - \frac{M}{4kT}\left(\boldsymbol{v} - 2\mathrm{i}D\boldsymbol{k}\right)^2 + \frac{3}{2}\right] f = \beta^{-1}\left(\lambda' + Dk^2\right) f. \tag{3.96}$$

Comparing Eq. (3.96) with Eq. (P3.16), we see that if $\widetilde{F}(\boldsymbol{v})$ is a solution of Eq. (P3.16) having eigenvalue λ, then $f(\boldsymbol{v}) = \widetilde{F}(\boldsymbol{v} - 2\mathrm{i}D\boldsymbol{k})$ is a solution of Eq. (3.96) with $\lambda' = \lambda - Dk^2$. With $F'(\boldsymbol{v}) = \sqrt{\phi(\boldsymbol{v})}f(\boldsymbol{v})$ and $\widetilde{F}(\boldsymbol{v}) = F(\boldsymbol{v})/\sqrt{\phi(\boldsymbol{v})}$, we have that

$$F'_n(\boldsymbol{v}) = \sqrt{\frac{\phi(\boldsymbol{v})}{\phi(\boldsymbol{v} - 2\mathrm{i}D\boldsymbol{k})}}F_n(\boldsymbol{v} - 2\mathrm{i}D\boldsymbol{k}) \tag{3.97}$$

is an eigenfunction of $\Omega + \mathrm{i}\boldsymbol{k}\cdot\boldsymbol{v}$ with eigenvalue $\lambda'_n = \lambda_n - Dk^2$. The orthonormality condition is the same as Eq. (3.85),

$$\int \mathrm{d}^3v \frac{1}{\phi(\boldsymbol{v})}F'_n(\boldsymbol{v})F'_m(\boldsymbol{v}) = \delta_{n,m}.$$

We therefore have the solution to Eq. (3.92),

$$\widetilde{P}(\boldsymbol{k}, \boldsymbol{v}, t) = \sum_n C'_n F'_n(\boldsymbol{v})\exp(\lambda'_n t), \tag{3.98}$$

where the coefficients C'_n are found from initial conditions, just as in Eq. (3.87). (The right side of Eq. (3.98) should indicate a dependence on $\boldsymbol{k}$, but it doesn't. F'_n, C'_n, and λ'_n each depend on $\boldsymbol{k}$.)

Consider a B-particle that at time $t = 0$ is at $\boldsymbol{r} = 0$ with velocity $\boldsymbol{v} = \boldsymbol{v}_0$, i.e., $P(\boldsymbol{r}, \boldsymbol{v}, 0) = \delta(\boldsymbol{r})\delta(\boldsymbol{v} - \boldsymbol{v}_0)$, implying $\widetilde{P}(\boldsymbol{k}, \boldsymbol{v}, 0) = \delta(\boldsymbol{v} - \boldsymbol{v}_0)$, in turn implying $C'_n = F'_n(\boldsymbol{v}_0)/\phi(\boldsymbol{v}_0)$. We can therefore write Eq. (3.98) as

$$\widetilde{P}(\boldsymbol{k}, \boldsymbol{v}, t|\boldsymbol{v}_0) = \frac{1}{\phi(\boldsymbol{v}_0)}\sum_n F'_n(\boldsymbol{v}_0)F'_n(\boldsymbol{v})\exp(\lambda'_n t), \tag{3.99}$$

the analog of Eq. (3.88). Equation (3.99) can be summed using Eq. (3.89). We find

$$\widetilde{P}(\boldsymbol{k}, \boldsymbol{v}, t|\boldsymbol{v}_0) = P(\boldsymbol{v}, t|\boldsymbol{v}_0)\mathrm{e}^{-Dk^2t}\exp\left[\beta^{-1}\tanh(\beta t/2)\left\{\mathrm{i}\boldsymbol{k}\cdot(\boldsymbol{v}_0 + \boldsymbol{v}) + 2Dk^2\right\}\right], \tag{3.100}$$

where $P(\boldsymbol{v}, t|\boldsymbol{v}_0)$ is given in Eq. (3.90). To find $P(\boldsymbol{r}, \boldsymbol{v}, t|\boldsymbol{v}_0)$ we have to evaluate the inverse Fourier transform, but the Fourier transform of a Gaussian is another Gaussian. We find

$$P(\boldsymbol{r}, \boldsymbol{v}, t|\boldsymbol{v}_0) = P(\boldsymbol{v}, t|\boldsymbol{v}_0)\frac{1}{(2\pi\sigma^2)^{3/2}}\exp\left[-\frac{1}{2\sigma^2}\left(\boldsymbol{r} - \beta^{-1}\tanh(\beta t/2)(\boldsymbol{v} + \boldsymbol{v}_0)\right)^2\right] \tag{3.101}$$

where $\sigma^2 \equiv 2D\left(t - 2\beta^{-1}\tanh(\beta t/2)\right)$.

3.4.4 Strong damping regime, Smoluchowski equation

Consider a system such that $\beta|\boldsymbol{v}| \gg |\dot{\boldsymbol{v}}|$. The friction in such a system is sufficiently strong that any accelerations produced by forces—random and external—are rapidly damped, resulting in B-particles having small accelerations. Under the approximation $\dot{\boldsymbol{v}} = 0$, the Langevin equation reduces to

$$\frac{\mathrm{d}\boldsymbol{r}}{\mathrm{d}t} = \frac{1}{\beta}\left(\boldsymbol{K}(\boldsymbol{r}) + \boldsymbol{A}(t)\right),$$

where $\boldsymbol{A}(t)$ is the fluctuating acceleration imposed by the medium. What is the form of the Fokker-Planck equation in this case? Because we're in a regime of small time variations in $\boldsymbol{v}$, we expect the velocity distribution to rapidly thermalize, even though the position distribution will remain far from equilibrium for a longer time. We seek therefore a Fokker-Planck equation for probability distributions depending only on $(\boldsymbol{r}, t)$.

We start with the Fokker-Planck equation (3.59) written in the form (restricting our analysis to one dimension),

$$\begin{aligned}\overline{\Omega}P(x,v,t) &\equiv \frac{\partial}{\partial v}\left(v + \frac{kT}{M}\frac{\partial}{\partial v}\right)P(x,v,t) \\ &= \beta^{-1}\left(\frac{\partial}{\partial t} + v\frac{\partial}{\partial x} + K\frac{\partial}{\partial v}\right)P(x,v,t) \equiv \beta^{-1}LP(x,v,t).\end{aligned} \tag{3.102}$$

where the first equality defines the operator $\overline{\Omega}$ (with $\beta\overline{\Omega} \equiv \Omega$ in Eq. (3.60)), and the last equality defines the operator L. We seek the small β^{-1} form of the Fokker-Planck equation.[49] We assume that an expansion in powers of β^{-1} can be developed:

$$P(x,v,t) = P^{(0)}(x,v,t) + \beta^{-1}P^{(1)}(x,v,t) + \beta^{-2}P^{(2)}(x,v,t) + \cdots. \tag{3.103}$$

Combining Eq. (3.103) with Eq. (3.102), we find the typical perturbative scheme where the term in the expansion at each order in β^{-1} is found from the terms preceding it,

$$\begin{aligned}\overline{\Omega}P^{(0)}(x,v,t) &= 0 \\ \overline{\Omega}P^{(i+1)}(x,v,t) &= LP^{(i)}(x,v,t). \qquad (i = 0,1,\dots)\end{aligned} \tag{3.104}$$

One might try to solve these equations by finding the Green function associated with $\overline{\Omega}$; such an approach, however, is complicated by the fact that $\overline{\Omega}$ has an eigenvector corresponding to zero eigenvalue. One has to ensure at each order that $LP^{(i)}$ is orthogonal to the null space of $\overline{\Omega}$ ([13, p259]). This program is illustrated in Exercise 3.36. We find to first order in β^{-1} (see Eq. (P3.29)),

$$\frac{\partial}{\partial t}P(x,t) = -\beta^{-1}\frac{\partial}{\partial x}\left[KP(x,t)\right] + D\frac{\partial^2}{\partial x^2}P(x,t), \tag{3.105}$$

where $P(x,t) \equiv \int P(x,v,t)\mathrm{d}v$.

Equation (3.105) is the *Smoluchowski equation* (derived in 1915), a generalized diffusion equation that includes a drift term associated with external forces. Note that it has the form of a continuity equation, implying the probability current density[50]

$$J = \beta^{-1}KP - D\frac{\partial P}{\partial x}. \tag{3.106}$$

There are convective as well as diffusive contributions to the probability current.

---

[49] We can't simply let $\beta \to \infty$ in Eq. (3.102) and be done with it because the solution $P(x,v,t)$ also depends on β.

[50] Equation (3.106) becomes a vector relation in three dimensions, $\boldsymbol{J} = \beta^{-1}\boldsymbol{K}P - D\boldsymbol{\nabla}P$.

3.4.5 Uniform field

For $\boldsymbol{K} = \boldsymbol{g} =$ constant, we have from Eq. (3.59)

$$\frac{\partial P}{\partial t} + \boldsymbol{v} \cdot \boldsymbol{\nabla}_r P + \boldsymbol{g} \cdot \boldsymbol{\nabla}_v P = \beta \boldsymbol{\nabla}_v \cdot \left(\boldsymbol{v} + \frac{kT}{M} \boldsymbol{\nabla}_v \right) P. \tag{3.107}$$

In allowing for the effects of gravity, the acceleration in a suspension medium must take buoyancy into account: $\boldsymbol{g} = (1 - \rho_0/\rho)\, \boldsymbol{g}_0$ where ρ, ρ_0 are the mass densities of B-particles and the surrounding fluid, $\boldsymbol{g}_0$ is the free-space gravitational acceleration at the earth's surface, and the coordinate system has been chosen so that the z-axis is directed upwards, opposite the direction of gravity. For definiteness, take $\rho > \rho_0$ so that $\boldsymbol{g} = -g\hat{\boldsymbol{z}}$ and the terminal velocity $\boldsymbol{v}_T = \boldsymbol{g}/\beta$ (see Section 3.1.3) is directed downward.

Because $\boldsymbol{g}$ is a constant vector, it can be placed on the right side of the equation,

$$\left(\frac{\partial}{\partial t} + \boldsymbol{v} \cdot \boldsymbol{\nabla}_r \right) P(\boldsymbol{r}, \boldsymbol{v}, t) = \beta \boldsymbol{\nabla}_v \cdot \left(\boldsymbol{v} - \boldsymbol{v}_T + \frac{kT}{M} \boldsymbol{\nabla}_v \right) P(\boldsymbol{r}, \boldsymbol{v}, t). \tag{3.108}$$

Equation (3.108) suggests working in a reference frame falling with the terminal velocity, because in that frame the effect of gravity on Brownian motion is eliminated.[51] This allows us to infer, for example, that Einstein's formula for the mean-square displacement (in three dimensions) generalizes to $\langle [\Delta \boldsymbol{r} - \boldsymbol{v}_T t]^2 \rangle = 6Dt$. Although beautiful physics, it's not terribly practical: A B-particle in a container cannot fall indefinitely.

To illustrate the effects of gravity, let's apply the Smoluchowski equation, which in a high-friction environment pertains to the phenomenon of sedimentation. With $\boldsymbol{K} = -g\hat{\boldsymbol{z}}$, Eq. (3.105) has the form

$$\frac{\partial}{\partial t} P(z,t) = (g/\beta) \frac{\partial}{\partial z} P(z,t) + D \frac{\partial^2}{\partial z^2} P(z,t) \tag{3.109}$$

Diffusion in the (x, y) plane takes place just as in the field-free case, and thus we restrict our attention to the z-direction. Equation (3.109) must be supplemented with appropriate boundary conditions. Suppose the B-particle is initially at $z = z_0$, implying that $\lim_{t \to 0} P(z,t) = \delta(z - z_0)$. Let the bottom of the container be the plane at $z = 0$, a boundary at which the normal component of the probability current must vanish, implying from Eq. (3.106) the mixed boundary condition[52]

$$\left. \frac{\partial}{\partial z} P(z,t) \right|_{z=0} = -\frac{Mg}{kT} P(0,t). \tag{3.110}$$

The solution to Eq. (3.109) with these boundary and initial conditions has been given by Chandrasekhar[82, p58], to which we refer.

SUMMARY

We introduced two methods for the dynamics of stochastic processes, the Langevin and Fokker-Planck equations, using Brownian motion as a theme.

- The Langevin equation is Newton's second law of motion involving two types of forces, the systematic force that affects the average motion of particles and the random force $F(t)$ representing interactions with microscopic degrees of freedom. We're unable to specify the form of $F(t)$ other than listing its statistical properties. The ensemble average vanishes

[51] Under $\boldsymbol{r} \to \boldsymbol{r}' = \boldsymbol{r} - \boldsymbol{v}_T t$ (Galilean transformation), $\boldsymbol{v} \to \boldsymbol{v}' = \boldsymbol{v} - \boldsymbol{v}_T$. In the "primed" frame, falling with the terminal velocity, Brownian motion is the same as in the field-free case. This is close to Einstein's equivalence principle that one cannot distinguish the effect of a uniform gravitational field from an accelerated frame in the absence of gravity[6, p187].

[52] We're using the term mixed boundary condition as it occurs in Sturm-Liouville theory; see [13, p58].

$\langle F(t)\rangle = 0$ and the second moment $\langle F(t)F(t')\rangle$ is determined by demanding concordance with the equipartition theorem. For the random force, we expect almost no correlation between impacts at different times; a common strategy is to take delta-correlated noise with $\langle F(t)F(t')\rangle = \Gamma\delta(t-t')$ with the strength Γ determined so that the equipartition theorem is satisfied.

- The motion of Brownian particles is inertial at short times and diffusive at long times.
- There is a connection between a system's resistive or damping process and the extent to which fluctuations are correlated [Eq. (3.7)]. Wherever there is damping there are fluctuations, and vice-versa.
- The Einstein relations, Eqs. (3.20) and (3.23), relate transport coefficients to temperature and other system parameters.
- The Fokker-Planck equation (either Eq. (3.36) or Eq. (3.37)) is a partial differential equation for time-dependent probabilities, either $P_1(y,t)$ or the transition probability $P_{1|1}(y,t|y_0,t_0)$. The stochastic process describing the velocity of a Brownian particle, Eq. (3.49), the Ornstein-Uhlenbeck process, is the only process that is simultaneously Markov, Gaussian, and stationary. Brownian motion in force fields was treated in Section 3.4.

EXERCISES

3.1 Chandrasekhar, in *Stochastic Problems in Physics and Astronomy*,[53] states, "Under normal conditions, in a liquid, a Brownian particle will suffer about 10^{21} collisions per second ...." Estimate the number of collisions per second suffered by a spherical B-particle of radius 1 μm suspended in water at room temperature. Do the calculation two ways.

a. Assume that water molecules of mass m have the thermal speed $\sqrt{kT/m}$.

b. That formula, however, pertains to gases, and liquids are not gases. The average interparticle separation in water (a condensed phase) is $a \approx 3\times 10^{-10}$ m (obtained from the density). Molecular motions in liquids are diffusive. An estimate for the speed of a diffusing particle can be had from Eq. (2.59), $(2D/a) = (a/\tau) \approx v$. The self-diffusion coefficient of water at room temperature is $\approx 2\times 10^{-9}$ m^2/s, implying $v \approx 10$ m/s. This estimate is an order of magnitude smaller than the first.

Using the lower estimate of the speed, a 1μm particle undergoes $\approx 10^{18}$ collisions per second. Chandrasekhar didn't specify a particle size.

3.2 a. Show that the following expression solves Eq. (3.1) for $t \geq 0$, where $V(t)$ is assumed to be known,

$$I(t) = I_0 \mathrm{e}^{-t/\tau_c} + \frac{1}{L}\int_0^t \mathrm{e}^{-(t-t')/\tau_c} V(t')\mathrm{d}t', \tag{P3.1}$$

and where I_0 is the current at an arbitrarily chosen instant of time (stationary processes have no unique origin in time). Use the Leibniz integral rule.

b. Define a time average $\overline{I} \equiv \lim_{T\to\infty}(1/T)\int_0^T I(t)\mathrm{d}t$. Show that $\overline{I} = 0$.

c. Equation (P3.1) holds for any initial condition. Positive values of I_0 are as equally likely as negative, and hence $\langle I_0\rangle = 0$. Show that $\langle I(t)\rangle = 0$. Hint: $\langle V(t)\rangle = 0$ for any time t.

[53] See Chandrasekhar[82], reprinted in Wax[44, pp3–91]. Chandrasekhar received the 1983 Nobel Prize in Physics.

3.3 Derive Eq. (3.3).

3.4 Suppose in Eq. (3.5) that I_0^2 is replaced with its value obtained from the equipartition theorem. In that case, what is the time evolution of $\langle I^2(t)\rangle$? Systems in equilibrium stay in equilibrium.

3.5 Derive Eq. (3.11) from Eq. (3.9). Hint: Take the ensemble average of Eq. (3.9); use Eqs. (2.21) and (3.10).

3.6 We formulated the Langevin equation in terms of scalar quantities, but it also applies to vectors. Treat the velocity $\boldsymbol{v}$ of a B-particle as a random variable subject to the stochastic differential equation

$$M\frac{\mathrm{d}\boldsymbol{v}}{\mathrm{d}t} = -\alpha\boldsymbol{v} + \boldsymbol{F}, \tag{P3.2}$$

where $\boldsymbol{F}$ is the random force considered a vector quantity. Equation (P3.2) implies three scalar equations, $M\mathrm{d}v_i/\mathrm{d}t = -\alpha v_i + F_i$, $i = 1, 2, 3$.

a. Show that the solution of Eq. (P3.2) with initial condition $\boldsymbol{v}(0) = \boldsymbol{v}_0$ is

$$\boldsymbol{v}(t) = \mathrm{e}^{-(\alpha/M)t}\left(\boldsymbol{v}_0 + \frac{1}{M}\int_0^t \mathrm{d}t'\mathrm{e}^{(\alpha/M)t'}\boldsymbol{F}(t')\right), \tag{P3.3}$$

the vector analog of Eq. (P3.1). The subensemble average $\langle\boldsymbol{F}\rangle_{\boldsymbol{v}_0} = 0$ for all $\boldsymbol{v}_0$, implying $\langle F_i\rangle_{\boldsymbol{v}_0} = 0$, $i = 1, 2, 3$. Thus, $\langle\boldsymbol{v}(t)\rangle_{\boldsymbol{v}_0} = \boldsymbol{v}_0\exp(-\alpha t/M)$, the analog of Eq. (3.11).

b. Use Eq. (P3.3) to show that

$$\begin{aligned}&\langle v_i(t_1)v_j(t_2)\rangle_{\boldsymbol{v}_0}\\ &= \mathrm{e}^{-(\alpha/M)(t_1+t_2)}\left[v_{0i}v_{0j} + \frac{1}{M^2}\int_0^{t_1}\int_0^{t_2}\mathrm{d}t'\mathrm{d}t''\mathrm{e}^{(\alpha/M)(t'+t'')}\langle F_i(t')F_j(t'')\rangle_{\boldsymbol{v}_0}\right].\end{aligned} \tag{P3.4}$$

Because the random force is independent of initial conditions, no harm would be done in erasing the subscript $\boldsymbol{v}_0$ on $\langle F_i(t')F_j(t'')\rangle_{\boldsymbol{v}_0}$, but let's leave it there. Note that the formula for $\langle v_i(t_1)v_j(t_2)\rangle_{\boldsymbol{v}_0}$ depends on t_1 and t_2 separately (instead of on their difference) because the values of $v_i(t_1)$ and $v_j(t_2)$ depend on when the velocity was $\boldsymbol{v}_0$, i.e., there is a preferred origin in time associated with the initial condition.

c. Equation (P3.4) indicates an average computed over a subensemble of systems having the same initial condition $\boldsymbol{v} = \boldsymbol{v}_0$. The full ensemble average is found by averaging over initial conditions. This is sometimes indicated as a double average $\langle\langle\ \rangle_{\boldsymbol{v}_0}\rangle$, a notation we won't adopt; we simply let $\langle\ \rangle$ denote the full ensemble average. In an average over all possible initial conditions, $\langle v_{0i}v_{0j}\rangle$ is zero for $i \neq j$ but nonzero when $i = j$,

$$\langle v_{0i}v_{0j}\rangle = \frac{1}{3}\delta_{ij}\langle v_0^2\rangle,$$

where we've used isotropy, $\langle v_{0x}^2\rangle = \langle v_{0y}^2\rangle = \langle v_{0z}^2\rangle$. We also have the generalization of Eq. (3.13),

$$\langle F_i(t')F_j(t'')\rangle = \delta_{ij}\Gamma\delta(t' - t''),$$

where the strength Γ is to be determined. Show from Eq. (P3.4) that

$$\langle v_i(t_1)v_j(t_2)\rangle = \delta_{ij}\left[\mathrm{e}^{-(\alpha/M)(t_1+t_2)}\left(\frac{1}{3}\langle v_0^2\rangle - \frac{\Gamma}{2\alpha M}\right) + \frac{\Gamma}{2\alpha M}\mathrm{e}^{-(\alpha/M)|t_1-t_2|}\right]. \tag{P3.5}$$

Stationarity requires that $\Gamma/(2\alpha M) = \langle v_0^2\rangle/3$, implying $\Gamma = 2\alpha kT$, just as in the scalar case, Eq. (3.14). The same result is obtained by setting $t_1 = t_2 \equiv t$ and letting $t \to \infty$,

$$\lim_{t\to\infty}\langle v_i^2(t)\rangle = \frac{\Gamma}{2\alpha M} = \langle v_i^2\rangle = \frac{kT}{M},$$

which is another form of the stationarity requirement. For the full ensemble average, we have a nice formula

$$\langle v_i(t_1)v_j(t_2)\rangle = \delta_{ij}\frac{kT}{M}\mathrm{e}^{-(\alpha/M)|t_1-t_2|}.$$

3.7 What are the dimensions of Γ, the strength of the random force autocorrelation function? Hint: What is the dimension of the Dirac delta function in Eq. (3.13)?

3.8 Derive Eq. (3.16). Note that the derivation requires the operations of taking a derivative and forming expectation values to commute; see Section 2.3.2.

3.9 Derive Eq. (3.17).

3.10 Derive Eq. (3.18). Hint: Expand all terms in Eq. (3.17) to $O(t^3)$.

3.11 Calculate the variance of displacements in Brownian motion from the Langevin equation.

a. Find the average displacement of a B-particle with initial speed v_0. Integrate Eq. (3.11) to show

$$\langle x(t) - x_0\rangle_{v_0} = v_0\tau_c\left(1 - \mathrm{e}^{-t/\tau_c}\right). \tag{P3.6}$$

Recognize that you're invoking Eq. (2.21), $\mathrm{d}\langle x(t)\rangle/\mathrm{d}t = \langle v\rangle$. Equation (P3.6) can be interpreted as the average distance traveled in time t with velocity $\langle v(t)\rangle_{v_0} = v_0\mathrm{e}^{-t/\tau_c}$. Note that $\langle x(t) - x_0\rangle_{v_0} \sim v_0\tau_c$ for $t \gg \tau_c$.

b. Now average over v_0 (see discussion in Exercise 3.6). Show that $\langle x(t) - x_0\rangle = 0$.

c. Average Eq. (3.17) over v_0.

i. Show that the variance is

$$\langle (x(t) - x_0)^2\rangle = \frac{2kTM}{\alpha^2}\left[\frac{t}{\tau_c} - 1 + \mathrm{e}^{-t/\tau_c}\right]. \tag{P3.7}$$

Hint: $\langle v_0^2\rangle = kT/M$. (We're using the scalar version of the Langevin equation.)

ii. Show from Eq. (P3.7) that

$$\langle (x(t) - x_0)^2\rangle \sim \begin{cases} \dfrac{kT}{M}t^2 & t \ll \tau_c \\ \dfrac{2kT}{\alpha}t. & t \gg \tau_c \end{cases} \tag{P3.8}$$

The variance of the diffusion process (obtained from the solution of the diffusion equation (2.61)), which grows linearly with time (as we see from Eq. (2.62), $\langle x^2(t)\rangle = 2Dt$, but also from the Gaussian form of Eq. (2.60)), is recovered in the theory of Brownian motion at long times (as we see from Eq. (P3.8)) when the Einstein relation Eq. (3.20) is invoked. Einstein was aware that his result ($\langle x^2(t)\rangle = 2Dt$) could not apply for short times. His reasoning is characteristically insightful:[48, p34] "The reason is that we have implicitly assumed in our development that the events during the time t are to be looked upon as phenomena independent of the events in the time immediately preceding. But this assumption becomes harder to justify the smaller the time t is chosen."

3.12 Show that Eq. (3.20) yields the correct dimensions for the diffusion coefficient.

3.13 In an experiment on the Brownian motion of a particle of radius 0.4 μm in a water-glycerine solution having viscosity $\nu = 0.0278$ Poise at a temperature of 18.8 °C, the observed x-component of the displacement in a 10 second interval was $\langle x^2(t)\rangle = 3.3 \times 10^{-12}$ m^2. A Poise is a non-SI unit of viscosity (named after Poiseuille), 1 Poise = 0.1 Pascal-second. Use these data to estimate the value of Boltzmann's constant. Hint: Use the Stokes formula.

3.14 In one of Perrin's experiments, Brownian motion of mastic grains (a type of plant resin) of radius 0.52 μm and mass 6.5×10^{-16} kg was observed in water of viscosity $\nu = 0.01$ Poise[62, p123]. Calculate the relaxation time $\tau_c \equiv M/\alpha$, where α is the friction coefficient. Use the Stokes formula. A: 6.6×10^{-9} s. Show that the ratio M/m of the mass of the B-particle to the mass of a water molecule is (in this case) $\approx 2 \times 10^{10}$, underscoring that Brownian motion is the cumulative effect of random weak impulses.

3.15 Consider a stochastic process $\{X(t), t \geq 0\}$ that has stationary independent increments (see Section 3.2) such that $\langle X(t)\rangle = 0$ for $t \geq 0$. This process has several useful properties.

a. Because $\langle X(t)\rangle = 0$, the second moment $\langle[X(t) - X(0)]^2\rangle \equiv f(t)$ is the variance. Show that

$$f(t_1 + t_2) = f(t_1) + f(t_2). \tag{P3.9}$$

Hint: $f(t_1 + t_2) = \langle\left[X(t_1 + t_2) - X(t_1) + X(t_1) - X(0)\right]^2\rangle$; use the statistical independence of increments and invoke stationarity. Equation (P3.9) is *Cauchy's functional equation* (see Aczel[58]), the only solution of which is a linear function $f(x) = \alpha x$ for constant α (proved by Cauchy in 1821). The variance therefore grows linearly with time, $f(t) = \alpha t$, where $\alpha > 0$ is a constant determined by the physics of the problem. For example, when $X(t)$ refers to the displacement of a B-particle, we infer from Eq. (2.62) that $\alpha = 2D$.

b. Show that the covariance

$$K(s,t) \equiv \langle X(s)X(t)\rangle = \alpha \min(s,t), \tag{P3.10}$$

where min denotes the minimum of s and t.
Hint: $\langle X(s)X(t)\rangle = \langle X(s)\left[X(t) - X(s) + X(s)\right]\rangle$.

3.16 Consider the differential form $\mathrm{d}v + \beta v \mathrm{d}t$ in Eq. (3.25) (Doob's form of the Langevin equation). Is it exact? (See [2, p6].) Show that $f(t) = \mathrm{e}^{\beta t}$ is an integrating factor.

3.17 Show that Eq. (3.29) with the choice $f(t) = \mathrm{e}^{\beta t}$ implies Eq. (3.26).

3.18 Derive Eq. (3.35).

3.19 Derive Eq. (3.37) from Eq. (3.36).

3.20 The Fokker-Planck equation associated with the Wiener-Levy process is the diffusion equation without drift. Calculate the transition moments for this process; see Eq. (3.34). A: $M_{2n+1} = 0$ $(n = 0, 1, \ldots)$, $M_2 = 2D$, $M_{2n, n\geq 2} = 0$.

3.21 Guided exercise: In this exercise, we show how the Fokker-Planck equation follows from the master equation, and then discuss some implications. From Eq. (2.66), reproduced here, erasing the subscript on $P_1(y,t)$, we have the master equation

$$\frac{\partial}{\partial t}P(y,t) = \int_{-\infty}^{\infty} \mathrm{d}x \left[W(y|x)P(x,t) - W(x|y)P(y,t)\right], \tag{2.66}$$

where the transition rate $W(y|x)$ is the probability per unit time of the transition $x \to y$.

a. The *jump length* $\xi \equiv y - x$ is the change in a random variable x in the transition $x \to y$. Make the change in variable $x = y - \xi$ and show that the master equation has the form

$$\frac{\partial}{\partial t}P(y,t) = \int_{-\infty}^{\infty} \mathrm{d}\xi \left[W(y|y-\xi)P(y-\xi,t) - W(y-\xi|y)P(y,t)\right].$$

b. Define a function of two arguments $g(a,b) \equiv W(a+b|b)$. Show that

$$\frac{\partial}{\partial t}P(y,t) = \int_{-\infty}^{\infty} \mathrm{d}\xi \left[g(\xi,y-\xi)P(y-\xi,t) - g(\xi,y)P(y,t)\right],$$

where in the second integral we've let $\xi \to -\xi$.

c. Let $u \equiv y - \xi$. Develop $g(\xi,u)P(u,t)$ in a power series in u about $u = y$ ($\xi = 0$),

$$g(\xi,u)P(u,t) = g(\xi,y)P(y,t) + \sum_{n=1}^{\infty} \frac{1}{n!}(-\xi)^n \left(\frac{\partial^n}{\partial u^n} \left[g(\xi,u)P(u,t)\right]\bigg|_{u=y} \right).$$

Show that

$$\frac{\partial}{\partial t}P(y,t) = \sum_{n=1}^{\infty} \frac{(-1)^n}{n!} \frac{\partial^n}{\partial y^n} \left[\alpha_n(y)P(y,t)\right], \tag{P3.11}$$

where

$$\alpha_n(y) \equiv \int_{-\infty}^{\infty} \xi^n g(\xi,y)\mathrm{d}\xi = \int_{-\infty}^{\infty} \mathrm{d}x(x-y)^n W(x|y) \tag{P3.12}$$

is the nth moment of the transition rate, the *jump moment.*

Comment: Equation (P3.11) is the *Kramers-Moyal expansion*; it appeared in Kramers[76] and was considerably improved upon by Moyal[83]. It provides an infinite-order differential operator to represent the time rate of change of the probability (as opposed to the integral operator of the master equation). It's formally equivalent to the master equation, and therefore not any easier to work with, but it suggests an approximation. If we truncate the expansion after $n = 2$, we recover the Fokker-Planck equation, (3.37). Obtaining the Fokker-Planck equation this way has been criticized by van Kampen[84][51]; a power series should have some small parameter to guarantee convergence and simply truncating the series is not a systematic approximation procedure. If we're to ignore the jump moments for higher n, the transition rate functions $W(x|y)$ should be sharply peaked for $x \approx y$, the requirement of small jump lengths. Clearly there are mathematical issues in passing from an integral to a differential operator. One must assume that certain partial derivatives exist as well as the convergence of the expansion. R.F. Pawula[85][86] showed that *if* the α_n defined by Eq. (P3.12) exist for all n, and if $\alpha_n = 0$ for some even n, then $\alpha_n = 0$ for all $n \geq 3$. By this theorem, it's logically inconsistent to retain more than two terms in the Kramers-Moyal expansion unless all terms are retained. In some ways, the Fokker-Planck equation is the best we can do in approximating the master equation. Pawula's theorem is discussed in Risken[67, Section 4.3].

3.22 Derive Eq. (3.41). Hint: $M_2 = \lim_{\Delta t \to 0}(1/\Delta t)\langle (v - v_0)^2 \rangle_{v_0}$. Use Eq. (3.11), Eq. (3.12) with $\tau = 0$, and delta-correlated noise, Eq. (3.13), with $\Gamma = 2kT\alpha$.

3.23 Show for a smooth function $f(x)$ such that $f(x) \to 0$ as $x \to \pm\infty$,

$$\int_{-\infty}^{\infty} \frac{\partial}{\partial x}\left[xf(x)\right] \mathrm{e}^{-\mathrm{i}qx}\mathrm{d}x = -q\frac{\partial}{\partial q}\int_{-\infty}^{\infty} f(x)\mathrm{e}^{-\mathrm{i}qx}\mathrm{d}x$$

$$\int_{-\infty}^{\infty} \frac{\partial^2 f(x)}{\partial x^2}\mathrm{e}^{-\mathrm{i}qx}\mathrm{d}x = -q^2\int_{-\infty}^{\infty} f(x)\mathrm{e}^{-\mathrm{i}qx}\mathrm{d}x.$$

Assume all "integrated parts" vanish as $x \to \pm\infty$.

3.24 Verify that the form of $\widetilde{P}(q,\tau)$ given in Eq. (3.47) solves the differential equation (3.46) for any function F.

3.25 Using the transition probability for Brownian motion, Eq. (3.49), verify that the transition moments M_1, M_2 from

$$M_n = \lim_{\Delta t \to 0} \frac{1}{\Delta t} \int_{-\infty}^{\infty} (v - v_0)^n P(v, \Delta t | v_0) \mathrm{d}v$$

are as given in Eqs. (3.40) and (3.41), and then show that to leading order $M_3 = O(\Delta t)^{5/2}$ for small Δt, and hence that $M_3 = 0$ in the limit. Hint: Hermite polynomials[54] have the integral representation[88, p338]

$$\int_{-\infty}^{\infty} x^n \mathrm{e}^{-(x-\beta)^2} \mathrm{d}x = (2\mathrm{i})^{-n} \sqrt{\pi} H_n(\mathrm{i}\beta),$$

where $H_1(z) = 2z$, $H_2(z) = 4z^2 - 2$, $H_3(z) = 8z^3 - 12z$, etc.

3.26 Fill in the steps between Eq. (3.48) and Eq. (3.49).

3.27 Show that the limit $\tau \to 0$ in Eq. (3.49) is a delta function. See Exercise 2.31.

3.28 Verify that with the proper substitutions Eq. (3.49) has precisely the form of a transition probability for stationary Gaussian processes, Eq. (2.108).

3.29 Find the stationary solution of the Fokker-Planck equation when $\boldsymbol{K}(\boldsymbol{r}) = -(1/M)\boldsymbol{\nabla}_r \psi(\boldsymbol{r})$.

a. Try the separation of variables technique. Let $P_{\text{st}}(\boldsymbol{r}, \boldsymbol{v}) = f(\boldsymbol{r}) g(\boldsymbol{v})$. Show that Eq. (3.59) reduces to

$$\boldsymbol{v} \cdot \left(\frac{1}{f} \boldsymbol{\nabla}_{\boldsymbol{r}} f \right) + \boldsymbol{K} \cdot \left(\frac{1}{g} \boldsymbol{\nabla}_{\boldsymbol{v}} g \right) = \frac{1}{g} \beta \boldsymbol{\nabla}_{\boldsymbol{v}} \cdot \left[\boldsymbol{v} g + \frac{kT}{M} \boldsymbol{\nabla}_{\boldsymbol{v}} g \right]. \tag{P3.13}$$

b. The terms in square brackets in Eq. (P3.13) generalize Eq. (3.43), which we obtained in finding the stationary solution of the single-variable Fokker-Planck equation, (3.42). Show that these terms vanish when $g(\boldsymbol{v}) = \exp(-Mv^2/(2kT))$. Hint: $\boldsymbol{\nabla}_{\boldsymbol{v}} v^2 = 2\boldsymbol{v}$.

c. With $g(\boldsymbol{v})$ so determined, $f(\boldsymbol{r})$ must be chosen so that the left side vanishes. Show that this requires

$$\boldsymbol{v} \cdot \left[\frac{1}{f} \boldsymbol{\nabla}_{\boldsymbol{r}} f - \frac{M}{kT} \boldsymbol{K} \right] = 0. \tag{P3.14}$$

If $\boldsymbol{K}(\boldsymbol{r}) = -(1/M)\boldsymbol{\nabla}_{\boldsymbol{r}} \psi(\boldsymbol{r})$, show that the terms in square brackets vanish if $f(\boldsymbol{r}) = \exp(-\psi(\boldsymbol{r})/kT)$.

Thus $P_{\text{st}}(\boldsymbol{r}, \boldsymbol{v}) \propto \exp(-\psi(\boldsymbol{r})/kT) \exp(-Mv^2/2kT)$ is the stationary solution of the Fokker-Planck equation, (3.59).

3.30 Consider the field-free Fokker-Planck equation for $P(\boldsymbol{r}, \boldsymbol{v}, t)$

$$\frac{\partial P}{\partial t} + \boldsymbol{v} \cdot \boldsymbol{\nabla}_{\boldsymbol{r}} P = \beta \boldsymbol{\nabla}_{\boldsymbol{v}} \cdot \left(\boldsymbol{v} + \frac{kT}{M} \boldsymbol{\nabla}_{\boldsymbol{v}} \right) P \tag{P3.15}$$

[54] Hermite polynomials are often listed in quantum mechanics texts. Or see Abramowitz and Stegun[87, p775].

in an unbounded system. The boundary conditions are that P vanishes for large $\boldsymbol{r}$ and $\boldsymbol{v}$. Define spatially dependent probability and current densities

$$\rho(\boldsymbol{r},t) \equiv \int \mathrm{d}^3 v P(\boldsymbol{r},\boldsymbol{v},t)$$

$$\boldsymbol{J}(\boldsymbol{r},t) \equiv \int \mathrm{d}^3 v \boldsymbol{v} P(\boldsymbol{r},\boldsymbol{v},t)$$

where the integrals are over all velocities.

a. Show that these definitions imply a continuity equation

$$\frac{\partial \rho}{\partial t} + \boldsymbol{\nabla}_{\boldsymbol{r}} \cdot \boldsymbol{J} = 0.$$

Hint: Use the divergence theorem along with the boundary condition.

b. With the average speed defined

$$\langle \boldsymbol{v} \rangle_t \equiv \int \mathrm{d}^3 x \mathrm{d}^3 v \boldsymbol{v} P(\boldsymbol{r},\boldsymbol{v},t),$$

show that the drag force law is obtained for the average,

$$\frac{\mathrm{d}}{\mathrm{d}t}\langle \boldsymbol{v} \rangle = -\beta \langle \boldsymbol{v} \rangle.$$

Hint: Use the Leibniz integral rule, $\mathrm{d}\langle \boldsymbol{v} \rangle/\mathrm{d}t = \int \mathrm{d}^3 x \mathrm{d}^3 \boldsymbol{v}\, (\partial P/\partial t)$. Recognize, using Eq. (P3.15), that one term is a spatial integration over a gradient and vanishes because of the boundary condition. For the second term, use integration by parts together with the boundary condition.

3.31 Find the adjoint of the Fokker-Planck operator in Eq. (3.60). Any operator Ω has associated with it a new operator $\Omega^\dagger$ (its adjoint) such that $\langle \Omega^\dagger f | g \rangle = \langle f | \Omega g \rangle$ for any functions f, g obeying appropriate boundary conditions (see [13, p56]), which in this case we require to decay sufficiently rapidly as $v \to \pm\infty$ that integrated parts vanish.
A: For $\Omega = \boldsymbol{\nabla}_{\boldsymbol{v}} \cdot (\boldsymbol{v} + (kT/M)\boldsymbol{\nabla}_{\boldsymbol{v}})$, $\Omega^\dagger = -\boldsymbol{v} \cdot \boldsymbol{\nabla}_{\boldsymbol{v}} + (kT/M)\nabla_v^2$.

3.32 Derive the form of $\widetilde{\Omega}$ in Eq. (3.81), given its definition in Eq. (3.79). Don't forget to let $\widetilde{\Omega}$ act on an unknown function.

3.33 We stated that with $\widetilde{\Omega}$ given in Eq. (3.81), the eigenvalue problem $\widetilde{\Omega}\widetilde{F} = \lambda \widetilde{F}$ has the form of the time-independent Schrödinger equation for the isotropic three-dimensional harmonic oscillator. Let's flesh that out.

a. In Cartesian coordinates the eigenvalue problem has the separable form

$$\left[\frac{kT}{M}\left(\frac{\partial^2}{\partial v_x^2} + \frac{\partial^2}{\partial v_y^2} + \frac{\partial^2}{\partial v_z^2}\right) - \frac{M}{4kT}\left(v_x^2 + v_y^2 + v_z^2\right) + \frac{3}{2}\right]\widetilde{F} = \frac{1}{\beta}\left(\lambda_x + \lambda_y + \lambda_z\right)\widetilde{F}. \tag{P3.16}$$

where $\lambda = \lambda_x + \lambda_y + \lambda_z$. Show that, with $\widetilde{F} = f(v_x)f(v_y)f(v_z)$, an isotropic solution requires that we need solve only a one-dimensional eigenvalue problem for $f = f(v)$,

$$\frac{kT}{M}f'' - \frac{M}{4kT}v^2 f = \left(\frac{\lambda}{\beta} - \frac{1}{2}\right)f. \tag{P3.17}$$

Introduce a dimensionless velocity $y = v/\alpha$, where the velocity scale α is to be determined. Show that with $\alpha = \sqrt{2kT/M}$, Eq. (P3.17) reduces to

$$f'' - y^2 f = 2\left(\frac{\lambda}{\beta} - \frac{1}{2}\right) f. \tag{P3.18}$$

b. With the substitution $f(y) = \exp(-y^2/2)H(y)$, show that Eq. (P3.18) reduces to

$$H'' - 2yH' - 2\left(\lambda/\beta\right) H = 0. \tag{P3.19}$$

Equation (P3.19) has polynomial solutions[55] when $-2(\lambda/\beta) = 2n$ for $n = 0, 1, 2, \ldots$, the Hermite polynomials, $H_n(y)$. We therefore have an equally spaced eigenvalue spectrum, $\lambda = -n\beta$. There is no "zero point" velocity here; this is a classical equation. Solutions of $\widetilde{\Omega}\widetilde{F} = \lambda\widetilde{F}$ occur in the form of a Gaussian multiplied by products of Hermite polynomials,

$$\widetilde{F}_{n_1,n_2,n_3} = A_{n_1,n_2,n_3} \exp\left(-v^2/(2\alpha^2)\right) H_{n_1}(v_1/\alpha) H_{n_2}(v_2/\alpha) H_{n_3}(v_3/\alpha), \tag{P3.20}$$

where $\lambda = -\beta(n_1 + n_2 + n_3)$ and A_{n_1,n_2,n_3} is a constant.

3.34 Show that Eq. (3.86) solves the field-free Fokker-Planck equation (3.77). Derive Eq. (3.87) for the expansion coefficient.

3.35 Derive Eq. (3.90).

3.36 Consider the strong-friction expansion of $P(x, v, t)$ in Eq. (3.103) for the Fokker-Planck equation $\overline{\Omega}P = \beta^{-1}LP$, where $\overline{\Omega}, L$ are defined in Eq. (3.102).

a. At zeroth order, $\overline{\Omega}P^{(0)}(x, v, t) = 0$ (see Eq. (3.104)) so that $P^{(0)}$ is an eigenvector of $\overline{\Omega}$ associated with zero eigenvalue. Show that

$$P^{(0)}(x, v, t) = \phi(x, t) \exp(-Mv^2/(2kT)), \tag{P3.21}$$

where $\phi(x, t)$ is an arbitrary (dimensionless) function of (x, t). As we show next, $\phi = \phi(x)$, so that $P^{(0)} = P^{(0)}(x, v)$ is stationary.

b. At first order in β^{-1} (the first order in which K appears), $\overline{\Omega}P^{(1)}(x, v, t) = LP^{(0)}(x, v, t)$. Show that

$$\overline{\Omega}P^{(1)}(x, v, t) = \left(\frac{\partial}{\partial t} + v\frac{\partial}{\partial x} - \frac{M}{kT}Kv\right) \phi(x, t) \exp(-Mv^2/(2kT)). \tag{P3.22}$$

One might consider solving Eq. (P3.22) for $P^{(1)}$ by finding the Green function for $\overline{\Omega}$ (it could happen). $\overline{\Omega}$, however, has an eigenvector with zero eigenvalue, which complicates the analysis. When this happens, a solution for $P^{(1)}$ exists if and only if the right side of Eq. (P3.22), $LP^{(0)}$, is orthogonal to the left eigenvector associated with zero eigenvalue (see [13, p260]), in this case a constant. Show that this requirement, met in this case by $\int_{-\infty}^{\infty} LP^{(0)} \mathrm{d}v = 0$, implies the solubility condition

$$\frac{\partial}{\partial t}\phi(x, t) = 0. \tag{P3.23}$$

A solution for $P^{(1)}(x, v, t)$ exists only if Eq. (P3.23) is obeyed, implying $\phi = \phi(x)$ or that $P^{(0)}$ is stationary. The stationary solution is independent of β.

[55] When λ does not meet the eigenvalue condition $\lambda = -\beta n$, the differential equation has an irregular singularity at infinity, and such solutions must be discarded as unphysical. See for example [13, p130].

c. With the time derivative absent from Eq. (P3.22), verify that

$$P^{(1)}(x,v,t) = \left[v\left(\frac{M}{kT}K\phi - \frac{\partial\phi}{\partial x}\right) + f(x,t)\right]\exp(-Mv^2/(2kT)), \tag{P3.24}$$

where $f(x,t)$ is another unknown function (that must have the dimension time^{-1}).

d. Working to one order higher, $P^{(2)}$ exists when $\int_{-\infty}^{\infty} LP^{(1)}\mathrm{d}v = 0$. Show that the associated solubility condition is

$$\frac{\partial}{\partial t}f(x,t) = -\frac{\partial}{\partial x}(K\phi) + \frac{kT}{M}\frac{\partial^2\phi}{\partial x^2}. \tag{P3.25}$$

e. Assemble the pieces and show that to first order in β^{-1},

$$P(x,v,t) = \exp(-Mv^2/(2kT))\left[\phi(x) + \beta^{-1}\left(\frac{Mv}{kT}K\phi - v\frac{\partial\phi}{\partial x} + f(x,t)\right)\right] + O(\beta^{-2}). \tag{P3.26}$$

Integrate $P(x,v,t)$ over v to find a probability distribution in (x,t):

$$P(x,t) = \int_{-\infty}^{\infty} P(x,v,t)\mathrm{d}v = \sqrt{\frac{2\pi kT}{M}}\left(\phi(x) + \beta^{-1}f(x,t)\right) + O(\beta^{-2}). \tag{P3.27}$$

f. Use Eq. (P3.25) and Eq. (P3.27) to show that

$$\frac{\partial}{\partial t}P(x,t) = \sqrt{\frac{2\pi kT}{M}}\beta^{-1}\left[-\frac{\partial}{\partial x}(K\phi) + \frac{kT}{M}\frac{\partial^2\phi}{\partial x^2}\right]. \tag{P3.28}$$

g. Recognize from Eq. (P3.27) that to lowest order $\phi(x) = \sqrt{M/(2\pi kT)}P(x,t) + O(\beta^{-1})$. Show that

$$\frac{\partial}{\partial t}P(x,t) = -\beta^{-1}\frac{\partial}{\partial x}[KP(x,t)] + D\frac{\partial^2}{\partial x^2}P(x,t), \tag{P3.29}$$

where we've used an Einstein relation.

CHAPTER 4

Kinetic theory: Boltzmann's approach to irreversibility

A common feature of macroscopic systems is their irreversible evolution toward thermal equilibrium, no matter how they're prepared in nonequilibrium states. Irreversibility is *built-in* to Markov processes.[1] Stochastic models provide phenomenological frameworks for describing the approach to equilibrium—and therein lies their justification—but they don't predict irreversibility, they're guided by it. Of the laws of physics, only the second law of thermodynamics speaks explicitly to irreversibility, wherein entropy is created in irreversible processes. Entropy, however, is not a microscopic property of matter, rendering microscopic treatments of irreversibility a challenging problem in theoretical physics.[2] We noted in Section 2.2 that the second law is not inherent in time-reversal-invariant equations of motion, but did anyone ever try? Ludwig Boltzmann sought, in the 1870s, to derive the second law as a consequence of microscopic physics. We'll identify the point in his analysis where time asymmetry is introduced through a non-mechanical, statistical assumption. Although he didn't succeed in deriving entropy from mechanics, his discoveries are widely used in modeling nonequilibrium systems. We develop Boltzmann's approach in this chapter, known as *kinetic theory*. Kinetic theory has a rich history; see Brush[89].

4.1 FROM MECHANICS TO STATISTICAL MECHANICS

Phase space for many-particle systems—Γ-space—is a mathematical space representing all possible mechanical states of a system, with each state associated with a point of the space. The time development of N particles in three spatial dimensions is governed by $6N$ first-order differential equations, Hamilton's equations of motion, Eq. (A.1). States are specified at an instant of time by $3N$ generalized coordinates $\{\boldsymbol{q}_i\}_{i=1}^N$ and $3N$ canonical momenta $\{\boldsymbol{p}_i\}_{i=1}^N$; Γ-space is $6N$-dimensional. Each state is associated with a vector[3] $\boldsymbol{\Gamma} \equiv (\boldsymbol{q}_1, \boldsymbol{p}_1, \ldots, \boldsymbol{q}_N, \boldsymbol{p}_N)$, its *system point* or its *phase point*, or, in the older literature, its phase. The evolution of the system can be visualized as a trajectory in Γ-space, $\boldsymbol{\Gamma}(t)$, the *phase trajectory*.

[1]Markov processes satisfy the SCK equation, (2.29), which presumes a one-way flow of time, $t_1 > t_2 > t_3$. The master equation follows from the SCK equation when transition probabilities are written as in Eq. (2.64), and the Fokker-Planck equation follows from the master equation (Kramers-Moyal expansion), or directly from the SCK equation, Eq. (3.32).

[2]Like temperature, one can't speak of the entropy of a single particle. What mechanical or electrodynamical property of matter is associated with entropy? Entropy is a property of equilibrium macroscopic systems as a whole.

[3]Γ-space is the direct product of the space of canonical momenta (of dimension $3N$) and the space of generalized coordinates (of dimension $3N$). The direct product of vector spaces U and V is a new vector space, $U \times V \equiv \{(u, v) | u \in U, v \in V\}$, with vector addition and scalar multiplication defined component wise. The dimension of $U \times V$ is the sum of the dimensions of U and V.

DOI: 10.1201/9781003512295-4

Precise knowledge of the system requires a precise specification of $6N$ microscopic quantities, quantities *we lack the ability to control.* Therein lies the transition from mechanics to statistical mechanics. To circumvent the fundamental problem of uncertainty in initial conditions, the concept of ensembles is introduced, collections of macroscopically identical systems; ensemble averages are averages over initial conditions consistent with constraints. Ensembles are represented by treating the point located by $\mathbf{\Gamma}$ as a random variable. Introduce a probability density $\rho(\mathbf{\Gamma},t)$ with $\rho(\mathbf{\Gamma},t)\mathrm{d}\mathbf{\Gamma}$ the probability at time t the phase point lies within $\mathrm{d}\mathbf{\Gamma} \equiv \mathrm{d}\boldsymbol{p}_1 \ldots \mathrm{d}\boldsymbol{p}_N \mathrm{d}\boldsymbol{q}_1 \ldots \mathrm{d}\boldsymbol{q}_N$ about $\mathbf{\Gamma}$. From $\rho(\mathbf{\Gamma},t)$, ensemble averages can be calculated. For $A(\mathbf{\Gamma})$ a Γ-space function, it's ensemble average is found from[4]

$$\langle A\rangle_t \equiv \int_\Gamma A(\mathbf{\Gamma})\rho(\mathbf{\Gamma},t)\mathrm{d}\mathbf{\Gamma}, \tag{4.1}$$

where unit normalization has been assumed, $\int_\Gamma \rho(\mathbf{\Gamma},t)\mathrm{d}\mathbf{\Gamma} = 1$.

How to find $\rho(\mathbf{\Gamma},t)$? By Liouville's theorem,[5] $\rho(\mathbf{\Gamma},t)$ is a constant of the motion:

$$\frac{\mathrm{d}}{\mathrm{d}t}\rho(\mathbf{\Gamma},t) \overset{\substack{\text{Liouville}\\ \text{theorem}\\ \downarrow}}{=} 0 \overset{\substack{\text{math}\\ \downarrow}}{=} \frac{\partial\rho}{\partial t} + \sum_{k=1}^{3N}\left(\frac{\partial\rho}{\partial p_k}\dot{p}_k + \frac{\partial\rho}{\partial q_k}\dot{q}_k\right) \overset{\substack{\text{Hamilton}\\ \text{equations}\\ \downarrow}}{=} \frac{\partial\rho}{\partial t} + \sum_{k=1}^{3N}\left(-\frac{\partial\rho}{\partial p_k}\frac{\partial H}{\partial q_k} + \frac{\partial\rho}{\partial q_k}\frac{\partial H}{\partial p_k}\right) \tag{4.2}$$

$$\equiv \frac{\partial\rho}{\partial t} + \mathrm{i}\Lambda\rho,$$

where Λ is the *Liouville operator* (a linear operator),

$$\Lambda(\cdot) \equiv \mathrm{i}\sum_{k=1}^{3N}\left(\frac{\partial H}{\partial q_k}\frac{\partial(\cdot)}{\partial p_k} - \frac{\partial H}{\partial p_k}\frac{\partial(\cdot)}{\partial q_k}\right) = \mathrm{i}\,[H,(\cdot)]_{\mathrm{P}}\,, \tag{4.3}$$

with $(\cdot)$ a placeholder for the Γ-space function that Λ acts on and $[H,(\cdot)]_{\mathrm{P}}$ the Poisson bracket, see Eq. (A.3). The factor of i (unit imaginary number) is included in the definition of Λ to make it Hermitian, $\Lambda^\dagger = \Lambda$; see [5, p335]. The Liouville operator can be written compactly as

$$\Lambda = \mathrm{i}\sum_{n=1}^{N}\left[(\boldsymbol{\nabla}_n H)\cdot\frac{\partial}{\partial \boldsymbol{p}_n} - \left(\frac{\partial H}{\partial \boldsymbol{p}_n}\right)\cdot\boldsymbol{\nabla}_n\right], \tag{4.4}$$

where $\boldsymbol{\nabla}_n \equiv \boldsymbol{\nabla}_{\boldsymbol{r}_n}$. Thus we have *Liouville's equation*, the fundamental equation of statistical mechanics,[6] on par with the status of Schrödinger's equation in quantum mechanics,

$$\mathrm{i}\frac{\partial}{\partial t}\rho(\mathbf{\Gamma},t) = \Lambda\rho(\mathbf{\Gamma},t). \tag{4.5}$$

It's solution defines a unitary time evolution operator $U(t) \equiv \mathrm{e}^{-\mathrm{i}\Lambda t}$ (see Exercise 6.1),

$$\rho(\mathbf{\Gamma},t) = U(t)\rho(\mathbf{\Gamma},0) = \exp\left(-\mathrm{i}\Lambda t\right)\rho(\mathbf{\Gamma},0), \tag{4.6}$$

where $\rho(\mathbf{\Gamma},0)$ is the state of the ensemble at $t=0$. States of mechanical systems are specified by points in Γ-space; *states of statistical mechanical systems are specified by weightings on Γ-space,* $\rho(\mathbf{\Gamma},0)$ with $\int \mathrm{d}\mathbf{\Gamma}\rho(\mathbf{\Gamma},0) = 1$.

---

[4]Note that $A(\mathbf{\Gamma})$ carries no time dependence; all time dependence is in the probability density. Equation (4.1) indicates to sample $A(\mathbf{\Gamma})$ at every point of Γ-space weighted by the probability $\rho(\mathbf{\Gamma},t)$ that the system point is at $\mathbf{\Gamma}$ at time t. This prescription is analogous to the Schrödinger picture of quantum mechanics where operators are fixed in time and the probability density is time dependent. See Appendix E.

[5]Liouville's theorem is a fundamental theorem of analytical mechanics. Its proof, omitted here (see [5, p47]), relies on the unit Jacobian of canonical transformations ([5, p331]) and that trajectories in Γ-space never cross ([5, p32]).

[6]In equilibrium, we have the *stationarity condition*, $\Lambda\rho_{\mathrm{eq}}(\mathbf{\Gamma}) = 0$; see [5, p47].

The eigenvalues of Λ are real (Hermiticity), with its eigenfunctions a complete orthogonal set (Sturm-Liouville theory[13, p66]), implying for an arbitrary initial condition $\rho(\boldsymbol{\Gamma}, 0)$ that the temporal behavior of $\rho(\boldsymbol{\Gamma}, t)$ is *oscillatory* with no decay to a unique state. *There is no irreversible decay inherent in the fundamental equations of motion*, qualitatively distinct from what we find in the solutions of the master equation (2.83) or the Fokker-Planck equation (3.49).

Let's find $\rho(\boldsymbol{\Gamma}, t)$ for free particles. Consider N identical noninteracting particles of mass m confined to a cubical box of edge length a. The Hamiltonian $H = (2m)^{-1} \sum_{n=1}^{N} p_n^2$ implies

$$\Lambda_N = -\mathrm{i} \sum_{n=1}^{N} \frac{\partial H}{\partial \boldsymbol{p}_n} \cdot \frac{\partial}{\partial \boldsymbol{r}_n} = -\mathrm{i} \sum_{n=1}^{N} \boldsymbol{v}_n \cdot \frac{\partial}{\partial \boldsymbol{r}_n}, \qquad \text{(free particles)} \tag{4.7}$$

where $\boldsymbol{v}_n \equiv \boldsymbol{p}_n / m$. The velocities $\boldsymbol{v}_n$ are constants of the motion (free particles). Moreover, Λ_N (in this case) is a sum of *one-body operators*, those that act on one particle only. To find the eigenvalues and eigenfunctions of Λ_N it suffices (in this case) to solve the one-body eigenproblem,

$$-\mathrm{i} \boldsymbol{v} \cdot \boldsymbol{\nabla} \psi(\boldsymbol{r}) = \lambda \psi(\boldsymbol{r}). \tag{4.8}$$

Equation (4.8) has solutions $\psi_{\boldsymbol{k}}(\boldsymbol{r}) = A \mathrm{e}^{\mathrm{i}\boldsymbol{k}\cdot\boldsymbol{r}}$, where A is a constant, with eigenvalues $\lambda_{\boldsymbol{k}} = \boldsymbol{v} \cdot \boldsymbol{k}$. For periodic boundary conditions, $\boldsymbol{k} = (2\pi/a)(n_1, n_2, n_3)$ for integer n_i. By normalizing to unity, $A = a^{-3/2}$. The eigenfunctions and eigenvalues of Λ_N in Eq. (4.7) have the form

$$\begin{aligned} \psi_{\{\boldsymbol{k}\}}(\boldsymbol{r}_1, \ldots, \boldsymbol{r}_N) &= \frac{1}{a^{3N/2}} \exp\left(\mathrm{i} \sum_{n=1}^{N} \boldsymbol{k}_n \cdot \boldsymbol{r}_n\right) \\ \lambda_{\{\boldsymbol{k}\}} &= \sum_{n=1}^{N} \boldsymbol{v}_n \cdot \boldsymbol{k}_n, \end{aligned} \tag{4.9}$$

where $\{\boldsymbol{k}\}$ denotes the collection of wave vectors, $\{\boldsymbol{k}\} \equiv (\boldsymbol{k}_1, \ldots, \boldsymbol{k}_N)$. There are $3N$ distinct eigenvalues implied by the notation $\{\boldsymbol{k}\}$. An arbitrary initial condition $\rho(\boldsymbol{\Gamma}, 0)$ can be expressed as a linear combination of eigenfunctions (complete set)

$$\rho(\boldsymbol{\Gamma}, 0) = \sum_{\{\boldsymbol{k}\}} A_{\{\boldsymbol{k}\}} \psi_{\{\boldsymbol{k}\}}(\boldsymbol{r}_1, \ldots, \boldsymbol{r}_N), \tag{4.10}$$

where the expansion coefficients $A_{\{\boldsymbol{k}\}}$ are functions of the momenta (see Exercise 4.4),

$$A_{\{\boldsymbol{k}\}}(\boldsymbol{p}_1, \ldots, \boldsymbol{p}_N) = \int \mathrm{d}\boldsymbol{r}_1 \cdots \mathrm{d}\boldsymbol{r}_N \psi^*_{\{\boldsymbol{k}\}}(\boldsymbol{r}_1, \ldots, \boldsymbol{r}_N) \rho(\boldsymbol{\Gamma}, 0). \tag{4.11}$$

Using Eq. (4.6), we have for free particles,

$$\rho(\boldsymbol{\Gamma}, t) = \frac{1}{a^{3N/2}} \sum_{\{\boldsymbol{k}\}} A_{\{\boldsymbol{k}\}}(\boldsymbol{p}_1, \ldots, \boldsymbol{p}_N) \exp\left(\mathrm{i} \sum_{n=1}^{N} \boldsymbol{k}_n \cdot (\boldsymbol{r}_n - \boldsymbol{v}_n t)\right), \tag{4.12}$$

a function of $6N$ constants of the motion, $\boldsymbol{p}_1, \ldots, \boldsymbol{p}_N$ and $\boldsymbol{r}_1 - \boldsymbol{v}_1 t, \ldots, \boldsymbol{r}_N - \boldsymbol{v}_N t$ (see Section A.1.2). The sum in Eq. (4.12) does not have a long-time limit.[7]

[7] Convergent trigonometric sums $\sum_n a_n \exp(\mathrm{i}\omega_n t)$ represent almost periodic functions (see Bohr[90]) which nearly repeat as time progresses and do not approach a limit as $t \to \infty$. (If the ω_n are integers, the sum represents a periodic function of period 2π.) If the ω_n are close together, one could try to approximate the sum by an integral, $\int a(\omega) \exp(\mathrm{i}\omega t) \mathrm{d}\omega$. For any reasonable function $a(\omega)$, the integral decays (irreversibly!) to zero as $t \to \infty$. Every discrete sum is almost periodic and every integral converges to zero as $t \to \infty$. Modeling irreversibility is tricky.

4.2 REDUCED PROBABILITY DISTRIBUTIONS

The solution of the Liouville equation is a joint probability density: For the N-body system to occupy state $\boldsymbol{\Gamma}$ at time t, particle 1 is in state $\boldsymbol{r}_1, \boldsymbol{p}_1 \equiv 1$, particle 2 is in state $\boldsymbol{r}_2, \boldsymbol{p}_2 \equiv 2$, and so on, at time t. To save writing, let $\mathrm{d}1 \equiv \mathrm{d}\boldsymbol{r}_1 \mathrm{d}\boldsymbol{p}_1, \mathrm{d}2 \equiv \mathrm{d}\boldsymbol{r}_2 \mathrm{d}\boldsymbol{p}_2, \ldots$ denote volume elements for each particle. Thus, $\rho(1,\ldots,N,t)\mathrm{d}1\cdots\mathrm{d}N$ is the probability at time t particle 1 is in d1 about state 1, particle 2 is in d2 about state 2, and so on. It turns out that the N-body distribution contains far more information than is necessary to calculate measurable quantities. For that reason, *reduced distributions* f_s are introduced for fewer particles (among the set of N particles)[8]

$$f_s(1,\ldots,s,t) \equiv \int \rho(1,\ldots,s,s+1,\ldots,N,t)\mathrm{d}(s+1)\cdots\mathrm{d}N, \tag{4.13}$$

where $\int f_s \mathrm{d}1\cdots\mathrm{d}s = 1$ for all s. Thus, f_1 (for example) is the probability of finding particle 1 in state 1, irrespective of the states of particles $2,\ldots,N$. The reduced distributions are quite useful. Yet if we can't solve the Liouville equation for $\rho(\boldsymbol{\Gamma},t)$, we can't calculate them from Eq. (4.13). The strategy for calculating these functions is treated in Section 4.3.

To see how reduced distributions arise, consider the average velocity of particle 1. From Eqs. (4.1) and (4.13),

$$\langle \boldsymbol{v}_1\rangle_t = \int \mathrm{d}\boldsymbol{\Gamma}\boldsymbol{v}_1\rho(\boldsymbol{\Gamma},t) = \int \mathrm{d}\boldsymbol{r}_1\mathrm{d}\boldsymbol{v}_1\boldsymbol{v}_1\mathrm{d}2\cdots\mathrm{d}N\rho(1,\ldots,N,t) = \int \mathrm{d}\boldsymbol{r}_1\mathrm{d}\boldsymbol{v}_1\boldsymbol{v}_1 f_1(\boldsymbol{r}_1,\boldsymbol{v}_1,t). \tag{4.14}$$

By this device, the phase-space variables of the other particles have been formally eliminated.[9] Sometimes we want the *l-particle velocity distribution* where spatial coordinates are eliminated,

$$\phi_l(\boldsymbol{v}_1,\ldots,\boldsymbol{v}_l,t) \equiv \int \mathrm{d}\boldsymbol{r}_1\cdots\mathrm{d}\boldsymbol{r}_l f_l(\boldsymbol{r}_1,\ldots,\boldsymbol{r}_l,\boldsymbol{v}_1,\ldots,\boldsymbol{v}_l,t), \tag{4.15}$$

in which case Eq. (4.14) can be written $\langle \boldsymbol{v}_1\rangle_t = \int \mathrm{d}\boldsymbol{v}_1\boldsymbol{v}_1\phi_1(\boldsymbol{v}_1,t)$. Sometimes we want the *spatial distribution* of l particles obtained by eliminating velocities,

$$n_l(\boldsymbol{r}_1,\ldots,\boldsymbol{r}_l,t) \equiv \int \mathrm{d}\boldsymbol{v}_1\cdots\mathrm{d}\boldsymbol{v}_l f_l(\boldsymbol{r}_1,\ldots,\boldsymbol{r}_l,\boldsymbol{v}_1,\ldots,\boldsymbol{v}_l,t). \tag{4.16}$$

The reduced distributions f_s pertain to subsystems of s *enumerated* particles, with $f_s\mathrm{d}1\cdots\mathrm{d}s$ the probability of finding the subsystem in the volumes $\mathrm{d}1\cdots\mathrm{d}s$ about the Γ-space points $(1,\ldots,s)$. A related quantity is the *s-tuple distribution*,[10] defined with a combinatoric factor,

$$F_s(1,\ldots,s,t) \equiv \frac{N!}{(N-s)!} f_s(1,\ldots,s,t). \tag{4.17}$$

Whereas f_s pertains to s labeled particles $(1,\ldots,s)$, F_s is the probability density of finding particles at points $(1,\ldots,s)$ without regard for *which* particles are counted among the s, with normalization $\int F_s\mathrm{d}1\cdots\mathrm{d}s = N!/(N-s)!$; F_s is a phase-space number density function for s-tuples.

[8] Reduced distribution functions were introduced in statistical mechanics by Yvon[91]. See Green[92] and Yvon[93]. Equation (4.13) holds for $1 \le s \le N-1$; for $s = N$, $f_N \equiv \rho$. The reader may notice an analogy with the hierarchy of joint probabilities introduced in Section 2.3. In that section, probabilities refer to events occurring at different times.

[9] We've worked with phase space spanned by $(\boldsymbol{r},\boldsymbol{v})$ in Eq. (4.14). When all particles have the same mass (a typical case), velocities are easier to work with than momenta. Be mindful of the dimensions of distribution functions.

[10] The notation for reduced and s-tuple distributions is not standardized; beware. The quantity $N!/(N-s)!$ is *not* the binomial coefficient $\binom{N}{s}$ (the number of ways of choosing s objects from N without regard to order); it counts the number of *distinct* ways of choosing s from $N > s$ objects. See Exercise 4.6. Combinatorics are reviewed in [5, Chapter 3].

4.3 DYNAMICS OF REDUCED DISTRIBUTIONS: THE HIERARCHY

To calculate reduced probability densities, we ask what their evolution equations are. It often happens that the same question is addressed independently by different researchers. As we now show, the reduced distribution functions satisfy a hierarchy of coupled dynamical equations known as the *BBGKY hierarchy* (or simply *the hierarchy*), after Bogoliubov (published in 1946, English translation in [94]), Born and Green,[95][96] Kirkwood,[97][98] and Yvon[91].

Consider N identical, structureless particles of mass m interacting through a two-body potential $\Phi_{ij} \equiv \Phi(|\boldsymbol{r}_i - \boldsymbol{r}_j|)$, with Hamiltonian

$$H(\boldsymbol{r}_1, \ldots, \boldsymbol{r}_N, \boldsymbol{p}_1, \ldots, \boldsymbol{p}_N) = \frac{1}{2m}\sum_{i=1}^{N} p_i^2 + \sum_{i=1}^{N}\sum_{j>i}^{N} \Phi_{ij}. \tag{4.18}$$

The potential energy term is written so that we never encounter a self-interaction, Φ_{ii}. There are N terms in the first sum in Eq. (4.18) and $\frac{1}{2}N(N-1)$ terms in the second. For convenience, we work with a modified Liouville operator, $L \equiv -\mathrm{i}\Lambda$. Referring to Eq. (4.3),

$$L_N = -\frac{1}{m}\sum_{k=1}^{N} \boldsymbol{p}_k \cdot \frac{\partial}{\partial \boldsymbol{r}_k} + \sum_{k=1}^{N}\sum_{i=1}^{N}\sum_{j>i}^{N} \frac{\partial \Phi_{ij}}{\partial \boldsymbol{r}_k} \cdot \frac{\partial}{\partial \boldsymbol{p}_k}. \tag{4.19}$$

The first term in Eq. (4.19), the noninteracting part of the operator, corresponds to Eq. (4.7); the remaining terms contain the effects of interactions. Of the terms in the interaction part of the operator, the only nonzero contributions are for $k = i$ and $k = j$ (Φ is a two-body interaction),

$$\frac{\partial \Phi_{ij}}{\partial \boldsymbol{r}_k} \cdot \frac{\partial}{\partial \boldsymbol{p}_k} = \delta_{k,i}\frac{\partial \Phi_{ij}}{\partial \boldsymbol{r}_i} \cdot \frac{\partial}{\partial \boldsymbol{p}_i} + \delta_{k,j}\frac{\partial \Phi_{ij}}{\partial \boldsymbol{r}_j} \cdot \frac{\partial}{\partial \boldsymbol{p}_j}.$$

But $\partial \Phi_{ij}/\partial \boldsymbol{r}_i = -\partial \Phi_{ij}/\partial \boldsymbol{r}_j$ from calculus (but which embodies Newton's third law), implying

$$\frac{\partial \Phi_{ij}}{\partial \boldsymbol{r}_k} \cdot \frac{\partial}{\partial \boldsymbol{p}_k} = \frac{\partial \Phi_{ij}}{\partial \boldsymbol{r}_i} \cdot \left(\delta_{k,i}\frac{\partial}{\partial \boldsymbol{p}_i} - \delta_{k,j}\frac{\partial}{\partial \boldsymbol{p}_j}\right).$$

Thus,

$$L_N = -\sum_{i=1}^{N} K_i + \sum_{i=1}^{N}\sum_{j>i}^{N} O_{ij}, \tag{4.20}$$

where K_i is a kinetic energy operator and O_{ij} contains the effects of interactions,

$$K_i \equiv \frac{1}{m}\boldsymbol{p}_i \cdot \frac{\partial}{\partial \boldsymbol{r}_i} \qquad O_{ij} \equiv \frac{\partial \Phi_{ij}}{\partial \boldsymbol{r}_i} \cdot \left(\frac{\partial}{\partial \boldsymbol{p}_i} - \frac{\partial}{\partial \boldsymbol{p}_j}\right) \equiv -\boldsymbol{F}_{ij} \cdot \left(\frac{\partial}{\partial \boldsymbol{p}_i} - \frac{\partial}{\partial \boldsymbol{p}_j}\right), \tag{4.21}$$

with $\boldsymbol{F}_{ij} = -\partial \Phi_{ij}/\partial \boldsymbol{r}_i$ the (internal) force experienced at $\boldsymbol{r}_i$ due to a particle at $\boldsymbol{r}_j$. The potential energy operator is symmetric, $O_{ij} = O_{ji}$. The kinetic energy operator K is associated with spatial gradients but the potential energy operator O is associated with momentum gradients.

To find an equation of motion for $s < N$ particles, decompose L_N into operators containing *only* the coordinates of the s particles, L_s, and those associated with the remaining $(N-s)$ particles,

$$L_N = L_s + L_{N-s}, \tag{4.22}$$

where (set $N = s$ in Eq. (4.20))

$$L_s = -\sum_{i=1}^{s} K_i + \sum_{i=1}^{s}\sum_{j>i}^{s} O_{ij}, \tag{4.23}$$

and

$$L_{N-s} \equiv -\sum_{j=s+1}^{N} K_j + \sum_{i=1}^{s}\sum_{j=s+1}^{N} O_{ij} + \sum_{i=s+1}^{N}\sum_{j>i}^{N} O_{ij}. \tag{4.24}$$

There are two types of interactions in L_{N-s}, those between the sets of s and $(N-s)$ particles and those among the $(N-s)$ particles. There are $\frac{1}{2}s(s+1)$ terms in L_s and $\frac{1}{2}\left[N(N+1)-s(s+1)\right]$ terms in L_{N-s}.

We seek an equation of motion for f_s. From Eq. (4.5) with $\Lambda = \mathrm{i}L$, we have, using Eq. (4.22),

$$\left(\frac{\partial}{\partial t} - L_s\right)\rho = L_{N-s}\rho = \left[-\sum_{j=s+1}^{N} K_j + \sum_{i=1}^{s}\sum_{j=s+1}^{N} O_{ij} + \sum_{i=s+1}^{N}\sum_{j>i}^{N} O_{ij}\right]\rho. \tag{4.25}$$

Integrate Eq. (4.25) over the Γ-space variables $(s+1,\ldots,N)$, leaving us with

$$\left(\frac{\partial}{\partial t} - L_s\right) f_s = -\int \sum_{i=1}^{s}\sum_{j=s+1}^{N} \boldsymbol{F}_{ij}\cdot\left(\frac{\partial}{\partial \boldsymbol{p}_i} - \frac{\partial}{\partial \boldsymbol{p}_j}\right)\rho \,\mathrm{d}(s+1)\cdots \mathrm{d}N. \tag{4.26}$$

The first and the third terms in square brackets in Eq. (4.25) vanish upon eliminating $s+1,\ldots,N$, when it's assumed that ρ vanishes at boundaries. Likewise the derivatives with respect to $\boldsymbol{p}_j$ in Eq. (4.26) contribute surface terms and vanish. We're left with

$$\begin{aligned}\left(\frac{\partial}{\partial t} - L_s\right) f_s &= -\sum_{i=1}^{s}\frac{\partial}{\partial \boldsymbol{p}_i}\cdot\int \sum_{j=s+1}^{N} \boldsymbol{F}_{ij}\rho \,\mathrm{d}(s+1)\cdots\mathrm{d}N \\ &= -(N-s)\sum_{i=1}^{s}\frac{\partial}{\partial \boldsymbol{p}_i}\cdot\int \boldsymbol{F}_{i,s+1} f_{s+1}\,\mathrm{d}(s+1),\end{aligned} \tag{4.27}$$

where the last step follows by symmetry (ρ is a totally symmetric function of its arguments).

The set of equations in (4.27) for $s=1,\ldots,N$, the BBGKY hierarchy, is not any easier to solve than the Liouville equation. To determine f_s requires that we know f_{s+1}, which in turn requires that we know f_{s+2} and so on. It might appear that little has been accomplished by this exercise. The virtue of the hierarchy is that it provides a means of introducing approximations so that the hierarchy is *truncated* at a given order.[11]

Note that the reduced distributions f_s are not constants of the motion along the phase trajectories of $s < N$ particles (the right side of Eq. (4.27) is nonzero) because of their interactions with the remaining $(N-s)$ particles. Consider the first equation of the hierarchy,

$$\left(\frac{\partial}{\partial t} - L_1\right) f_1(\boldsymbol{r}_1,\boldsymbol{p}_1,t) = -(N-1)\frac{\partial}{\partial \boldsymbol{p}_1}\cdot\int \boldsymbol{F}_{12} f_2(\boldsymbol{r}_1,\boldsymbol{p}_1,\boldsymbol{r}_2,\boldsymbol{p}_2)\mathrm{d}2. \tag{4.28}$$

The rate of change of $f_1(t)$ is affected by the force $\boldsymbol{F}_{12}$ exerted on 1 by 2, weighted by the probability of finding the pair at phase points 1 and 2. If the net force exerted on 1 is aligned with $\boldsymbol{p}_1$, the change in $f_1(1,t)$ is *negative* as the particle gains speed (make sure you understand that). When we single out for consideration a group of $s < N$ particles, f_s is not a constant of the motion; Liouville's theorem holds when Newton's third law can be applied between all particles.

It's often easier to work with the hierarchy cast in terms of s-tuple distributions. Multiply Eq. (4.27) for $s=1$ by N and that for $s=2$ by $N(N-1)$ to find[12]

$$\begin{aligned}\left(\frac{\partial}{\partial t} - L_1\right)F_1(1) &= -\frac{\partial}{\partial \boldsymbol{p}_1}\cdot\int \boldsymbol{F}_{12}F_2(1,2)\mathrm{d}2 \\ \left(\frac{\partial}{\partial t} - L_2\right)F_2(1,2) &= -\frac{\partial}{\partial \boldsymbol{p}_1}\cdot\int \boldsymbol{F}_{13}F_3(1,2,3)\mathrm{d}3 - \frac{\partial}{\partial \boldsymbol{p}_2}\cdot\int \boldsymbol{F}_{23}F_3(1,2,3)\mathrm{d}3.\end{aligned} \tag{4.29}$$

[11]Theoretical physics is the art of approximation.

[12]Apologies for notation: Bold $\boldsymbol{F}_{ij}$ is the internal force on particle i from particle j; unbold F_s is an s-tuple distribution.

4.4 HYDRODYNAMICS AND THE HIERARCHY

Liouville's equation is not suited for practical calculations; working in $6N$-dimensions is impossible for $N \approx 10^{23}$. Maxwell and Boltzmann worked with one-particle probability densities $f_1(\boldsymbol{r}, \boldsymbol{p}, t)$, and Boltzmann in particular derived an evolution equation for $f_1(\boldsymbol{r}, \boldsymbol{p}, t)$. Before taking up that all-important topic (see Section 4.5), we stop to consider the question of how nonequilibrium statistical mechanics connects microscopic with macroscopic. We had to address the analogous question in statistical mechanics, which must reproduce the laws of thermodynamics—the macroscopic theory of equilibrium—a demand met by the simple step of relating the parameter β in the stationary distribution $\rho(E) \equiv Z^{-1}\mathrm{e}^{-\beta E}\Omega(E)$ to the absolute temperature,[13] $\beta = (kT)^{-1}$ (see [5, p92]). There is no comparably simple way in nonequilibrium theory of connecting microscopic with macroscopic because of the variety of phenomena and the complexity of evolution processes; the connection must be established in stages. In the sense that $\beta = (kT)^{-1}$ ensures the second law of thermodynamics is reproduced by statistical mechanics, which macroscopic phenomena must the nonequilibrium theory get right? In analogy with quantum mechanics, what is the "hydrogen atom" of nonequilibrium statistical mechanics? The equations of hydrodynamics embody local conservation of mass, momentum, and energy—physical principles the theory must obviously retain. Kinetic theory is tasked with deriving the equations of hydrodynamics from first principles and of calculating transport coefficients in terms of molecular parameters.

In this section, we show, first, that energy conservation is contained in the first two equations of the hierarchy, without having to know the three-particle distribution. We derive balance equations for mass, momentum, and energy from the hierarchy, implying the equations of hydrodynamics have a microscopic basis.[14] We thereby obtain expressions for the pressure tensor and heat flux (introduced in Chapter 1) in terms of molecular variables. Finally, we find the long-wavelength, low-frequency dynamical modes of simple fluids from the equations of hydrodynamics.[15]

4.4.1 Densities macroscopic and microscopic

The basic fields of hydrodynamics are densities: mass, momentum, and energy. There are two uses of the term density, however, depending on the level of description. Macroscopically, matter is seen as a continuum, whereas microscopically (nanometer length scales) it appears as a collection of discrete particles moving under the influence of mutual interactions. The natural mathematical representation of physical quantities in the first case is with continuous functions (classical fields), with their spatiotemporal behavior determined by partial differential equations. At the microscopic level, systems are described in terms of many particle dynamics. Fundamentally motion is governed by quantum mechanics, but classical mechanics is often a good approximation.

Classical microscopic densities are possible because (classically) one can say either there is a particle at position $\boldsymbol{r}$ or there is not. We define a microscopic mass density function through the use of Dirac delta functions,

$$\hat{\rho}(\boldsymbol{r}) \equiv m \sum_{\alpha=1}^{N} \delta(\boldsymbol{r} - \boldsymbol{r}_\alpha). \tag{4.30}$$

Here $\boldsymbol{r}$ is a parameter, an arbitrary position in the system, $\boldsymbol{r}_\alpha$ is the canonical position variable of the αth-particle, and the "hat" on $\hat{\rho}$ indicates microscopic. Equation (4.30) specifies an irregularly fluctuating quantity and is a function only in the sense that delta functions are functions.[16] Yet it defines a density: $\int \hat{\rho}(\boldsymbol{r})\mathrm{d}\boldsymbol{r} = Nm$.

---

[13] And if we redefine the partition function to include a factor of $N!$ for identical particles.

[14] Balance equations were introduced in Chapter 1 phenomenologically, based on what seems obvious from experience.

[15] A simple fluid is a fluid with only a single chemical component.

[16] The Dirac delta function $\delta(x)$ is a *generalized function* such that $\int_{-\infty}^{\infty} \delta(x)F(x) = F(0)$, for F a smooth function. No ordinary function has this property; $\delta(x)$ has meaning only "inside" integrals. See Lighthill[99] or [13, p66].

Let's evaluate the ensemble average of $\hat{\rho}$ at $(\boldsymbol{r}, t)$ [see Eq. (4.1)]:

$$\begin{aligned}
\rho(\boldsymbol{r}, t) \equiv \langle \hat{\rho}(\boldsymbol{r}) \rangle_t &= m \sum_{\alpha=1}^{N} \int \mathrm{d}\boldsymbol{r}_1 \mathrm{d}\boldsymbol{v}_1 \cdots \mathrm{d}\boldsymbol{r}_\alpha \mathrm{d}\boldsymbol{v}_\alpha \delta(\boldsymbol{r} - \boldsymbol{r}_\alpha) \cdots \mathrm{d}\boldsymbol{r}_N \mathrm{d}\boldsymbol{v}_N \rho(1, \ldots, N, t) \\
&= m \sum_{\alpha=1}^{N} \int \mathrm{d}\boldsymbol{r}_\alpha \mathrm{d}\boldsymbol{v}_\alpha \delta(\boldsymbol{r} - \boldsymbol{r}_\alpha) f_1(\boldsymbol{r}_\alpha, \boldsymbol{v}_\alpha, t) = m \sum_{\alpha=1}^{N} \int \mathrm{d}\boldsymbol{v}_\alpha f_1(\boldsymbol{r}, \boldsymbol{v}_\alpha, t) \\
&= Nm \int \mathrm{d}\boldsymbol{v} f_1(\boldsymbol{r}, \boldsymbol{v}, t) = m \int \mathrm{d}\boldsymbol{v} F_1(\boldsymbol{r}, \boldsymbol{v}, t),
\end{aligned} \tag{4.31}$$

where we've used that $\boldsymbol{v}_\alpha$ is a dummy variable and Eq. (4.17).[17] Apart from the factor of mass, $\rho(\boldsymbol{r}, t)$ is the zeroth velocity moment of $F_1(\boldsymbol{r}, \boldsymbol{v}, t)$. The macroscopic density $\rho(\boldsymbol{r}, t)$ is associated with the same total mass as the microscopic density $\hat{\rho}(\boldsymbol{r})$, $\int \mathrm{d}\boldsymbol{r} \rho(\boldsymbol{r}, t) = Nm$; $\rho(\boldsymbol{r}, t)$ "smoothes out" the delta functions of $\hat{\rho}$.

A local, microscopic momentum density (or mass flux) is defined similarly:

$$\hat{\boldsymbol{g}}(\boldsymbol{r}) \equiv m \sum_\alpha \boldsymbol{v}_\alpha \delta(\boldsymbol{r} - \boldsymbol{r}_\alpha). \tag{4.32}$$

Repeating analogous steps, the ensemble average of $\hat{\boldsymbol{g}}$ at $(\boldsymbol{r}, t)$, is

$$\boldsymbol{g}(\boldsymbol{r}, t) \equiv \langle \hat{\boldsymbol{g}}(\boldsymbol{r}) \rangle_t = m \int \boldsymbol{v} F_1(\boldsymbol{r}, \boldsymbol{v}, t) \mathrm{d}\boldsymbol{v}. \tag{4.33}$$

The momentum density $\boldsymbol{g}(\boldsymbol{r}, t)$ is the first velocity moment of $F_1(\boldsymbol{r}, \boldsymbol{v}, t)$. From the mass density $\rho(\boldsymbol{r}, t)$, a velocity field $\boldsymbol{u}(\boldsymbol{r}, t)$, the *local velocity*, can be inferred from the mass flux,

$$\boldsymbol{g}(\boldsymbol{r}, t) = \rho(\boldsymbol{r}, t) \frac{1}{\rho(\boldsymbol{r}, t)} \boldsymbol{g}(\boldsymbol{r}, t) = \rho(\boldsymbol{r}, t) \frac{\int \boldsymbol{v} F_1(\boldsymbol{r}, \boldsymbol{v}, t) \mathrm{d}\boldsymbol{v}}{\int F_1(\boldsymbol{r}, \boldsymbol{v}, t) \mathrm{d}\boldsymbol{v}} \equiv \rho(\boldsymbol{r}, t) \boldsymbol{u}(\boldsymbol{r}, t). \tag{4.34}$$

Thus, $\boldsymbol{u}(\boldsymbol{r}, t)$ is the mean velocity at $(\boldsymbol{r}, t)$.[18]

Likewise we define the local energy density,[19]

$$\hat{\varepsilon}(\boldsymbol{r}) \equiv \sum_\alpha \left[\tfrac{1}{2} m v_\alpha^2 + \tfrac{1}{2} \sum_{\beta \neq \alpha} \Phi(\boldsymbol{r}_\alpha, \boldsymbol{r}_\beta) \right] \delta(\boldsymbol{r} - \boldsymbol{r}_\alpha), \tag{4.35}$$

with ensemble average at $(\boldsymbol{r}, t)$,

$$\begin{aligned}
\varepsilon(\boldsymbol{r}, t) \equiv \langle \hat{\varepsilon}(\boldsymbol{r}) \rangle_t &= \frac{1}{2} m \int \mathrm{d}\boldsymbol{v} v^2 F_1(\boldsymbol{r}, \boldsymbol{v}, t) + \frac{1}{2} \int \mathrm{d}\boldsymbol{v} \mathrm{d}\boldsymbol{r}' \mathrm{d}\boldsymbol{v}' F_2(\boldsymbol{r}, \boldsymbol{v}, \boldsymbol{r}', \boldsymbol{v}', t) \Phi(\boldsymbol{r}, \boldsymbol{r}') \\
&\equiv \varepsilon^K(\boldsymbol{r}, t) + \varepsilon^\Phi(\boldsymbol{r}, t).
\end{aligned} \tag{4.36}$$

The kinetic energy density ε^K is the second velocity moment of F_1; the potential energy density ε^Φ, although not a moment, results from a spatial weighting of F_2 by the two-body potential $\Phi(\boldsymbol{r}, \boldsymbol{r}')$.

The prescription just developed, of a correspondence through ensemble averages between microscopic functions $(\hat{\rho}, \hat{\boldsymbol{g}}, \hat{\varepsilon})$ and macroscopic densities $(\rho, \boldsymbol{g}, \varepsilon)$, works for mechanical quantities associated with one or more particles. The correspondence is a mapping between phase space and physical space. Not every macroscopic quantity can be so represented, however—macroscopic quantities exist that cannot be expressed as the average of a dynamical function weighted by a distribution function. Entropy is not attached to the dynamics of single particles; it's associated with the system as a whole. The densities of mass, momentum, and energy, however, suffice to establish balance equations for those quantities.

---

[17] These manipulations rely on the symmetry of joint probabilities; see Section 2.3.1.

[18] The quantity $F_1(\boldsymbol{r}, \boldsymbol{v}, t)/[\int F_1(\boldsymbol{r}, \boldsymbol{v}, t) \mathrm{d}\boldsymbol{v}]$ defines a local ($\boldsymbol{r}$-dependent) velocity probability distribution.

[19] Note that $\hat{\varepsilon}(\boldsymbol{r})$ is the total energy density; we'll soon define the internal energy used in thermodynamics. A microscopic form of the internal energy density is given in Chapter 6.

4.4.2 Balance equations from the hierarchy

Balance equations for mass, momentum, and energy follow from moments of the first two equations of the hierarchy.

4.4.2.1 *Mass conservation*

Equations (4.31) and (4.33) imply that the continuity equation holds at every $(\boldsymbol{r}, t)$,

$$\frac{\partial}{\partial t}\rho(\boldsymbol{r},t) + \boldsymbol{\nabla}\cdot\boldsymbol{g}(\boldsymbol{r},t) \equiv m\int \mathrm{d}\boldsymbol{v}\left(\frac{\partial}{\partial t} + \boldsymbol{v}\cdot\boldsymbol{\nabla}\right)F_1(\boldsymbol{r},\boldsymbol{v},t) = 0, \tag{4.37}$$

where the second equality follows from the zeroth velocity moment of the first equation of the hierarchy when distribution functions are assumed to vanish at far-off surfaces in velocity space. Using Eq. (P4.1), one sees that mass conservation holds in the presence of external forces.

4.4.2.2 *Momentum balance, pressure tensor* $\mathbf{P}$

With mass conservation implied by the zeroth moment, let's take the first velocity moment of the first equation of the hierarchy. Operate on Eq. (P4.1) with $\int \boldsymbol{v}\mathrm{d}\boldsymbol{v}$:

$$\int \boldsymbol{v}\left(\frac{\partial}{\partial t} + \boldsymbol{v}\cdot\boldsymbol{\nabla}_{\boldsymbol{r}} + \frac{\boldsymbol{F}}{m}\cdot\boldsymbol{\nabla}_{\boldsymbol{v}}\right)F_1(\boldsymbol{r},\boldsymbol{v},t)\mathrm{d}\boldsymbol{v} \tag{4.38}$$
$$= \frac{1}{m}\int \mathrm{d}\boldsymbol{r}'\mathrm{d}\boldsymbol{v}'[\boldsymbol{\nabla}_{\boldsymbol{r}}\Phi(\boldsymbol{r},\boldsymbol{r}')\cdot\boldsymbol{\nabla}_{\boldsymbol{v}}F_2(\boldsymbol{r},\boldsymbol{v},\boldsymbol{r}',\boldsymbol{v}',t)]\boldsymbol{v}\mathrm{d}\boldsymbol{v}.$$

This equation simplifies with the identities derived in Exercise 4.12; we find

$$\frac{\partial}{\partial t}\boldsymbol{g}(\boldsymbol{r},t) + m\int \boldsymbol{v}\left(\boldsymbol{v}\cdot\boldsymbol{\nabla}_{\boldsymbol{r}}\right)F_1(\boldsymbol{r},\boldsymbol{v},t)\mathrm{d}\boldsymbol{v}$$
$$+ \int \mathrm{d}\boldsymbol{v}\mathrm{d}\boldsymbol{r}'\mathrm{d}\boldsymbol{v}'F_2(\boldsymbol{r},\boldsymbol{v},\boldsymbol{r}',\boldsymbol{v}',t)\boldsymbol{\nabla}_{\boldsymbol{r}}\Phi(\boldsymbol{r},\boldsymbol{r}') = \frac{1}{m}\rho(\boldsymbol{r},t)\boldsymbol{F}(\boldsymbol{r}). \tag{4.39}$$

On the left of Eq. (4.39) we have the rate of change of momentum density $\partial\boldsymbol{g}/\partial t$ and on the right the external force density, $\rho(\boldsymbol{r},t)\boldsymbol{F}(\boldsymbol{r})/m$. To interpret the other terms, we note that momentum is conserved in the absence of external forces and thus we expect Eq. (4.39) for $\boldsymbol{F} = 0$ to be in the form of a balance equation,

$$\frac{\partial}{\partial t}g_i(\boldsymbol{r},t) + \sum_{j=1}^{3}\frac{\partial}{\partial \boldsymbol{r}_j}P_{ij}(\boldsymbol{r},t) = 0, \qquad (i = 1,2,3) \tag{4.40}$$

with P_{ij} elements of a momentum flux tensor $\mathbf{P}$, also known as the *pressure tensor*.[20] Let's see how that comes about. Readers uninterested in the details should jump to Eq. (4.49).

The vector field $\boldsymbol{I}(\boldsymbol{r}) \equiv \int \mathrm{d}\boldsymbol{v}\mathrm{d}\boldsymbol{r}'\mathrm{d}\boldsymbol{v}'F_2(\boldsymbol{r},\boldsymbol{v},\boldsymbol{r}',\boldsymbol{v}',t)\boldsymbol{\nabla}_{\boldsymbol{r}}\Phi(\boldsymbol{r},\boldsymbol{r}')$ is the force density at $\boldsymbol{r}$ arising from interactions with other system particles. Such a force would vanish if F_2 were spherically symmetric in the separation $\Delta\boldsymbol{r} \equiv \boldsymbol{r}' - \boldsymbol{r}$ [if Φ is spherically symmetric,[21] $\Phi(\boldsymbol{r},\boldsymbol{r}') = \Phi(|\boldsymbol{r}' - \boldsymbol{r}|)$]. The quantity $\boldsymbol{I}(\boldsymbol{r})$ is thus a measure of *inhomogeneities* in the system. To bring that out, we work with a generalization of $\boldsymbol{I}$ involving an integration over a new spatial variable $\boldsymbol{r}_1$,

$$\boldsymbol{I}(\boldsymbol{r}) = \int \mathrm{d}\boldsymbol{r}'h(\boldsymbol{r},\boldsymbol{r}')\boldsymbol{\nabla}_{\boldsymbol{r}}\Phi(\boldsymbol{r},\boldsymbol{r}') = \int \mathrm{d}\boldsymbol{r}'\mathrm{d}\boldsymbol{r}_1 h(\boldsymbol{r}_1,\boldsymbol{r}')\delta(\boldsymbol{r}_1 - \boldsymbol{r})\boldsymbol{\nabla}_{\boldsymbol{r}_1}\Phi(\boldsymbol{r}_1,\boldsymbol{r}'), \tag{4.41}$$

[20] The dimension of $\mathbf{P}$ is, equivalently, pressure, energy density, or momentum flux (see Exercise 4.14). The stress tensor $\mathbf{T}$ introduced in Eq. (1.17) can be used synonymously with $\mathbf{P}$; there's a minus sign difference in definition, $\mathbf{T} = -\mathbf{P}$.

[21] The dominant part of most intermolecular interactions is spherically symmetric.

where $h(\boldsymbol{r},\boldsymbol{r}') \equiv \int \mathrm{d}\boldsymbol{v}\mathrm{d}\boldsymbol{v}' F_2(\boldsymbol{r},\boldsymbol{v},\boldsymbol{r}',\boldsymbol{v}',t)$. As a step we'll use more than once, change variables $\boldsymbol{r}' \to \boldsymbol{r}_1$, $\boldsymbol{r}_1 \to \boldsymbol{r}'$, use $\boldsymbol{\nabla}_{\boldsymbol{r}}\Phi(|\boldsymbol{r}-\boldsymbol{r}'|) = -\boldsymbol{\nabla}_{\boldsymbol{r}'}\Phi(|\boldsymbol{r}-\boldsymbol{r}'|)$, and symmetry[22] $h(\boldsymbol{r}_1,\boldsymbol{r}') = h(\boldsymbol{r}',\boldsymbol{r}_1)$ to write

$$\boldsymbol{I}(\boldsymbol{r}) = \frac{1}{2}\int \mathrm{d}\boldsymbol{r}_1\mathrm{d}\boldsymbol{r}' h(\boldsymbol{r}_1,\boldsymbol{r}')\left[\delta(\boldsymbol{r}_1-\boldsymbol{r}) - \delta(\boldsymbol{r}'-\boldsymbol{r})\right]\boldsymbol{\nabla}_{\boldsymbol{r}_1}\Phi(\boldsymbol{r}_1,\boldsymbol{r}'). \tag{4.42}$$

In this form, $\boldsymbol{I}(\boldsymbol{r})$ is a measure of anisotropy in the internal force field.[23] Is there a way to handle the difference of two delta functions? Delta functions, although highly singular, nevertheless have derivatives when defined "inside" integrals (as in Eq. (4.42)).[24] For that reason, delta functions can be treated as analytic functions possessing Taylor series:

$$\delta(\boldsymbol{r}-\boldsymbol{r}_1) = \delta(\boldsymbol{r}-\boldsymbol{r}') - (\boldsymbol{r}_1-\boldsymbol{r}')\cdot\frac{\partial}{\partial\boldsymbol{r}}\delta(\boldsymbol{r}-\boldsymbol{r}') + \frac{1}{2}(\boldsymbol{r}_1-\boldsymbol{r}')(\boldsymbol{r}_1-\boldsymbol{r}'):\frac{\partial^2}{\partial\boldsymbol{r}^2}\delta(\boldsymbol{r}-\boldsymbol{r}') - \cdots,$$

where we've used dyadic notation and that $(\partial/\partial x)\delta(x-y) = -(\partial/\partial y)\delta(x-y)$ (the delta function is even; see [13, p106]). Thus, for the terms in square brackets in Eq. (4.42),

$$\delta(\boldsymbol{r}-\boldsymbol{r}_1) - \delta(\boldsymbol{r}-\boldsymbol{r}') = -(\boldsymbol{r}_1-\boldsymbol{r}')\cdot\frac{\partial}{\partial\boldsymbol{r}}\left[D_{\boldsymbol{r}}\delta(\boldsymbol{r}-\boldsymbol{r}')\right], \tag{4.43}$$

where $D_{\boldsymbol{r}} \equiv 1-(1/2)(\boldsymbol{r}_1-\boldsymbol{r}')\cdot\boldsymbol{\nabla}_{\boldsymbol{r}}+\cdots+(1/n!)\left[-(\boldsymbol{r}_1-\boldsymbol{r}')\cdot\boldsymbol{\nabla}_{\boldsymbol{r}}\right]^{n-1}+\cdots$. Equation (4.43) is due to Irving and Kirkwood.[100] Other approaches utilize Fourier transforms[101]. Combine Eq. (4.43) with Eq. (4.42) and take $D_{\boldsymbol{r}} = 1$ for simplicity,[25]

$$\boldsymbol{I}(\boldsymbol{r}) = \frac{1}{2}\int \mathrm{d}\boldsymbol{r}_1\mathrm{d}\boldsymbol{v}\mathrm{d}\boldsymbol{r}'\mathrm{d}\boldsymbol{v}' F_2(\boldsymbol{r}_1,\boldsymbol{v},\boldsymbol{r}',\boldsymbol{v}',t)\left[-(\boldsymbol{r}_1-\boldsymbol{r}')\cdot\boldsymbol{\nabla}_{\boldsymbol{r}}\delta(\boldsymbol{r}-\boldsymbol{r}')\right]\boldsymbol{\nabla}_{\boldsymbol{r}_1}\Phi(\boldsymbol{r}_1,\boldsymbol{r}'). \tag{4.44}$$

From Eq. (4.40), we expect $\boldsymbol{I}$ to be in the form of the divergence of a second-rank tensor, $\boldsymbol{I} \equiv \boldsymbol{\nabla}\cdot\mathbf{P}^{\Phi}$ (Φ indicates the contribution of intermolecular forces), with i^{th} vector component $I_i = \left[\boldsymbol{\nabla}\cdot\mathbf{P}^{\Phi}\right]_i \equiv \sum_j \partial P_{ij}^{\Phi}/\partial r_j$; see Eq. (1.21). We infer the tensor elements from Eq. (4.44) after a few steps,

$$P_{ij}^{\Phi}(\boldsymbol{r},t) = -\frac{1}{2}\int \mathrm{d}\boldsymbol{v}\mathrm{d}\boldsymbol{r}'\mathrm{d}\boldsymbol{v}' F_2(\boldsymbol{r},\boldsymbol{v},\boldsymbol{r}',\boldsymbol{v}',t)[\boldsymbol{\nabla}_{\boldsymbol{r}}\Phi(\boldsymbol{r},\boldsymbol{r}')]_i(\boldsymbol{r}-\boldsymbol{r}')_j. \tag{4.45}$$

$\mathbf{P}^{\Phi}(\boldsymbol{r},t)$ is symmetric ($P_{ij}^{\Phi} = P_{ji}^{\Phi}$) when $\Phi(\boldsymbol{r},\boldsymbol{r}')$ is spherically symmetric; see Exercise 4.15.

The remaining term in Eq. (4.39) is already in the form of the divergence of a tensor:

$$m\int \boldsymbol{v}(\boldsymbol{v}\cdot\boldsymbol{\nabla})F_1(\boldsymbol{r},\boldsymbol{v},t)\mathrm{d}\boldsymbol{v} = \sum_i \frac{\partial}{\partial r_i}\int m\boldsymbol{v}v_i F_1(\boldsymbol{r},\boldsymbol{v},t)\mathrm{d}\boldsymbol{v} \equiv \boldsymbol{\nabla}\cdot\mathbf{P}^{K},$$

with

$$P_{ij}^{K}(\boldsymbol{r},t) = m\int v_i v_j F_1(\boldsymbol{r},\boldsymbol{v},t)\mathrm{d}\boldsymbol{v}, \tag{4.46}$$

where K indicates the kinetic contribution to $\mathbf{P}$. The manifestly symmetric tensor field $\mathbf{P}^{K}(\boldsymbol{r},t)$ is associated with the flux of particle momentum, averaged over the one-particle distribution.

---

[22] A general property of joint phase-space probability distributions, $F_2(\boldsymbol{r}_1,\boldsymbol{v}_1,\boldsymbol{r}_2,\boldsymbol{v}_2,t) = F_2(\boldsymbol{r}_2,\boldsymbol{v}_2,\boldsymbol{r}_1,\boldsymbol{v}_1,t)$.

[23] We can also write the integral, with $\boldsymbol{r}' = \boldsymbol{r}+\Delta\boldsymbol{r}$, $\boldsymbol{I}(\boldsymbol{r}) = \frac{1}{2}\int \mathrm{d}(\Delta\boldsymbol{r})\boldsymbol{\nabla}_{\boldsymbol{r}}\Phi(|\Delta\boldsymbol{r}|)\left[h(\boldsymbol{r},\boldsymbol{r}+\Delta\boldsymbol{r}) - h(\boldsymbol{r},\boldsymbol{r}-\Delta\boldsymbol{r})\right]$.

[24] The nth-derivative of $\delta(x)$, $\delta^{(n)}(x)$, has the property $\int_{-\infty}^{\infty}\delta^{(n)}(x)F(x)\mathrm{d}x = (-1)^n F^{(n)}(0)$ for any smooth function $F(x)$; see Lighthill[99, p19].

[25] Expressions featuring the full expansion with $D_{\boldsymbol{r}} \neq 1$ are given in Irving and Kirkwood[100]. Massignon[102, p32] avoids these expansions and provides a more careful presentation on the use of delta function techniques.

We now separate the contributions to $\mathbf{P}^K$ into those from the mean velocity $\boldsymbol{u}(\boldsymbol{r},t)$ and from "random" velocities measured relative to the mean—*thermal velocities*.[26] Using dyadic notation, add and subtract $\boldsymbol{u}$,

$$\begin{aligned}\mathbf{P}^K(\boldsymbol{r},t) &= m\int \boldsymbol{v}\boldsymbol{v}F_1(\boldsymbol{r},\boldsymbol{v},t)\mathrm{d}\boldsymbol{v} = m\int(\boldsymbol{v}-\boldsymbol{u}+\boldsymbol{u})(\boldsymbol{v}-\boldsymbol{u}+\boldsymbol{u})F_1(\boldsymbol{r},\boldsymbol{v},t)\mathrm{d}\boldsymbol{v}\\ &= \rho(\boldsymbol{r},t)\boldsymbol{u}(\boldsymbol{r},t)\boldsymbol{u}(\boldsymbol{r},t) + m\int(\boldsymbol{v}-\boldsymbol{u})(\boldsymbol{v}-\boldsymbol{u})F_1(\boldsymbol{r},\boldsymbol{v},t)\mathrm{d}\boldsymbol{v}\\ &\equiv \rho(\boldsymbol{r},t)\boldsymbol{u}(\boldsymbol{r},t)\boldsymbol{u}(\boldsymbol{r},t) + \widetilde{\mathbf{P}}^K(\boldsymbol{r},t),\end{aligned} \tag{4.47}$$

with

$$\widetilde{P}^K_{ij}(\boldsymbol{r},t) = m\int(\boldsymbol{v}-\boldsymbol{u})_i(\boldsymbol{v}-\boldsymbol{u})_jF_1(\boldsymbol{r},\boldsymbol{v},t)\mathrm{d}\boldsymbol{v}. \tag{4.48}$$

We define the pressure tensor $\mathbf{P}$ without the convection term $\rho\boldsymbol{u}\boldsymbol{u}$, $\mathbf{P}\equiv\widetilde{\mathbf{P}}^K+\mathbf{P}^\Phi$.

Equation (4.39) is therefore a momentum balance equation [see Eq. (1.20) noting that $\mathbf{T}=-\mathbf{P}$],

$$\frac{\partial \boldsymbol{g}}{\partial t} + \boldsymbol{\nabla}\cdot(\rho\boldsymbol{u}\boldsymbol{u}+\mathbf{P}) = \frac{1}{m}\rho\boldsymbol{F}. \tag{4.49}$$

The term $\boldsymbol{\nabla}\cdot\rho\boldsymbol{u}\boldsymbol{u}$ represents the rate of change in local momentum due to the mean motion of the system; $\boldsymbol{\nabla}\cdot\mathbf{P}$ represents the rate of momentum change by means other than average flow, from random motions about the mean velocity ($\widetilde{\mathbf{P}}^K$) or internal forces ($\mathbf{P}^\Phi$).

A stress system is *hydrostatic* if an element of area always has a stress normal to itself for any surface orientation. The hydrostatic pressure, a scalar field, is defined as a third of the trace of $\mathbf{P}$,

$$\begin{aligned}P(\boldsymbol{r},t)\equiv\frac{1}{3}\operatorname{Tr}\mathbf{P}(\boldsymbol{r},t) =& \frac{m}{3}\int(\boldsymbol{v}-\boldsymbol{u})^2F_1(\boldsymbol{r},\boldsymbol{v},t)\mathrm{d}\boldsymbol{v}\\ &-\frac{1}{6}\int\mathrm{d}\boldsymbol{v}\mathrm{d}\boldsymbol{r}'\mathrm{d}\boldsymbol{v}'F_2(\boldsymbol{r},\boldsymbol{v},\boldsymbol{r}',\boldsymbol{v}',t)\boldsymbol{\nabla}_{\boldsymbol{r}}\Phi(\boldsymbol{r},\boldsymbol{r}')\cdot(\boldsymbol{r}-\boldsymbol{r}').\end{aligned} \tag{4.50}$$

There's a kinetic contribution to the pressure (present in all systems, significant in gases) and a contribution from forces associated with two-body interactions (important in liquids).[27]

4.4.2.3 Energy balance total and internal, heat flux $\boldsymbol{J}_Q$

Operate on Eq. (P4.1) with $\frac{1}{2}m\int\mathrm{d}\boldsymbol{v}v^2$ (second velocity moment) to derive a balance equation for kinetic energy,

$$\frac{\partial}{\partial t}\varepsilon^K + \boldsymbol{\nabla}\cdot\boldsymbol{J}^K = -\int\mathrm{d}\boldsymbol{v}\mathrm{d}\boldsymbol{r}'\mathrm{d}\boldsymbol{v}'F_2(\boldsymbol{r},\boldsymbol{v},\boldsymbol{r}',\boldsymbol{v}',t)\boldsymbol{v}\cdot\boldsymbol{\nabla}_{\boldsymbol{r}}\Phi(\boldsymbol{r},\boldsymbol{r}')+\frac{1}{m}\rho\boldsymbol{F}\cdot\boldsymbol{u}, \tag{4.51}$$

where we have the kinetic energy flux (third velocity moment of F_1),

$$\boldsymbol{J}^K(\boldsymbol{r},t)\equiv\frac{m}{2}\int\boldsymbol{v}v^2F_1(\boldsymbol{r},\boldsymbol{v},t)\mathrm{d}\boldsymbol{v}. \tag{4.52}$$

Equation (4.51) is a work-energy theorem: The rate at which kinetic energy changes is balanced by the rate at which forces do work, internal and external.

Now operate on Eq. (4.29) (the second equation of the hierarchy) with $\frac{1}{2}\int\mathrm{d}\boldsymbol{v}\mathrm{d}\boldsymbol{r}'\mathrm{d}\boldsymbol{v}'\Phi(\boldsymbol{r},\boldsymbol{r}')$:

$$\frac{\partial}{\partial t}\varepsilon^\Phi(\boldsymbol{r},t)+\frac{1}{2}\int\mathrm{d}\boldsymbol{v}\mathrm{d}\boldsymbol{r}'\mathrm{d}\boldsymbol{v}'\Phi(\boldsymbol{r},\boldsymbol{r}')\left(\boldsymbol{v}\cdot\frac{\partial}{\partial\boldsymbol{r}}+\boldsymbol{v}'\cdot\frac{\partial}{\partial\boldsymbol{r}'}\right)F_2(\boldsymbol{r},\boldsymbol{v},\boldsymbol{r}',\boldsymbol{v}',t)=0. \tag{4.53}$$

[26] What we're calling thermal velocities, astronomers refer to as *peculiar velocities*.

[27] Equation (4.50) justifies the reasoning adopted by van der Waals that pressure is lowered by attractive interatomic forces; see [5, p191]. Conversely, pressure is increased by repulsive interactions, such as degeneracy pressure; see [5, p149].

Contributions from O_{12} in L_2 and the source terms on the right of Eq. (4.29) vanish in arriving at Eq. (4.53); all velocity derivatives lead to vanishing surface terms when integrated over $\mathrm{d}\boldsymbol{v}\mathrm{d}\boldsymbol{v}'$. To simplify Eq. (4.53), use the identity $\Phi\partial F_2/\partial\boldsymbol{r} = \partial(F_2\Phi)/\partial\boldsymbol{r} - F_2\partial\Phi/\partial\boldsymbol{r}$. We then have the balance equation for potential energy:

$$\frac{\partial}{\partial t}\varepsilon^{\Phi} + \boldsymbol{\nabla}\cdot\boldsymbol{J}^{\Phi} = \frac{1}{2}\int \mathrm{d}\boldsymbol{v}\mathrm{d}\boldsymbol{r}'\mathrm{d}\boldsymbol{v}'F_2(\boldsymbol{r},\boldsymbol{v},\boldsymbol{r}',\boldsymbol{v}',t)\left[\boldsymbol{v}\cdot\boldsymbol{\nabla}_{\boldsymbol{r}} + \boldsymbol{v}'\cdot\boldsymbol{\nabla}_{\boldsymbol{r}'}\right]\Phi(\boldsymbol{r},\boldsymbol{r}'), \tag{4.54}$$

where the potential energy flux is

$$\boldsymbol{J}^{\Phi}(\boldsymbol{r},t) \equiv \frac{1}{2}\int \mathrm{d}\boldsymbol{v}\mathrm{d}\boldsymbol{r}'\mathrm{d}\boldsymbol{v}'F_2(\boldsymbol{r},\boldsymbol{v},\boldsymbol{r}',\boldsymbol{v}',t)\Phi(\boldsymbol{r},\boldsymbol{r}')\boldsymbol{v}. \tag{4.55}$$

By adding Eqs. (4.51) and (4.54) and integrating over space, one can show that

$$\int \mathrm{d}\boldsymbol{r}\left[\frac{\partial}{\partial t}\left(\varepsilon^K + \varepsilon^{\Phi}\right) + \boldsymbol{\nabla}\cdot\left(\boldsymbol{J}^K + \boldsymbol{J}^{\Phi}\right)\right] = \frac{1}{m}\int \mathrm{d}\boldsymbol{r}\rho\boldsymbol{u}\cdot\boldsymbol{F}, \tag{4.56}$$

implying local energy balance (see Exercise 4.16),

$$\frac{\partial}{\partial t}\varepsilon + \boldsymbol{\nabla}\cdot\boldsymbol{J}_{\varepsilon} = \frac{1}{m}\rho\boldsymbol{u}\cdot\boldsymbol{F}, \tag{4.57}$$

where $\varepsilon \equiv \varepsilon^K + \varepsilon^{\Phi}$ and $\boldsymbol{J}_{\varepsilon} \equiv \boldsymbol{J}^K + \boldsymbol{J}^{\Phi}$. In the absence of external forces, *energy conservation is an exact consequence of the first two equations of the hierarchy* (without having to know F_3).

Internal energy, the province of thermodynamics, is a component of the total energy. It can be isolated by separating the contributions to $\varepsilon^K, \boldsymbol{J}^K, \boldsymbol{J}^{\Phi}$ into those from the mean motion and from thermal velocities (as we did with $\mathbf{P}^K$ in Eq. (4.47)). It can be shown that:

$$\begin{aligned}
\varepsilon^K &= \frac{1}{2}\rho u^2 + \frac{m}{2}\int(\boldsymbol{v}-\boldsymbol{u})^2F_1(\boldsymbol{r},\boldsymbol{v},t)\mathrm{d}\boldsymbol{v} \equiv \frac{1}{2}\rho u^2 + \widetilde{\varepsilon}^K;\\
\boldsymbol{J}^K &= \boldsymbol{u}\cdot\widetilde{\mathbf{P}}^K + \boldsymbol{u}\left(\frac{1}{2}\rho u^2 + \widetilde{\varepsilon}^K\right) + \frac{m}{2}\int(\boldsymbol{v}-\boldsymbol{u})(\boldsymbol{v}-\boldsymbol{u})^2F_1(\boldsymbol{r},\boldsymbol{v},t)\mathrm{d}\boldsymbol{v}\\
&\equiv \boldsymbol{u}\cdot\widetilde{\mathbf{P}}^K + \boldsymbol{u}\left(\frac{1}{2}\rho u^2 + \widetilde{\varepsilon}^K\right) + \widetilde{\boldsymbol{J}}^K;\\
\boldsymbol{J}^{\Phi} &= \boldsymbol{u}\varepsilon^{\Phi} + \frac{1}{2}\int \mathrm{d}\boldsymbol{v}\mathrm{d}\boldsymbol{r}'\mathrm{d}\boldsymbol{v}'(\boldsymbol{v}-\boldsymbol{u})\Phi(\boldsymbol{r},\boldsymbol{r}')F_2(\boldsymbol{r},\boldsymbol{v},\boldsymbol{r}'\boldsymbol{v}',t) \equiv \boldsymbol{u}\varepsilon^{\Phi} + \widetilde{\boldsymbol{J}}^{\Phi}.
\end{aligned} \tag{4.58}$$

Internal energy, denoted $\widetilde{\varepsilon}$, is the sum of the average thermal kinetic energy and potential energy,[28]

$$\widetilde{\varepsilon} \equiv \widetilde{\varepsilon}^K + \varepsilon^{\Phi} = \varepsilon - \frac{1}{2}\rho u^2. \tag{4.59}$$

The flux of internal energy, traditionally called the heat flux [see Eq. (1.30)], is defined

$$\boldsymbol{J}_Q \equiv \widetilde{\boldsymbol{J}}^K + \widetilde{\boldsymbol{J}}^{\Phi}. \tag{4.60}$$

By combining Eqs. (4.58)–(4.60) with Eq. (4.57), we have a rewrite of the energy balance equation,

$$\frac{\partial}{\partial t}\left(\frac{1}{2}\rho u^2 + \widetilde{\varepsilon}\right) + \boldsymbol{\nabla}\cdot\left[\boldsymbol{J}_Q + \boldsymbol{u}\cdot\widetilde{\mathbf{P}}^K + \boldsymbol{u}\left(\frac{1}{2}\rho u^2 + \widetilde{\varepsilon}\right)\right] = \frac{1}{m}\rho\boldsymbol{u}\cdot\boldsymbol{F}. \tag{4.61}$$

The total energy flux has contributions from heat, work, and convected energy.

---

[28] In Chapter 1 we denoted the internal energy density ρu. We use $\widetilde{\varepsilon}$ here to avoid confusion with the average speed u.

A balance equation for $\frac{1}{2}\rho u^2$ follows by projecting the momentum balance equation onto $\boldsymbol{u}$,

$$\boldsymbol{u}\cdot\frac{\partial}{\partial t}(\rho\boldsymbol{u})+\boldsymbol{u}\cdot\boldsymbol{\nabla}\cdot(\mathbf{P}+\rho\boldsymbol{u}\boldsymbol{u})=\frac{1}{m}\rho\boldsymbol{u}\cdot\boldsymbol{F}, \tag{4.62}$$

which is equivalent to [see Exercise 4.18]

$$\frac{\partial}{\partial t}\left(\frac{1}{2}\rho u^2\right)+\boldsymbol{\nabla}\cdot\left(\frac{1}{2}\rho u^2\boldsymbol{u}+\boldsymbol{u}\cdot\mathbf{P}\right)=\mathbf{P}{:}\boldsymbol{\nabla}\boldsymbol{u}+\frac{1}{m}\rho\boldsymbol{u}\cdot\boldsymbol{F}. \tag{4.63}$$

By subtracting Eq. (4.63) from Eq. (4.61), we arrive at a balance equation for internal energy,

$$\frac{\partial\widetilde{\varepsilon}}{\partial t}+\boldsymbol{\nabla}\cdot\left(\boldsymbol{u}\widetilde{\varepsilon}+\boldsymbol{J}_Q-\boldsymbol{u}\cdot\mathbf{P}^{\Phi}\right)=-\mathbf{P}{:}\boldsymbol{\nabla}\boldsymbol{u}. \tag{4.64}$$

Note that 1) internal energy does not couple to external forces and 2) convection of internal energy occurs naturally in the theory and we have a microscopic expression for the heat flux [see Eq. (4.60)], physical effects introduced phenomenologically in Chapter 1.[29] We see explicitly that internal energy is not conserved; the source term $-\mathbf{P}{:}\boldsymbol{\nabla}\boldsymbol{u}$ represents a conversion of internal into kinetic energy, it's the work done by the fluid against pressure forces.

4.4.3 Microscopic balance equations

Balance equations for $(\rho, \boldsymbol{g}, \varepsilon)$ result from moments of the hierarchy, equations in the form of relations among ensemble averages. Yet, conservation of mass, momentum, and energy hold for each member of an ensemble, implying that *balance equations don't require ensemble averaging to be true*. In this section, we show that the balance equations for $(\rho, \boldsymbol{g}, \varepsilon)$ follow from averages of balance equations for the microscopic densities $(\hat{\rho}, \hat{\boldsymbol{g}}, \hat{\varepsilon})$. These results will be used in Chapter 6.

Differentiate Eq. (4.30) with respect to time and Eq. (4.32) with respect to space, leaving us with $(\boldsymbol{v}_\alpha\equiv\dot{\boldsymbol{r}}_\alpha)$,[30]

$$\frac{\partial\hat{\rho}}{\partial t}=m\sum_\alpha(-\dot{\boldsymbol{r}}_\alpha\cdot\boldsymbol{\nabla}_{\boldsymbol{r}})\delta(\boldsymbol{r}-\boldsymbol{r}_\alpha)=-\boldsymbol{\nabla}\cdot\hat{\boldsymbol{g}}. \tag{4.65}$$

The time rate of change of $\hat{\rho}$ is balanced by the negative divergence of $\hat{\boldsymbol{g}}$. The ensemble average of Eq. (4.65) reproduces Eq. (4.37); $\langle\hat{\rho}(\boldsymbol{r})\rangle_t=\rho(\boldsymbol{r},t)$ and $\langle\hat{\boldsymbol{g}}(\boldsymbol{r})\rangle_t=\boldsymbol{g}(\boldsymbol{r},t)$.

Microscopic momentum balance follows by differentiating Eq. (4.32) with respect to time:

$$\frac{\partial\hat{g}_i}{\partial t}=m\sum_\alpha\left[v_{\alpha,i}\left(-\boldsymbol{v}_\alpha\cdot\boldsymbol{\nabla}\right)+\dot{v}_{\alpha,i}\right]\delta(\boldsymbol{r}-\boldsymbol{r}_\alpha)\equiv-\sum_j\frac{\partial}{\partial x_j}\hat{P}^K_{ij}+\hat{f}_i, \tag{4.66}$$

where $\hat{P}^K_{ij}(\boldsymbol{r})\equiv m\sum_\alpha v_{\alpha,i}v_{\alpha,j}\delta(\boldsymbol{r}-\boldsymbol{r}_\alpha)$ are microscopic components of $\mathbf{P}^K$ $[\langle\hat{P}^K_{ij}(\boldsymbol{r})\rangle_t=P^K_{ij}(\boldsymbol{r},t)]$ and $\hat{f}_i(\boldsymbol{r})\equiv m\sum_\alpha\dot{v}_{\alpha,i}\delta(\boldsymbol{r}-\boldsymbol{r}_\alpha)$ are components of a microscopic force density, $\hat{\boldsymbol{f}}$. In vector form,

$$\frac{\partial\hat{\boldsymbol{g}}}{\partial t}+\boldsymbol{\nabla}\cdot\hat{\mathbf{P}}^K=\hat{\boldsymbol{f}}. \tag{4.67}$$

In identifying $\hat{\boldsymbol{f}}$ with force, we've used Newton's second law wherein acceleration is associated with force when observed from inertial reference frames.[31] The force could be external or internal in origin, however; let's write $\hat{\boldsymbol{f}}=\hat{\boldsymbol{f}}^{\text{int}}+\hat{\boldsymbol{f}}^{\text{ext}}$.

---

[29] Internal energy is motivated in thermodynamics as the kinetic energy of thermal agitations and the potential energy associated with interatomic forces, a picture that emerges here from microscopic dynamics. Heat flux was defined in Chapter 1 as "whatever's left over" in accounting for different types of energies, with no analytic way of characterizing it other than giving it a name. Here have an expression for $\boldsymbol{J}_Q(\boldsymbol{r},t)$ starting from first principles (Liouville's equation).

[30] To show that $(\partial/\partial t)\delta(\boldsymbol{r}-\boldsymbol{r}_\alpha)=-(\boldsymbol{v}_\alpha\cdot\boldsymbol{\nabla}_{\boldsymbol{r}})\delta(\boldsymbol{r}-\boldsymbol{r}_\alpha)$, note that $\delta(\boldsymbol{r})\equiv\delta(x)\delta(y)\delta(z)$; see [13, p261].

[31] We've implicitly used inertial frames in our development. A careful discussion of inertial frames is given in [6, p4].

For internal forces derivable from two-body interactions $\Phi(\boldsymbol{r}_\alpha, \boldsymbol{r}_\beta)$ ($\beta \neq \alpha$ in these sums),

$$\hat{f}_i^{\text{int}}(\boldsymbol{r}) \equiv \sum_{\alpha,\beta} \delta(\boldsymbol{r}-\boldsymbol{r}_\alpha)\left[-\boldsymbol{\nabla}_{\boldsymbol{r}_\alpha}\Phi(\boldsymbol{r}_\alpha,\boldsymbol{r}_\beta)\right]_i = \frac{1}{2}\sum_{\alpha,\beta}\left[\delta(\boldsymbol{r}_\alpha-\boldsymbol{r})-\delta(\boldsymbol{r}_\beta-\boldsymbol{r})\right]\left[-\boldsymbol{\nabla}_{\boldsymbol{r}_\alpha}\Phi(\boldsymbol{r}_\alpha,\boldsymbol{r}_\beta)\right]_i$$
$$= -\frac{1}{2}\sum_{\alpha,\beta}\left[-\boldsymbol{\nabla}_{\boldsymbol{r}_\alpha}\Phi(\boldsymbol{r}_\alpha,\boldsymbol{r}_\beta)\right]_i(\boldsymbol{r}_\alpha-\boldsymbol{r}_\beta)\cdot\frac{\partial}{\partial \boldsymbol{r}}\delta(\boldsymbol{r}-\boldsymbol{r}_\beta) \equiv -\sum_j \frac{\partial}{\partial r_j}\hat{P}_{ij}^\Phi = -\left[\boldsymbol{\nabla}\cdot\hat{\mathbf{P}}^\Phi\right]_i,$$

where the second equality follows as the discrete version of Eq. (4.42) (through similar steps), the third uses Eq. (4.43) with $D_{\boldsymbol{r}} = 1$, and $\hat{P}_{ij}^\Phi$ are microscopic components of $\mathbf{P}^\Phi$ [$\langle \hat{P}_{ij}^\Phi(\boldsymbol{r})\rangle_t = P_{ij}^\Phi(\boldsymbol{r},t)$],

$$\hat{P}_{ij}^\Phi(\boldsymbol{r}) \equiv -\frac{1}{2}\sum_{\alpha,\beta}\left[\boldsymbol{\nabla}_{\boldsymbol{r}_\alpha}\Phi(\boldsymbol{r}_\alpha,\boldsymbol{r}_\beta)\right]_i(\boldsymbol{r}_\alpha-\boldsymbol{r}_\beta)_j\delta(\boldsymbol{r}-\boldsymbol{r}_\alpha). \tag{4.68}$$

Thus, $\hat{\boldsymbol{f}}^{\text{int}} = -\boldsymbol{\nabla}\cdot\hat{\mathbf{P}}^\Phi$, implying microscopic momentum balance,

$$\frac{\partial \hat{\boldsymbol{g}}}{\partial t} + \boldsymbol{\nabla}\cdot\hat{\mathbf{P}} = \hat{\boldsymbol{f}}^{\text{ext}}, \tag{4.69}$$

where $\hat{\mathbf{P}} \equiv \hat{\mathbf{P}}^K + \hat{\mathbf{P}}^\Phi$ and $\hat{\boldsymbol{f}}^{\text{ext}}(\boldsymbol{r}) \equiv \sum_\alpha \boldsymbol{F}^{\text{ext}}(\boldsymbol{r}_\alpha)\delta(\boldsymbol{r}-\boldsymbol{r}_\alpha)$ is the microscopic force density associated with external forces; $\langle \hat{\boldsymbol{f}}^{\text{ext}}(\boldsymbol{r})\rangle_t = (1/m)\rho(\boldsymbol{r},t)\boldsymbol{F}^{\text{ext}}(\boldsymbol{r})$.

Microscopic energy balance follows by differentiating Eq. (4.35) with respect to time,

$$\frac{\partial}{\partial t}\hat{\varepsilon}(\boldsymbol{r}) = \sum_\alpha\left[\frac{1}{2}mv_\alpha^2 + \frac{1}{2}\sum_{\beta\neq\alpha}\Phi(\boldsymbol{r}_\alpha,\boldsymbol{r}_\beta)\right](-\boldsymbol{v}_\alpha\cdot\boldsymbol{\nabla})\,\delta(\boldsymbol{r}-\boldsymbol{r}_\alpha)$$
$$+\sum_\alpha \delta(\boldsymbol{r}-\boldsymbol{r}_\alpha)\left[m\boldsymbol{v}_\alpha\cdot\dot{\boldsymbol{v}}_\alpha + \frac{1}{2}\sum_{\beta\neq\alpha}(\boldsymbol{v}_\alpha+\boldsymbol{v}_\beta)\cdot\boldsymbol{\nabla}_{\boldsymbol{r}_\alpha}\Phi(\boldsymbol{r}_\alpha,\boldsymbol{r}_\beta)\right]. \tag{4.70}$$

Through familiar steps, Eq. (4.70) has the form of a balance equation,

$$\frac{\partial \hat{\varepsilon}}{\partial t} + \boldsymbol{\nabla}\cdot\hat{\boldsymbol{J}}_\varepsilon = \hat{w}, \tag{4.71}$$

where $\hat{\boldsymbol{J}}_\varepsilon$ is the total energy flux and $\hat{w}$ is the power per volume expended by external forces,

$$\hat{\boldsymbol{J}}_\varepsilon(\boldsymbol{r}) \equiv \sum_\alpha\left[\boldsymbol{v}_\alpha\left(\frac{1}{2}mv_\alpha^2 + \frac{1}{2}\sum_{\beta\neq\alpha}\Phi(\boldsymbol{r}_\alpha,\boldsymbol{r}_\beta)\right)\right.$$
$$\left.+\frac{1}{4}\sum_{\beta\neq\alpha}(\boldsymbol{v}_\alpha+\boldsymbol{v}_\beta)\cdot\boldsymbol{\nabla}_{\boldsymbol{r}_\alpha}\Phi(\boldsymbol{r}_\alpha,\boldsymbol{r}_\beta)\,(\boldsymbol{r}_\alpha-\boldsymbol{r}_\beta)\right]\delta(\boldsymbol{r}-\boldsymbol{r}_\alpha)$$
$$\hat{w}(\boldsymbol{r}) \equiv \sum_\alpha \boldsymbol{v}_\alpha\cdot\boldsymbol{F}^{\text{ext}}(\boldsymbol{r}_\alpha)\delta(\boldsymbol{r}-\boldsymbol{r}_\alpha). \tag{4.72}$$

One can show[103, p200] that $\langle \hat{\boldsymbol{J}}_\varepsilon(\boldsymbol{r})\rangle_t = \boldsymbol{J}_\varepsilon(\boldsymbol{r},t)$ and $\langle \hat{w}(\boldsymbol{r})\rangle_t = (1/m)\rho(\boldsymbol{r},t)\boldsymbol{u}(\boldsymbol{r},t)\cdot\boldsymbol{F}^{\text{ext}}(\boldsymbol{r})$.

4.4.4 Hydrodynamics: Adding constitutive relations to conservation laws

The equations of hydrodynamics in their most basic form are the balance equations for mass, momentum, and internal energy. We collect them here,

$$\frac{\partial}{\partial t}\rho + \boldsymbol{\nabla}\cdot(\rho\boldsymbol{u}) = 0$$
$$\frac{\partial}{\partial t}\rho\boldsymbol{u} + \boldsymbol{\nabla}\cdot(\rho\boldsymbol{u}\boldsymbol{u}+\mathbf{P}) = 0$$
$$\frac{\partial}{\partial t}\widetilde{\varepsilon} + \boldsymbol{\nabla}\cdot(\boldsymbol{u}\widetilde{\varepsilon}+\boldsymbol{J}) = -\mathbf{P}{:}\boldsymbol{\nabla}\boldsymbol{u}, \tag{4.73}$$

where $\boldsymbol{J}$ includes all contributions to the heat flux. These equations are incomplete, however. To be practical, they must be supplemented with constitutive relations between the fluxes $(\mathbf{P}, \boldsymbol{J})$ and the fields $(\rho, \boldsymbol{u}, \widetilde{\varepsilon})$. We have *definitions* of $(\mathbf{P}, \boldsymbol{J})$ involving the one and two-particle distributions but such expressions are formal unless we can calculate the reduced distributions (the job of kinetic theory). The traditional way of closing the equations of hydrodynamics is to parameterize $\mathbf{P}$ and $\boldsymbol{J}$.

In equilibrium, spatially uniform[32] and time independent, $\boldsymbol{u} = 0$, $\boldsymbol{J} = 0$, and $\mathbf{P}$ is diagonal, $P_{ij} = P\delta_{ij}$, with P the hydrostatic pressure. For out-of-equilibrium systems, $\boldsymbol{J} \neq 0$ and $\mathbf{P}$ has off-diagonal elements, *effects associated with deviations from uniformity*. Write $\mathbf{P}$ so that the contributions of inhomogeneities add to the equilibrium form,

$$P_{ij} = P\delta_{ij} + \Pi_{ij}, \tag{4.74}$$

where the tensor elements Π_{ij} model momentum fluxes mediated by inhomogeneities. Deviations from uniformity are, in a first approximation, represented by linear functions of gradients in the local fields[33] $(\rho, \boldsymbol{u}, T)$. We take Π_{ij} to be a linear combination of velocity gradients,

$$\Pi_{ij} \equiv \sum_{l,m} A_{ij}^{lm} \nabla_l u_m, \tag{4.75}$$

where A_{ij}^{lm} are expansion coefficients. Any second-rank tensor $\mathbf{B}$ can be written as a sum of symmetric traceless, antisymmetric, and diagonal tensors:

$$B_{ij} = \tfrac{1}{2}\left(B_{ij} + B_{ji} - \tfrac{2}{3}\delta_{ij}\operatorname{Tr} B\right) + \tfrac{1}{2}\left(B_{ij} - B_{ji}\right) + \tfrac{1}{3}\delta_{ij}\operatorname{Tr} B. \tag{4.76}$$

Because we require $\mathbf{P}$ to be symmetric (conservation of angular momentum, Exercise 4.19), the simplest form of the expansion in Eq. (4.75) is that of Eq. (4.76),

$$\Pi_{ij} = -\eta\left(\nabla_i u_j + \nabla_j u_i - \tfrac{2}{3}\delta_{ij}\boldsymbol{\nabla}\cdot\boldsymbol{u}\right) - \zeta\delta_{ij}\boldsymbol{\nabla}\cdot\boldsymbol{u} \tag{4.77}$$

where η and ζ are the coefficients of shear and bulk viscosity.[34] Likewise we represent heat flux in terms of gradients,

$$\boldsymbol{J} = -\lambda\boldsymbol{\nabla}\rho - \kappa\boldsymbol{\nabla}T, \tag{4.78}$$

where κ is the thermal conductivity and λ has no standard name. We show presently that $\lambda = 0$.

For simple fluids the entropy source has the form[35]

$$\sigma_S = -\frac{1}{T^2}\boldsymbol{J}\cdot\boldsymbol{\nabla}T - \frac{1}{T}\boldsymbol{\Pi}{:}\boldsymbol{\nabla}\boldsymbol{u}; \tag{4.79}$$

Combining Eqs. (4.77)–(4.79), we find

$$\sigma_S = \frac{\lambda}{T^2}\boldsymbol{\nabla}\rho\cdot\boldsymbol{\nabla}T + \frac{\kappa}{T^2}\left|\boldsymbol{\nabla}T\right|^2 + \frac{\eta}{2T}\sum_{ij}\left[\nabla_j u_i + \nabla_i u_j - \tfrac{2}{3}\delta_{ij}\boldsymbol{\nabla}\cdot\boldsymbol{u}\right]^2 + \frac{3\zeta}{T}\left(\boldsymbol{\nabla}\cdot\boldsymbol{u}\right)^2. \tag{4.80}$$

Thus, the inequality $\sigma_S \geq 0$ implies $\kappa \geq 0, \eta \geq 0, \zeta \geq 0$. These sign requirements are as general as the thermodynamic stability conditions, $C_V \geq 0$ and $\beta_T \geq 0$; see [2, p48] or [5, p26]. Because $\boldsymbol{\nabla}\rho\cdot\boldsymbol{\nabla}T$ can be of either sign, we set $\lambda = 0$ to have a theory manifestly consistent with the second law of thermodynamics.[36] The hydrodynamics of simple fluids involves three transport coefficients,

---

[32] By uniform we mean isotropic and homogeneous—all directions in space equivalent with no preferred spatial location. These concepts are distinct: Isotropy does not necessarily imply homogeneity nor does homogeneity necessarily imply isotropy; see [6, p45]. Homogeneity is implied if the system is isotropic at every point in space.

[33] The equations in (4.73) have $(\rho, \boldsymbol{u}, \widetilde{\varepsilon})$ as the basic fields. We'll use thermodynamics to relate T to $\widetilde{\varepsilon}$.

[34] Equation (4.77) is the same as Eq. (1.18) up to a minus sign; the sign difference between stress and pressure tensors.

[35] Equation (4.79) differs from Eq. (1.34) in the sign of $\boldsymbol{\Pi}$; compare Eqs. (1.19) and (4.74) noting that $\mathbf{T} = -\mathbf{P}$.

[36] In relativistic fluid dynamics, $\lambda \neq 0$ and is not independent of κ; see Landau and Lifshitz[15, p506].

(κ, η, ζ). Note that, because the precise forms of the pressure tensor and heat flux have been subsumed into the symbols $\mathbf{P}$ and $\boldsymbol{J}$, the equations of hydrodynamics obtained from the conservation laws in (4.73) hold independently of the microscopic forms of $\mathbf{P}$ and $\boldsymbol{J}$. See Exercise 4.32.

Combining the parameterized expressions for $\mathbf{P}, \boldsymbol{J}$ with the equations in (4.73), we arrive at a form of the hydrodynamic equations (see Exercise 4.20),

$$\begin{aligned}\frac{\partial \rho u_i}{\partial t} &= -\sum_j \nabla_j(\rho u_i u_j) - \nabla_i P + \eta \nabla^2 u_i + (\zeta + \tfrac{1}{3}\eta)\nabla_i(\boldsymbol{\nabla}\cdot\boldsymbol{u}) \\ \frac{\partial \widetilde{\varepsilon}}{\partial t} &= -\boldsymbol{\nabla}\cdot(\boldsymbol{u}\widetilde{\varepsilon}) - P\boldsymbol{\nabla}\cdot\boldsymbol{u} + \kappa\nabla^2 T + \eta\sum_{ij}\left[\nabla_i u_j + \nabla_j u_i\right]\nabla_i u_j + (\zeta - \tfrac{2}{3}\eta)(\boldsymbol{\nabla}\cdot\boldsymbol{u})^2,\end{aligned} \tag{4.81}$$

two equations in the five unknowns $\rho, \boldsymbol{u}, P, \widetilde{\varepsilon}, T$ (three equations including the continuity equation). The number of variables can be reduced assuming local equilibrium with the concomitant local equations of state $P(\boldsymbol{r}) = P[T(\boldsymbol{r}), \rho(\boldsymbol{r})], \widetilde{\varepsilon}(\boldsymbol{r}) = \widetilde{\varepsilon}[T(\boldsymbol{r}), \rho(\boldsymbol{r})]$ having the same functional form as in global equilibrium, $P = P(T, \rho), \widetilde{\varepsilon} = \widetilde{\varepsilon}(T, \rho)$. That leaves $\rho, \boldsymbol{u}, T$ as the basic variables.[37] We now further assume that *temporal* variations of $P, \widetilde{\varepsilon}$ occur through the variations of[38] $T(\boldsymbol{r}, t), \rho(\boldsymbol{r}, t)$, e.g., $P(\boldsymbol{r}, t) = P[T(\boldsymbol{r}, t), \rho(\boldsymbol{r}, t)]$. With these assumptions,

$$\boldsymbol{\nabla} P = (\partial P/\partial T)_\rho \boldsymbol{\nabla} T + (\partial P/\partial \rho)_T \boldsymbol{\nabla}\rho, \tag{4.82}$$

where the thermodynamic derivatives have known connections with measurable quantities; see Exercise 4.22. For variations of internal energy, express $\widetilde{\varepsilon}$ as the product of mass density ρ and specific internal energy: let[39] $\widetilde{\varepsilon} \to \rho(\rho^{-1}\widetilde{\varepsilon}) \equiv \rho\widetilde{u}$. We require the space and time derivatives of $\widetilde{u}$,

$$\left\{\begin{array}{c}\boldsymbol{\nabla}\\ \partial_t\end{array}\right\}\widetilde{u} = \left(\frac{\partial \widetilde{u}}{\partial T}\right)_\rho \left\{\begin{array}{c}\boldsymbol{\nabla}\\ \partial_t\end{array}\right\} T + \left(\frac{\partial \widetilde{u}}{\partial \rho}\right)_T \left\{\begin{array}{c}\boldsymbol{\nabla}\\ \partial_t\end{array}\right\}\rho, \tag{4.83}$$

where $\partial_t \equiv \partial/\partial t$. It's straightforward to show that (see Exercise 4.22)

$$\rho^2\left(\frac{\partial \widetilde{u}}{\partial \rho}\right)_T = -T\left(\frac{\partial P}{\partial T}\right)_\rho + P \qquad \left(\frac{\partial \widetilde{u}}{\partial T}\right)_\rho = c_v, \tag{4.84}$$

where c_v is the specific heat at constant volume, the heat capacity C_V per mass.

Combining these expressions with Eq. (4.81), we find, including the continuity equation,

$$\begin{aligned}\partial_t\rho &= -\,\boldsymbol{u}\cdot\boldsymbol{\nabla}\rho - \rho\boldsymbol{\nabla}\cdot\boldsymbol{u} \\ \partial_t u_i &= -\,(\boldsymbol{u}\cdot\boldsymbol{\nabla})u_i - \frac{1}{\rho}\left[\left(\frac{\partial P}{\partial T}\right)_\rho \nabla_i T + \left(\frac{\partial P}{\partial \rho}\right)_T \nabla_i\rho\right] + \frac{1}{\rho}\left[\eta\nabla^2 u_i + (\zeta + \tfrac{1}{3}\eta)\,\nabla_i\,(\boldsymbol{\nabla}\cdot\boldsymbol{u})\right] \\ \partial_t T &= -\,\boldsymbol{u}\cdot\boldsymbol{\nabla}T - \frac{T}{\rho c_v}\left(\frac{\partial P}{\partial T}\right)_\rho \boldsymbol{\nabla}\cdot\boldsymbol{u} \\ &\quad + \frac{1}{\rho c_v}\left[\kappa\nabla^2 T + \eta\sum_{ij}\left[\nabla_i u_j + \nabla_j u_i\right]\nabla_i u_j + (\zeta - \tfrac{2}{3}\eta)(\boldsymbol{\nabla}\cdot\boldsymbol{u})^2\right].\end{aligned} \tag{4.85}$$

We now have five equations in the five fields $\rho(\boldsymbol{r}, t), \boldsymbol{u}(\boldsymbol{r}, t), T(\boldsymbol{r}, t)$. These are the most general hydrodynamic equations[40] consistent with the assumption of local thermodynamic equilibrium and our parameterizations of $\mathbf{P}$ and $\boldsymbol{J}$.

[37] State variables are not independent of each other; much of the mathematical apparatus of thermodynamics is devoted to exposing their interrelations. We're free to assume the existence of functional relations among state variables[2, p12].

[38] This seemingly strong assumption is justified in the *normal solutions* of the Boltzmann equation; see Section 4.8

[39] Apologies for notation. In Section 1.2 we denoted the specific internal energy u (a common notation) and in this chapter we used $\boldsymbol{u}$ for the mean local velocity (also a common notation). Here we indicate specific internal energy $\widetilde{u}$.

[40] One might ask: What *is* hydrodynamics? Is it the set of equations in (4.73), or those in (4.85)? Practically speaking, hydrodynamics = conservation laws + constitutive relations, the equations in (4.85).

4.4.5 Linearized hydrodynamics, normal modes

The equations in (4.85) are quite complicated and have a rich variety of solutions because of their nonlinear character. Our purpose is not to provide a detailed account of fluid dynamics (see Batchelor[12] or Landau and Lifshitz[15]), but to develop the essential features that contribute to our understanding of kinetic theory. There is a regime in which these equations simplify, where ρ and T deviate slightly from their equilibrium values. We indicate this with the scheme

$$\begin{aligned} \rho(\boldsymbol{r},t) &= \rho + \delta\rho(\boldsymbol{r},t) \\ \boldsymbol{u}(\boldsymbol{r},t) &= 0 + \delta\boldsymbol{u}(\boldsymbol{r},t) \\ T(\boldsymbol{r},t) &= T + \delta T(\boldsymbol{r},t). \end{aligned} \tag{4.86}$$

Small deviations $(\delta\rho, \delta T)$ are expected to generate small velocity variations $\delta\boldsymbol{u}$ around the equilibrium value of zero, and vice versa. Substituting (4.86) in (4.85) and keeping terms linear in small quantities, we have[41]

$$\begin{aligned} \partial_t\delta\rho &= -\,\rho\boldsymbol{\nabla}\cdot\delta\boldsymbol{u} \\ \partial_t\delta\boldsymbol{u} &= -\frac{1}{\rho}\left[\left(\frac{\partial P}{\partial T}\right)_\rho \boldsymbol{\nabla}\delta T + \left(\frac{\partial P}{\partial \rho}\right)_T \boldsymbol{\nabla}\delta\rho\right] + \frac{1}{\rho}\left[\eta\nabla^2\delta\boldsymbol{u} + \left(\zeta + \tfrac{1}{3}\eta\right)\boldsymbol{\nabla}\left(\boldsymbol{\nabla}\cdot\delta\boldsymbol{u}\right)\right] \\ \partial_t\delta T &= -\frac{T}{\rho c_v}\left(\frac{\partial P}{\partial T}\right)_\rho \boldsymbol{\nabla}\cdot\delta\boldsymbol{u} + \frac{\kappa}{\rho c_v}\nabla^2\delta T. \end{aligned} \tag{4.87}$$

These are the equations of *linearized hydrodynamics*.[42] A key result is that they contain all the transport coefficients of the nonlinear theory. The problem of determining transport coefficients is therefore well-posed for linear systems.

We seek normal-mode solutions of these equations which (as with any normal-mode analysis) can be used to represent arbitrary small amplitude deviations from equilibrium. For simplicity we consider a system infinite in spatial extent to avoid the complication of boundary conditions. We seek solutions in the form

$$\begin{aligned} \delta\rho(\boldsymbol{r},t) &= \rho_{\boldsymbol{k}}\exp(\mathrm{i}\boldsymbol{k}\cdot\boldsymbol{r} + \omega t) \\ \delta\boldsymbol{u}(\boldsymbol{r},t) &= \boldsymbol{u}_{\boldsymbol{k}}\exp(\mathrm{i}\boldsymbol{k}\cdot\boldsymbol{r} + \omega t) \\ \delta T(\boldsymbol{r},t) &= T_{\boldsymbol{k}}\exp(\mathrm{i}\boldsymbol{k}\cdot\boldsymbol{r} + \omega t), \end{aligned} \tag{4.88}$$

where we allow that ω can be complex-valued and there's no restriction on $\boldsymbol{k}$ (infinite system). The amplitudes $\rho_{\boldsymbol{k}}, \boldsymbol{u}_{\boldsymbol{k}}, T_{\boldsymbol{k}}$ are Fourier transforms of $\delta\rho, \delta\boldsymbol{u}, \delta T$. Substituting (4.88) in (4.87) with $\partial_t \to \omega$ and $\boldsymbol{\nabla} \to \mathrm{i}\boldsymbol{k}$, we have a system of linear homogeneous equations,

$$\begin{aligned} \omega\rho_{\boldsymbol{k}} + \mathrm{i}\rho\boldsymbol{k}\cdot\boldsymbol{u}_{\boldsymbol{k}} &= 0 \\ \mathrm{i}B\boldsymbol{k}\rho_{\boldsymbol{k}} + \left(\omega + \phi k^2\right)\boldsymbol{u}_{\boldsymbol{k}} + \chi\boldsymbol{k}(\boldsymbol{k}\cdot\boldsymbol{u}_{\boldsymbol{k}}) + \mathrm{i}A\boldsymbol{k}T_{\boldsymbol{k}} &= 0 \\ \mathrm{i}C\boldsymbol{k}\cdot\boldsymbol{u}_{\boldsymbol{k}} + \left(\omega + \sigma k^2\right)T_{\boldsymbol{k}} &= 0, \end{aligned} \tag{4.89}$$

where A, B, C denote equilibrium properties and ϕ, χ, σ denote nonequilibrium properties,

$$A \equiv \frac{1}{\rho}\left(\frac{\partial P}{\partial T}\right)_\rho \qquad B \equiv \frac{1}{\rho}\left(\frac{\partial P}{\partial \rho}\right)_T \qquad C \equiv \frac{T}{\rho c_v}\left(\frac{\partial P}{\partial T}\right)_\rho$$

$$\phi \equiv \frac{\eta}{\rho} \qquad \chi \equiv \frac{1}{\rho}\left(\zeta + \tfrac{1}{3}\eta\right) \qquad \sigma \equiv \frac{\kappa}{\rho c_v}.$$

[41] The term $\kappa/(\rho c_v)$ in the temperature equation is known as the *thermal diffusivity*, sometimes denoted D_T. The ratio η/ρ is known as the *kinematic viscosity*, with η itself the dynamic viscosity. Note that $\kappa/(\rho c_v)$ and η/ρ have the same dimensions as the diffusion coefficient, length^2 / time.

[42] We note that another fundamental theory of physics featuring nonlinear field equations, the general theory of relativity, has a linearized regime with numerous physical consequences, such as gravitational waves[6, Chapter 18].

Do the equations of linearized hydrodynamics admit solutions in the form of (4.88)? Not for arbitrary ω. For a square system of linear homogeneous equations to have nontrivial solutions, the determinant of the coefficient matrix must vanish, a requirement that imposes a connection between ω and k, the *dispersion relations*, $\omega_i = \omega_i(k)$, $i = 1, \ldots, 5$. A full 5×5 coefficient matrix is implied by the equations in (4.89). Those equations simplify when the coordinate system is oriented with one axis parallel to $\boldsymbol{k}$. We're free to do that without loss of generality; such transformations do not affect eigenvalues. Let $\boldsymbol{k} = k\hat{\boldsymbol{x}}$ with $k_y = k_z = 0$. In that case, we seek the characteristic polynomial associated with

$$\begin{vmatrix} \omega & \mathrm{i}\rho k & 0 & 0 & 0 \\ \mathrm{i}Bk & \omega + (\phi+\chi)k^2 & 0 & 0 & \mathrm{i}Ak \\ 0 & 0 & \omega + \phi k^2 & 0 & 0 \\ 0 & 0 & 0 & \omega + \phi k^2 & 0 \\ 0 & \mathrm{i}Ck & 0 & 0 & \omega + \sigma k^2 \end{vmatrix} = 0. \tag{4.90}$$

Expanding the determinant, the allowed frequencies are the roots of the quintic function in ω,

$$\left(\omega + \phi k^2\right)^2 \left\{\omega^3 + \omega^2 k^2(\phi + \chi + \sigma) + \omega[\sigma(\phi + \chi)k^4 + k^2(B\rho + AC)] + \rho B \sigma k^4\right\} = 0.$$

Clearly there is a doubly degenerate root (which we refer to as ω_3, ω_4),

$$\omega_3 = \omega_4 = -\phi k^2 = -(\eta/\rho)k^2. \tag{4.91}$$

The remaining allowed frequencies are found from the roots of the cubic function in curly braces. That (arduous) path can be avoided by observing that all roots coalesce to zero as $k \to 0$. Because we're interested in the slowest processes, we want only the small-k form of $\omega(k)$. Rather than seek exact expressions, let's try to find approximate solutions,

$$\omega_i(k) = a_i k + b_i k^2 + O(k^3). \qquad (i = 1, 2, 5) \tag{4.92}$$

The unknown coefficients (a_i, b_i) are determined by matching powers of $\omega^n k^m$ for $n + m \leq 4$ when Eq. (4.92) is substituted in the left side of

$$(\omega - \omega_1)(\omega - \omega_2)(\omega - \omega_5) = \omega^3 + \omega^2 k^2(\phi + \chi + \sigma) + \omega[\sigma(\phi + \chi)k^4 + k^2(B\rho + AC)] + \rho B \sigma k^4. \tag{4.93}$$

By this procedure (see Exercise 4.23) we find the long-wavelength forms shown in Table 4.1. The

Table 4.1 Frequencies of long-wavelength hydrodynamic modes

Dispersion relation	Mode
$\omega_{1,2} = \pm \mathrm{i} c_s k - \Gamma k^2 + O(k^3)$	Sound
$\omega_{3,4} = -\dfrac{\eta}{\rho} k^2$	Shear
$\omega_5 = -\dfrac{\kappa}{\rho c_p} k^2 + O(k^3)$	Thermal

complex roots $\omega_{1,2}$ describe propagation in two opposite directions (along and against $\boldsymbol{k}$, the *longitudinal direction*), at the speed of sound[43]

$$c_s \equiv \sqrt{\left(\frac{\partial P}{\partial \rho}\right)_T \frac{c_p}{c_v}}, \tag{4.94}$$

[43] Equation (4.94) is equivalent to $c_s = \sqrt{(\partial P/\partial \rho)_S}$, the adiabatic sound speed (see [2, p24]). There is no dissipation (constant entropy) associated with undamped sound propagation (set transport coefficients to zero in Eq. (4.80)).

with damping coefficient

$$\Gamma \equiv \frac{1}{2\rho}\left[\frac{4}{3}\eta + \zeta + \kappa\left(\frac{1}{c_v} - \frac{1}{c_p}\right)\right] . \tag{4.95}$$

The sound attenuation rate $\Gamma > 0$ has contributions from the viscosity η, the bulk viscosity ζ, and the heat conductivity κ. The roots $\omega_{3,4,5}$ are purely damped excitations.

4.5 THE BOLTZMANN EQUATION, MOLECULAR CHAOS

The issue with the hierarchy isn't lack of computational resources, as it is with Liouville's equation, it's that it isn't closed. The simplest possibility would be for the hierarchy to be truncated at $s = 1$. A closed evolution equation for the one-particle distribution f_1 is important enough to get its own name, a *kinetic equation*,[44] which has the form[45]

$$\left(\frac{\partial}{\partial t} - L_1\right) f_1(\boldsymbol{r}, \boldsymbol{p}, t) = \left(\frac{\partial}{\partial t} + \underbrace{(\boldsymbol{p}/m)\cdot\boldsymbol{\nabla}}_{\text{free streaming}}\right) f_1(\boldsymbol{r}, \boldsymbol{p}, t) \equiv D_v f_1(\boldsymbol{r}, \boldsymbol{p}, t) = \underbrace{J[f_1]}_{\substack{\text{collision}\\ \text{operator}}} , \tag{4.96}$$

where D_v is the convective derivative ($\boldsymbol{v} = \boldsymbol{p}/m$). The operator $L_1 = -(\boldsymbol{p}/m)\cdot\boldsymbol{\nabla}$ appears in the first equation of the hierarchy [see Eq. (4.29)] and is associated with the motion of independent particles. The *collision operator* $J[f_1]$ (a functional[46]) is associated with the rate of change of f_1 due to particle interactions, referred to as *collisions*. The right side of Eq. (4.96) effectively represents the rest of the hierarchy; it therefore can't be found as an exact consequence of the equations of motion, it must be devised from approximations that capture the physics of particle trajectories modified by interactions. An extreme approximation to Eq. (4.96), neglecting J, would apply to systems of such low density that encounters between molecules are so rare as to be negligible. Although there *are* gases like that,[47] we consider the kinetic equation established by Boltzmann, the *Boltzmann transport equation*, that treats the more representative problem of particle interactions through collisions.[48]

4.5.1 Derivation

Assume that 1) the intermolecular potential is *short ranged*—there is a distance r_0 beyond which the potential rapidly approaches zero—and 2) the system is sufficiently *dilute* that the average distance between particles far exceeds r_0. For the average number density $n \equiv N/V$, the *diluteness criterion*[49] is $nr_0^3 \ll 1$. Under these conditions particles move with constant velocities $\boldsymbol{v}$ until they encounter other particles closely enough that they deflect over short time intervals $\tau_c \approx r_0/v$ (the microscopic *collision time*, see Exercise 4.25) after which free motion resumes with velocities $\boldsymbol{v}'$. Simultaneous encounters of three or more particles occur so infrequently as to be negligible. Two-body encounters can be seen as discrete events confined to a small volume r_0^3, occurring over a short time τ_c, in which changes in *free-state* velocities occur almost abruptly, $\boldsymbol{v} \to \boldsymbol{v}'$; see Fig. 4.1.

---

[44] Although the Fokker-Planck equation (3.37) is a closed equation in the one-particle distribution, it's a linear differential equation. Kinetic equations are closed *nonlinear* equations for f_1; see Section 4.5.2.5.

[45] Equation (4.96) ignores the effects of external forces on the streaming (left) side of the equation; see Exercise 4.7.

[46] Functionals map functions to numbers, often by way of integrals, $J[y] \equiv \int F(y(x))dx$ where F is known. In quantum mechanics we encounter linear functionals—*bras* in Dirac notation; in kinetic theory functionals are often nonlinear.

[47] A *Knudsen gas* is one in which the mean free path exceeds the size of the container, a state of matter important in nanometer-scale systems[104].

[48] As one soon discovers, Boltzmann's written output is voluminous; English translations of select articles are available in Brush[89]. A valuable resource is *Lectures on Gas Theory*[105], described by M. Kac[42, p261] as "one of the greatest books in the history of exact sciences." Other references are [106, Chapter 3] and [107, Chapter 4]. The unadorned term "Boltzmann equation" is ambiguous—it could refer to the Boltzmann entropy formula or the Boltzmann factor, $\exp(-\beta E)$, or to a kinetic equation—hence the more precise term Boltzmann transport equation.

[49] The same criterion appears in the equilibrium theory of dilute gases, in the virial expansion[5, Section 6.2].

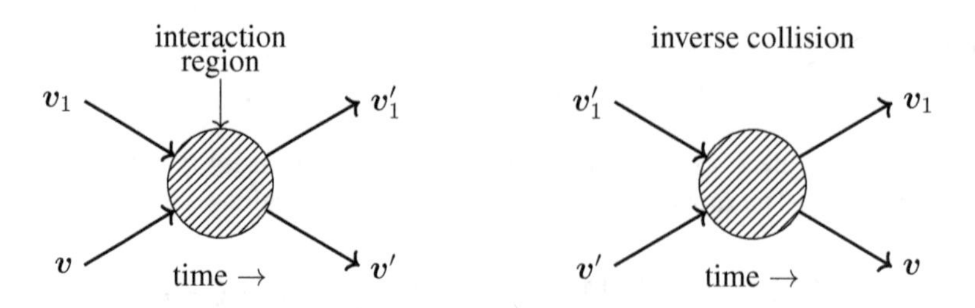

Figure 4.1 Two-body collisions mediated by short-range forces. Left: Direct collision. Right: Inverse collision (role of initial and final states interchanged).

Under these assumptions, the dynamics of dilute-gas particles reduce to a series of two-body encounters, and two-body problems (for motion in central force fields) reduce to the dynamics of single particles (see Appendix C). Because before and after-collision velocities pertain to free particles, and because all gas particles have the same mass, for particles of velocities $\boldsymbol{v}, \boldsymbol{v}_1$ interacting to produce velocities $\boldsymbol{v}', \boldsymbol{v}_1'$ we have the conservation laws,

$$\boldsymbol{v} + \boldsymbol{v}_1 = \boldsymbol{v}' + \boldsymbol{v}_1' \qquad v^2 + v_1^2 = v'^2 + v_1'^2. \tag{4.97}$$

The four scalar equations implied by (4.97) are not sufficient to determine the six scalar quantities comprising the after-collision velocities; a knowledge of the force law is required for a unique solution.[50] Expressions for $\boldsymbol{v}', \boldsymbol{v}_1'$ that solve the equations in (4.97) can be written

$$\boldsymbol{v}' = \frac{1}{2}\left(\boldsymbol{v} + \boldsymbol{v}_1 - g\hat{\boldsymbol{\epsilon}}\right) \qquad \boldsymbol{v}_1' = \frac{1}{2}\left(\boldsymbol{v} + \boldsymbol{v}_1 + g\hat{\boldsymbol{\epsilon}}\right), \tag{4.98}$$

where (for now) $\hat{\boldsymbol{\epsilon}}$ is an arbitrary unit vector[51] and $g \equiv |\boldsymbol{g}|$ denotes the magnitude of the pre-collision relative velocity,

$$\boldsymbol{g} \equiv \boldsymbol{v}_1 - \boldsymbol{v}. \tag{4.99}$$

The effect of an elastic collision is to rigidly rotate the relative velocity (see Fig. 4.2, use Eq. (4.98)),

$$\boldsymbol{g}' \equiv \boldsymbol{v}_1' - \boldsymbol{v}' = g\hat{\boldsymbol{\epsilon}}. \tag{4.100}$$

Thus, $\hat{\boldsymbol{\epsilon}}$ is aligned with $\boldsymbol{g}'$; to determine $\hat{\boldsymbol{\epsilon}}$ we must solve for the dynamical motion using the force law between particles. The calculation of the scattering angle is reviewed in Appendix C.

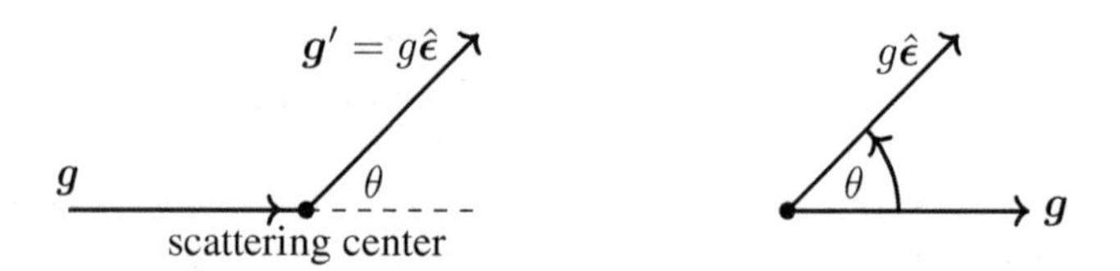

Figure 4.2 Elastic scattering. The relative velocity $\boldsymbol{g}$ prior to the collision is rotated through the scattering angle θ to the relative velocity after the collision $\boldsymbol{g}'$, with $|\boldsymbol{g}'| = |\boldsymbol{g}|$.

The Boltzmann equation, an equation of motion for $f(\boldsymbol{r}, \boldsymbol{v}, t)$, the number density at phase point[52] $(\boldsymbol{r}, \boldsymbol{v})$ at time t, can be derived by analyzing how the number of particles in state $(\boldsymbol{r}, \boldsymbol{v})$

[50] Only in one dimension do conservation laws determine motion. In d-dimensions, there are $d+1$ scalar equations in (4.97), yet there are $2d$ scalar unknowns in the after-collision velocities; $d+1 = 2d$ only for $d = 1$.

[51] The arbitrariness of $\hat{\epsilon}$ underscores the indeterminism of the conservation laws in specifying the motion.

[52] As noted in footnote 9 of this chapter, we often work with a phase space spanned by $\boldsymbol{r}, \boldsymbol{v}$. We also drop the subscript on f_1. It's conceptually simpler in the derivation of the Boltzmann equation to let $f(\boldsymbol{r}, \boldsymbol{v}, t)\mathrm{d}\boldsymbol{r}\mathrm{d}\boldsymbol{v}$ denote the *number* of

changes in time, from flows at velocity $\boldsymbol{v}$ or from changes in $\boldsymbol{v}$ through collisions. The particles that flow to $(\boldsymbol{r}, \boldsymbol{v})$ are those that Δt seconds ago[53] were in state $(\boldsymbol{r} - \boldsymbol{v}\Delta t, \boldsymbol{v})$. The change in particle number at $(\boldsymbol{r}, \boldsymbol{v})$ by means of streaming is therefore[54]

$$[f(\boldsymbol{r} - \boldsymbol{v}\Delta t, \boldsymbol{v}, t) - f(\boldsymbol{r}, \boldsymbol{v}, t)]\, \mathrm{d}\boldsymbol{r}\mathrm{d}\boldsymbol{v} \approx -\Delta t\, (\boldsymbol{v} \cdot \boldsymbol{\nabla} f)\, \mathrm{d}\boldsymbol{r}\mathrm{d}\boldsymbol{v}. \tag{4.101}$$

Dividing by Δt (to give us a rate), we see that the streaming term in Eq. (4.96) (with $\boldsymbol{p} = m\boldsymbol{v}$) represents a balance between gains and losses from state $(\boldsymbol{r}, \boldsymbol{v})$ by means of flow.

Let's apply gain-loss reasoning to the collision term, the new wrinkle in the Boltzmann equation.[55] Keeping our emphasis on the rates of change of particle numbers in $(\boldsymbol{r}, \boldsymbol{v})$, we write

$$\mathrm{d}\boldsymbol{r}\mathrm{d}\boldsymbol{v}\left(\frac{\partial}{\partial t} + \boldsymbol{v} \cdot \boldsymbol{\nabla}\right) f(\boldsymbol{r}, \boldsymbol{v}, t) = (G - L)\, \mathrm{d}\boldsymbol{r}\mathrm{d}\boldsymbol{v}, \tag{4.102}$$

where G and L are gain and loss rates per μ-space volume at $(\boldsymbol{r}, \boldsymbol{v})$. Collisions affect the rate of change of $f(\boldsymbol{r}, \boldsymbol{v}, t)$ in two ways: 1) among particles with velocity $\boldsymbol{v}$, some undergo collisions, resulting in a loss from state $(\boldsymbol{r}, \boldsymbol{v})$, and 2) among particles with velocities $\boldsymbol{v}_1', \boldsymbol{v}'$, some undergo collisions such that one of the particles has final velocity $\boldsymbol{v}$, a gain. We symbolize the loss collision $(\boldsymbol{v}_1, \boldsymbol{v}) \to (\boldsymbol{v}_1', \boldsymbol{v}')$ and the gain collision $(\boldsymbol{v}_1', \boldsymbol{v}') \to (\boldsymbol{v}_1, \boldsymbol{v})$; see Fig. 4.1. The (real-space) volume containing particles of impact parameters between b and $b + \mathrm{d}b$ (see Fig. C.6) and velocity[56] $g = |\boldsymbol{v}_1 - \boldsymbol{v}|$ that will scatter in Δt seconds is $2\pi b\mathrm{d}bg\Delta t$. Thus the number of particles about to encounter a scatterer is $2\pi b\mathrm{d}bg\Delta t f(\boldsymbol{r}, \boldsymbol{v}_1, t)\mathrm{d}\boldsymbol{v}_1$ and there are $f(\boldsymbol{r}, \boldsymbol{v}, t)\mathrm{d}\boldsymbol{r}\mathrm{d}\boldsymbol{v}$ scatterers within $\mathrm{d}\boldsymbol{v}\mathrm{d}\boldsymbol{r}$ of $(\boldsymbol{r}, \boldsymbol{v})$. The total loss rate from collisions is therefore

$$L\mathrm{d}\boldsymbol{r}\mathrm{d}\boldsymbol{v} = \left(\int \mathrm{d}\boldsymbol{v}_1 2\pi b\mathrm{d}bg f(\boldsymbol{r}, \boldsymbol{v}_1, t) f(\boldsymbol{r}, \boldsymbol{v}, t)\right)\mathrm{d}\boldsymbol{r}\mathrm{d}\boldsymbol{v}. \tag{4.103}$$

The gain term is handled similarly. In the *inverse collision* $(\boldsymbol{v}_1', \boldsymbol{v}') \to (\boldsymbol{v}_1, \boldsymbol{v})$, the roles of initial and final states are interchanged from the direct collision (see Fig. 4.1).[57] Applying the reasoning leading to Eq. (4.103) to the inverse collision, we have the total gain rate from collisions,

$$\begin{aligned} G\mathrm{d}\boldsymbol{r}\mathrm{d}\boldsymbol{v} &= \left(\int \mathrm{d}\boldsymbol{v}'\mathrm{d}\boldsymbol{v}_1' 2\pi b\mathrm{d}bg' f(\boldsymbol{r}, \boldsymbol{v}_1', t) f(\boldsymbol{r}, \boldsymbol{v}', t)\right)\mathrm{d}\boldsymbol{r} \\ &= \left(\int \mathrm{d}\boldsymbol{v}_1 2\pi b\mathrm{d}bg f(\boldsymbol{r}, \boldsymbol{v}_1', t) f(\boldsymbol{r}, \boldsymbol{v}', t)\right)\mathrm{d}\boldsymbol{r}\mathrm{d}\boldsymbol{v}, \end{aligned} \tag{4.104}$$

where we've applied the *integral invariant of Poincaré*,[108, p73] $\mathrm{d}\boldsymbol{v}_1'\mathrm{d}\boldsymbol{v}' = \mathrm{d}\boldsymbol{v}_1\mathrm{d}\boldsymbol{v}$ and we've set $g' = g$ (elastic scattering).

Combining Eqs. (4.102)–(4.104), we arrive at the Boltzmann transport equation,[58]

$$\left(\frac{\partial}{\partial t} + \boldsymbol{v} \cdot \boldsymbol{\nabla}\right) f(\boldsymbol{r}, \boldsymbol{v}, t) = \int \mathrm{d}\boldsymbol{v}_1 2\pi b\mathrm{d}bg \left[f(\boldsymbol{r}, \boldsymbol{v}_1', t) f(\boldsymbol{r}, \boldsymbol{v}', t) - f(\boldsymbol{r}, \boldsymbol{v}_1, t) f(\boldsymbol{r}, \boldsymbol{v}, t)\right]. \tag{4.105}$$

It can be written equivalently in terms of the cross section $\sigma(\Omega)$ for scattering into the differential solid angle $\mathrm{d}\Omega$ about Ω (see Eq. (C.6)),

$$\left(\frac{\partial}{\partial t} + \boldsymbol{v} \cdot \boldsymbol{\nabla}\right) f(\boldsymbol{r}, \boldsymbol{v}, t) = \int \mathrm{d}\boldsymbol{v}_1 \mathrm{d}\Omega \sigma(\Omega) |\boldsymbol{v}_1 - \boldsymbol{v}| \left[f(\boldsymbol{r}, \boldsymbol{v}_1', t) f(\boldsymbol{r}, \boldsymbol{v}', t) - f(\boldsymbol{r}, \boldsymbol{v}_1, t) f(\boldsymbol{r}, \boldsymbol{v}, t)\right]. \tag{4.106}$$

---

particles within $\mathrm{d}\boldsymbol{r}\mathrm{d}\boldsymbol{v}$ of $(\boldsymbol{r}, \boldsymbol{v})$ at time t, rather than probability. It's a matter of normalization—probability densities or particle densities. By "at" $(\boldsymbol{r}, \boldsymbol{v})$, we mean within $\mathrm{d}\boldsymbol{r}\mathrm{d}\boldsymbol{v}$ of $(\boldsymbol{r}, \boldsymbol{v})$.

[53] We leave the time interval Δt unspecified for now. We'll see that it must satisfy inequality (4.107).

[54] The particles in state $(\boldsymbol{r}, \boldsymbol{v})$ at time t flow to $(\boldsymbol{r} + \boldsymbol{v}\Delta t, \boldsymbol{v})$ in the next Δt seconds.

[55] The master equation (2.66) also has the gain-loss form.

[56] Figure C.6 is in the reference frame of the particle with velocity $\boldsymbol{v}$; see also Fig. C.5.

[57] The inverse collision is not the time-reversed collision, which in our notation would be $(-\overline{\boldsymbol{v}}', -\boldsymbol{v}') \to (-\overline{\boldsymbol{v}}, -\boldsymbol{v})$. In the older literature, the inverse collision is called the *restituting collision*.

[58] In an external force field $\boldsymbol{F}$ there is an additional term on the left side, $(\boldsymbol{F}/m) \cdot \partial/\partial \boldsymbol{v}$; see Exercise 4.7.

The collision operator $J[f]$ is the expression on the right side of Eq. (4.106). Boltzmann's equation is a nonlinear integro-differential equation. The variables $(\boldsymbol{r}, \boldsymbol{v}, t)$ are specified on the left side of the equation, we're summing over $\boldsymbol{v}_1$, and $(\boldsymbol{v}_1', \boldsymbol{v}')$ are determined by $(\boldsymbol{v}_1, \boldsymbol{v})$ through the scattering cross section (see Appendix C). We cannot in the space of this book consider applications (see Cercignani[109]). It's used in disparate fields of research: astrophysics,[110] neutron transport,[111] traffic flow,[112] and charge transport in solids,[113] to name just a few. *It's one of the fundamental equations of nonequilibrium statistical mechanics.* Despite its history, it remains far from fully understood.[59] We offer some remarks.

4.5.2 Comments

4.5.2.1 Derivation

Our derivation, although physically motivated, is ad hoc and not part of a systematic scheme. It would be illuminating if the Boltzmann equation could be derived from the first equation of the BBGKY hierarchy, where the assumptions underlying such a derivation would be expected to present themselves explicitly. Attempts have been made by Bogoliubov,[94] Kirkwood,[98] Grad,[115] and Prigogine[116, Section 2.5]. We lack the space to analyze these approaches; see Liboff[117].

4.5.2.2 Dilute gases

The Boltzmann equation applies to dilute gases. The diluteness criterion $nr_0^3 \ll 1$ is sufficiently embedded in the concepts underlying the derivation, that relaxing this assumption would require a major revamping of the ideas involved. It ensures that collisions are localized events in space and time; it allows us to conceive of particle interactions as a succession of binary collisions and to calculate the velocities $(\boldsymbol{v}_1', \boldsymbol{v}')$ from the solution of a two-body problem. The extension to dense gases has occupied the attention of many researchers, and we can only refer to reviews: Cohen[118], Dorfman and van Beijeren[119], Mazenko and Yip[120], and Brush[121].

4.5.2.3 Irreversibility, molecular chaos assumption

The Boltzmann equation is not time-reversal symmetric. The left side of the equation reverses sign under $t \to -t$ and the right side does not.[60,61] Thus, if $f(\boldsymbol{r}, \boldsymbol{v}, t)$ is a solution of the Boltzmann equation, $f(\boldsymbol{r}, -\boldsymbol{v}, -t)$ is not, in contrast to solutions of the Liouville equation (see Exercise 4.3). Although we see in the finished product that the Boltzmann equation is not time-reversal symmetric, where in its derivation did we depart from the laws of mechanics?

In loss collisions $(\boldsymbol{v}_1, \boldsymbol{v}) \to (\boldsymbol{v}_1', \boldsymbol{v}')$, $\boldsymbol{v}_1$ can be chosen independently of $\boldsymbol{v}$; the colliding particles are uncorrelated. In gain collisions, the velocities of the colliding particles are constrained to satisfy the conservation law $\boldsymbol{v}' + \boldsymbol{v}_1' = \boldsymbol{v}_1 + \boldsymbol{v}$; one can't freely vary $\boldsymbol{v}_1'$ and $\boldsymbol{v}'$. *Collisions create correlations.* Yet we have in the gain term Eq. (4.104) a product of one-particle distribution functions befitting uncorrelated particles.[62] It's at this point we've departed from the laws of mechanics: $\boldsymbol{v}_1', \boldsymbol{v}'$ follow deterministically from the force law, yet we're treating them as statistically

[59] See the articles in [114] on such basic questions as the mathematical existence of solutions to the Boltzmann equation.

[60] We found the same in the Fokker-Planck equation (3.59) with Ω a linear operator.

[61] The scattering cross section is the same for direct, inverse, and time-reversed collisions. Equality of cross sections for the direct and time-reverse collisions follows from the time-reversal invariance of the equations of motion; the conservation laws in Eq. (4.97) are the same if velocities are replaced with their time-reversed versions. Equality of cross sections for the inverse and direct collisions follows because one can read the conservation laws from left to right, or from right to left. The direct, inverse, and time-reverse collisions are kinematically equivalent.

[62] The number of pairs of particles located simultaneously at different phase points is specified by the two-body distribution function $f_2(\boldsymbol{r}, \boldsymbol{v}, \boldsymbol{r}, \boldsymbol{v}', t)$. In setting $f_2(\boldsymbol{r}, \boldsymbol{v}, \boldsymbol{r}, \boldsymbol{v}', t) = f_1(\boldsymbol{r}, \boldsymbol{v}, t) f_1(\boldsymbol{r}, \boldsymbol{v}', t)$, correlations are ignored.

independent.[63] Boltzmann referred to uncorrelated after-collision particles the "molecular disordered state,"[105, pp40–42] known today as the *molecular chaos assumption*,[64] a statistical rather than a mechanical assumption.[65] Boltzmann therefore did not derive irreversibility from time-symmetric physics, it was introduced into the theory as an artifact of ignoring correlations. Despite its extra-mechanical nature, Boltzmann's equation makes predictions in good agreement with experimental results. There's no question that the Boltzmann equation captures the physics of dilute gases.

4.5.2.4 *Markovian character, spatial inhomogeneity*

The Boltzmann equation has irreversibility in common with Markov processes; is it Markovian? That requires we say something about the time Δt (upon which no constraints have been placed up to now). It must be shorter than the *relaxation time* τ_r, the time over which $f(\boldsymbol{r}, \boldsymbol{v}, t)$ changes appreciably, and it must be longer than the collision time τ_c, the time to turn $\boldsymbol{g}$ into $g\hat{\boldsymbol{e}}$ (see Fig. 4.2),

$$\tau_c \ll \Delta t \ll \tau_r. \tag{4.107}$$

The Boltzmann equation can't apply to times shorter than the microscopic time τ_c. The diluteness criterion (underlying the theory) implies a clear separation in time scales: $nr_0^3 \ll 1 \implies \tau_c/\tau_r \ll 1$. The Boltzmann equation is not defined in the opposite limit $\tau_r \ll \tau_c$.

When, in analyzing the collision $(\boldsymbol{v}_1, \boldsymbol{v}) \to (\boldsymbol{v}_1', \boldsymbol{v}')$, we counted the number of particles of velocity $\boldsymbol{v}_1$ about to collide with particles of velocity $\boldsymbol{v}$ (at $\boldsymbol{r}$), we assumed that $f(\boldsymbol{r}, \boldsymbol{v}_1, t)$ does not change much during Δt (because $\Delta t \ll \tau_r$) and hence we used $f(\boldsymbol{r}, \boldsymbol{v}_1, t)$ evaluated at the same time as the number of scatterers, $f(\boldsymbol{r}, \boldsymbol{v}, t)\mathrm{d}\boldsymbol{r}\mathrm{d}\boldsymbol{v}$. If $\Delta t \approx \tau_r$, we should have used the number of particles at the phase point $(\boldsymbol{r}, \boldsymbol{v}_1)$ at an earlier time $t - \tau$, where τ is the time it takes for particles with velocity $\boldsymbol{v}_1$ to reach the scatterer. In that case, the rate of change of $f(\boldsymbol{r}, \boldsymbol{v}, t)$ would depend on its instantaneous value as well as its previous history—a non-Markovian process. The condition $\Delta t \ll \tau_r$ ensures that the rate of change of $f(\boldsymbol{r}, \boldsymbol{v}, t)$ depends only on the time t, and in that sense the Boltzmann equation is Markovian.

These considerations apply spatially as well, that the distribution function not vary appreciably over the range of distances traveled by particles in time Δt. Consider that $2\pi b\mathrm{d}bg\Delta t$ represents the "$\mathrm{d}\boldsymbol{r}$" associated with $f(\boldsymbol{r}, \boldsymbol{v}_1, t)\mathrm{d}\boldsymbol{v}_1$ to determine the number of particles about to encounter a scatterer. If $f(\boldsymbol{r}, \boldsymbol{v}_1, t)$ varied appreciably over the distance $g\Delta t$, we would have to evaluate the distribution function at the point where the colliding particle comes from, resulting in a collision operator *delocalized* in space, in essence "spatially non-Markovian." *The Boltzmann equation is local in space and time*: All distribution functions are evaluated at the same $(\boldsymbol{r}, t)$.

4.5.2.5 *Nonlinearity*

The Boltzmann equation is nonlinear; the collision term scales quadratically, $J[\lambda f] = \lambda^2 J[f]$ (where λ is a constant) and the streaming terms scale linearly. As opposed to the hierarchy, where the time rate of change of f_1 depends on the two-particle distribution function f_2, and so on, through the molecular chaos assumption *the entire hierarchy has been replaced by the collision operator* to produce a single equation nonlinear in f_1. *Kinetic equations are necessarily nonlinear*; collisions between particles are treated in terms of products of one-particle distributions.

We note that $J[f]$ as a nonlinear operator can be represented in terms of a bilinear operator[66]

$$Q(f, g) \equiv \frac{1}{2} \int \mathrm{d}\boldsymbol{v}_1 \mathrm{d}\Omega\sigma(\Omega) \left|\boldsymbol{v}_1 - \boldsymbol{v}\right| \left(f_1' g' + f' g_1' - f_1 g - f g_1\right), \tag{4.108}$$

---

[63] For repulsive interactions (hard core potentials), particles push each other away. The probability of finding two particles close together (post-collision) is smaller than the product of probabilities of finding a single particle anywhere in the system.

[64] Also called the *Stosszahlansatz* (introduced by the Ehrenfests[122]), the "assumption about the number of collisions."

[65] To be clear, the molecular chaos assumption is a violation of mechanics.

[66] The operator $Q(f, g)$ is bilinear: $Q(\lambda f, g) = \lambda Q(f, g)$ and $Q(a + b, g) = Q(a, g) + Q(b, g)$; see Exercise 4.27.

where $f_1' \equiv f(\boldsymbol{r},\boldsymbol{v}_1',t)$, $f' \equiv f(\boldsymbol{r},\boldsymbol{v}',t)$, $f_1 \equiv f(\boldsymbol{r},\boldsymbol{v}_1,t)$, $f \equiv f(\boldsymbol{r},\boldsymbol{v},t)$; likewise for g. For $g = f$, $Q(f,f) = J[f]$. Clearly, $Q(f,g) = Q(g,f)$. We'll use the bilinear form of J in Section 4.8.

4.6 BOLTZMANN'S H-THEOREM, CONNECTION WITH ENTROPY

In 1872 Boltzmann published, in the same article, two celebrated results: the kinetic equation for $f(\boldsymbol{r},\boldsymbol{v},t)$ and that a quantity traditionally called[67,68]

$$H(t) \equiv \int \mathrm{d}\boldsymbol{r}\mathrm{d}\boldsymbol{v} f(\boldsymbol{r},\boldsymbol{v},t) \ln\left[f(\boldsymbol{r},\boldsymbol{v},t)\right] \tag{4.109}$$

never increases as time progresses, the so-called *H-theorem*,

$$\frac{\mathrm{d}}{\mathrm{d}t} H(t) \leq 0. \tag{4.110}$$

The H-theorem has far-reaching consequences.[69]

4.6.1 Proof of the H-theorem

Because the normalization is fixed $[(\mathrm{d}/\mathrm{d}t)\int f(\boldsymbol{r},\boldsymbol{v},t)\mathrm{d}\boldsymbol{r}\mathrm{d}\boldsymbol{v} = \int(\partial f/\partial t)\mathrm{d}\boldsymbol{r}\mathrm{d}\boldsymbol{v} = 0]$, we have from Eq. (4.109),

$$\frac{\mathrm{d}}{\mathrm{d}t} H(t) = \int \frac{\partial f}{\partial t}\left(1 + \ln f\right) \mathrm{d}\boldsymbol{r}\mathrm{d}\boldsymbol{v} = \int \frac{\partial f}{\partial t} \ln f \mathrm{d}\boldsymbol{r}\mathrm{d}\boldsymbol{v}. \tag{4.111}$$

Let's write the Boltzmann equation in the form

$$\frac{\partial}{\partial t} f(\boldsymbol{r},\boldsymbol{v},t) = -\left(\boldsymbol{v}\cdot\boldsymbol{\nabla} + \boldsymbol{a}\cdot\boldsymbol{\nabla}_{\boldsymbol{v}}\right) f(\boldsymbol{r},\boldsymbol{v},t) + J[f], \tag{4.112}$$

where we've included in the streaming term the acceleration ($\boldsymbol{a} \equiv \boldsymbol{F}/m$) caused by external forces (see Exercise 4.7). Equation (4.111) can then be written schematically

$$\frac{\mathrm{d}}{\mathrm{d}t} H(t) = \left(\frac{\mathrm{d}H}{\mathrm{d}t}\right)_{\text{streaming}} + \left(\frac{\mathrm{d}H}{\mathrm{d}t}\right)_{\text{collisions}}. \tag{4.113}$$

The streaming term vanishes, $(\mathrm{d}H/\mathrm{d}t)_{\text{streaming}} = 0$ (see Exercise 4.28), and thus $H(t)$ *changes only from collisions*,

$$\frac{\mathrm{d}}{\mathrm{d}t} H(t) = \int \mathrm{d}\boldsymbol{r}\mathrm{d}\boldsymbol{v} J[f] \ln f. \tag{4.114}$$

The theorem follows if we can show the integrand in Eq. (4.114) is never positive.

We do that by making use of the symmetry properties of $J[f]$. Consider an integral of the product of $J[f(\boldsymbol{v})]$ with an arbitrary function of $\boldsymbol{v}$, $\psi(\boldsymbol{v})$. That operation specifies a linear operator on ψ associated with $J[f]$,

$$\mathcal{L}(\psi) \equiv \int \mathrm{d}\boldsymbol{v} J[f(\boldsymbol{v})]\psi(\boldsymbol{v}) = \int \mathrm{d}\boldsymbol{v}\mathrm{d}\boldsymbol{v}_1\mathrm{d}\Omega\sigma(\Omega)|\boldsymbol{v}_1 - \boldsymbol{v}|\left[f(\boldsymbol{v}_1')f(\boldsymbol{v}') - f(\boldsymbol{v}_1)f(\boldsymbol{v})\right]\psi(\boldsymbol{v}). \tag{4.115}$$

---

[67] Boltzmann used E (for entropy) in 1872, but changed it to H in 1895, which has become standard. It's said that H is intended as the capital Greek letter η (used by Gibbs and others to symbolize entropy); see Brush[89, p182]. In 1937, S. Chapman[123] wrote (on Boltzmann changing E to H): "This use of H must have seemed mysterious to many generations of students, and it would be interesting to know whether any reader can account for its use or give an earlier instance of it."

[68] One should be wary of equations featuring logarithms of dimensional quantities. Transcendental functions can only be functions of dimensionless variables. The quantity f, whether particle or probability density, is not dimensionless. The only way such equations can be valid is if there are other terms in the equation (perhaps not displayed) that lead to logarithms of dimensionless quantities. A constant can be added to Eq. (4.109) that fixes this issue; see Eq. (4.130).

[69] R.C. Tolman devoted an entire chapter of *The Principles of Statistical Mechanics*[124, Chapter 6] to the H-theorem, referring to it as "among the greatest achievements of physical science." Strong praise.

The interchange of dummy variables $\boldsymbol{v} \leftrightarrow \boldsymbol{v}_1$ doesn't alter the value of the integral, which we indicate as $\mathcal{L}(\psi) = \mathcal{L}(\psi_1)$, where $\psi_1 \equiv \psi(\boldsymbol{v}_1)$. The change of dummy variables $\boldsymbol{v}, \boldsymbol{v}_1 \to \boldsymbol{v}', \boldsymbol{v}_1'$ effects the inverse transformation of the dependent variables $\boldsymbol{v}', \boldsymbol{v}_1' \to \boldsymbol{v}, \boldsymbol{v}_1$ (use $\mathrm{d}\boldsymbol{v}_1'\mathrm{d}\boldsymbol{v}' = \mathrm{d}\boldsymbol{v}_1\mathrm{d}\boldsymbol{v}$, $g' = g$, and the invariance of the cross section; see footnote 61 on page 112) leading to $\mathcal{L}(\psi') = -\mathcal{L}(\psi)$, where $\psi' \equiv \psi(\boldsymbol{v}')$. Under $\boldsymbol{v}' \leftrightarrow \boldsymbol{v}_1'$, $\mathcal{L}(\psi') = \mathcal{L}(\psi_1')$. By linearity, therefore,

$$\mathcal{L}(\psi) = \tfrac{1}{4}\mathcal{L}\left(\psi + \psi_1 - \psi' - \psi_1'\right). \tag{4.116}$$

Equation (4.116) expresses a basic symmetry of $J[f]$, one that we'll use again, several times.

Combining Eqs. (4.116) and (4.114) and using $\psi = \ln f$, we find

$$\frac{\mathrm{d}}{\mathrm{d}t}H(t) = \frac{1}{4}\int \mathrm{d}\boldsymbol{r}\mathrm{d}\boldsymbol{v}\mathrm{d}\boldsymbol{v}_1\mathrm{d}\Omega\sigma(\Omega)|\boldsymbol{v}_1 - \boldsymbol{v}|\ln\left(\frac{ff_1}{f'f_1'}\right)(f'f_1' - ff_1) \le 0, \tag{4.117}$$

the non-positivity of which follows from the inequality $(x-y)\ln(y/x) \le 0$ for positive x, y. The integrand is therefore nonpositive for any distribution functions (which are individually positive). Solutions of the Boltzmann equation have the property that, regardless of initial conditions, $f(\boldsymbol{r}, \boldsymbol{v}, t)$ evolves in such a way that $H(t)$ never increases as time progresses.

4.6.2 Equilibrium, collisional invariants, Maxwell-Boltzmann distribution

Although $H(t)$ can't increase, it also can't decrease indefinitely; it's bounded from below.[70] A bounded monotonic function must approach a limit, the time-invariant state $\mathrm{d}H/\mathrm{d}t = 0$, *which we identify with equilibrium.* From Eq. (4.117) we see that $\mathrm{d}H/\mathrm{d}t = 0$ is achieved if and only if

$$f(\boldsymbol{r}, \boldsymbol{v}', t)f(\boldsymbol{r}, \boldsymbol{v}_1', t) = f(\boldsymbol{r}, \boldsymbol{v}, t)f(\boldsymbol{r}, \boldsymbol{v}_1, t). \tag{4.118}$$

The state of $\mathrm{d}H/\mathrm{d}t = 0$ is therefore one in which *all collisions are in detailed balance*, in which loss and gain collisions $(\boldsymbol{v}, \boldsymbol{v}_1) \to (\boldsymbol{v}', \boldsymbol{v}_1')$ and $(\boldsymbol{v}', \boldsymbol{v}_1') \to (\boldsymbol{v}, \boldsymbol{v}_1)$ occur with equal frequency, *for all velocities*. Equilibrium is thus a special *dynamical* state.[71] Note that Eq. (4.118) holds for any $(\boldsymbol{r}, t)$; the Boltzmann equation is local in $(\boldsymbol{r}, t)$ (see Section 4.5.2.4).

To find the distribution f_{eq} satisfying Eq. (4.118), note that, as a consequence of Eqs. (4.115) and (4.116), and the conservation laws Eq. (4.97), we have, *for any* f,

$$\int \mathrm{d}\boldsymbol{v}J[f]\begin{Bmatrix} 1 \\ \boldsymbol{v} \\ v^2 \end{Bmatrix} = 0. \tag{4.119}$$

Quantities $\psi(\boldsymbol{v})$ such that $\psi_1 + \psi = \psi_1' + \psi'$ (those for which $\int \mathrm{d}\boldsymbol{v}J[f]\psi(\boldsymbol{v}) = 0$) are said to be *collisional invariants* or *summational invariants*, quantities preserved in elastic collisions—particle number, momentum, and kinetic energy. A key result in kinetic theory, on which much depends, and a proof of which we omit, is that $(1, \boldsymbol{v}, v^2)$ are *the only linearly independent collisional invariants.*[72] For future use, we denote the collisional invariants ψ_i, $i = 1, 2, 3, 4, 5$, for $(1, v_x, v_y, v_z, v^2)$.

---

[70] For $H(t)$ not to have a lower bound ($H(t) \to -\infty$), the integral in Eq. (4.109) would have to diverge, an assumption that leads to a contradiction. Although $\ln f \to -\infty$ as $v \to \infty$ ($f \to 0$), the integral $\int \frac{1}{2}mv^2 f(\boldsymbol{v})\mathrm{d}\boldsymbol{v}$ representing the kinetic energy in the gas must exist on physical grounds. For the integral for H to diverge, $-\ln f$ must tend to ∞ faster than v^2, implying that f tends to zero more rapidly than $\exp(-v^2)$, for which the integral for H exists, a contradiction! Assuming it diverges implies that it converges. A finite minimum of H is consistent with finite kinetic energy.

[71] Detailed balance is a microscopic, dynamical characterization of equilibrium; see Sections 2.8.2 and 3.4.3.1 for detailed balance in the master equation and the Fokker-Planck equation. Compare with the picture of equilibrium provided by traditional thermodynamics, the timeless state in which a system's macroscopic properties appear not to be changing[2, p4].

[72] Jeans[125, p111] wrote that it's *obvious* these are the only collisional invariants, yet a proof is not trivial. See Sommerfeld[126, p307] or Harris[127, p57]. The invariants ψ_i may be multiplied by arbitrary functions of $(\boldsymbol{r}, t)$; the Boltzmann equation is local in $(\boldsymbol{r}, t)$. It's obvious these are the only collisional invariants in the sense that assuming the opposite leads to unphysical conclusions. Besides the trivial invariant $\psi_1 = 1$, the four remaining invariants impose four constraints on the after-collision velocities $(\boldsymbol{v}', \boldsymbol{v}_1')$, leaving the unit vector $\hat{\boldsymbol{\epsilon}}$ unspecified (see Section 4.5.1). If there were additional independent invariants, after-collision velocities could be determined without knowing $\hat{\boldsymbol{\epsilon}}$; no need for $\boldsymbol{F} = m\boldsymbol{a}$.

4.6.2.1 Spatially uniform systems

The state specified by Eq. (4.118) is equivalent to $\ln f'_{\text{eq}} + \ln f'_{\text{eq},1} = \ln f_{\text{eq}} + \ln f_{\text{eq},1}$. Thus, $\ln f_{\text{eq}}$ is a collisional invariant, implying it's a linear combination of the fundamental invariants,

$$\ln f_{\text{eq}}(\boldsymbol{v}) = A + \boldsymbol{B} \cdot \boldsymbol{v} + Cv^2, \tag{4.120}$$

where the coefficients $A, \boldsymbol{B}, C$ (five quantities) are constants (for spatially uniform systems).[73] Equation (4.120) implies an equivalent parameterization as a Gaussian[74] (see Exercise 4.33),

$$f_{\text{eq}}(\boldsymbol{v}) = K \exp\left(-(m\beta/2)\left(\boldsymbol{v} - \boldsymbol{u}\right)^2 \right). \tag{4.121}$$

This expression, the Maxwell-Boltzmann velocity distribution,[75] shows that gas in equilibrium, a quiescent system macroscopically, is, microscopically, a collection of atoms having a range of speeds at every point $\boldsymbol{r}$, the distribution of which has been verified experimentally. The five parameters $K, \boldsymbol{u}, \beta$ in Eq. (4.121) are determined by demanding that $f_{\text{eq}}(\boldsymbol{v})$ be consistent with known macroscopic information. The quantity K is found from particle number,

$$N = \int \mathrm{d}\boldsymbol{r}\mathrm{d}\boldsymbol{v} f_{\text{eq}}(\boldsymbol{v}) = V \int \mathrm{d}\boldsymbol{v} f_{\text{eq}}(\boldsymbol{v}) \implies K = n\left(\frac{m\beta}{2\pi}\right)^{3/2}, \tag{4.122}$$

with $n \equiv N/V$; $\boldsymbol{u}$ is the average velocity,

$$N\langle \boldsymbol{v}\rangle = \int \mathrm{d}\boldsymbol{r}\mathrm{d}\boldsymbol{v}\, \boldsymbol{v} f_{\text{eq}}(\boldsymbol{v}) = N\boldsymbol{u} \implies \boldsymbol{u} = \langle \boldsymbol{v}\rangle, \tag{4.123}$$

and β is related to the average energy[76]

$$\frac{1}{2}mN\langle(\boldsymbol{v} - \langle\boldsymbol{v}\rangle)^2\rangle = \frac{1}{2}m\int \mathrm{d}\boldsymbol{r}\mathrm{d}\boldsymbol{v}(\boldsymbol{v} - \langle\boldsymbol{v}\rangle)^2 f_{\text{eq}}(\boldsymbol{v}) = \frac{3}{2}\frac{N}{\beta} = \frac{3}{2}NkT \implies \beta = \frac{1}{kT}, \tag{4.124}$$

where we've invoked $\frac{3}{2}NkT$ as the energy of an N-particle ideal gas[77] at absolute temperature T[5, p21]. Note that temperature (a thermodynamic quantity) has been introduced in Eq. (4.124) as proportional to the distribution of velocities relative to the mean velocity.[78] Temperature is thus a measure of the width of the equilibrium velocity distribution,[79]

$$kT = \frac{1}{3}m\langle(\boldsymbol{v} - \boldsymbol{u})^2\rangle. \tag{4.125}$$

If the distribution is sharply peaked about the mean, temperature is small; a wide distribution implies high temperature. The velocity $\boldsymbol{u}$ is not particularly relevant here because it can be transformed away

[73] Boltzmann gave a proof of Eq. (4.120) in [105, p139]. A simpler proof was devised by Gronwall[128] in 1915.

[74] Gaussians are completely specified by their first three moments (when normalization is counted as the zeroth moment); see Section 2.9.1. These line up with the three parameters $K, \boldsymbol{u}, \beta$.

[75] Maxwell derived the form of Eq. (4.121) in 1860 for a gas at rest. His argument, although of historical interest, was not mathematically rigorous. Boltzmann put the distribution on a much firmer footing using the H-theorem. There is no one way to derive the Maxwell-Boltzmann distribution; see [5, Section 4.1.2.7], *Getting to Boltzmann: A discussion.* We attempted in [5, Section 4.1.2] a probabilistic derivation based on the central limit theorem.

[76] Entirely translational for a dilute gas of atoms with short-ranged interactions undergoing elastic collisions.

[77] One could object we're using the equipartition theorem of statistical mechanics (see [5, p97]) and hence we're assuming the Maxwell-Boltzmann distribution, the form of which we're trying to establish from kinetic theory. One can take as experimental input that the heat capacity C_V of noble gases (which approximate ideal gases) is $\frac{3}{2}Nk$ and hence $U = \frac{3}{2}NkT$.

[78] Does the temperature of a container of gas increase if it's placed on a train moving at constant speed? If that were the case, we'd have a way of detecting absolute motion which the theory of relativity says you can't do (see [6, Chapter 1]).

[79] In thermodynamics, equilibrium is attained when the intensive variables of a system (associated with conserved quantities) are equal to their environmental counterparts, i.e., temperature is set by the environment[2, p45]. Equation (4.125) is an intrinsic definition of absolute temperature in terms of the statistical properties of a system in thermal equilibrium.

(Galilean transformation) by choosing a reference frame in which $\boldsymbol{u} = 0$. Temperature is a Galilean invariant, as we see from Eq. (4.125). Note that when the Maxwell-Boltzmann distribution is used in Eq. (4.50), we recover the ideal gas equation of state, $P = nkT$.

Equation (4.118) is equivalent to $J[f_{\text{eq}}] = 0$. *The Maxwell-Boltzmann distribution is invariant under collisions*;[80] equilibrium is a special dynamical mode in which all collisions are in detailed balance, a state the H-theorem guarantees is always attained. Because the streaming terms in the Boltzmann equation vanish for uniform systems, $J[f_{\text{eq}}] = 0$ implies $\partial f_{\text{eq}}/\partial t = 0$. Establishing *and maintaining* equilibrium is effected by collisions, and collisions require interparticle forces. A collection of noninteracting particles would never come to equilibrium; *irreversibility requires interactions*. Ideal systems are a convenient fiction not occurring in nature.[81]

4.6.2.2 Inhomogeneous systems, local Maxwellian

For inhomogeneous systems, the streaming terms in the Boltzmann equation can be nonzero even when $J[f] = 0$. Systems can satisfy detailed balance—a condition in velocity space—at each point of configuration space.[82] The general Boltzmann equation for systems in detailed balance is

$$\left(\frac{\partial}{\partial t} + \boldsymbol{v} \cdot \frac{\partial}{\partial \boldsymbol{r}} + \boldsymbol{a} \cdot \frac{\partial}{\partial \boldsymbol{v}}\right) f(\boldsymbol{r}, \boldsymbol{v}, t) = 0. \tag{4.126}$$

Let's try a stationary solution of Eq. (4.126) using our favorite method, separation of variables. Generalize Eq. (4.121) to $f(\boldsymbol{r}, \boldsymbol{v}) = K(\boldsymbol{r})\text{e}^{-(m\beta/2)(\boldsymbol{v}-\boldsymbol{u})^2}$, with all $\boldsymbol{r}$-dependence carried by $K(\boldsymbol{r})$. Assuming conservative forces with $m\boldsymbol{a} = -\boldsymbol{\nabla}\phi(\boldsymbol{r})$, Eq. (4.126) implies (under this assumption)

$$\boldsymbol{v} \cdot [\boldsymbol{\nabla} K(\boldsymbol{r}) + \beta K(\boldsymbol{r}) \boldsymbol{\nabla}\phi(\boldsymbol{r})] - \beta K(\boldsymbol{r}) \boldsymbol{u} \cdot \boldsymbol{\nabla}\phi(\boldsymbol{r}) = 0. \tag{4.127}$$

The only way Eq. (4.127) can be satisfied with the form $K(\boldsymbol{r}) = K_0 \text{e}^{-\beta\phi(\boldsymbol{r})}$ (see Eq. (3.61)) is if $\boldsymbol{u} = 0$ (see Uhlenbeck and Ford[107, p86]). This isn't a serious restriction; there is no global average velocity in an inhomogeneous system. We can define a local density,

$$n(\boldsymbol{r}) \equiv \int \text{d}\boldsymbol{v} f(\boldsymbol{r}, \boldsymbol{v}) = K_0 \text{e}^{-\beta\phi(\boldsymbol{r})} \int \text{d}\boldsymbol{v}\, \text{e}^{-\frac{1}{2}m\beta v^2} = K_0 \left(\frac{2\pi}{m\beta}\right)^{3/2} \text{e}^{-\beta\phi(\boldsymbol{r})} \equiv n_0 \text{e}^{-\beta\phi(\boldsymbol{r})}, \tag{4.128}$$

where n_0 is the density where $\phi = 0$. Thus, $K_0 = n_0(m\beta/2\pi)^{3/2}$ (compare with Eq. (4.122)). Equation (4.128) is the *barometric formula*.

For the most general solution of $J[f] = 0$, return to Eq. (4.120) and let $A, \boldsymbol{B}, C$ be arbitrary functions of $(\boldsymbol{r}, t)$. In that case we have the generalization of Eq. (4.121),

$$f_{\text{LM}}(\boldsymbol{r}, \boldsymbol{v}, t) \equiv n(\boldsymbol{r}, t) \left(\frac{m\beta(\boldsymbol{r}, t)}{2\pi}\right)^{3/2} \exp\left[-\frac{1}{2} m\beta(\boldsymbol{r}, t) \left(\boldsymbol{v} - \boldsymbol{u}(\boldsymbol{r}, t)\right)^2\right]. \tag{4.129}$$

Although the *form* of Eq. (4.129) solves Eq. (4.126) ($J[f] = 0$), the functions $n(\boldsymbol{r}, t), \boldsymbol{u}(\boldsymbol{r}, t), \beta(\boldsymbol{r}, t)$ must be found from the streaming side of the Boltzmann equation, (4.126). Equation (4.129) with

[80] $H(t)$ changes only through collisions and $\text{d}H/\text{d}t = 0$ in equilibrium.

[81] Systems composed of noninteracting components (ideal systems) are used in statistical mechanics to approximate the equilibrium properties of real systems[5, Chapter 5]. Equilibrium is specified by the *values* of state variables and not by *how* they have come to have their values. Exact differentials, widely used in thermodynamics, are predicated on the history independence of equilibrium[2, p5]. Interactions are required for systems to come to equilibrium (no matter how long equilibration takes); once in equilibrium, however, most systems have regimes in which interactions can be neglected. For example, all gases (in equilibrium!) behave ideally at sufficiently low pressure[129].

[82] Single-particle phase space (μ-space) is the direct product of configuration space and velocity space; Eq. (4.118) can be satisfied at every point of $\boldsymbol{r}$-space. The collision operator acts in $\boldsymbol{v}$-space only; one can achieve detailed balance in $\boldsymbol{v}$-space ($J[f] = 0$) while evolution remains in $\boldsymbol{r}$-space. Irreversibility occurs in $\boldsymbol{v}$-space (collisions); streaming is reversible.

space and time-dependent parameters $n, \boldsymbol{u}, \beta$ is the *local Maxwellian*,[83] which provides a theoretical foundation for the concept of local thermodynamic equilibrium. We return to inhomogeneous systems and local Maxwellians in Section 4.8 and in Chapter 6.

4.6.3 Connection with the second law, Gibbs and Boltzmann entropies

The H-theorem shows that associated with solutions of the Boltzmann equation is a quantity $H(t)$ that decreases monotonically to a lower bound, the state of the gas in which $\mathrm{d}H/\mathrm{d}t = 0$. We therefore have, in the framework of the Boltzmann equation, a model of nonequilibrium behavior in which a dilute gas irreversibly comes to equilibrium regardless of initial state, in agreement with observational experience. Even though we know something about the mathematical properties of $H(t)$, we don't know what it represents. Entropy suggests itself,[84] which *increases* until equilibrium is achieved, when it has the maximum value it can have subject to constraints. Is the value of H in the state $\mathrm{d}H/\mathrm{d}t = 0$, call it H_∞, proportional to entropy? Is $-kH_\infty = S$? We'll see how that equality can be arranged, but first we must address the issue raised in footnote 68 on page 114.

Referring to Eq. (4.109), redefine H so that the argument of the logarithm is dimensionless, and let's do that using the equilibrium distribution for spatially uniform systems,

$$H_\infty \equiv \int \mathrm{d}\boldsymbol{r}\mathrm{d}\boldsymbol{v} f_{\text{eq}}(\boldsymbol{v}) \ln\left(f_{\text{eq}}(\boldsymbol{v})/\widetilde{f}\right), \tag{4.130}$$

where $\widetilde{f}$ is a constant (to be determined) having the dimensions of f_{eq}. Using Eq. (4.121),

$$H_\infty = VK \int \mathrm{d}\boldsymbol{v}\, \mathrm{e}^{-\frac{1}{2}m\beta v^2} \ln\left((K/\widetilde{f})\mathrm{e}^{-\frac{1}{2}m\beta v^2}\right) = -N\left[\frac{3}{2} + \ln\left(\widetilde{f}\frac{V}{N}\left(\frac{2\pi}{m\beta}\right)^{3/2}\right)\right], \tag{4.131}$$

where we've integrated over $\boldsymbol{r}$ to produce the volume V. How to choose $\widetilde{f}$? Appeal to experiment!

The *Sackur-Tetrode equation* is an experimentally tested formula for the entropy of an ideal gas,[2, p118]

$$S = Nk\left[\frac{5}{2} + \ln\left(\frac{V}{N\lambda_T^3}\right)\right] = Nk\ln\left(\mathrm{e}^{5/2}\frac{V}{N\lambda_T^3}\right), \tag{4.132}$$

where the *thermal wavelength* $\lambda_T \equiv h/\sqrt{2\pi mkT}$. Note Planck's constant in λ_T; the entropy of even the simplest macroscopic system, the ideal gas, cannot be calculated using classical physics.[85] Classical theories miss the mark in two ways: the Heisenberg uncertainty principle which forces a discretization of μ-space into cells of size[86] h^3, and the recognition that permutations of identical

---

[83] When $n, \boldsymbol{u}, \beta$ are constants, as in Eq. (4.121), it's referred to as a *global Maxwellian* or an *absolute Maxwellian*.

[84] Entropy and $H(t)$ evolve through irreversible processes; they have that in common. Boltzmann wrote in 1872:[89, p263] "H must approach a minimum value and remain constant thereafter, and the corresponding final value of f will be the Maxwell distribution. Since H is closely related to the thermodynamic entropy in the final equilibrium state, our result is equivalent to a proof that the entropy must always increase or remain constant, and thus provides a microscopic interpretation of the second law of thermodynamics." Class assignment: Was Boltzmann justified in making this claim?

[85] In 1982, Feynman[130] quipped: "Nature isn't classical, dammit, ..." Although true, one might think that thermodynamics is far from the quantum realm. Not so with entropy. It can be shown (see [2, Section 3.7]) that for entropy to be extensive, there must be an intrinsic quantity associated with material particles having the dimension of action, which experiment shows is Planck's constant[2, Section 7.4]. We noted in [2] the many instances in which thermodynamics anticipates quantum physics (see the index of [2] *thermodynamics, anticipates quantum physics*). One can't scratch the surface of thermodynamics without finding quantum mechanics lurking beneath. Equation (4.132) shows there are two independent lengths in an ideal gas: the average distance between particles $(V/N)^{1/3}$ and the thermal wavelength λ_T, the average de Broglie wavelength of the gas particles[5, p174]. Particles behave independently when $V/N \gg \lambda_T^3$; when $V/N \ll \lambda_T^3$ they interact strongly and their quantum natures manifest. Quantum aspects of matter feature in calculations of entropy, but not in calculations of internal energy and other quantities obtained in statistical mechanics from logarithmic derivatives, e.g., $U = -\partial \ln Z/\partial\beta$.

[86] For the Sackur-Tetrode equation to agree with experiment, we must use h, not $\hbar$ nor $\frac{1}{2}\hbar$; see [2, p110]. The Sackur-Tetrode formula (circa 1912) is to our knowledge the first instance in the logical development of physics of associating

particles are not distinct states.[87] Single-particle theories lead to a factor of $\frac{3}{2}$ in expressions for entropy. If the permutation symmetry of collections of identical particles is not taken into account, one will not obtain the experimentally verified factor of $\frac{5}{2}$ in Eq. (4.132).[88]

Equating (4.131) (multiplied by $-k$) with Eq. (4.132) implies a value of $\widetilde{f}$ so that $-kH_{\infty} = S$,

$$\widetilde{f} = \mathrm{e}m^3/h^3. \tag{4.133}$$

The factor of e in $\widetilde{f}$ (the *indistinguishability factor*) changes $\frac{3}{2}$ in Eq. (4.131) to $\frac{5}{2}$. The factor of m^3 occurs because we're working with a μ-space spanned by $\boldsymbol{r}, \boldsymbol{v}$; it would be unity if spanned by $\boldsymbol{r}, \boldsymbol{p}$. Choosing $\widetilde{f}$ as in Eq. (4.133),[89] we have a connection between entropy and the velocity distribution of gas particles,[90]

$$S = -k \int \mathrm{d}\boldsymbol{r}\mathrm{d}\boldsymbol{v} f_{\mathrm{eq}}(\boldsymbol{v}) \ln \left(\frac{h^3}{\mathrm{e}m^3} f_{\mathrm{eq}}(\boldsymbol{v}) \right). \tag{4.134}$$

We note that 75 years after the H-theorem, *information* was discovered as a generalization of entropy having the same "P-log P" form as the H-function, where the same symbol is used, $H \equiv -K \sum_i p_i \ln p_i$ (K is a positive constant and the p_i are known probabilities).[91]

Entropy calculated from the one-particle distribution is known as the *Boltzmann entropy*,[92]

$$S_B \equiv -k \int f_1 \ln f_1 \mathrm{d}\boldsymbol{r}\mathrm{d}\boldsymbol{v} \equiv -kH_B, \tag{4.135}$$

with H_B the long-time value of Eq. (4.109). Gibbs introduced a similar, yet different quantity, the *Gibbs entropy*, calculated from the N-body distribution, $f_N(\boldsymbol{r}_1, \boldsymbol{v}_1, \ldots, \boldsymbol{r}_N, \boldsymbol{v}_N)$,[93]

$$S_G \equiv -k \int f_N \ln f_N \mathrm{d}\boldsymbol{r}_1 \mathrm{d}\boldsymbol{v}_1 \cdots \mathrm{d}\boldsymbol{r}_N \mathrm{d}\boldsymbol{v}_N \equiv -kH_G, \tag{4.136}$$

where H_G denotes a Gibbs H-function. With $S_G = -k\langle \ln f_N \rangle$ as the statistical definition of entropy, the laws of thermodynamics are reproduced in the canonical ensemble (see [5, pp92–96]),

Planck's constant with material particles. Before that point in the history of physics, Planck's constant was associated only with photons. The thermal wavelength presages the de Broglie wavelength.

[87] Noted by Gibbs in 1902[131, p187].

[88] The difference between $\frac{5}{2}$ and $\frac{3}{2}$ might seem small, but it's *huge*. In a change of entropy $\Delta S = Nk$, the number of states accessible to the system is increased by a multiplicative factor of e^N.

[89] Although $\widetilde{f} = \mathrm{e}m^3/h^3$ ensures that the ideal-gas entropy is correctly reproduced by $-kH_{\infty}$, it's a kludge designed to get the right answer from a single-particle theory. Single-particle theories treat the gas as N particles in six-dimensional μ-space, rather than, as is required, a system of identical particles in $6N$-dimensional Γ-space; see [2, Section 7.6].

[90] In thermodynamics, entropy is a function of state variables, $S = Nkf(U/N, V/N)$. In statistical mechanics, entropy is found from the partition function, $S = k\partial(T \ln Z)/\partial T$, where Z is a function of the system Hamiltonian. The Hamiltonian is microscopic in that it contains all information about system components and their interactions, yet the view of the system afforded by Z (the number of energy states available in equilibrium at temperature T) is static. Equation (4.134) is a connection between entropy and a *dynamical* property of system components, the distribution of velocities in equilibrium. We were led to Eq. (4.134) through the H-theorem, a dynamical approach. Note that $S \to \infty$ if $h \to 0$; S is finite because $h \neq 0$. Entropy is another quantity saved from the ultraviolet catastrophe by the introduction of Planck's constant.

[91] C. Shannon published a landmark paper in 1948, "A Mathematical Theory of Communication,"[132] which was soon thereafter reprinted in book form[133]. We're not trying to imply that information was copied from the H-theorem; Shannon proposed a set of axioms to be satisfied by any definition of information and then showed that information (also called the *Shannon entropy*) $H = -K \sum_i p_i \ln p_i$ is the only function meeting those requirements. See [2, Chapter 12]. Information applies to any system for which its probability distribution is known; entropy applies to systems for which the probability distribution characterizes thermal equilibrium. Entropy is information, but information isn't necessarily entropy.

[92] A *second* use of the term Boltzmann entropy (besides $S = k \ln W$). In deriving the Boltzmann equation we took f to be a number density such that $\int f(\boldsymbol{r}, \boldsymbol{v})\mathrm{d}\boldsymbol{r}\mathrm{d}\boldsymbol{v} = N$. S_B is therefore extensive; see Eq. (4.131). If, however, f is taken to be a probability density with $\int f(\boldsymbol{r}, \boldsymbol{v})\mathrm{d}\boldsymbol{r}\mathrm{d}\boldsymbol{v} = 1$, we should define $S_B = -Nk \int f \ln f \mathrm{d}\boldsymbol{r}\mathrm{d}\boldsymbol{v}$.

[93] Gibbs did not explicitly write down Eq. (4.136); he introduced an equivalent quantity, $\overline{\eta} \equiv \int \eta \mathrm{e}^{\eta} \mathrm{d}p_1 \cdots \mathrm{d}q_N$, the average of η, where $\mathrm{e}^{\eta} = P$[131, Chapter 11]. Gibbs entropy is not always included in statistical physics texts. See Feynman[3, p28] or Kittel[24, p54]. Entropy, as we see from Eq. (4.136), is not the mean value of a mechanical quantity, but rather is a measure of the uncertainty in the systems of the ensemble.

an important consistency requirement.[94] If the Gibbs definition preserves the structure of thermodynamic relations in statistical mechanics, and the Boltzmann definition reproduces the entropy of the ideal gas, there must be a relation between them. E.T. Jaynes[134] studied the two entropies (see also [135]) and derived the inequality $H_B \leq H_G$, with equality for independent particles.[95] The customary entropy formula attributed to Boltzmann,[96] $S = k \ln W$ (which applies to the microcanonical ensemble), is consistent with the Gibbs formula $S = -k\langle \ln f_N \rangle$ (canonical ensemble) and is derivable from it; the formula for S_G reduces to $k \ln W$ in the microcanonical ensemble.[97] The Gibbs formula makes use of N-particle phase space (Γ-space), as does a correct calculation of W for identical particles[2, p112].

We've been led into this discussion of entropy—a property of equilibrium systems—because of its possible connection with the long-time value of $H(t)$. The Gibbs entropy has the property of being *a constant of the motion for N-particle systems*, a conclusion that follows from general principles of classical mechanics[98] or from direct calculation (see Exercise 4.40). Thus, S_G exhibits a required property of state variables, time invariance. Of course S_B is also time invariant, but that's because $J[f_{\text{eq}}] = 0$ is a property of the Boltzmann equation (which pertains to dilute gases and is based on the molecular chaos assumption); S_G remains constant under the microscopic dynamics of systems of interacting particles of any density. Although the H-theorem shows the tendency of dilute gases to irreversibly come to equilibrium with the Maxwell velocity distribution, that does not constitute a proof of the second law of thermodynamics, which applies to all physical systems, not just the ideal gas.[99] It's not enough to have a bounded, monotonic mathematical function $H(t)$; it must have a relation with experimentally measured quantities.

4.6.4 Coarse graining and loss of information

Of broader interest is that although the Gibbs H-function remains constant under the time-reversal-invariant microdynamics, the Boltzmann H-function evolves irreversibly under the Boltzmann equation. This qualitative difference illustrates the effects of *coarse graining*, the replacement

---

[94] Setting up statistical mechanics is guided methodologically by the requirement that it be the simplest mechanical model consistent with the laws of thermodynamics[5, Section 2.5]. This demand is met by identifying macroscopically measurable quantities with appropriate ensemble averages, as in $U = \langle H \rangle$ where H is the Hamiltonian.

[95] To quote Jaynes:[134] "...the Gibbs formula gives the correct entropy, as defined in phenomenological thermodynamics, while the Boltzmann H expression is correct only in the case of an ideal gas." Jaynes showed that S_G satisfies the Clausius relation $\Delta S_G = \int \mathrm{d}Q/T$ (but not S_B) for general systems (those having interacting components). The quantity S_B is the entropy of an ideal system of the same density and temperature as the actual system. The difference between S_B and S_G is not negligible for systems in which interparticle forces have observable effects on thermodynamic properties. Interactions of course affect the equation of state and the free energy; see [5, Sections 6.1–6.3].

[96] There is debate whether Boltzmann ever wrote the formula attributed to him explicitly in the form $S = k \ln W$ (apparently due to Planck). The closest he came to writing $S = k \ln W$ seems to be in [105, p74].

[97] Whereas $S = k \ln W$ applies to isolated systems of fixed (U, V, N) (microcanonical ensemble), S_G applies to closed systems in thermal contact with their surroundings, of fixed (T, V, N) (canonical ensemble). Thus S_G is more general than $k \ln W$, yet it reduces to that expression in the microcanonical ensemble. There, each member of the ensemble has a number of states Ω, all of which are equally likely with probability $P_r = 1/\Omega$, the principle of *equal a priori probabilities*[5, p116]. Using the Gibbs formula for discrete states, $S_G = -k \sum_{r=1}^{\Omega} [(1/\Omega) \ln(1/\Omega)] = k \ln \Omega$.

[98] Classical motion can be described as the continuous unfolding of an infinitesimal canonical transformation generated by the Hamiltonian[5, p334]. An arbitrary phase-space function at time $t = 0$, $A(\boldsymbol{p}(0), \boldsymbol{q}(0))$, is mapped (under the natural motion in phase space) into the same function of the transformed variables $A(\boldsymbol{p}(t), \boldsymbol{q}(t))$ at time t, and the Jacobian of canonical transformations is unity (see [5, p332]). Thus, $S_G(t_1) = S_G(t_2)$ whether $t_1 > t_2$ or $t_1 < t_2$.

[99] By the way, what *is* the second law of thermodynamics? There is no one, all-encompassing way of stating the second law, which has diverse, interrelated implications (see [2, Chapter 9]). Students (and others) often leap to "entropy" as the second law. Entropy is best viewed as a *consequence* of the second law, with the second law itself the recognition of irreversibility. It's remarkable how much physics follows from the observation that heat does not spontaneously flow from cold to hot, the simplest form of the second law. As noted by P.W. Bridgman[136] (1946 Nobel Prize in Physics), "There have been nearly as many formulations of the second law as there have been discussions of it."

of N-body dynamics with the effective dynamics of fewer degrees of freedom.[100] In this case, the one-particle distribution is obtained by eliminating almost all degrees of freedom of the gas, $f_1(\boldsymbol{r}_1, \boldsymbol{v}_1) \equiv \int \mathrm{d}\boldsymbol{r}_2 \mathrm{d}\boldsymbol{v}_2 \cdots \mathrm{d}\boldsymbol{r}_N \mathrm{d}\boldsymbol{v}_N f_N(\boldsymbol{r}_1, \boldsymbol{v}_1, \ldots, \boldsymbol{r}_N, \boldsymbol{v}_N)$. In 1894, E.P. Culverwell[137] asked: "Will someone say exactly what the H-theorem proves?" A reply can be given when the connection between entropy and information is brought out. Entropy is a measure of the microstates of a system consistent with the macrostate specified by state variables (other than entropy)[2, p171]. Entropy represents the *missing information* about a system that can't be accounted for in macroscopic descriptions. Entropy as missing information, what could *potentially* be known about a system, is an appealing idea, but can it be made precise? Can information be quantified in a way suitable for physical theories? As the term is used in everyday life, information connotes a subjective experience, yet we're suggesting a connection between something nominally qualitative (information) with something physical (entropy).[101] Is information physical? In 1961, R. Landauer[138] published a seminal article, "Irreversibility and Heat Generation in the Computing Process," proposing a link between information and the thermodynamics of its manifestation in physical systems. Landauer's thesis, in brief, is that indeed, *information is physical*, that information is stored in physical media, is manipulated by physical means, and is subject to the laws of physics.[102] *Landauer's principle* is that entropy increases when information is destroyed (as in clearing a memory register), an experimentally verified prediction[140][141][142]. We can now respond to Culverwell's question. The H-theorem tells us there is a loss of information in a dilute gas passing from nonequilibrium (low probability) configurations to the state of equilibrium (the most probable state of macroscopic systems; see [5, pp363–365]). There is no proof that coarse-grained entropy increases in time, just as there is no proof of the second law, yet Landauer's principle is a useful way to think about it.[103]

4.6.5 The reversibility paradox, not

An objection to the H-theorem was raised soon after it was published (*Loschmidt's paradox* or the *reversibility paradox*), that it should not be possible to derive irreversible phenomena from time-symmetric physics. Consider in the evolution of a system from time t_0 to t_1 to t_2 ($t_0 < t_1 < t_2$) there is an associated decrease in H with time. Suppose it would be possible at time t_1 to reverse the velocity of every particle, $\boldsymbol{v}_i \to -\boldsymbol{v}_i$. For systems governed by time-reversal-invariant dynamics, for every state of the system with particles having velocities $\{\boldsymbol{v}_i\}$, there is a time-reversed state with velocities $\{-\boldsymbol{v}_i\}$, and thus the value of H would be unchanged by this process. Before the time reversal at $t = t_1$, $\partial H/\partial t < 0$, but after $\partial H/\partial t > 0$, a contradiction.

The paradox resolves itself when the flaw in the argument is noticed: The dynamics described by the H-theorem is not time-reversal invariant! Pre-collision velocities of dilute-gas particles are uncorrelated, but after-collision particles become correlated through the conservation laws, Eq. (4.97),

[100] Coarse graining is also used in statistical mechanics. In Ginzburg-Landau theory, a system of spins on a lattice $\{\sigma_i\}$ is replaced with a continuous spin-density field $S(\boldsymbol{r})$[5, p295]. A field description constitutes a coarse graining, a view of the system from sufficiently large distances that it appears continuous. The renormalization group studies how the effective interactions between coarse-grained degrees of freedom change over progressively larger length scales[5, Chapter 8].

[101] The ability to quantify information is why information theory is so widely used in scientific research. Information (as the term is used in information theory) is similar to entropy, with their commonality being missing information. (Information is thus missing information. Got it?) The definition of information $H = -K\sum_i p_i \ln p_i$ takes into account the element of surprise one experiences in learning of events whose occurrence was not previously certain. Which statement conveys more information: 1) The sun rose this morning; 2) The sun exploded this morning. There's no news in the first statement; the sun rises everyday with high probability. The second statement, however, is real news: It's unexpected (low probability). The more likely an event, the less is the information conveyed in learning of its occurrence.

[102] Landauer wrote in 1999:[139] "Information is inevitably inscribed in a physical medium. It is not an abstract entity. It can be denoted by a hole in a punched card, by the orientation of a nuclear spin, or by the pulses transmitted by a neuron. The quaint notion that information has an existence independent of its physical manifestation is still seriously advocated. This concept, very likely, has its roots in the fact that we were aware of mental information long before we realized that it, too, utilized real physical degrees of freedom."

[103] Landauer's principle is not independent of the second law of thermodynamics[2, p184].

a fact ignored in deriving Boltzmann's equation (molecular chaos assumption). If the velocities of all particles are reversed, collisions occur in the reverse sense, but the colliding particles would not be uncorrelated (if the dynamics is time symmetric), and the value of $\partial H/\partial t$ found previously no longer applies.

4.7 COLLISION FREQUENCY, MEAN FREE PATH

We can infer the rate at which particles collide using the arguments leading to the collision operator. Divide Eq. (4.103) (the *total* rate of loss collisions[104]) by $f(\boldsymbol{r}, \boldsymbol{v}, t)$ to yield the *per-particle* rate at $(\boldsymbol{r}, t)$ that particles of velocity $\boldsymbol{v}$ undergo collisions, $\nu(\boldsymbol{r}, \boldsymbol{v}, t) \equiv \int \mathrm{d}\boldsymbol{v}_1 \mathrm{d}\Omega \sigma(\Omega)|\boldsymbol{v}_1 - \boldsymbol{v}| f(\boldsymbol{r}, \boldsymbol{v}_1, t)$. This result is too general; let's use $f_0(\boldsymbol{v}) = n(m/2\pi kT)^{3/2} \exp(-mv^2/2kT)$, the equilibrium distribution for spatially uniform systems: $\nu(v) = \int \mathrm{d}\boldsymbol{v}_1 \mathrm{d}\Omega \sigma(\Omega)|\boldsymbol{v}_1 - \boldsymbol{v}| f_0(\boldsymbol{v}_1)$. Treat gas particles as *hard spheres* that undergo elastic scattering when their surfaces come into contact,[105] when their centers are separated by the diameter D. The total scattering cross section is the area blocked out by a sphere of radius D, $\sigma_T = \pi D^2$. Take σ_T outside the integral; for hard spheres, therefore, $\nu(v) = \sigma_T \int \mathrm{d}\boldsymbol{v}_1 |\boldsymbol{v}_1 - \boldsymbol{v}| f_0(\boldsymbol{v}_1)$. Evaluating the integral (see Exercise 4.41), we find

$$\nu(v) = \nu_0 \left[\mathrm{e}^{-\xi^2} + (\xi^{-1} + 2\xi) \int_0^\xi \mathrm{e}^{-t^2} \mathrm{d}t \right], \tag{4.137}$$

where $\xi \equiv v/v_{\mathrm{mp}}$, with $v_{\mathrm{mp}} \equiv \sqrt{2kT/m}$ the most probable speed of gas particles,[5, p129] and $\nu_0 \equiv n\sigma_T\sqrt{2kT/(m\pi)}$. The collision frequency of a particle moving through an equilibrium gas thus depends on its speed. For particles at rest (for small x, $\int_0^x \exp(-t^2)\mathrm{d}t = x - \frac{1}{3}x^3 + \cdots$), $\lim_{v\to 0} \nu(v) = 2\nu_0 = n\sigma_T\overline{v}$, where $\overline{v} \equiv \sqrt{8kT/(m\pi)}$ is the mean speed of particles in an equilibrium gas. As v increases, $\nu(v)$ increases monotonically, becoming linear, $\nu(v) \overset{v\to\infty}{\sim} n\sigma_T v$.

The inverse of $\nu(v)$ provides the time between successive collisions for particles of speed v. The distance between successive collisions, $l(v) \equiv v/\nu(v)$, ranges from zero for the slowest atoms to the value for the fastest

$$l \equiv \lim_{v\to\infty} l(v) = 1/(n\sigma_T). \tag{4.138}$$

The limiting value $l(\infty)$ is called the *mean free path*.[106] Note that it's independent of particle mass.

Example. Atomic diameters range from 50 to 500 pm (picometers);[107] take $D = 0.1$ nm as representative. At standard temperature and pressure[108] $n = 2.65 \times 10^{25}$ m^{-3}, implying from Eq. (4.138) a mean free path $l = 1.2 \times 10^{-6}$ m. The interatomic separation can be estimated from $d \approx n^{-1/3} = 3.35 \times 10^{-9}$ m. The average distance between collisions is thus 360 times the average distance between atoms—a dilute gas. Taking D to be the range of the interaction, $nD^3 = 2.65 \times 10^{-5}$, satisfying the diluteness criterion $nr_0^3 \ll 1$. For 10^7 Pa pressure, the molar volume is 0.227 liters (assuming an ideal gas), with a mean free path 1.2×10^{-8} m. In this case, $d = 7.2 \times 10^{-10}$ m, implying $l = 17d$. At higher pressures, the mean free path becomes on the order of the interatomic separation, making the molecular chaos assumption difficult to justify.

---

[104] For every direct collision, there is a loss scattering event.

[105] This is our first use of the hard-sphere approximation. Expositions of kinetic theory often treat gas particles as hard spheres from the outset.

[106] Paths taken by rigid molecules between successive collisions are called *free paths*. (For non-rigid molecules, scattering events have no definite beginning and end, rendering the concept of free path problematic.) Equation (4.138) is widely used for the mean free path, but there are other expressions in the literature; see Chapman and Cowling[106, p88]. A heuristic derivation of Eq. (4.138) consists of finding the length l of a tube of cross-sectional area A in a gas of density n such that for molecules traveling parallel to the tube one is guaranteed to encounter a scatterer, when $lAn\sigma_T = A$.

[107] Atomic radii are widely tabulated. See for example Slater[143].

[108] From 22.7 liters per mole of an ideal gas at standard pressure and temperature, a good number to remember.

4.8 NORMAL SOLUTIONS OF THE BOLTZMANN EQUATION

Hydrodynamic conservation laws, as found from balance equations for mass, momentum, and energy [see Eq. (4.73)], are incomplete without constitutive relations for the momentum and heat fluxes. To obtain a closed set of equations for the hydrodynamic fields density $\rho(\boldsymbol{r},t)$, velocity $\boldsymbol{u}(\boldsymbol{r},t)$, and temperature $T(\boldsymbol{r},t)$ [see Eq. (4.85)], it's assumed that: 1) fluxes can be parameterized in terms of local gradients [see Eqs. (4.77), (4.78)], and 2) local equations of state such as $P(\boldsymbol{r}) = P[T(\boldsymbol{r}),\rho(\boldsymbol{r})]$ extend to nonequilibrium states such that $P(\boldsymbol{r},t) = P[T(\boldsymbol{r},t),\rho(\boldsymbol{r},t)]$ [see Eqs. (4.82), (4.83)]. These ideas are (as we show) validated in *normal solutions* of the Boltzmann equation in which spatiotemporal variations occur through a functional dependence on hydrodynamic fields,

$$f(\boldsymbol{r},\boldsymbol{v},t) = f(\{\rho_i(\boldsymbol{r},t)\},\boldsymbol{v}), \tag{4.139}$$

where $\{\rho_i\}$ is the set of the *conserved moments* of f associated with the collisional invariants $\psi_i(\boldsymbol{v})$ (see Exercise 4.43),

$$\rho_i(\boldsymbol{r},t) \equiv m\int \psi_i(\boldsymbol{v})f(\boldsymbol{r},\boldsymbol{v},t)\mathrm{d}\boldsymbol{v}. \tag{4.140}$$

From Eqs. (4.31), (4.34), (4.36) and (4.58), $\rho_1 = \rho$, $\rho_{2,3,4} = \rho u_{x,y,z}$, and $\rho_5 = \rho u^2 + 2\tilde{\varepsilon}^K$. As an example, the local Maxwellian Eq. (4.129) is a normal solution. With normal solutions, not only are $\mathbf{P}, \boldsymbol{J}$ determined by macroscopic fields, so are averages of any function of the velocity. Under what conditions do normal solutions occur?

4.8.1 Hilbert's theorem, hydrodynamic regime

Consider Boltzmann's equation as an initial-value problem: One solves for $f(t>0)$ given $f(t=0)$. It would be unrealistic to expect that we could ever have such a detailed knowledge of nonequilibrium systems as to know $f(t=0)$ precisely. We noted in Section 2.1.1 that if one knew all the moments of a probability distribution, but not the distribution itself, one could in principle construct the distribution from the moments. That would require knowledge of an infinite collection of moments—also unrealistic. A result of Hilbert is then remarkable that if f (as a solution of the Boltzmann equation) has a power series representation, f is uniquely determined for $t>0$ by the values at $t=0$ of only *five* moments, the conserved moments $\rho_i(\boldsymbol{r},0)$. We outline the proof.

4.8.1.1 Time scales in the Boltzmann equation

Flows and collisions, the processes by which distribution functions change in the Boltzmann equation, occur at different rates. From Eq. (4.106) we see that $[J[f]] = [n\sigma\bar{v}f] \equiv [f/\tau_r]$ (square brackets denote dimension), where $\tau_r = l/\bar{v}$ (relaxation time) is the time scale associated with the collision operator (l is the mean free path, $\bar{v}$ is the mean speed). For $\bar{v} = 10^3$ m s^{-1} and $l = 10^{-6}$ m, $\tau_r = 10^{-9}$ s. For the collision time $\tau_c = D/\bar{v}$ (D is the range of the inter-particle potential), the time for a scattering event to occur, we have for $\bar{v} = 10^3$ m s^{-1} and $D = 10^{-10}$ m, $\tau_c = 10^{-13}$ s. Thus, $\tau_c \ll \tau_r$, as in (4.107). In uniform systems, $\partial f/\partial t = 0$ when $J[f] = 0$ (when detailed balance is achieved). In nonuniform systems, the streaming terms can be nonzero when $J[f] = 0$; detailed balance—a condition in velocity space—can be satisfied at each point of position space. We can infer a time scale associated with the spatial inhomogeneities that drive flows, $\tau_h \equiv L_h/\bar{v}$, where L_h can be estimated from the gradient, $L_h^{-1} \equiv |\boldsymbol{\nabla} f|/f$. There is no typical value of L_h, which depends on experimental conditions, but it can be considered macroscopic. For $L_h = 10^{-2}$ m and $\bar{v} = 10^3$ m s^{-1}, $\tau_h = 10^{-5}$ s.

Systems featuring a clear separation of length scales $D \ll l \ll L_h$, implying the separation of time scales $\tau_c \ll \tau_r \ll \tau_h$, are said to be in the *hydrodynamic regime*. To compare time scales, let $\delta \equiv \tau_r/\tau_h$. With $\tau_r/\tau_h = (l/\bar{v})/(L_h/\bar{v}) = l/L_h$, we have $\delta = l/L_h \equiv \mathrm{Kn}$, the *Knudsen number*. Systems with $\mathrm{Kn} \ll 1$ behave as a continuum, with many collisions in time τ_h, $\sigma n\bar{v}\tau_h \gg 1$, the *collision-dominated regime*. Solutions of the Boltzmann equation that we develop (and the equations of hydrodynamics) are valid for small Knudsen numbers. Boltzmann's equation can then be written in *scaled form*,

$$\frac{\partial f}{\partial t} + \boldsymbol{v} \cdot \nabla f \equiv D_v f = \frac{1}{\delta} Q(f, f), \qquad (\delta \ll 1) \tag{4.141}$$

with t dimensionless, D_v the convective derivative, and Q the bilinear form of J, Eq. (4.108).

4.8.1.2 The Hilbert expansion

Hilbert posited a power series solution of the Boltzmann equation [in the form of (4.141)],[109]

$$f(\boldsymbol{r}, \boldsymbol{v}, t) = \sum_{n=0}^{\infty} f^{(n)}(\boldsymbol{r}, \boldsymbol{v}, t) \delta^n. \tag{4.142}$$

Combining Eqs. (4.142) and (4.141), we find by matching powers[110] of δ, the system of equations,

$$\begin{aligned} Q(f^{(0)}, f^{(0)}) &= 0 \\ 2Q(f^{(1)}, f^{(0)}) &= D_v f^{(0)} \\ 2Q(f^{(n)}, f^{(0)}) &= D_v f^{(n-1)} - \sum_{m=1}^{n-1} Q(f^{(m)}, f^{(n-m)}). \qquad (n > 1) \end{aligned} \tag{4.143}$$

Thus, when 1) Boltzmann's equation can be written as in (4.141) and 2) the series in Eq. (4.142) exists, the nonlinear integro-differential equation for f can be replaced with a homogeneous integral equation for $f^{(0)}$ together with an infinite system of linear integral equations of the second kind for $f^{(n\geq 1)}$. Appendix D is a review of integral equations.

The leading term $f^{(0)}$ is the solution of $J[f^{(0)}] = 0$, which by the H-theorem is uniquely given by a local Maxwellian, Eq. (4.129), which we write

$$f^{(0)}(\boldsymbol{r}, \boldsymbol{v}, t) = n^{(0)} \left(m/(2\pi k T^{(0)}) \right)^{3/2} \exp\left(-m/(2kT^{(0)})(\boldsymbol{v} - \boldsymbol{u}^{(0)})^2 \right). \tag{4.144}$$

The fields $n^{(0)}(\boldsymbol{r}, t), \boldsymbol{u}^{(0)}(\boldsymbol{r}, t), T^{(0)}(\boldsymbol{r}, t)$ (left unspecified momentarily) are the moments of $f^{(0)}$:

$$\left\{ \begin{matrix} n^{(0)} \\ n^{(0)} \boldsymbol{u}^{(0)} \\ 3n^{(0)} kT^{(0)}/m \end{matrix} \right\} = \int f^{(0)}(\boldsymbol{r}, \boldsymbol{v}, t) \left\{ \begin{matrix} 1 \\ \boldsymbol{v} \\ (\boldsymbol{v} - \boldsymbol{u}^{(0)})^2 \end{matrix} \right\} \mathrm{d}\boldsymbol{v}. \tag{4.145}$$

The quantity $f^{(0)}$ is not necessarily associated with an actual state of the system; $(n^{(0)}, \boldsymbol{u}^{(0)}, T^{(0)})$, the moments of $f^{(0)}$, are not necessarily equal to $(n, \boldsymbol{u}, T)$, the moments of f. As we'll show, consistency requires that $(n^{(0)}, \boldsymbol{u}^{(0)}, T^{(0)})$ satisfy the Euler equations of fluid dynamics.[111]

4.8.1.3 The linearized collision operator

The left side of each equation in (4.143) for $n \geq 1$ involves $2Q(f^{(n)}, f^{(0)})$. Suppose $f = f^{(0)} + h$. Because of bilinearity, $Q(f^{(0)} + h, f^{(0)}) = Q(f^{(0)}, f^{(0)}) + Q(h, f^{(0)}) = Q(h, f^{(0)})$, which specifies

[109] Published in 1912 as the last chapter of Hilbert's treatise on integral equations[144] and reprinted in [145]. There is an English translation in Brush[121, pp89–101]. In Hilbert's treatment, there is no physical basis for δ; it's a formal parameter only. Hilbert took the form of Boltzmann's equation in (4.141) as an *ansatz* without physical motivation.

[110] Equating coefficients of like powers of δ on the two sides of the equation follows from the uniqueness of power series.

[111] The Euler equations are the equations of hydrodynamics for ideal fluids, which we take to be the five equations in (4.85) with $\eta = \zeta = \kappa = 0$. There are no dissipative effects in the Euler equations. Some authors refer to the Euler equation (singular) as the Navier-Stokes equation (1.22) with $\eta = \zeta = 0$; see Landau and Lifshitz[15, p3].

a linear operator on h ($f^{(0)}$ is fixed), L, the *linearized collision operator*. Using Eq. (4.108) and the detailed balance condition Eq. (4.118), we find:

$$\begin{aligned} L(h(\boldsymbol{v})) &\equiv 2Q(h, f^{(0)}) \qquad (4.146)\\ &= \int \mathrm{d}\boldsymbol{v}_1 \mathrm{d}\Omega\sigma(\Omega)|\boldsymbol{v}_1 - \boldsymbol{v}| f^{(0)}(\boldsymbol{v}) f^{(0)}(\boldsymbol{v}_1)\left[\left(\frac{h}{f^{(0)}}\right)_1' + \left(\frac{h}{f^{(0)}}\right)' - \left(\frac{h}{f^{(0)}}\right)_1 - \frac{h}{f^{(0)}}\right], \end{aligned}$$

where the subscript 1 indicates evaluated at $\boldsymbol{v}_1$ and the prime indicates post-collision quantities (see Fig. 4.1).[112] We therefore have the *linearized Boltzmann equation* (see Exercise 4.46),

$$\frac{\partial h}{\partial t} + \boldsymbol{v} \cdot \boldsymbol{\nabla} h = L(h), \qquad (4.147)$$

simpler than Boltzmann's equation in that it's linear, yet it's still an integro-differential equation for which there is no standard mathematical theory.

By combining Eq. (4.146) with the result of Exercise 4.29, we have for functions $k(\boldsymbol{v})$ and $h(\boldsymbol{v})$,

$$\begin{aligned} \int \mathrm{d}\boldsymbol{v} \left(f^{(0)}(\boldsymbol{v})\right)^{-1} k^*(\boldsymbol{v}) L(h(\boldsymbol{v})) = -\frac{1}{4}\int & \mathrm{d}\boldsymbol{v}\mathrm{d}\boldsymbol{v}_1 \mathrm{d}\Omega\sigma(\Omega)|\boldsymbol{v}_1 - \boldsymbol{v}| f^{(0)}(\boldsymbol{v}) f^{(0)}(\boldsymbol{v}_1) \\ &\times \left[\left(\frac{k}{f^{(0)}}\right)_1' + \left(\frac{k}{f^{(0)}}\right)' - \left(\frac{k}{f^{(0)}}\right)_1 - \frac{k}{f^{(0)}}\right]^* \\ &\times \left[\left(\frac{h}{f^{(0)}}\right)_1' + \left(\frac{h}{f^{(0)}}\right)' - \left(\frac{h}{f^{(0)}}\right)_1 - \frac{h}{f^{(0)}}\right]. \qquad (4.148) \end{aligned}$$

We now use Dirac notation and treat functions $f(\boldsymbol{v})$ as elements $|f\rangle$ of a Hilbert space of square-integrable functions with inner product defined with respect to the weighting function[113] $(f^{(0)})^{-1}$,

$$\langle k|h\rangle \equiv \int \mathrm{d}\boldsymbol{v} \left(f^{(0)}(\boldsymbol{v})\right)^{-1} k^*(\boldsymbol{v}) h(\boldsymbol{v}). \qquad (4.149)$$

Thus, L is Hermitian, $\langle k|L|h\rangle = \langle h|L|k\rangle^*$, and negative semi-definite, $\langle h|L|h\rangle \leq 0$. For $L|\phi_n\rangle = \lambda_n|\phi_n\rangle$, the eigenvalues are real and nonpositive,[114] $\lambda_n \leq 0$. The case of $\lambda_n = 0$ occurs if and only if $h = f^{(0)}(\boldsymbol{v})\psi_i(\boldsymbol{v})$, where ψ_i is one of the five collisional invariants (see Section 4.6.2). The zero eigenvalue is precisely five-fold degenerate.

4.8.1.4 Existence of terms in the expansion

The next equation in (4.143), $L(f^{(1)}) = D_v f^{(0)}$, has a solution for $f^{(1)}$ if and only if the *solvability conditions* are satisfied, that the source term $D_v f^{(0)}$ have no projection onto the null space of L (the subspace spanned by zero-eigenvalue eigenfunctions),[115]

$$\langle f^{(0)}\psi_i|D_v f^{(0)}\rangle = \int \mathrm{d}\boldsymbol{v}\psi_i(\boldsymbol{v}) D_v f^{(0)}(\boldsymbol{r}, \boldsymbol{v}, t) = 0. \qquad (i = 1, \ldots, 5) \qquad (4.150)$$

---

[112] Note that the factor of $f^{(0)}(\boldsymbol{v})$ could come outside the integral in Eq. (4.146), a step we'll take later.

[113] A Hilbert space is a complete inner product space[13, p43]; an inner product must be specified, and inner products can be defined with weighting functions[13, Chapter 2]. We encountered the same form of inner product in our treatments of the master equation transition matrix and the Fokker-Planck equation, Eqs. (2.82) and (3.85).

[114] The master equation matrix and the Fokker-Planck operator have nonpositive eigenvalues; see Exercises 2.45 and 3.33.

[115] For solutions of linear inhomogeneous equations to exist, the source terms must have no projection onto the null space of the operator. This is true of differential equations ([72, p356], [13, p259]), systems of algebraic equations,[72, p6] and integral equations (see Appendix D). The linearized collision operator L can be written $L = -\nu I + K$, with ν the collision frequency, I the identity, and K a symmetric kernel, the *Hilbert decomposition* (establishing this result occupies approximately 25% of Hilbert's article).[146, p197][147, p58][127, p61]. Thus, $L(g) = 0$ is represented by a homogeneous Fredholm equation of the second kind, allowing us to make use of associated theorems for the solutions of such equations.

The five equations implied by (4.150) are related to the five Euler equations of fluid dynamics; the moments of $f^{(0)}$ listed in Eq. (4.145) satisfy the Euler equations. For $L(f^{(n+1)})$ in Eq. (4.143), we have the associated solvability conditions

$$\int d\boldsymbol{v}\psi_i(\boldsymbol{v})\left[D_v f^{(n)} - \sum_{m=1}^{n} Q(f^{(m)}, f^{(n+1-m)})\right] = 0.$$

For collisional invariants ψ_j, $\int \psi_j(\boldsymbol{v})Q(f^{(m)}, f^{(n+1-m)})d\boldsymbol{v}$ vanishes identically for all (n, m) (see Exercise 4.29). For the terms in the Hilbert expansion to exist[116] we therefore require, for $n \geq 0$,

$$\int d\boldsymbol{v}\psi_i(\boldsymbol{v})D_v f^{(n)}(\boldsymbol{r}, \boldsymbol{v}, t) = 0. \qquad (i = 1, \ldots, 5) \tag{4.151}$$

With the particular solutions to Eq. (4.143) denoted $\bar{f}^{(n)}$ (assumed known), we have the general solutions

$$f^{(n)}(\boldsymbol{r}, \boldsymbol{v}, t) = \bar{f}^{(n)}(\boldsymbol{r}, \boldsymbol{v}, t) + \sum_{j=1}^{5} \gamma_j^{(n)}(\boldsymbol{r}, t) f^{(0)}(\boldsymbol{r}, \boldsymbol{v}, t)\psi_j(\boldsymbol{v}), \tag{4.152}$$

where the $\gamma_j^{(n)}(\boldsymbol{r}, t)$ are expansion coefficients. These are uniquely determined by the Fredholm condition (see the Fredholm theorem on page 210 or [72, p116]),

$$\int \bar{f}^{(n)}(\boldsymbol{r}, \boldsymbol{v}, t)\psi_i(\boldsymbol{v})d\boldsymbol{v} = 0. \qquad (i = 1, \ldots, 5) \tag{4.153}$$

By combining Eq. (4.152) with the solvability conditions Eq. (4.151), and when Eq. (4.153) is invoked, we find the equations determining the $\gamma_j^{(n)}$,

$$\frac{\partial}{\partial t}\left[\sum_{j=1}^{5} a_{ij}(\boldsymbol{r}, t)\gamma_j^{(n)}(\boldsymbol{r}, t)\right] + \sum_{k=1}^{3} \frac{\partial}{\partial x_k}\left[\sum_{j=1}^{5} b_{ijk}(\boldsymbol{r}, t)\gamma_j^{(n)}(\boldsymbol{r}, t) + c_{ik}^{(n)}(\boldsymbol{r}, t)\right] = 0, \tag{4.154}$$

where

$$\begin{aligned} \begin{Bmatrix} a_{ij}(\boldsymbol{r}, t) \\ b_{ijk}(\boldsymbol{r}, t) \end{Bmatrix} &\equiv \int d\boldsymbol{v} f^{(0)}(\boldsymbol{r}, \boldsymbol{v}, t)\psi_i(\boldsymbol{v})\psi_j(\boldsymbol{v}) \begin{Bmatrix} 1 \\ v_k \end{Bmatrix} \\ c_{ik}^{(n)}(\boldsymbol{r}, t) &\equiv \int d\boldsymbol{v} \bar{f}^{(n)}(\boldsymbol{r}, \boldsymbol{v}, t)\psi_i(\boldsymbol{v})v_k. \end{aligned} \tag{4.155}$$

The quantities a_{ij}, b_{ijk} are known; the $c_{ik}^{(n)}$ presume the particular solutions $\bar{f}^{(n)}$. We have for each n, five inhomogeneous first-order hyperbolic partial differential equations in the five unknowns $\gamma_j^{(n)}(\boldsymbol{r}, t)$. Solutions to these equations are determined when the initial data $\gamma_j^{(n)}(\boldsymbol{r}, 0)$ are specified.

By combining Eq. (4.142) with Eq. (4.140):

$$\rho_i(\boldsymbol{r}, t) = m\sum_{n=0}^{\infty}\left(\int f^{(n)}(\boldsymbol{r}, \boldsymbol{v}, t)\psi_i(\boldsymbol{v})d\boldsymbol{v}\right)\delta^n \equiv \sum_{n=0}^{\infty} \rho_i^{(n)}(\boldsymbol{r}, t)\delta^n. \tag{4.156}$$

The expansion of the distribution function therefore implies associated expansions for the moments. Using Eqs. (4.152), (4.153), (4.155), we find

$$\rho_i^{(n)}(\boldsymbol{r}, t) = m\sum_{j=1}^{5} a_{ij}(\boldsymbol{r}, t)\gamma_j^{(n)}(\boldsymbol{r}, t). \tag{4.157}$$

Solutions for $f^{(n)}(\boldsymbol{r}, \boldsymbol{v}, t)$ in Eq. (4.152) are therefore determined by specifying $\rho_i^{(n)}(\boldsymbol{r}, 0)$ as initial data. That's not the content of Hilbert's theorem, however; it's the starting point.

[116] To speak of the existence of the Hilbert expansion (not the existence of its terms $f^{(n)}$), the issue of convergence should be investigated, of which not much is known; see [146, Section 11.4].

4.8.1.5 Hilbert's result and discussion

It would appear that every time a new term $f^{(n)}$ is added to the expansion $f = \sum_k f^{(k)}\delta^k$, five new arbitrary functions $\rho_i^{(n)}(\boldsymbol{r},0)$ would have to be found. Hilbert showed, to the contrary, that arbitrary functions appear in the expression for $f(\boldsymbol{r},\boldsymbol{v},t)$ in such a way that it contains only five arbitrary functions. For this he relied on the uniqueness of power series, that no matter how you've found the solution to a problem in the form of a power series, once you have it, it's *the* series.

Consider that we're free to choose[117] initial data such that

$$\begin{aligned}\rho_i^{(0)}(\boldsymbol{r},0) &\equiv m\int \mathrm{d}\boldsymbol{v} f^{(0)}(\boldsymbol{r},\boldsymbol{v},0)\psi_i(\boldsymbol{v}) = \rho_i(\boldsymbol{r},0)\\ \rho_i^{(n)}(\boldsymbol{r},0) &\equiv m\int \mathrm{d}\boldsymbol{v} f^{(n)}(\boldsymbol{r},\boldsymbol{v},0)\psi_i(\boldsymbol{v}) = 0. \qquad (n\geq 1)\end{aligned} \tag{4.158}$$

The conserved moments of $f^{(0)}$ are equated to those of f, with moments of $f^{(n\geq 1)}$ defined to be zero. We therefore have Hilbert's result of a power series solution for f depending only on the initial data $\rho_i(\boldsymbol{r},0)$.[118]

Hilbert's theorem raises questions, physical and mathematical. Why does the solution involve only five moments of f, whereas an exact solution would require the full specification of f at $t = 0$? The answer is that Boltzmann's equation doesn't make sense for very short times. In the regime $\delta \ll 1$ (collision frequency far exceeds the rate of flow processes), local equilibrium is established so rapidly that the ρ_i basically don't change from their initial values. By writing Eq. (4.141) in the form $\delta D_v f = J[f]$, we see that it reduces to $J[f] = 0$ if the limit $\delta \to 0$ is naively taken, the solution of which is a Gaussian (local Maxwellian) parameterized by the five moments associated with collisional invariants (see Section 4.6.2.1). Gaussians *have* higher moments, they're just not independent of the lower moments, the normalization, the mean, and the variance.

Hilbert's analysis assumes the limit $\delta \to 0$ can be taken, of which there are issues. Physically, $\tau_r \to 0$ for arbitrarily large densities, a regime not in the scope of Boltzmann's equation based on $\tau_r \gg \tau_c$. We therefore have a *singular perturbation* ($\delta = 0$ is fundamentally different from the "neighboring theory" for $\delta \ll 1$) which can't be handled with the usual perturbation methods (see Bender and Orszag[148]). In Hilbert's theory the moments $\rho_i^{(0)}$ must satisfy the Euler equations. It tries, therefore, to construct the correct moments ρ_i through a perturbation theory based on ideal fluids. Singular-perturbative phenomena in fluid dynamics (see for example Van Dyke[149]) involve nonanalytic deviations from ideal-fluid behavior, where a perturbation δ can't be expanded around $\delta = 0$. The hope of perturbation theory is that the more the terms included in Eq. (4.156), the better one approximates the fields ρ_i, a possibility precluded by Eq. (4.158) in which the moments of $f^{(0)}$ are equated with those of f in one fell swoop. The reliance of Hilbert's method on expansions for the conserved moments is therefore problematic. We won't dwell further on these issues. The merit of Hilbert's treatment is that it suggests the possibility of normal solutions.

4.8.2 Chapman-Enskog method

Between 1911 and 1917, S. Chapman and D. Enskog independently published articles on deriving the equations of hydrodynamics from Boltzmann's equation.[119] Their method is similar to, yet generalizes Hilbert's treatment in a way that leads to testable predictions for transport coefficients.[120]

[117] Guessing is a valid method of problem solving.

[118] We still have the particular solutions in Eq. (4.152). One might wonder what the problem is that we've solved. We have a solution $f(\boldsymbol{r},\boldsymbol{v},t)$ of the Boltzmann equation that's non-negative, vanishes for $|\boldsymbol{v}| \to \infty$, and is finite and continuous for all times t.

[119] Brush[121] contains reprints of Chapman's articles and an English translation of Enskog's doctoral dissertation. Chapman and Cowling[106, Chapter 7] present a general treatment of the Enskog method.

[120] Grad[115][150][151] examined the Hilbert and Chapman-Enskog methods in detail and concluded they should both ultimately lead to identical results.

Normal solutions are assumed from the outset, $f(\boldsymbol{r},\boldsymbol{v},t) = f(\boldsymbol{v},\{\rho_i(\boldsymbol{r},t)\})$, implying from Eq. (4.141),

$$\sum_{i=1}^{5} \frac{\partial f}{\partial \rho_i}\frac{\partial \rho_i}{\partial t} + \boldsymbol{v}\cdot\boldsymbol{\nabla} f = \frac{1}{\delta}Q(f,f). \qquad (\delta \ll 1) \tag{4.159}$$

A key step is to recognize that for the conserved moments (see Exercise 4.43),

$$\frac{\partial \rho_i}{\partial t} = -m\boldsymbol{\nabla}\cdot\int \psi_i(\boldsymbol{v})\boldsymbol{v}f(\boldsymbol{v},\{\rho_k(\boldsymbol{r},t)\})\mathrm{d}\boldsymbol{v} \equiv \mathcal{D}_i(\{\rho_k(\boldsymbol{r},t)\}). \tag{4.160}$$

The *time* derivatives of macroscopic quantities are expressed in terms of *spatial* derivatives! We therefore have Boltzmann's equation for normal solutions (in which time is formally eliminated),

$$\sum_{i=1}^{5} \frac{\partial f}{\partial \rho_i}\mathcal{D}_i(\{\rho_j\}) + \boldsymbol{v}\cdot\boldsymbol{\nabla} f = \frac{1}{\delta}Q(f,f). \qquad (\delta \ll 1) \tag{4.161}$$

As with Hilbert, expand the distribution function, Eq. (4.142). We don't, however, expand the conserved moments to avoid nonanalyticities associated with singular perturbations. To enforce the no-expansion treatment of these moments, the conditions in Eq. (4.158) are assumed,

$$\begin{aligned}\rho_i(\boldsymbol{r},t) &= m\int \psi_i(\boldsymbol{v})f^{(0)}(\boldsymbol{r},\boldsymbol{v},t)\mathrm{d}\boldsymbol{v}\\ 0 &= m\int \psi_i(\boldsymbol{v})f^{(n>0)}(\boldsymbol{r},\boldsymbol{v},t)\mathrm{d}\boldsymbol{v}.\end{aligned} \tag{4.162}$$

We then introduce expansions of the time-derivative functions,

$$\mathcal{D}_i(\{\rho_j\}) = \sum_{n=0}^{\infty} \mathcal{D}_i^{(n)}(\{\rho_j\})\delta^n, \tag{4.163}$$

where

$$\mathcal{D}_i^{(n)}(\{\rho_j\}) \equiv -m\boldsymbol{\nabla}\cdot\int \psi_i(\boldsymbol{v})\boldsymbol{v}f^{(n)}(\boldsymbol{v},\{\rho_j(\boldsymbol{r},t)\})\mathrm{d}\boldsymbol{v}. \tag{4.164}$$

By combining the expansions for f and $\mathcal{D}_i$, Eqs. (4.142) and (4.163), with Eq. (4.161), we find the generalization of Hilbert's scheme (see Eq. (4.143); L is the linear collision operator),

$$\begin{aligned}&J[f^{(0)}] = 0\\ &L(f^{(1)}) = \sum_{i=1}^{5}\frac{\partial f^{(0)}}{\partial \rho_i}\mathcal{D}_i^{(0)}(\{\rho_j\}) + \boldsymbol{v}\cdot\boldsymbol{\nabla} f^{(0)}\\ &L(f^{(n)}) = \sum_{k=0}^{n-1}\sum_{i=1}^{5}\left[\frac{\partial f^{(k)}}{\partial \rho_i}\mathcal{D}_i^{(n-k-1)}(\{\rho_j\}) + \boldsymbol{v}\cdot\boldsymbol{\nabla} f^{(n-1)}\right] - \sum_{k=1}^{n-1}Q(f^{(k)},f^{(n-k)}). \qquad (n>1)\end{aligned} \tag{4.165}$$

The solvability conditions must be satisfied. We require, for $n \geq 1$ and $j = 1,\ldots,5$,

$$m\int \mathrm{d}\boldsymbol{v}\psi_j(\boldsymbol{v})\left[\sum_{k=0}^{n-1}\sum_{i=1}^{5}\left[\frac{\partial f^{(k)}}{\partial \rho_i}\mathcal{D}_i^{(n-k-1)}(\{\rho_j\}) + \boldsymbol{v}\cdot\boldsymbol{\nabla} f^{(n-1)}\right] - \sum_{k=1}^{n-1}Q(f^{(k)},f^{(n-k)})\right] = 0.$$

The integral associated with the third term vanishes identically (see Section 4.8.1.4). For the first term, the time derivatives $\mathcal{D}_i^{(l)}$ involve only $(\boldsymbol{r},t)$ and can come outside the integral. Thus,

$$\begin{aligned}m\int \mathrm{d}\boldsymbol{v}\psi_j(\boldsymbol{v})\sum_{k=0}^{n-1}\sum_{i=1}^{5}\frac{\partial f^{(k)}}{\partial \rho_i}\mathcal{D}_i^{(n-k-1)} &= \sum_{k=0}^{n-1}\sum_{i=1}^{5}\mathcal{D}_i^{(n-k-1)}\frac{\partial}{\partial \rho_i}m\int \mathrm{d}\boldsymbol{v}\psi_j(\boldsymbol{v})f^{(k)}(\boldsymbol{r},\boldsymbol{v},t)\\ = \sum_{k=0}^{n-1}\sum_{i=1}^{5}\mathcal{D}_i^{(n-k-1)}\frac{\partial}{\partial \rho_i}\rho_j(\boldsymbol{r},t)\delta_{k,0} &= \sum_{k=0}^{n-1}\sum_{i=1}^{5}\mathcal{D}_i^{(n-k-1)}\delta_{i,j}\delta_{k,0} = \mathcal{D}_j^{(n-1)},\end{aligned}$$

where we've used Eq. (4.162). For the solvability conditions to hold, we must have

$$\mathcal{D}_j^{(n-1)} = -m\boldsymbol{\nabla}\cdot\int \mathrm{d}\boldsymbol{v}\psi_j(\boldsymbol{v})\boldsymbol{v}f^{(n-1)}(\boldsymbol{r},\boldsymbol{v},t),$$

precisely Eq. (4.164). The solvability conditions are identities in the Chapman-Enskog theory.

The solution of the first equation in (4.165) is the local Maxwellian $f^{(0)}$, Eq. (4.144), with moments given in Eq. (4.145) but with superscripts erased, $(n,\boldsymbol{u},T)$, because of Eq. (4.162). The momentum and heat fluxes associated with $f^{(0)}$ are $\mathbf{P}=nkT\mathbf{I}$ and $\boldsymbol{J}=0$; there are no dissipative effects at lowest order in the Chapman-Enskog expansion.[121]

Rewrite the second equation in (4.165),

$$\frac{1}{f^{(0)}}L(f^{(1)}) = \sum_{i=1}^{5}\frac{\partial \ln f^{(0)}}{\partial \rho_i}\left(\mathcal{D}_i^{(0)} + \boldsymbol{v}\cdot\boldsymbol{\nabla}\rho_i\right), \tag{4.166}$$

which follows because $f^{(0)}$ depends on $\boldsymbol{r}$ only through $\{\rho_i(\boldsymbol{r},t)\}$. To find the derivatives with respect to ρ_i we must change variables, $\{\rho_i\}\to(n,\boldsymbol{u},T)$; $f^{(0)}$ is a function of the fields $(n,\boldsymbol{u},T)$.[122] From Eq. (4.144) (where $l=x,y,z$),

$$\frac{\partial \ln f^{(0)}}{\partial n} = \frac{1}{n}, \qquad \frac{\partial \ln f^{(0)}}{\partial u_l} = \frac{m}{kT}(\boldsymbol{v}-\boldsymbol{u})_l, \qquad \frac{\partial \ln f^{(0)}}{\partial T} = \frac{1}{T}\left(\frac{m}{2kT}(\boldsymbol{v}-\boldsymbol{u})^2 - \frac{3}{2}\right),$$

and from Eq. (4.162) with $\psi_i = \{1, v_x, v_y, v_z, v^2\}$,[123]

$$\begin{aligned}
\rho_1 &= mn \equiv \rho &&\Longrightarrow n = \rho_1/m\\
\rho_{2,3,4} &= mnu_{x,y,z} &&\Longrightarrow u_{x,y,z} = \rho_{2,3,4}/\rho_1\\
\rho_5 &= 3nkT + mnu^2 &&\Longrightarrow T = \frac{m}{3k}\left(\frac{\rho_5}{\rho_1} - \frac{1}{\rho_1^2}\sum_{l=2,3,4}\rho_l^2\right).
\end{aligned} \tag{4.167}$$

Using the chain rule, we find the required derivatives:

$$\begin{aligned}
\frac{\partial \ln f^{(0)}}{\partial \rho_1} &= \frac{1}{\rho}\left[\frac{5}{2} - \frac{1}{2}\frac{m}{kT}v^2 + \frac{1}{6}\left(\frac{m}{kT}\right)^2 u^2(\boldsymbol{v}-\boldsymbol{u})^2\right];\\
\frac{\partial \ln f^{(0)}}{\partial \rho_l} &= \frac{1}{\rho}\left[\frac{m}{kT}\boldsymbol{v} - \frac{1}{3}\left(\frac{m}{kT}\right)^2 \boldsymbol{u}(\boldsymbol{v}-\boldsymbol{u})^2\right]_l;\\
\frac{\partial \ln f^{(0)}}{\partial \rho_5} &= \frac{1}{\rho}\left[-\frac{1}{2}\frac{m}{kT} + \frac{1}{6}\left(\frac{m}{kT}\right)^2(\boldsymbol{v}-\boldsymbol{u})^2\right].
\end{aligned} \tag{4.168}$$

We also need $\mathcal{D}_i^{(0)} + \boldsymbol{v}\cdot\boldsymbol{\nabla}\rho_i$. From Eqs. (4.164) and (4.167):[124]

$$\begin{aligned}
\mathcal{D}_1^{(0)} + \boldsymbol{v}\cdot\boldsymbol{\nabla}\rho_1 &= -\boldsymbol{\nabla}\cdot(\rho\boldsymbol{u}) + \boldsymbol{v}\cdot\boldsymbol{\nabla}\rho;\\
\mathcal{D}_l^{(0)} + \boldsymbol{v}\cdot\boldsymbol{\nabla}\rho_l &= -\left[\boldsymbol{\nabla}\cdot(\rho\boldsymbol{u}\boldsymbol{u})\right]_l - \nabla_l(nkT) + \boldsymbol{v}\cdot\boldsymbol{\nabla}(\rho u_l);\\
\mathcal{D}_5^{(0)} + \boldsymbol{v}\cdot\boldsymbol{\nabla}\rho_5 &= -\boldsymbol{\nabla}\cdot\left[\boldsymbol{u}(5nkT + \rho u^2)\right] + \boldsymbol{v}\cdot\boldsymbol{\nabla}(3nkT + \rho u^2).
\end{aligned} \tag{4.169}$$

---

[121] What's in a name? In kinetic theory, the first equation in Eq. (4.165) for $f^{(0)}$ is confusingly referred to as the first Chapman-Enskog approximation, with the second equation, for $f^{(1)}$, the second approximation. This nomenclature differs from other branches of physics where $f^{(0)}$ would be referred to as the zeroth approximation, with $f^{(1)}$ first order.

[122] The equations of hydrodynamics are also given in terms of $(n,\boldsymbol{u},T)$ or $(\rho\equiv mn,\boldsymbol{u},T)$; see Eq. (4.85).

[123] Some authors take $\psi_5 = \frac{1}{2}v^2$. Scalings of ψ_i are immaterial: $(\partial f/\partial\rho_i)\mathcal{D}_i$ is invariant under $\psi_i\to\lambda\psi_i$.

[124] Set $P = nkT$ and $\widetilde{\varepsilon} = \frac{3}{2}nkT$ to recover the Euler equations. See Exercise 4.49.

Finally, using Eqs. (4.168) and (4.169), we find from Eq. (4.166), after considerable algebra,[125]

$$\frac{1}{f^{(0)}}L(f^{(1)}) = \frac{1}{T}\left(\frac{my^2}{2kT} - \frac{5}{2}\right)\boldsymbol{y}\cdot\boldsymbol{\nabla}T + \frac{m}{kT}\overset{\circ}{\boldsymbol{y}\boldsymbol{y}}\mathbf{:D}, \tag{4.170}$$

where: $\boldsymbol{y} \equiv \boldsymbol{v} - \boldsymbol{u}$; $\overset{\circ}{\mathrm{w}}$ indicates a traceless (or *non-divergent*) tensor, $\overset{\circ}{\mathrm{w}} \equiv \mathrm{w} - \left(\frac{1}{3}\sum_i w_{ii}\right)\mathbf{I}$ (notation of Chapman and Cowling[106, p15]); and $\mathbf{D}$ is the *rate-of-strain tensor* with elements $D_{ij} \equiv \frac{1}{2}\left(\nabla_i u_j + \nabla_j u_i\right)$. Terms proportional to the density gradient $\boldsymbol{\nabla} n$ vanish identically.

The structure of L suggests a solution in the form $f^{(1)}(\boldsymbol{v}) = f^{(0)}(\boldsymbol{v})\Phi(\boldsymbol{v})$; see Eq. (4.146). With that substitution, we have from the linearized Boltzmann equation (4.147),

$$\left(\frac{\partial}{\partial t} + \boldsymbol{v}\cdot\boldsymbol{\nabla}\right)\Phi(\boldsymbol{r},\boldsymbol{v},t) = \frac{1}{f^{(0)}}L(f^{(0)}\Phi) \equiv n\sigma\sqrt{\frac{2kT}{m}}\mathcal{I}(\Phi). \tag{4.171}$$

We've introduced $n\sigma\sqrt{2kT/m} = \overline{v}/l = \tau_r^{-1}$ to make $\mathcal{I}(\Phi)$ dimensionless. With the dimensionless variable $\boldsymbol{\xi} \equiv \sqrt{m/(2kT)}\boldsymbol{y}$, $f^{(0)}(\boldsymbol{v})\mathrm{d}\boldsymbol{v} \to (n/\pi^{3/2})\exp(-\xi^2)\mathrm{d}\boldsymbol{\xi}$. The quantity σ (with dimension area) can be found from the total cross section, $\sigma \equiv \int \mathrm{d}\Omega\sigma(\Omega)$. Let $\widetilde{\sigma} \equiv \sigma(\Omega)/\sigma$ denote a dimensionless cross section. Also, let $|\boldsymbol{v}_1 - \boldsymbol{v}| \to \sqrt{(2kT/m)}g$, where g is dimensionless. Thus,

$$\begin{aligned}\mathcal{I}(\Phi(\boldsymbol{v})) &\equiv \frac{1}{n\sigma}\sqrt{\frac{m}{2kT}}\int \mathrm{d}\boldsymbol{v}_1\mathrm{d}\Omega\sigma(\Omega)|\boldsymbol{v}_1 - \boldsymbol{v}|f^{(0)}(\boldsymbol{v}_1)\left[\Phi(\boldsymbol{v}_1') + \Phi(\boldsymbol{v}') - \Phi(\boldsymbol{v}_1) - \Phi(\boldsymbol{v})\right]\\ &= \frac{1}{\pi^{3/2}}\int \mathrm{d}\boldsymbol{\xi}_1\mathrm{e}^{-\xi_1^2}\mathrm{d}\Omega\widetilde{\sigma}(\Omega)g\left[\Phi(\boldsymbol{\xi}_1') + \Phi(\boldsymbol{\xi}') - \Phi(\boldsymbol{\xi}_1) - \Phi(\boldsymbol{\xi})\right].\end{aligned} \tag{4.172}$$

By expressing the right side of Eq. (4.170) in terms of dimensionless velocities, we find

$$n\sigma\mathcal{I}(\Phi) = \left(\xi^2 - \frac{5}{2}\right)\boldsymbol{\xi}\cdot\boldsymbol{\nabla}\ln T + 2\overset{\circ}{\boldsymbol{\xi}\boldsymbol{\xi}}\mathbf{:}\widetilde{\mathbf{D}}, \tag{4.173}$$

where $\widetilde{\mathbf{D}} \equiv \sqrt{m/(2kT)}\mathbf{D}$ is the rate of strain tensor expressed in terms of a dimensionless velocity. Note that $n\sigma = l^{-1}$, the inverse mean free path. If one pulled out the length scale L_h associated with gradients on the right side of Eq. (4.173), we would have the Knudsen number $\delta = l/L_h$. We can (if we want) set $l = 1$ and work in length units of the mean free path.

We can guess the form of Φ by noting that the left side of Eq. (4.173) is linear in Φ but the right side is linear in spatial derivatives. We can therefore build Φ out of a scalar product of $\boldsymbol{\nabla}T$ with a vector $\boldsymbol{A}(\xi)$ and the contraction of $\widetilde{\mathbf{D}}$ with a second-rank tensor,[126] $\mathbf{B}(\xi)$. Let's try

$$\Phi = l\boldsymbol{A}\cdot\boldsymbol{\nabla}\ln T + 2l\mathbf{B:}\widetilde{\mathbf{D}} + a_1 + \boldsymbol{a}\cdot\boldsymbol{v} + a_5 v^2, \tag{4.174}$$

where we've included the complementary solution. The dimensionless quantities $\boldsymbol{A}, \mathbf{B}$ are solutions of:

$$\mathcal{I}(\boldsymbol{A}) = \left(\xi^2 - \frac{5}{2}\right)\boldsymbol{\xi}; \qquad \mathcal{I}(\mathbf{B}) = \overset{\circ}{\boldsymbol{\xi}\boldsymbol{\xi}}. \tag{4.175}$$

The solvability conditions for these integral equations are satisfied identically; see Exercise 4.50. The inhomogeneous terms in Eq. (4.175) involve $\boldsymbol{\xi}$, implying $\boldsymbol{A}, \mathbf{B}$ are functions of $\boldsymbol{\xi}$. We can write

$$\boldsymbol{A}(\xi) = \mathcal{A}(\xi)\boldsymbol{\xi}, \tag{4.176}$$

where $\mathcal{A}$ is a scalar function. We see that $\mathbf{B}$ is traceless, $\sum_i \mathcal{I}(B_{ii}) = 0$; it's also symmetric, $\mathcal{I}(B_{ij} - B_{ji}) = 0$. Thus we can write, where $\mathcal{B}$ is a scalar function,

$$\mathbf{B}(\xi) = \mathcal{B}(\xi)\overset{\circ}{\boldsymbol{\xi}\boldsymbol{\xi}}. \tag{4.177}$$

[125] Group together terms associated with gradients of $(n, \boldsymbol{u}, T)$, a task Harris[127, p90] refers to as "gruesome."

[126] With $\boldsymbol{A}$ and $\mathbf{B}$, we're using the notation of Chapman and Cowling[106, p123], the standard reference on this material.

The quantities $(a_1, \boldsymbol{a}, a_5)$ in Eq. (4.174) are uniquely determined by the Fredholm requirements,

$$\int f^{(0)}(\boldsymbol{v})\psi_i(\boldsymbol{v})\left[\boldsymbol{A}\cdot\boldsymbol{\nabla}\ln T + 2\mathbf{B}{:}\widetilde{\mathbf{D}} + a_1 + \boldsymbol{a}\cdot\boldsymbol{v} + a_5 v^2\right]\mathrm{d}\boldsymbol{v} = 0. \tag{4.178}$$

With $\psi_i(\boldsymbol{v}) = (1, \boldsymbol{v}, v^2)$, Eq. (4.178) implies the three equations

$$\int f^{(0)}(\boldsymbol{v})\left(a_1 + a_5 v^2\right)\mathrm{d}\boldsymbol{v} = 0$$

$$\int f^{(0)}(\boldsymbol{v})\left(\frac{1}{T}\mathcal{A}(y)\boldsymbol{\nabla}T + \boldsymbol{a}\right)y^2\mathrm{d}\boldsymbol{v} = 0$$

$$\int f^{(0)}(\boldsymbol{v})\left(a_1 + a_5 v^2\right)v^2\mathrm{d}\boldsymbol{v} = 0.$$

From the first and third equations, $a_1 = a_5 = 0$, while the middle indicates $\boldsymbol{a}$ is proportional to $\boldsymbol{\nabla}T$, implying $\boldsymbol{a}\cdot\boldsymbol{v}$ can be absorbed into $\boldsymbol{A}\cdot\boldsymbol{\nabla}T$. We can take $\boldsymbol{a} = 0$ if we impose the constraint

$$\int \mathrm{d}\boldsymbol{v} f^{(0)}(\boldsymbol{v})\mathcal{A}(y)y^2 = 0, \tag{4.179}$$

an equation that plays an important role.

Thus, assuming Eq. (4.179), we have the *form* of $f^{(1)}$,

$$f^{(1)}(\boldsymbol{r},\boldsymbol{v},t) = lf^{(0)}(\boldsymbol{r},\boldsymbol{v},t)\left[\boldsymbol{A}\cdot\boldsymbol{\nabla}\ln T(\boldsymbol{r},t) + 2\mathbf{B}{:}\widetilde{\mathbf{D}}(\boldsymbol{r},t)\right]. \tag{4.180}$$

We'd be done if we could account for $\boldsymbol{A}$ and $\mathbf{B}$. In practice, no attempt is made at directly solving the equations in (4.175) because (as we'll see) only certain integrals of $\boldsymbol{A}, \mathbf{B}$ are of interest, integrals that turn out to be easier to evaluate than finding $\boldsymbol{A}, \mathbf{B}$ themselves. Clearly $f^{(1)}$ is a normal solution, a function of the hydrodynamic fields $n(\boldsymbol{r},t), \boldsymbol{u}(\boldsymbol{r},t), T(\boldsymbol{r},t)$. Although normal solutions are assumed in the Chapman-Enskog method, we see that $f^{(1)}$ involves gradients of $T(\boldsymbol{r},t)$ and $\boldsymbol{u}(\boldsymbol{r},t)$. As we show, $f^{(1)}$ leads to expressions for $\boldsymbol{J}$ and $\mathbf{P}$ having the form assumed in Section 4.4.4. Thus, $f^{(0)}$ is associated with the Euler equations and $f^{(1)}$ with the Navier-Stokes equations. At the next order, $f^{(2)}$ leads to the *Burnett equations*[152]. Higher-order terms may not give reliable improvements over the Navier-Stokes equations because the Chapman-Enskog expansion does not always converge[153]. The expansion is thought to be asymptotic to solutions of the Boltzmann equation[151] and thus truncating at low order may give accurate results. The interpretation of the Burnett equations remains a research topic[154]. We won't dwell further on the existence of normal solutions. Normal solutions are expected for small Knudsen numbers, when the time between successive collisions is small compared to the time scale of variations in macroscopic properties.

4.9 CHAPMAN-ENSKOG THEORY OF TRANSPORT COEFFICIENTS

Before seeking $\boldsymbol{A}$ and $\mathbf{B}$ in Eq. (4.180), we develop expressions for the fluxes $\boldsymbol{J}$ and $\mathbf{P}$ assuming $\boldsymbol{A}$ and $\mathbf{B}$ are known.

4.9.1 The heat and momentum fluxes associated with $f^{(1)}$

Using Eq. (4.180) for $f^{(1)}$ in Eq. (4.58), we have the heat flux,

$$\begin{aligned}\boldsymbol{J} &= \frac{m}{2}\int \mathrm{d}\boldsymbol{v} f^{(1)}(\boldsymbol{v})\boldsymbol{y}y^2 = \frac{m}{2\sigma\pi^{3/2}}\left(\frac{2kT}{m}\right)^{3/2}\int \mathrm{d}\boldsymbol{\xi}\mathrm{e}^{-\xi^2}\boldsymbol{\xi}\xi^2\mathcal{A}(\xi)\boldsymbol{\xi}\cdot\boldsymbol{\nabla}\ln T\\ &= \frac{k}{3\sigma\pi^{3/2}}\sqrt{\frac{2kT}{m}}\boldsymbol{\nabla}T\int \mathrm{d}\boldsymbol{\xi}\mathrm{e}^{-\xi^2}\mathcal{A}(\xi)\xi^4 = \frac{k}{3\sigma\pi^{3/2}}\sqrt{\frac{2kT}{m}}\boldsymbol{\nabla}T\int \mathrm{d}\boldsymbol{\xi}\mathrm{e}^{-\xi^2}\mathcal{A}(\xi)\left(\xi^4 - \frac{5}{2}\xi^2\right),\end{aligned}$$

where we've used the result of Exercise 4.51 and Eq. (4.179). The last step was not taken randomly. Using Eqs. (4.175) and (4.176), we have

$$\boldsymbol{J} = \left(\frac{k}{3\sigma\pi^{3/2}} \sqrt{\frac{2kT}{m}} \int \mathrm{d}\boldsymbol{\xi} \mathrm{e}^{-\xi^2} \boldsymbol{A}(\xi) \cdot \mathcal{I}(\boldsymbol{A}) \right) \boldsymbol{\nabla} T.$$

Thus, $\boldsymbol{J}$ is in the form of Fourier's law, $\boldsymbol{J} = -\kappa \boldsymbol{\nabla} T$, with the thermal conductivity

$$\kappa = -\frac{k}{3\sigma\pi^{3/2}} \sqrt{\frac{2kT}{m}} \int \mathrm{d}\boldsymbol{\xi} \mathrm{e}^{-\xi^2} \boldsymbol{A}(\xi) \cdot \mathcal{I}(\boldsymbol{A}) \equiv \frac{k}{3\sigma\pi^{3/2}} \sqrt{\frac{2kT}{m}} \left[\boldsymbol{A}, \boldsymbol{A} \right]. \tag{4.181}$$

The integral represented by brackets $[\boldsymbol{A}, \boldsymbol{A}]$ is non-negative (see Exercise 4.52) and thus $\kappa \geq 0$.

With Eq. (4.180) in Eq. (4.48) we have the momentum flux (where we restore dimensions to $\mathbf{D}$),

$$\begin{aligned}
\mathbf{P} &= m \int \mathrm{d}\boldsymbol{v} f^{(0)}(\boldsymbol{v}) \boldsymbol{y}\boldsymbol{y}\Phi(\boldsymbol{v}) = \frac{2m}{\sigma\pi^{3/2}} \sqrt{\frac{2kT}{m}} \int \mathrm{d}\boldsymbol{\xi} \mathrm{e}^{-\xi^2} \boldsymbol{\xi}\boldsymbol{\xi} \mathcal{B}(\xi) \overset{\circ}{\boldsymbol{\xi}\boldsymbol{\xi}} : \mathbf{D} \\
&= \frac{2m}{5\sigma\pi^{3/2}} \sqrt{\frac{2kT}{m}} \overset{\circ}{\mathbf{D}} \int \mathrm{d}\boldsymbol{\xi} \mathrm{e}^{-\xi^2} \mathcal{B}(\xi) \left(\overset{\circ}{\boldsymbol{\xi}\boldsymbol{\xi}} : \overset{\circ}{\boldsymbol{\xi}\boldsymbol{\xi}} \right) = \left(\frac{2m}{5\sigma\pi^{3/2}} \sqrt{\frac{2kT}{m}} \int \mathrm{d}\boldsymbol{\xi} \mathrm{e}^{-\xi^2} \mathbf{B} : \mathcal{I}(\mathbf{B}) \right) \overset{\circ}{\mathbf{D}},
\end{aligned}$$

where we've used an integral theorem from Chapman and Cowling[106, p22] and Eqs. (4.175), (4.177). Thus $\mathbf{P} = -2\eta \overset{\circ}{\mathbf{D}}$, Newton's law of viscosity,[127] with transport coefficient

$$\eta = -\frac{m}{5\sigma\pi^{3/2}} \sqrt{\frac{2kT}{m}} \int \mathrm{d}\boldsymbol{\xi} \mathrm{e}^{-\xi^2} \mathbf{B} : \mathcal{I}(\mathbf{B}) \equiv \frac{m}{5\sigma\pi^{3/2}} \sqrt{\frac{2kT}{m}} \left[\mathbf{B}, \mathbf{B} \right]. \tag{4.182}$$

The dimensionless integral represented by $[\mathbf{B}, \mathbf{B}]$ is non-negative and hence $\eta \geq 0$.

4.9.2 Calculating A and B

We've succeeded in relating transport coefficients to molecular properties through the integrals $[\boldsymbol{A}, \boldsymbol{A}]$ and $[\mathbf{B}, \mathbf{B}]$, each involving the linear collision operator which depends on particle interactions through the scattering cross section. It's traditional in kinetic theory to examine various models of particle interactions (see Chapman and Cowling[106, Chapter 10]), a task we're largely going to sidestep. We mention one case, however, that of *Maxwell molecules*, particles assumed to interact with a repulsive force varying with the inverse fifth power of the intermolecular separation. Many aspects of the Boltzmann equation simplify in this case, and for that reason calculations are made assuming Maxwell molecules even if they have little physical significance.

In particular, the eigenproblem $\mathcal{I}\psi = \lambda\psi$ associated with the linear collision operator $\mathcal{I}$ can be solved exactly for Maxwell molecules[155, Chapter 4]. We know in general that $\mathcal{I}$ is Hermitian with real and nonpositive eigenvalues, with different eigenfunctions orthogonal to each other, and that there are five eigenfunctions corresponding to zero eigenvalue, a consequence of the five collisional invariants. It's also isotropic in velocity space ($\mathcal{I}$ commutes with the rotation operator defined on that space) implying its eigenfunctions have the form $\psi_{rlm}(\boldsymbol{v}) = R_{rl}(v) Y_{lm}(\theta, \phi)$, where the spherical harmonics Y_{lm} are functions of the polar angles (θ, ϕ) of $\boldsymbol{v}$ with respect to an arbitrary direction. The eigenvalues λ_{rl} are independent of m and are at least $(2l + 1)$-fold degenerate (the spherical harmonics are defined for $l = 0, 1, 2, \ldots$ and $m = 0, \pm 1, \ldots, \pm l$). In our application to Maxwell molecules, the index r takes integer values,[128] $r = 0, 1, 2, \ldots$.

---

[127]"Pulling" a factor of two out of Eq. (4.77) allows us to express the rate-of-strain tensor $\mathbf{D}$ as symmetric and traceless.

[128]The spectrum of $\mathcal{I}$ is discrete with no accumulation points for purely repulsive power-law potentials proportional to r^{-n} with $n > 2$[156]. It's assumed the spectrum remains discrete for more general intermolecular forces that are not strictly repulsive, such as the Lennard-Jones potential.

The "radial functions" R_{rl} for Maxwell molecules have the form,[155, Chapter 4]

$$R_{rl}(\xi) = \xi^l L_r^{(l+1/2)}(\xi^2), \tag{4.183}$$

where $L_n^{(\alpha)}$ denotes an associated Laguerre polynomial[129] of degree $n = 0, 1, 2, \ldots$, defined by

$$L_n^{(\alpha)}(x) \equiv \sum_{m=0}^{n} \frac{(-x)^m}{m!(n-m)!} \frac{\Gamma(n+\alpha+1)}{\Gamma(m+\alpha+1)}, \tag{4.184}$$

where $\alpha > -1$ is a real number and $\Gamma(x)$ is the Γ-function (see [13, Appendix C]). Some special cases are $L_0^{(\alpha)}(x) = 1$ and $L_1^{(\alpha)}(x) = \alpha + 1 - x$. The set of functions $\{\text{e}^{-x/2} x^{\alpha/2} L_n^{(\alpha)}(x)\}|_{n=0}^{\infty}$ is complete on the interval $0 \leq x < \infty$ (Szego[78, p104]) and have the orthogonality property[130]

$$\int_0^\infty \text{e}^{-x} x^\alpha L_n^{(\alpha)}(x) L_m^{(\alpha)}(x) \text{d}x = \frac{1}{n!} \Gamma(n+\alpha+1) \delta_{n,m}. \tag{4.185}$$

4.9.2.1 Formal evaluation of $\boldsymbol{A}$ and κ

Thus we have explicit expressions for the eigenfunctions of the linear collision operator associated with Maxwell molecules—a restricted class of particles admittedly, yet which have the important property of being a complete orthogonal set. A basis set is a basis set, which can be used to represent $\boldsymbol{A}, \mathbf{B}$ for general intermolecular potentials. We start by assuming that $\mathcal{A}(\xi)$ can be expanded in a convergent series of the form

$$\mathcal{A}(\xi) = \sum_{m=0}^{\infty} a_m L_m^{(\alpha)}(\xi^2),$$

where α is to be determined. As usual, the expansion coefficients are isolated utilizing orthogonality. We find

$$a_n = \frac{2n!}{\Gamma(n+\alpha+1)} \int_0^\infty \text{d}\xi \text{e}^{-\xi^2} \xi^{2\alpha+1} L_n^{(\alpha)}(\xi^2) \mathcal{A}(\xi).$$

Of course, we don't know $\mathcal{A}(\xi)$, but we did impose a constraint on it in Eq. (4.179), $\int \text{d}\boldsymbol{\xi} \text{e}^{-\xi^2} \xi^2 \mathcal{A}(\xi) = 0$, equivalent to $\int_0^\infty \text{d}\xi \text{e}^{-\xi^2} \xi^4 \mathcal{A}(\xi) = 0$ by isotropy. By comparing with the orthogonality condition, we see that $a_0 = 0$ ($L_0^{(\alpha)} = 1$) if we take $\alpha = \frac{3}{2}$, a choice we adopt.[131]

Thus,

$$\boldsymbol{A} = \mathcal{A}(\xi)\boldsymbol{\xi} = \sum_{m=1}^{\infty} a_m L_m^{(3/2)}(\xi^2)\boldsymbol{\xi} \equiv \sum_{m=1}^{\infty} a_m \boldsymbol{a}^{(m)}. \tag{4.186}$$

We see the expansion in associated Laguerre polynomials has an immediate payoff: From Eq. (4.175), $\mathcal{I}(\boldsymbol{A}) = (\xi^2 - \frac{5}{2})\boldsymbol{\xi} = -L_1^{(3/2)}(\xi^2)\boldsymbol{\xi} = -\boldsymbol{a}^{(1)}$. Using the bracket introduced in Eq. (4.181), consider

$$\alpha_q \equiv [\boldsymbol{a}^{(q)}, \boldsymbol{A}] = 4\pi \int_0^\infty \text{d}\xi \text{e}^{-\xi^2} \xi^4 L_q^{(3/2)}(\xi^2) L_1^{(3/2)}(\xi^2) = \frac{15}{4} \pi^{3/2} \delta_{q,1}. \tag{4.187}$$

With Eq. (4.187), we have $[\boldsymbol{A}, \boldsymbol{A}] = (15/4)\pi^{3/2} a_1$. Because $[\boldsymbol{A}, \boldsymbol{A}] \geq 0$ (Exercise 4.52), we infer that $a_1 \geq 0$. By combining Eq. (4.186) with the definition of α_q in Eq. (4.187), we have

$$\alpha_q = \sum_{m=1}^{\infty} a_m [\boldsymbol{a}^{(q)}, \boldsymbol{a}^{(m)}] \equiv \sum_{m=1}^{\infty} a_m a_{qm}, \qquad (q = 1, 2, \ldots) \tag{4.188}$$

---

[129] In kinetic theory these are often referred to as *Sonine polynomials*, an obsolete reference to associated Laguerre polynomials; see Whittaker and Watson[157, p352].

[130] An equivalent orthogonality expression follows from the substitution $x = y^2$ in Eq. (4.185), in which case we have $\int_0^\infty \text{d}y \text{e}^{-y^2} y^{2\alpha+1} L_n^{(\alpha)}(y^2) L_m^{(\alpha)}(y^2) = \Gamma(n+\alpha+1)\delta_{n,m}/(2n!)$.

[131] By taking $\alpha = \frac{3}{2}$, the constraint in Eq. (4.179) is equivalent to $a_0 = 0$.

where $a_{qm} = a_{mq}$; see Exercise 4.52. The $\boldsymbol{a}^{(m)}$ are known and thus the a_{mq} can be considered known. Equation (4.188) comprises an infinite set of equations for the infinite collection of coefficients[132] a_m. Solutions to infinite sets of equations (when they exist) can be found through a method of successive approximations. Define the kth approximant, $\boldsymbol{A}^{(k)} \equiv \sum_{m=1}^{k} a_m^{(k)} \boldsymbol{a}^{(m)}$, where

$$\sum_{m=1}^{k} a_m^{(k)} a_{mq} = \alpha_q. \qquad (q = 1, 2, \ldots, k) \tag{4.189}$$

This scheme follows from neglecting terms in these expansions with $m > k$. It's assumed that $a_m^{(k)} \to a_m$ and $\boldsymbol{A}^{(k)} \to \boldsymbol{A}$ as $k \to \infty$. Using Eq. (4.187), we have from Cramer's rule

$$a_m^{(k)} = \frac{15}{4}\pi^{3/2} \frac{\Delta_{1m}^{(k)}}{\Delta^{(k)}}, \qquad (m = 1, 2, \ldots, k) \tag{4.190}$$

where $\Delta^{(k)}$ denotes the determinant of the square symmetric matrix $[a_{mq}]$ $(m, q = 1, 2, \ldots, k)$, and Δ_{1m} is the cofactor of a_{1m} in $\Delta^{(k)}$. From Eq. (4.181), $\kappa \propto [\boldsymbol{A}, \boldsymbol{A}] \propto a_1$, and thus

$$\kappa = \frac{5k}{4\sigma}\sqrt{\frac{2kT}{m}}a_1 = \frac{5k}{4\sigma}\sqrt{\frac{2kT}{m}} \lim_{k\to\infty}\left(\frac{\Delta_{11}^{(k)}}{\Delta^{(k)}}\right). \tag{4.191}$$

4.9.2.2 Formal evaluation of $\mathbf{B}$ *and* η

Because $\mathbf{B} = \mathcal{B}(\xi)\overset{\circ}{\boldsymbol{\xi}\boldsymbol{\xi}}$ (see Eq. (4.177)), we assume it can be expanded as

$$\mathbf{B} = \sum_{m=1}^{\infty} b_m \mathbf{b}^{(m)}, \tag{4.192}$$

where

$$\mathbf{b}^{(m)} \equiv L_{m-1}^{(5/2)}(\xi^2)\overset{\circ}{\boldsymbol{\xi}\boldsymbol{\xi}}. \tag{4.193}$$

The reason for $\alpha = 5/2$ will become apparent, and the shift in indices is for convenience, so that we don't end up referring to the $(0, 0)$ element of a matrix. Define, using the bracket in Eq. (4.182),

$$\begin{aligned}\beta_q \equiv [\mathbf{b}^{(q)}, \mathbf{B}] &= -\int \mathrm{d}\boldsymbol{\xi} \mathrm{e}^{-\xi^2}\mathbf{b}^{(q)}{:}\mathcal{I}(\mathbf{B}) = -\int \mathrm{d}\boldsymbol{\xi}\mathrm{e}^{-\xi^2} L_{q-1}^{(5/2)}(\xi^2)\overset{\circ}{\boldsymbol{\xi}\boldsymbol{\xi}}{:}\overset{\circ}{\boldsymbol{\xi}\boldsymbol{\xi}} \\ &= -\frac{8\pi}{3}\int_0^{\infty} \mathrm{d}\xi \mathrm{e}^{-\xi^2}\xi^6 L_{q-1}^{(5/2)}(\xi^2) L_0^{(5/2)}(\xi^2) = -\frac{5}{2}\pi^{3/2}\delta_{q,1},\end{aligned}$$

where we've used Eq. (4.175) and $\overset{\circ}{\boldsymbol{\xi}\boldsymbol{\xi}}{:}\overset{\circ}{\boldsymbol{\xi}\boldsymbol{\xi}} = \frac{2}{3}\xi^4$ (Chapman and Cowling[106, p18]). We infer that $b_1 \leq 0$ because $[\mathbf{B}, \mathbf{B}] \geq 0$. And, because $\eta \propto [\mathbf{B}, \mathbf{B}] \propto |b_1|$, we have, by the reasoning leading to Eq. (4.191),

$$\eta = \frac{m}{2\sigma}\sqrt{\frac{2kT}{m}}|b_1| = \frac{m}{2\sigma}\sqrt{\frac{2kT}{m}} \lim_{k\to\infty}\left|\frac{\Delta_{11}^{(k)}}{\Delta^{(k)}}\right|, \tag{4.194}$$

where $\Delta^{(k)}$ denotes the determinant of the square symmetric matrix $b_{mq} \equiv [\boldsymbol{b}^{(m)}, \boldsymbol{b}^{(q)}]$ $(m, q = 1, 2, \ldots, k)$ and $\Delta_{11}^{(k)}$ is the cofactor of b_{11} in $\Delta^{(k)}$.

Explicit expressions for (κ, η) for various molecular models are given in Chapman and Cowling[106, Chapter 10]. Comparisons with experiment would require the actual form of the intermolecular potential. One way by which that information can be inferred is through the second virial coefficient[5, Section 6.2]. Detailed calculations made this way are in excellent agreement with experiment[158, Section 8.4].

[132] It's obvious but worth stating that to know $\boldsymbol{A}$ we require the coefficients a_m.

SUMMARY

An introduction to kinetic theory was presented, a more microscopic approach to irreversibility than stochastic methods.

- We started with Liouville's theorem on the phase-space distribution function $\rho(\mathbf{\Gamma}, t)$, where $\mathbf{\Gamma} \equiv (\boldsymbol{q}_1, \boldsymbol{p}_1, \ldots, \boldsymbol{q}_N, \boldsymbol{p}_N)$ denotes a vector in Γ-space (N-particle phase space), with $\rho(\mathbf{\Gamma}, t)\mathrm{d}\mathbf{\Gamma}$ the probability at time t that the phase point lies within $\mathrm{d}\mathbf{\Gamma} \equiv \mathrm{d}\boldsymbol{p}_1 \ldots \mathrm{d}\boldsymbol{p}_N \mathrm{d}\boldsymbol{q}_1 \ldots \mathrm{d}\boldsymbol{q}_N$ about $\mathbf{\Gamma}$. Nonequilibrium ensemble averages are constructed from $\rho(\mathbf{\Gamma}, t)$, $\langle A \rangle_t \equiv \int A(\mathbf{\Gamma})\rho(\mathbf{\Gamma}, t)\mathrm{d}\mathbf{\Gamma}$, where $A(\mathbf{\Gamma})$ is a Γ-space function. By Liouville's theorem, $\rho(\mathbf{\Gamma}, t)$ is a constant of the motion, $\mathrm{d}\rho(\mathbf{\Gamma}, t)/\mathrm{d}t = 0$, implying Liouville's equation, $\mathrm{i}\partial\rho(\mathbf{\Gamma}, t)/\partial t = \Lambda\rho(\mathbf{\Gamma}, t)$, where Λ is the Liouville operator, Eq. (4.4). $\rho(\mathbf{\Gamma}, t)$ contains far more information than necessary to calculate measurable quantities; for that reason reduced probabilities are introduced, $f_s(1, \ldots, s; t) \equiv \int \rho(1, \ldots, s, s+1, \ldots, N; t)\mathrm{d}(s+1)\cdots \mathrm{d}N$. The quantity f_1 for example is the probability of finding particle 1 in state 1, irrespective of the states of particles $2, \ldots, N$. Macroscopic properties can be calculated from f_1 and f_2 (see Section 4.4). Yet if we can't solve Liouville's equation for $\rho(\mathbf{\Gamma}, t)$, we can't calculate reduced distributions from their definition. The reduced distribution functions satisfy a hierarchy of coupled dynamical equations, the BBGKY hierarchy, Eq. (4.27), where to determine f_s requires that we know f_{s+1}, which in turn requires that we know f_{s+2}, and so on. The challenge is to devise approximations so that the hierarchy is truncated at a given order.

- Boltzmann derived a closed evolution equation for f_1, one of the fundamental equations of nonequilibrium statistical mechanics. Before taking that up in Section 4.5, we examined in detail the equations of hydrodynamics in Section 4.4, foundational material for constructing solutions of the Boltzmann equation in Section 4.8. The basic fields of hydrodynamics are densities: mass, $\rho(\boldsymbol{r}, t) = m \int \mathrm{d}\boldsymbol{v} f_1(\boldsymbol{r}, \boldsymbol{v}, t)$; momentum, $\boldsymbol{g}(\boldsymbol{r}, t) = m \int \mathrm{d}\boldsymbol{v}\boldsymbol{v} f_1(\boldsymbol{r}, \boldsymbol{v}, t) = \rho(\boldsymbol{r}, t)\boldsymbol{u}(\boldsymbol{r}, t)$, with $\boldsymbol{u}(\boldsymbol{r}, t)$ the mean velocity at $(\boldsymbol{r}, t)$; and energy, $\varepsilon(\boldsymbol{r}, t) = \varepsilon^K(\boldsymbol{r}, t) + \varepsilon^\Phi(\boldsymbol{r}, t)$, with the kinetic energy density $\varepsilon^K(\boldsymbol{r}, t) = \frac{1}{2}m \int \mathrm{d}\boldsymbol{v}v^2 f_1(\boldsymbol{r}, \boldsymbol{v}, t)$ and $\varepsilon^\Phi(\boldsymbol{r}, t) = \frac{1}{2}\int \mathrm{d}\boldsymbol{v}\mathrm{d}\boldsymbol{r}'\mathrm{d}\boldsymbol{v}' f_2(\boldsymbol{r}, \boldsymbol{v}, \boldsymbol{r}', \boldsymbol{v}', t)\Phi(\boldsymbol{r}, \boldsymbol{r}')$ the potential energy density, where $\Phi(\boldsymbol{r}, \boldsymbol{r}')$ is the intermolecular potential energy function. The quantities $(\rho, \boldsymbol{g}, \varepsilon^K)$ represent the zeroth, first, and second velocity moments of the one-particle probability distribution f_1; ε^Φ results from a weighting of the two-particle distribution f_2 by $\Phi(\boldsymbol{r}, \boldsymbol{r}')$. These relations follow from ensemble averages of microscopic expressions for mass, momentum, and energy density, $(\hat{\rho}, \hat{\boldsymbol{g}}, \hat{\varepsilon})$; see Section 4.4.1. The correspondence between ensemble averages of microscopic functions $(\hat{\rho}, \hat{\boldsymbol{g}}, \hat{\varepsilon})$ and macroscopic densities $(\rho, \boldsymbol{g}, \varepsilon)$ works for mechanical quantities associated with one or more particles. Not every macroscopic quantity can be so represented. Entropy is not attached to the dynamics of single particles; it's a property of the system as a whole.

  Balance equations for mass, momentum, and energy (introduced phenomenologically in Chapter 1) can be derived from moments of the first two equations of the hierarchy. The continuity equation results from the zeroth velocity moment of the first equation of the hierarchy and the momentum balance equation is derived from its first velocity moment, thereby providing us with a microscopic expression for the pressure tensor $\mathbf{P}$. There are two contributions to $\mathbf{P}$: kinetic, $\mathbf{P}^K(\boldsymbol{r}, t) \equiv m \int \mathrm{d}\boldsymbol{v} f_1(\boldsymbol{r}, \boldsymbol{v}, t)\boldsymbol{v}\boldsymbol{v}$, and potential (when we set $D_r = 1$ in Eq. (4.43)), $\mathbf{P}^\Phi(\boldsymbol{r}, t) \equiv -\frac{1}{2}\int \mathrm{d}\boldsymbol{v}\mathrm{d}\boldsymbol{r}'\mathrm{d}\boldsymbol{v}' f_2(\boldsymbol{r}, \boldsymbol{v}, \boldsymbol{r}', \boldsymbol{v}', t)\boldsymbol{\nabla}_{\boldsymbol{r}}\Phi(\boldsymbol{r}, \boldsymbol{r}')(\boldsymbol{r} - \boldsymbol{r}')$, where we've used dyadic notation. We separate the contributions to $\mathbf{P}^K$ into those from the local mean velocity $\boldsymbol{u}(\boldsymbol{r}, t)$ and random velocities measured relative to the mean, $\mathbf{P}^K = \rho\boldsymbol{u}\boldsymbol{u} + \widetilde{\mathbf{P}}^K$, with $\widetilde{\mathbf{P}}^K(\boldsymbol{r}, t) \equiv m \int (\boldsymbol{v} - \boldsymbol{u})(\boldsymbol{v} - \boldsymbol{u}) f_1(\boldsymbol{r}, \boldsymbol{v}, t)\mathrm{d}\boldsymbol{v}$. The pressure tensor $\mathbf{P} \equiv \widetilde{\mathbf{P}}^K + \mathbf{P}^\Phi$. The hydrostatic pressure $P(\boldsymbol{r}, t) \equiv \frac{1}{3}\mathrm{Tr}\,\mathbf{P}(\boldsymbol{r}, t)$, a scalar field, has two contributions, $P(\boldsymbol{r}, t) = \frac{m}{3}\int \mathrm{d}\boldsymbol{v}(\boldsymbol{v} - \boldsymbol{u})^2 f_1(\boldsymbol{r}, \boldsymbol{v}, t) - \frac{1}{6}\int \mathrm{d}\boldsymbol{v}\mathrm{d}\boldsymbol{r}'\mathrm{d}\boldsymbol{v}' f_2(\boldsymbol{r}, \boldsymbol{v}, \boldsymbol{r}', \boldsymbol{v}', t)\boldsymbol{\nabla}_{\boldsymbol{r}}\Phi(\boldsymbol{r}, \boldsymbol{r}') \cdot (\boldsymbol{r} - \boldsymbol{r}')$, kinetic, present in all systems but significant in gases, and from internal forces associated with two-body interactions, significant in liquids.

The energy balance equation (4.57) is derived from the first and second equations of the hierarchy, in a way that requires no knowledge of f_3. Energy conservation is an exact consequence of the first two equations of the hierarchy. There are contributions to the energy flux $\boldsymbol{J}_\varepsilon$ associated with kinetic and potential energies, $\boldsymbol{J}_\varepsilon \equiv \boldsymbol{J}^K + \boldsymbol{J}^\Phi$, with $\boldsymbol{J}^K(\boldsymbol{r},t) = \frac{m}{2}\int \mathrm{d}\boldsymbol{v}\boldsymbol{v}v^2 f_1(\boldsymbol{r},\boldsymbol{v},t)$ (third velocity moment of f_1), and $\boldsymbol{J}^\Phi(\boldsymbol{r},t) = \frac{1}{2}\int \mathrm{d}\boldsymbol{v}\mathrm{d}\boldsymbol{r}'\mathrm{d}\boldsymbol{v}' f_2(\boldsymbol{r},\boldsymbol{v},\boldsymbol{r}',\boldsymbol{v}',t)\Phi(\boldsymbol{r},\boldsymbol{r}')\boldsymbol{v}$. Internal energy, the province of thermodynamics, can be isolated from the total energy by separating the contributions to $\varepsilon^K, \boldsymbol{J}^K, \boldsymbol{J}^\Phi$ into those from the mean motion and velocities specified relative to the mean (as with $\mathbf{P}^K$). In Eq. (4.58) it's shown that: $\varepsilon^K = \frac{1}{2}\rho u^2 + \widetilde{\varepsilon}^K$ where $\widetilde{\varepsilon}^K(\boldsymbol{r},t) \equiv \frac{m}{2}\int \mathrm{d}\boldsymbol{v} f_1(\boldsymbol{r},\boldsymbol{v},t)(\boldsymbol{v}-\boldsymbol{u})^2$; $\boldsymbol{J}^K = \boldsymbol{u}\cdot\widetilde{\mathbf{P}}^K + \boldsymbol{u}(\frac{1}{2}\rho u^2 + \widetilde{\varepsilon}^K) + \widetilde{\boldsymbol{J}}^K$ with $\widetilde{\boldsymbol{J}}^K(\boldsymbol{r},t) \equiv \frac{m}{2}\int \mathrm{d}\boldsymbol{v} f_1(\boldsymbol{r},\boldsymbol{v},t)(\boldsymbol{v}-\boldsymbol{u})(\boldsymbol{v}-\boldsymbol{u})^2$; and $\boldsymbol{J}^\Phi = \boldsymbol{u}\varepsilon^\Phi + \widetilde{\boldsymbol{J}}^\Phi$ with $\widetilde{\boldsymbol{J}}^\Phi(\boldsymbol{r},t) \equiv \frac{1}{2}\int \mathrm{d}\boldsymbol{v}\mathrm{d}\boldsymbol{r}'\mathrm{d}\boldsymbol{v}' f_2(\boldsymbol{r},\boldsymbol{v},\boldsymbol{r}',\boldsymbol{v}',t)(\boldsymbol{v}-\boldsymbol{u})\Phi(\boldsymbol{r},\boldsymbol{r}')$. Internal energy $\widetilde{\varepsilon} \equiv \widetilde{\varepsilon}^K + \varepsilon^\Phi$ is the average random kinetic energy and the average potential energy. Internal energy flux (heat flux) $\boldsymbol{J}_Q \equiv \widetilde{\boldsymbol{J}}^K + \widetilde{\boldsymbol{J}}^\Phi$. A balance equation for the mean kinetic energy follows by projecting the momentum balance equation onto $\boldsymbol{u}$, Eq. (4.63). By subtracting that equation from the energy balance equation, we arrive at a balance equation for internal energy, Eq. (4.64).

- Balance equations for $\rho, \boldsymbol{g}, \varepsilon$ follow from moments of the equations of the hierarchy in the form of relations among ensemble averages. Yet conservation of mass, momentum, and energy hold for every member of an ensemble, implying that balance equations don't require ensemble averaging to be true. It's shown in Section 4.4.3 that the balance equations for $\rho, \boldsymbol{g}, \varepsilon$ follow from ensemble averages of balance equations for $\hat{\rho}, \hat{\boldsymbol{g}}, \hat{\varepsilon}$.

- The equations of hydrodynamics in their most basic form are the balance equations for mass, momentum, and internal energy; see Eq. (4.73). These equations must be supplemented with constitutive relations between the fluxes $(\mathbf{P}, \boldsymbol{J})$ and the fields $(\rho, \boldsymbol{u}, \widetilde{\varepsilon})$. For systems in thermal equilibrium, spatially uniform and time independent, $\boldsymbol{u} = 0$, $\boldsymbol{J} = 0$, and $\mathbf{P}$ is diagonal, $P_{ij} = P\delta_{ij}$, with P the hydrostatic pressure. For nonequilibrium systems, $\boldsymbol{J} \neq 0$ and $\mathbf{P}$ has off-diagonal elements, effects associated with deviations from uniformity. Write $\mathbf{P}$ so that nonequilibrium contributions add to the equilibrium form, $P_{ij} = P\delta_{ij} + \Pi_{ij}$, where the tensor elements Π_{ij} model momentum fluxes mediated by inhomogeneities. Deviations from uniformity are, in a first approximation, represented by linear functions of gradients in local fields. We take Π_{ij} to be a linear combination of velocity gradients, $\Pi_{ij} = \sum_{lm} A_{ij}^{lm}\nabla_l u_m$. The simplest form of Π_{ij}, Eq. (4.77), involves two parameters (η, ζ), the coefficients of shear and bulk viscosity. In the same way, heat flux is modeled as proportional to temperature gradients $\boldsymbol{J} = -\kappa\boldsymbol{\nabla}T$, with κ the thermal conductivity. The non-negativity of the entropy source, $\sigma_S \geq 0$, implies $\kappa \geq 0, \eta \geq 0, \zeta \geq 0$; see Eq. (4.80). Combining the parameterized expressions for $\mathbf{P}, \boldsymbol{J}$ with the equations in (4.73), we arrive at a form of the hydrodynamic equations in (4.81), two equations (three with the continuity equation) in five unknowns, $(\rho, \boldsymbol{u}, P, \widetilde{\varepsilon}, T)$. The number of independent variables can be reduced by assuming local thermodynamic equilibrium, that local equations of state $P(\boldsymbol{r}) = P[T(\boldsymbol{r}), \rho(\boldsymbol{r})]$, $\widetilde{\varepsilon}(\boldsymbol{r}) = \widetilde{\varepsilon}[T(\boldsymbol{r}), \rho(\boldsymbol{r})]$ have the same functional form as in global equilibrium, $P = P(T,\rho)$, $\widetilde{\varepsilon} = \widetilde{\varepsilon}(T,\rho)$, leaving $(\rho, \boldsymbol{u}, T)$ as independent variables. We assume that temporal variations of $P, \widetilde{\varepsilon}$ occur through variations of $T(\boldsymbol{r},t), \rho(\boldsymbol{r},t)$, e.g., $P(\boldsymbol{r},t) = P[T(\boldsymbol{r},t), \rho(\boldsymbol{r},t)]$ (an assumption justified in Section 4.8). We arrive in (4.85) at a closed set of nonlinear equations among the five hydrodynamic fields $\rho(\boldsymbol{r},t), \boldsymbol{u}(\boldsymbol{r},t), T(\boldsymbol{r},t)$, the most general hydrodynamic equations consistent with the assumptions of local thermodynamic equilibrium and our parameterizations of $\mathbf{P}$ and $\boldsymbol{J}$. We found normal-mode solutions of the linearized version of these equations, (4.87), with the long-wavelength form of dispersion relations shown in Table 4.1.

- The Boltzmann equation, (4.106), is a nonlinear integro-differential equation for the one-body distribution $f(\boldsymbol{r},\boldsymbol{v},t)$ in which change occurs in two ways: streaming and collisions.

In the absence of intermolecular forces, f would be unchanged along particle trajectories, $\mathrm{d}f(\boldsymbol{r},\boldsymbol{v},t)/\mathrm{d}t \equiv (\partial/\partial t + \boldsymbol{v}\cdot\nabla_{\boldsymbol{r}} + \boldsymbol{a}\cdot\nabla_{\boldsymbol{v}})f = 0$, with $\boldsymbol{a} \equiv \dot{\boldsymbol{v}}$. The Boltzmann equation adds a term specifying the rate at which f changes due to particle interactions, $\mathrm{d}f/\mathrm{d}t = J[f]$, where the collision operator $J[f]$ is a nonlinear functional of f; see the right side of Eq. (4.106). As $J[f]$ in essence represents the rest of the hierarchy, it can't be found as an exact consequence of the equations of motion; it must be devised from approximations that capture the physics of particle trajectories modified by interactions (collisions).

The derivation of $J[f]$ relies on 1) short-range forces (there is a length r_0 beyond which the intermolecular potential rapidly approaches zero) and 2) dilute systems (average inter-particle separation far exceeds r_0). The diluteness criterion is $nr_0^3 \ll 1$, with $n \equiv N/V$. Under these conditions, particles move with constant velocities $\boldsymbol{v}$ until they encounter other particles closely enough that they deflect over the collision time $\tau_c \equiv r_0/v$, after which free motion resumes with velocities $\boldsymbol{v}'$. The time τ_c is microscopic, with $\tau_c \approx 10^{-13}$s representative. Two-body encounters can be seen as discrete events confined to a small volume r_0^3, in which changes in free-state velocities occur almost abruptly, $\boldsymbol{v} \to \boldsymbol{v}'$, over the short time τ_c.

The Boltzmann equation is not time-reversal symmetric. The streaming terms reverse sign under $t \to -t$ but the collision term does not. If $f(\boldsymbol{r},\boldsymbol{v},t)$ is a solution of the Boltzmann equation, $f(\boldsymbol{r},-\boldsymbol{v},-t)$ is not (in contrast to solutions of the Liouville equation). Irreversibility is associated with the collision operator. Where in its derivation did we depart from the laws of mechanics? Collisions effect change in two ways: 1) among particles with velocity $\boldsymbol{v}$, some undergo collisions, resulting in a loss from $(\boldsymbol{r},\boldsymbol{v})$; 2) among particles with velocities $\boldsymbol{v}_1', \boldsymbol{v}'$, some undergo collisions such that one of the particles has velocity $\boldsymbol{v}$, a gain. In loss collisions $(\boldsymbol{v}_1,\boldsymbol{v}) \to (\boldsymbol{v}_1',\boldsymbol{v}')$, $\boldsymbol{v}_1$ can be chosen independently of $\boldsymbol{v}$; the colliding particles are uncorrelated. In gain collisions $(\boldsymbol{v}_1',\boldsymbol{v}') \to (\boldsymbol{v}_1,\boldsymbol{v})$, the velocities of the colliding particles are constrained to conserve momentum, $\boldsymbol{v}' + \boldsymbol{v}_1' = \boldsymbol{v}_1 + \boldsymbol{v}$; we can't freely vary $\boldsymbol{v}_1'$ and $\boldsymbol{v}'$. It's here we've departed from the laws of mechanics: $\boldsymbol{v}_1', \boldsymbol{v}'$ follow deterministically from force laws, yet Boltzmann assumed they're statistically independent, the so-called molecular chaos assumption. Despite its extra-mechanical nature, the Boltzmann equation makes predictions about dilute gases in good agreement with experiment.

The Boltzmann equation is nonlinear; the collision term scales quadratically, $J[\lambda f] = \lambda^2 J[f]$ with the streaming terms linear. The nonlinear operator $J[f]$ can be represented in terms of a bilinear operator $Q(f,g)$ such that $J[f] = Q(f,f)$; see Eq. (4.108). This form of the collision operator is used in more advanced work; Sections 4.8, 4.9.

- Boltzmann's H-theorem, an important consequence of the Boltzmann equation, states that the quantity $H(t) \equiv \int \mathrm{d}\boldsymbol{r}\mathrm{d}\boldsymbol{v} f(\boldsymbol{r},\boldsymbol{v},t)\ln[f(\boldsymbol{r},\boldsymbol{v},t)]$ never increases in time, $\mathrm{d}H(t)/\mathrm{d}t \le 0$, when $f(\boldsymbol{r},\boldsymbol{v},t)$ is a solution of Boltzmann's equation. The proof relies on the symmetries of $J[f]$; see Section 4.6.1. Regardless of initial conditions, solutions of the Boltzmann equation evolve in a way that $H(t)$ never increases. In that regard, $-H(t)$ behaves as a proxy for entropy S, and for the ideal gas the equality $\lim_{t\to\infty} H(t) = -S/k$ can be arranged. More generally, $H(t)$ reflects the loss of information associated with nonequilibrium initial states as the system equilibrates. The connection between $H(t)$ and the second law of thermodynamics is discussed in Sections 4.6.3 and 4.6.4.

  Although $H(t)$ can't increase in time, it also can't decrease indefinitely—it's bounded from below; see Section 4.6.2. It approaches a limit, the time-invariant state $\mathrm{d}H/\mathrm{d}t = 0$, which we identify with equilibrium, a state achieved if and only if detailed balance holds, $f(\boldsymbol{r},\boldsymbol{v}',t)f(\boldsymbol{r},\boldsymbol{v}_1',t) = f(\boldsymbol{r},\boldsymbol{v},t)f(\boldsymbol{r},\boldsymbol{v}_1,t)$, Eq. (4.118), where $\boldsymbol{v}' + \boldsymbol{v}_1' = \boldsymbol{v} + \boldsymbol{v}_1$; see Fig. 4.1. Equilibrium is a special dynamical state that occurs when gain and loss collisions $(\boldsymbol{v}',\boldsymbol{v}_1') \to (\boldsymbol{v},\boldsymbol{v}_1)$ and $(\boldsymbol{v},\boldsymbol{v}_1) \to (\boldsymbol{v}',\boldsymbol{v}_1')$ occur with equal frequency for all velocities $\boldsymbol{v}$. Taking the logarithm of the detailed balance condition, we have for the equilibrium distribution $\ln f'_{\text{eq}} + \ln f'_{\text{eq},1} = \ln f_{\text{eq}} + \ln f_{\text{eq},1}$. Quantities ψ such that $\psi' + \psi'_1 = \psi + \psi_1$ are termed collisional invariants, quantities preserved in elastic collisions, such as particle number, momentum, and kinetic energy. A key result, on which much of the theory depends, is that ψ_i, $i = 1, 2, 3, 4, 5$ for $(1, v_x, v_y, v_z, v^2)$ are the only linearly independent collisional invariants. The logarithm of the equilibrium distribution can therefore be expressed as a linear combination of fundamental invariants, $\ln f_{\text{eq}}(\boldsymbol{v}) = A + \boldsymbol{B} \cdot \boldsymbol{v} + Cv^2$, where $A, \boldsymbol{B}, C$ are constants for spatially uniform systems. It has an equivalent parameterization as a Gaussian, $f_{\text{eq}}(\boldsymbol{v}) = K \exp(-(m\beta/2)(\boldsymbol{v} - \boldsymbol{u})^2)$, the Maxwell-Boltzmann velocity distribution, Eq. (4.121). The constants can be related to known macroscopic information, the zeroth, first, and second velocity moments; $f_{\text{eq}}(\boldsymbol{v}) = n\,(m\beta/2\pi)^{3/2} \exp[-\frac{1}{2}m\beta\,(\boldsymbol{v} - \boldsymbol{u})^2]$, with $\beta = (1/kT)$. For inhomogeneous systems, the streaming terms can be nonzero even though detailed balance has been achieved. Systems can satisfy detailed balance, a condition in velocity space, at each point in configuration space. The most general solution of $J[f] = 0$ is to let the quantities $A, \boldsymbol{B}, C$ (constants for uniform systems) be arbitrary functions of $(\boldsymbol{r}, t)$. In that case, we have the generalization of the velocity distribution, $f(\boldsymbol{r}, \boldsymbol{v}, t) = n(\boldsymbol{r}, t)\,(m\beta(\boldsymbol{r}, t)/2\pi)^{3/2} \exp[-\frac{1}{2}m\beta(\boldsymbol{r}, t)\,(\boldsymbol{v} - \boldsymbol{u}(\boldsymbol{r}, t))^2]$, Eq. (4.129), known as a local Maxwellian function.

- We derived an expression for the collision rate of particles in an equilibrium gas, Eq. (4.137). An asymptotic form of this expression yields the customary formula for the mean free path, Eq. (4.138), $l = (n\sigma_T)^{-1}$, with σ_T the total scattering cross section.

- As noted in Section 4.4.4, the hydrodynamic conservation laws are incomplete without constitutive relations for the momentum and heat fluxes. We developed such relations by parameterizing local gradients with transport coefficients. To pass from the equations of hydrodynamics in (4.81) to their form in (4.85), we had to assume that local equations of state such as $P(\boldsymbol{r}) = P[T(\boldsymbol{r}), \rho(\boldsymbol{r})]$ extend to nonequilibrium systems such that $P(\boldsymbol{r}, t) = P\,[T(\boldsymbol{r}, t), \rho(\boldsymbol{r}, t)]$. In so-called normal solutions of the Boltzmann equation (see Section 4.8), spatiotemporal variations occur through a functional dependence on hydrodynamic fields, $f(\boldsymbol{r}, \boldsymbol{v}, t) = f(\boldsymbol{v}, \{\rho_i(\boldsymbol{r}, t)\})$, where $\rho_i(\boldsymbol{r}, t) \equiv m \int \psi_i(\boldsymbol{v}) f(\boldsymbol{r}, \boldsymbol{v}, t) \mathrm{d}\boldsymbol{v}$ denote the conserved moments of f associated with the collisional invariants $\psi_i(\boldsymbol{v})$. Normal solutions provide microscopic justification for the assumptions made in Section 4.4.4 that would otherwise be considered phenomenological. We presented in Section 4.8 Hilbert's theorem on the existence of normal solutions and the Chapman-Enskog method for constructing them. Normal solutions are expected for small Knudsen numbers, the collision-dominated hydrodynamic regime wherein the time between successive collisions is small compared to the time scale of variations in macroscopic properties. In this way, we have a self-consistent construction of kinetic theory that relies on the properties of solutions to the Boltzmann equation. Kinetic theory must be able to derive the equations of hydrodynamics from first principles (which embody local conservation of mass, momentum, and energy), yet the construction of hydrodynamics would be phenomenological in introducing transport coefficients as unknown parameters. We outlined the Chapman-Enskog method for calculating transport coefficients in Section 4.9.

EXERCISES

4.1 Does the Liouville operator (see Eq. (4.4)) obey the product rule of differential calculus? For functions $f(\boldsymbol{r}, \boldsymbol{p}), g(\boldsymbol{r}, \boldsymbol{p})$, is $\Lambda(fg) = f\Lambda g + g\Lambda f$? For simplicity, let $N = 1$.

4.2 The N-body distribution $\rho(\boldsymbol{\Gamma}, t)$ satisfies the Liouville equation $\partial\rho(\boldsymbol{\Gamma}, t)/\partial t = -\mathrm{i}\Lambda\rho(\boldsymbol{\Gamma}, t)$, where Λ is the Liouville operator. Show that phase-space functions $A(\boldsymbol{\Gamma})$ satisfy a closely related dynamical equation, $\mathrm{d}A(\boldsymbol{\Gamma})/\mathrm{d}t = \mathrm{i}\Lambda A(\boldsymbol{\Gamma})$ (assuming $\partial A/\partial t = 0$). The equation of motion for $\rho(\boldsymbol{\Gamma}, t)$ is special in that it derives from Liouville's theorem. The time dependence of $A(\mathbf{\Gamma})$ is implicit through the evolution of $\mathbf{\Gamma}$, $A(\mathbf{\Gamma}(t))$, where $\mathbf{\Gamma}(t)$ is the image of $\mathbf{\Gamma}(t=0)$ under the natural motion in phase space. That is, $\mathbf{\Gamma}(t=0)$, signifying a set of variables satisfying Hamilton's equations of motion at time $t=0$, is mapped into $\mathbf{\Gamma}(t)$, a set of variables satisfying Hamilton's equations at time t.

4.3 Show that the solution of the Liouville equation is time-reversal symmetric, i.e., if $\rho(\mathbf{\Gamma}, t)$ is a solution to the Liouville equation, then so is the time-reversed function $\rho(\widetilde{\mathbf{\Gamma}}, \tau)$, where $\tau \equiv -t$ and $\widetilde{\mathbf{\Gamma}}$ is the time-reversed version of phase-space vector $\mathbf{\Gamma}$. Hint: $\widetilde{\Lambda} = -\Lambda$.

4.4 a. Derive Eq. (4.11). First show the orthonormality of the eigenfunctions in Eq. (4.9),

$$\int \mathrm{d}\boldsymbol{r}_1 \ldots \mathrm{d}\boldsymbol{r}_N \psi^*_{\{\boldsymbol{k}'\}}(\boldsymbol{r}_1, \ldots, \boldsymbol{r}_N)\psi_{\{\boldsymbol{k}\}}(\boldsymbol{r}_1, \ldots, \boldsymbol{r}_N) = \delta_{\{\boldsymbol{k}\},\{\boldsymbol{k}'\}} \equiv \delta_{\boldsymbol{k}_1,\boldsymbol{k}'_1} \cdots \delta_{\boldsymbol{k}_N,\boldsymbol{k}'_N}.$$

b. Derive Eq. (4.12).

4.5 Show that the reduced distribution functions are normalized to unity, $\int f_s \mathrm{d}1 \cdots \mathrm{d}s = 1$. What is the dimension of f_s? What are the dimensions of ϕ_l and n_l in Eqs. (4.15) and (4.16)?

4.6 Among three particles there are $3!/1!$ distinct arrangements of two particles. Enumerate the possibilities. The s-tuple distribution counts all *distinct* arrangements of s particles out of N.

4.7 Add a term to the Hamiltonian in Eq. (4.18) representing the effects of external forces, $\sum_{i=1}^{N} \phi(\boldsymbol{r}_i)$, where $\phi(\boldsymbol{r})$ is a one-body field coupling to every particle (such as electrostatic potential energy), with $\boldsymbol{F}(\boldsymbol{r}) = -\boldsymbol{\nabla}\phi(\boldsymbol{r})$ the associated force at position $\boldsymbol{r}$.

a. Derive the Liouville operator $L_N = -\mathrm{i}\Lambda_N$. A: Add $-\sum_{n=1}^{N} \boldsymbol{F}(\boldsymbol{r}_n)\cdot\partial/\partial\boldsymbol{p}_n$ to Eq. (4.20).

b. Write down expressions for L_1 and L_2. Remember where you put these for future use.

$$L_1 = -\frac{\boldsymbol{p}}{m}\cdot\frac{\partial}{\partial\boldsymbol{r}} - \boldsymbol{F}(\boldsymbol{r})\cdot\frac{\partial}{\partial\boldsymbol{p}}$$

$$L_2 = -\left(\frac{\boldsymbol{p}_1}{m}\cdot\frac{\partial}{\partial\boldsymbol{r}_1} + \boldsymbol{F}(\boldsymbol{r}_1)\cdot\frac{\partial}{\partial\boldsymbol{p}_1}\right) - \left(\frac{\boldsymbol{p}_2}{m}\cdot\frac{\partial}{\partial\boldsymbol{r}_2} + \boldsymbol{F}(\boldsymbol{r}_2)\cdot\frac{\partial}{\partial\boldsymbol{p}_2}\right) + O_{12}.$$

4.8 Show that the interaction operator in Eq. (4.21) is symmetric, $O_{ij} = O_{ji}$.

4.9 Show that L_s has $\frac{1}{2}s(s+1)$ terms and L_{N-s} has $\frac{1}{2}\left[N(N+1) - s(s+1)\right]$ terms.

4.10 Show that Eq. (4.28) is dimensionally correct.

4.11 Write down the first equation of the hierarchy for a system on which an external force field $F(\boldsymbol{r})$ acts. Answer:

$$\left(\frac{\partial}{\partial t} + \boldsymbol{v}\cdot\boldsymbol{\nabla}_r + \frac{\boldsymbol{F}}{m}\cdot\boldsymbol{\nabla}_v\right) F_1(\boldsymbol{r}, \boldsymbol{v}, t) = \frac{1}{m}\int \boldsymbol{\nabla}_r\Phi(\boldsymbol{r}, \boldsymbol{r}')\cdot\boldsymbol{\nabla}_v F_2(\boldsymbol{r}, \boldsymbol{v}, \boldsymbol{r}', \boldsymbol{v}', t)\mathrm{d}\boldsymbol{r}'\mathrm{d}\boldsymbol{v}'. \tag{P4.1}$$

Verify that this equation is dimensionally correct.

4.12 In this exercise we derive two identities that follow from integration by parts, and are better relegated to an exercise than devoting space to them in the main text. Assume integrated parts vanish at infinitely distant surfaces in velocity space.

a. For a constant vector $\boldsymbol{A}$ and $f(\boldsymbol{r}, \boldsymbol{v}, t)$ the one-particle distribution, show that

$$\int \mathrm{d}\boldsymbol{v} \left[\boldsymbol{A} \cdot \boldsymbol{\nabla}_{\boldsymbol{v}} f(\boldsymbol{r}, \boldsymbol{v}, t)\right] \boldsymbol{v} = -\frac{1}{m}\rho(\boldsymbol{r}, t)\boldsymbol{A},$$

where $\rho(\boldsymbol{r}, t)$ is defined in Eq. (4.31).

b. Show that

$$\int \mathrm{d}\boldsymbol{r}'\mathrm{d}\boldsymbol{v}'\mathrm{d}\boldsymbol{v}[\boldsymbol{\nabla}_{\boldsymbol{r}}\Phi(\boldsymbol{r}, \boldsymbol{r}') \cdot \boldsymbol{\nabla}_{\boldsymbol{v}} F_2(\boldsymbol{r}, \boldsymbol{v}, \boldsymbol{r}', \boldsymbol{v}', t)]\boldsymbol{v} =$$
$$- \int \mathrm{d}\boldsymbol{r}'\mathrm{d}\boldsymbol{v}'\mathrm{d}\boldsymbol{v}[\boldsymbol{\nabla}_{\boldsymbol{r}}\Phi(\boldsymbol{r}, \boldsymbol{r}')]F_2(\boldsymbol{r}, \boldsymbol{v}, \boldsymbol{r}', \boldsymbol{v}', t),$$

where Φ is the interparticle potential and F_2 is the two-body distribution (properties of those functions are not used in the derivation other than F_2 vanishing as $v \to \infty$).

4.13 Derive Eq. (4.39) from Eq. (4.38). Make use of the results derived in the previous exercise.

4.14 Show the dimensional equivalence of force per area, energy density, and momentum flux.

4.15 For spherically symmetric interatomic potentials $\Phi(|\boldsymbol{r} - \boldsymbol{r}'|)$, show that the potential energy part of the pressure tensor $\mathbf{P}^{\Phi}$ is symmetric, i.e., $P^{\Phi}_{ij} = P^{\Phi}_{ji}$. See Eq. (4.45). Hint: Show that

$$\frac{\partial}{\partial \boldsymbol{r}_i}\Phi(|\boldsymbol{r} - \boldsymbol{r}'|) = \frac{(\boldsymbol{r}_i - \boldsymbol{r}'_i)}{|\boldsymbol{r} - \boldsymbol{r}'|}\Phi',$$

where $\Phi' \equiv \mathrm{d}\Phi/\mathrm{d}(|\boldsymbol{r} - \boldsymbol{r}'|)$.

4.16 Derive Eq. (4.56). You should find that the right sides of Eqs. (4.51) and (4.54) cancel. If an integral vanishes for all possible regions of integration, the integrand vanishes everywhere. This reasoning doesn't hold for definite integrals, e.g., $\int_{-1}^{1} \sin x \mathrm{d}x = 0$ does not imply that $\sin x$ vanishes identically on $[-1, 1]$. If however $\int_E f(x)\mathrm{d}x = 0$ for any measurable set E, then $f = 0$.

4.17 Derive the results in Eq. (4.58).

4.18 In this exercise we work through the ingredients that go into the derivation of Eq. (4.63).

a. Using calculus, verify the identity

$$\boldsymbol{u} \cdot \frac{\partial}{\partial t}(\rho \boldsymbol{u}) = \frac{1}{2}\frac{\partial}{\partial t}(\rho u^2) + \frac{1}{2}u^2\frac{\partial \rho}{\partial t}.$$

Combine with the continuity equation to show that

$$\boldsymbol{u} \cdot \frac{\partial}{\partial t}(\rho \boldsymbol{u}) = \frac{1}{2}\frac{\partial}{\partial t}(\rho u^2) - \frac{1}{2}u^2 \boldsymbol{\nabla} \cdot (\rho \boldsymbol{u}).$$

b. From Eq. (1.24), $\boldsymbol{u} \cdot \boldsymbol{\nabla} \cdot \mathbf{P} = \boldsymbol{\nabla} \cdot (\mathbf{u} \cdot \mathbf{P}) - \mathbf{P}{:}\boldsymbol{\nabla}\mathbf{u}$ and $\boldsymbol{u} \cdot [\boldsymbol{\nabla} \cdot \rho \boldsymbol{u}\boldsymbol{u}] = \boldsymbol{\nabla} \cdot (\boldsymbol{u} \cdot \rho \boldsymbol{u}\boldsymbol{u}) - \rho \boldsymbol{u}\boldsymbol{u}{:}\boldsymbol{\nabla}\boldsymbol{u}$, where the contraction of second-rank tensors $\mathbf{A}{:}\mathbf{B} \equiv \sum_{ij} A_{ij}B_{ij}$. Show that:

i. $\boldsymbol{u} \cdot \rho \boldsymbol{u}\boldsymbol{u} = \rho u^2 \boldsymbol{u}$;
ii. $\rho \boldsymbol{u}\boldsymbol{u}{:}\boldsymbol{\nabla}\boldsymbol{u} = \frac{1}{2}\rho \boldsymbol{u} \cdot \boldsymbol{\nabla} u^2$;
iii. $\boldsymbol{\nabla} \cdot (\rho u^2 \boldsymbol{u}) = u^2 \boldsymbol{\nabla} \cdot (\rho \boldsymbol{u}) + \rho \boldsymbol{u} \cdot \boldsymbol{\nabla} u^2$.

c. Use these results to show that Eq. (4.62) implies Eq.(4.63).

4.19 In Section 4.4.3, balance equations for mass, momentum, and energy were derived, but not for angular momentum. Here we establish that orbital angular momentum is conserved when the pressure tensor is symmetric. We define an orbital angular momentum density tensor as the moment of the linear momentum density $\hat{g}(\boldsymbol{r})$ (see Eq. (4.32)),

$$\hat{L}_{ij}(\boldsymbol{r}) \equiv x_i\hat{g}_j(\boldsymbol{r}) - x_j\hat{g}_i(\boldsymbol{r}) = m\sum_{\alpha}\left[x_i v_{\alpha,j} - x_j v_{\alpha,i}\right]\delta(\boldsymbol{r}-\boldsymbol{r}_\alpha), \tag{P4.2}$$

where x_i are the coordinates of $\boldsymbol{r}$. Note that $\hat{L}_{ij} = -\hat{L}_{ji}$; angular momentum is described by an antisymmetric tensor, a property it inherits from the vector cross product, $\boldsymbol{L} = \boldsymbol{r}\times\boldsymbol{p}$.

a. Show in the absence of external forces that

$$\frac{\partial}{\partial t}\hat{L}_{ij} = -\sum_k \frac{\partial}{\partial x_k}\hat{M}_{kij} + \hat{P}_{ij} - \hat{P}_{ji},$$

where the angular momentum flux tensor $\hat{M}_{kij}$ is the moment of the linear momentum flux (pressure tensor),

$$\hat{M}_{kij} \equiv x_i\hat{P}_{kj} - x_j\hat{P}_{ki},$$

where $\hat{\mathbf{P}} = \hat{\mathbf{P}}^K + \hat{\mathbf{P}}^\Phi$ is the full pressure tensor; see Eq. (4.69). Note that $\hat{M}_{kij}$, the elements of a third-rank tensor, are antisymmetric in the second and third indices. Orbital angular momentum is conserved when $\hat{\mathbf{P}}$ is symmetric.

b. Show that $\hat{M}_{kij}$ has the dimension of angular momentum flux.

4.20 Show that the divergence of the pressure tensor [see Eqs. (4.74) and (4.77)] has the form

$$[\boldsymbol{\nabla}\cdot\mathbf{P}]_i = \nabla_i P - (\zeta + \tfrac{1}{3}\eta)\nabla_i(\boldsymbol{\nabla}\cdot\boldsymbol{u}) - \eta\nabla^2 u_i,$$

where P is the hydrostatic pressure.

4.21 Derive Eq. (4.80). Hint: Symmetrize the velocity gradient in the contraction with the "η" part of the tensor; let $\nabla_j u_i \to \frac{1}{2}(\nabla_j u_i + \nabla_i u_j)$. The pressure tensor is symmetric, and we have a sum over the indices (i,j). Add and subtract $\frac{2}{3}\delta_{ij}\boldsymbol{\nabla}\cdot\boldsymbol{u}$.

4.22 The thermal expansivity α and isothermal compressibility β_T measure fractional changes in V that occur with changes in T and P: $\alpha \equiv (1/V)(\partial V/\partial T)_P$ and $\beta_T \equiv -(1/V)(\partial V/\partial P)_T$ (see [2, p19]). The quantity α can be of either sign, and β_T is always positive.

a. Show that the derivatives required in Eq. (4.82) are given by

$$\left(\frac{\partial P}{\partial T}\right)_\rho = \frac{\alpha}{\beta_T} \qquad \left(\frac{\partial P}{\partial \rho}\right)_T = \frac{1}{\rho\beta_T}.$$

b. In thermodynamics, the derivative $(\partial U/\partial V)_T$ (which is not easily measured), can be related to measurable quantities through the relation (see [2, p20])

$$\left(\frac{\partial U}{\partial V}\right)_T = T\left(\frac{\partial P}{\partial T}\right)_V - P.$$

Derive the related expression in terms of specific quantities in Eq. (4.83). Hint: Show that (where M is the total mass)

$$\left(\frac{\partial U}{\partial V}\right)_T = \left[\frac{\partial(U/M)}{\partial(V/M)}\right]_T \equiv \left(\frac{\partial \tilde{u}}{\partial \rho^{-1}}\right)_T = -\rho^2\left(\frac{\partial \tilde{u}}{\partial \rho}\right)_T.$$

4.23 Calculate the entries in Table 4.1 by the strategy of substituting Eq. (4.92) in Eq. (4.93).

a. Show by matching powers of $\omega^n k^m$ at third order $(\omega^0 k^3, \omega k^2, \omega^2 k)$ that $a_1 a_2 a_5 = 0$, $a_1 a_2 + a_2 a_5 + a_5 a_1 = B\rho + AC$, and $a_1 + a_2 + a_5 = 0$.

b. Show at fourth order $(\omega^0 k^4, \omega k^3, \omega^2 k^2)$ that we have $a_1 b_2 a_5 + a_2 b_5 a_1 + a_5 b_1 a_2 = -\rho B \sigma$, $a_1 b_2 + b_1 a_2 + a_2 b_5 + b_2 a_5 + a_5 b_1 + b_5 a_1 = 0$, and $b_1 + b_2 + b_5 = -(\phi + \chi + \sigma)$.

c. We don't expect the thermal mode to have a propagating component. Show that setting $a_5 = 0$ implies $a_2 = -a_1$ and $b_2 = b_1$, what we expect of longitudinal sound modes propagating in opposite directions with the same speed and the same damping coefficient.

d. Show that $B\rho + AC = (\partial P/\partial \rho)_T (c_p/c_v)$, where c_p, c_v are the specific heat capacities at constant pressure and volume. Show from thermodynamics that $\alpha^2 T/(\rho \beta_T) = c_p - c_v$. See [2, p20] and use the results of Exercise 4.22.

e. Show that $b_5 = -\kappa/(\rho c_p)$. Note the factor of c_p; that's not a misprint. Show that

$$b_1 = -\frac{1}{2\rho}\left[\frac{4}{3}\eta + \zeta + \kappa\left(\frac{1}{c_v} - \frac{1}{c_p}\right)\right].$$

From thermodynamics $c_p > c_v$ (see [2, p20]); b_1 is always negative.

4.24 Show that kinetic equations have the form [see Eq. (4.96)]

$$\left(\frac{\partial}{\partial t} + \frac{\boldsymbol{p}}{m} \cdot \frac{\partial}{\partial \boldsymbol{r}} + \boldsymbol{F} \cdot \frac{\partial}{\partial \boldsymbol{p}}\right) f_1(\boldsymbol{r}, \boldsymbol{p}, t) = J[f_1]. \tag{P4.3}$$

There are two sources of change to the momentum dependence of $f_1(\boldsymbol{r}, \boldsymbol{p}, t)$—that generated by external forces and that due to particle collisions. Compare Eq. (P4.3) with Eq. (3.59), the Fokker-Planck equation for a Brownian particle in an external force field.

4.25 Estimate the collision time $\tau_c = r_0/v$ for dilute gases at room temperature. For the interaction range, use the Lennard-Jones length parameter for noble gases, $r_0 \approx 3 \times 10^{-10}$ m,[159, p398] and for the speed, the rms speed in gases, $\sqrt{3kT/m}$. Contrast this time with an estimate from the uncertainty principle, $\tau_c = \hbar/(kT)$.

4.26 Show that the velocities $\boldsymbol{v}_1', \boldsymbol{v}_2'$ in Eq. (4.98) solve Eq. (4.97) when $\boldsymbol{g}$ is given in Eq. (4.99).

4.27 a. Show that the bilinear form of the collision operator $Q(f, g)$, Eq. (4.108), is indeed a bilinear form. See footnote 38 on page 9.

b. Show that $J[f + g] = J[f] + J[g] + 2Q(f, g)$. This indicates that J is nonlinear.

4.28 Show that $(\mathrm{d}H/\mathrm{d}t)_{\text{streaming}} = 0$. See Eq. (4.113). Integrate by parts and assume f vanishes at the boundaries of μ-space; $f = 0$ outside the region occupied by the gas and $f \to 0$ at an infinitely distant surface in velocity space.

4.29 Consider the integral over $\boldsymbol{v}$ of the product of the bilinear form of the collision operator $Q(f, g)$ with an arbitrary function of $\boldsymbol{v}$, $\psi(\boldsymbol{v})$, $I \equiv \int \mathrm{d}\boldsymbol{v} Q(f, g)\psi(\boldsymbol{v})$. Show that

$$I = -\frac{1}{8}\int \mathrm{d}\boldsymbol{v}\mathrm{d}\boldsymbol{v}_1 \mathrm{d}\Omega \sigma(\Omega)|\boldsymbol{v}_1 - \boldsymbol{v}| \left[f(\boldsymbol{v}_1')g(\boldsymbol{v}') + f(\boldsymbol{v}')g(\boldsymbol{v}_1') - f(\boldsymbol{v}_1)g(\boldsymbol{v}) - f(\boldsymbol{v})g(\boldsymbol{v}_1)\right]$$
$$\times \left[\psi(\boldsymbol{v}') + \psi(\boldsymbol{v}_1') - \psi(\boldsymbol{v}) - \psi(\boldsymbol{v}_1)\right].$$

Use the same steps leading to Eq. (4.116).

4.30 Fill in the details leading to the H-theorem inequality (4.117), and to Eq. (4.119).

4.31 Consider the function $f(x, y) \equiv (x - y)\ln(y/x)$, where (x, y) are positive. Show for $y = x + \delta$, where $|\delta| \ll x$, that $f \approx -\delta^2/x$. Note that f vanishes independently of the sign of δ.

4.32 What is the form of the hydrodynamic conservation laws associated with the Boltzmann equation? Take velocity moments of the Boltzmann equation and use Eq. (4.119) together with any results from Section 4.4.2. You should find the same conservation laws as in Eqs. (4.37), (4.49), and (4.57), except they include only the kinetic parts of the pressure tensor and the energy flux. Long-range internal forces are ignored in the Boltzmann equation, with elastic scattering modeled by the Boltzmann collision operator. Because the precise forms of $\mathbf{P}, \boldsymbol{J}$ were left unspecified, hydrodynamic equations derived from the Boltzmann equation are the same as those in Section 4.4.4.

4.33 Find expressions for the quantities $K, \boldsymbol{u}, \beta$ in Eq. (4.121) in terms of $A, \boldsymbol{B}, C$ in Eq. (4.120). A: $K = \exp[A - B^2/(4C)], \boldsymbol{u} = -(1/2C)\boldsymbol{B}, \beta = (-2/m)C$.

4.34 Show explicitly that the Maxwell-Boltzmann distribution in Eq. (4.121) satisfies the detailed balance condition, Eq. (4.118), for any value of $\boldsymbol{u}$. Hint: Use Eq. (4.97). Does the local Maxwellian, Eq. (4.129), satisfy Eq. (4.118)?

4.35 What is the dimension of the parameter K in Eq. (4.122)?

4.36 Find the kinetic pressure tensor and heat flux vector associated with the Maxwell-Boltzmann distribution, Eq. (4.121). A: $\widetilde{P}^K_{ij} = nkT\delta_{ij}$ and $\widetilde{\boldsymbol{J}}^K = 0$.

4.37 Derive Eq. (4.127).

4.38 Evaluate the integral in Eq. (4.131).

4.39 Show that the Sackur-Tetrode formula for the entropy of an ideal gas, Eq. (4.132), can be written in the general form expected from thermodynamics, $S = Nkf(U/N, V/N)$.

4.40 We stated that the Gibbs entropy is a constant of the motion as a consequence of general principles of classical dynamics. Use the Liouville equation to verify explicitly from Eq. (4.136) that

$$\frac{\mathrm{d}}{\mathrm{d}t} S_G = 0.$$

Hint: The Liouville operator is self-adjoint, $\Lambda^\dagger = \Lambda$.

4.41 Derive Eq. (4.137) starting from $\nu(v) = \sigma_T \int \mathrm{d}\boldsymbol{v}_1 |\boldsymbol{v}_1 - \boldsymbol{v}| f_0(\boldsymbol{v}_1)$. Change variables with $\boldsymbol{u} \equiv \boldsymbol{v}_1 - \boldsymbol{v}$. Let $\boldsymbol{v}$ define the polar axis of the coordinate system. Use the integral $\int_0^\infty x^2 \mathrm{e}^{-x^2} \sinh(\gamma x) = (\gamma/4) + (\sqrt{\pi}/8)(2 + \gamma^2)\exp(\gamma^2/4)\,\mathrm{erf}(\gamma/2)$ (Gradshteyn and Ryzhik[88, p365]), where the error function $\mathrm{erf}(x) \equiv (2/\sqrt{\pi}) \int_0^x \mathrm{e}^{-t^2} \mathrm{d}t$. Note that $\lim_{x\to\infty} \mathrm{erf}(x) = 1$.

4.42 Estimate the collision frequency of an atom in an argon gas in thermal equilibrium, under temperature and pressure conditions of your choice. Compare your answer with estimates of the collision frequency in Brownian motion; see Exercise 3.1.

4.43 Show that the moments of the solution of Boltzmann's equation associated with the collisional invariants satisfy conservation laws. Starting from Eq. (4.140), show that

$$\frac{\partial \rho_i}{\partial t} + m\boldsymbol{\nabla} \cdot \int \psi_i(\boldsymbol{v})\boldsymbol{v} f(\boldsymbol{r}, \boldsymbol{v}, t)\mathrm{d}\boldsymbol{v} = 0.$$

Thus $m \int \boldsymbol{v}\psi_i(\boldsymbol{v}) f(\boldsymbol{r}, \boldsymbol{v}, t)\mathrm{d}\boldsymbol{v}$ is a current associated with the invariants.

4.44 Derive the system of equations in (4.143).

4.45 Derive the linearized collision operator, Eq. (4.146).

4.46 Derive the linearized Boltzmann equation, Eq. (4.147). Start with the Boltzmann equation $[(\partial/\partial t) + \boldsymbol{v}\cdot\boldsymbol{\nabla}]f = Q(f,f)$ and let $f = f^{(0)} + h$, where $Q(f^{(0)}, f^{(0)}) = 0$; show its equivalence with

$$\frac{\partial h}{\partial t} + \boldsymbol{v}\cdot\boldsymbol{\nabla} h = L(h) + Q(h,h),$$

where L is defined in Eq. (4.146). Equation (4.147) follows by ignoring the nonlinear term $Q(h,h)$. It's obvious but worth noting that $Q(f,g)$ has no projection on the null space of L.

4.47 Derive Eq. (4.148).

4.48 Derive Eq. (4.166).

4.49 Using results from Section 4.8.2, show that the Euler equations are obtained from the lowest order time derivatives $\mathcal{D}_j^{(0)}, j = 1, \ldots 5$. Show that (with $i = x, y, z \equiv 2, 3, 4$):

$$\frac{\partial \rho}{\partial t} = \mathcal{D}_1^{(0)}; \qquad \frac{\partial \rho u_i}{\partial t} = \mathcal{D}_i^{(0)}; \qquad \frac{\partial \widetilde{\varepsilon}}{\partial t} = \frac{1}{2}\mathcal{D}_5^{(0)} - \sum_i u_i \mathcal{D}_i^{(0)} + \frac{1}{2}u^2 \mathcal{D}_1^{(0)}.$$

Let $P = nkT$ and $\widetilde{\varepsilon} = \frac{3}{2}nkT$.

4.50 Show that the solvability conditions associated with Eq. (4.175) are satisfied. Show that

$$\int \mathrm{d}\boldsymbol{v} f^{(0)}(\boldsymbol{v})\psi_i(\boldsymbol{v})\left(\frac{my^2}{2kT} - \frac{5}{2}\right)y_j = 0, \qquad \int \mathrm{d}\boldsymbol{v} f^{(0)}(\boldsymbol{v})\psi_i(\boldsymbol{v})\left(y_k y_l - \frac{1}{3}y^2 \delta_{k,l}\right) = 0.$$

4.51 For $F = F(y)$ a scalar function of the vector $\boldsymbol{y}$, and for $\boldsymbol{D}$ a constant vector, show that

$$\int F(y)\boldsymbol{y}\,(\boldsymbol{y}\cdot\boldsymbol{D})\,\mathrm{d}\boldsymbol{y} = \frac{1}{3}\boldsymbol{D}\int F(y)y^2 \mathrm{d}\boldsymbol{y}.$$

Hint: By symmetry, $\int y_x^2 F(y)\mathrm{d}\boldsymbol{y} = \int y_y^2 F(y)\mathrm{d}\boldsymbol{y} = \int y_z^2 F(y)\mathrm{d}\boldsymbol{y}$.

4.52 Verify that the bracket defined in Eq. (4.181) is never negative. It's best to show this for two scalar functions F, G; the vector case then follows. Let

$$\begin{aligned}[F,G] &\equiv -\int f^{(0)}(\boldsymbol{v})F(\boldsymbol{v})\mathcal{I}(G)\,\mathrm{d}\boldsymbol{v} \\ &= -\int \mathrm{d}\boldsymbol{v}\mathrm{d}\boldsymbol{v}_1 \mathrm{d}\Omega\sigma(\Omega)|\boldsymbol{v}_1 - \boldsymbol{v}| f^{(0)}(\boldsymbol{v}) f^{(0)}(\boldsymbol{v}_1) F(\boldsymbol{v})\left[G(\boldsymbol{v}_1') + G(\boldsymbol{v}') - G(\boldsymbol{v}_1) - G(\boldsymbol{v})\right],\end{aligned}$$

where $\mathcal{I}$ is defined in Eq. (4.172). Bracket notations have long been used in kinetic theory to symbolize complicated expressions, going back to Enskog's thesis at least. Using the arguments leading to Eq. (4.116), show that

$$\begin{aligned}[F,G] =& \frac{1}{4}\int \mathrm{d}\boldsymbol{v}\mathrm{d}\boldsymbol{v}_1 \mathrm{d}\Omega\sigma(\Omega)|\boldsymbol{v}_1 - \boldsymbol{v}| f^{(0)}(\boldsymbol{v}_1) f^{(0)}(\boldsymbol{v}) \\ &\times \left[F(\boldsymbol{v}_1') + F(\boldsymbol{v}') - F(\boldsymbol{v}_1) - F(\boldsymbol{v})\right]\left[G(\boldsymbol{v}_1') + G(\boldsymbol{v}') - G(\boldsymbol{v}_1) - G(\boldsymbol{v})\right].\end{aligned}$$

Thus, $[F,F] \geq 0$. Note that $[F,G] = [G,F]$.

CHAPTER 5

Weakly coupled systems: Landau-Vlasov theories

THE Boltzmann equation applies to dilute gases having short-ranged interactions. Particle collisions in such systems are well localized in space and time, allowing us to model the dynamics as a succession of binary scattering events caused primarily by the repulsive part of the intermolecular potential. Many gases have potentials featuring repulsive interactions at short distances, such as the Lennard-Jones potential (see [5, p181]). Here we consider a different type of system, *weakly coupled*, in which the interaction potential is everywhere small relative to kinetic energies.[1] Plasmas approximate weakly coupled gases (and the plasma state is the most abundant form of visible matter in the universe). In this chapter, we consider the application of kinetic theory to plasmas. We start with the *Landau equation*[160], the kinetic equation for homogeneous weakly coupled gases.

5.1 HOMOGENEOUS WEAK COUPLING: THE LANDAU EQUATION

We can interpret $\sigma(\Omega)|\boldsymbol{v}_1-\boldsymbol{v}|$ in the Boltzmann equation as the rate w at which particles of velocities $(\boldsymbol{v},\boldsymbol{v}_1)$ transition to $(\boldsymbol{v}',\boldsymbol{v}_1')$, and thus, for spatially homogeneous systems (suppressing $\boldsymbol{r},t$),

$$\frac{\partial}{\partial t}f(\boldsymbol{v}) = \int \mathrm{d}\boldsymbol{v}_1\mathrm{d}\Omega w(\boldsymbol{v},\boldsymbol{v}_1;\boldsymbol{v}',\boldsymbol{v}_1')\left[f(\boldsymbol{v}_1')f(\boldsymbol{v}') - f(\boldsymbol{v}_1)f(\boldsymbol{v})\right].$$

Writing the transition rate in the form $w(\boldsymbol{v},\boldsymbol{v}_1;\boldsymbol{v}',\boldsymbol{v}_1')$ is misleading, however. The four arguments $(\boldsymbol{v},\boldsymbol{v}_1,\boldsymbol{v}',\boldsymbol{v}_1')$ are not independent; there's a constraint among them, implying w is a function of three vector variables. To show that, introduce the velocity deviation vector $\boldsymbol{\Delta}$ with $\boldsymbol{v}'=\boldsymbol{v}+\boldsymbol{\Delta}$ and $\boldsymbol{v}_1'=\boldsymbol{v}_1-\boldsymbol{\Delta}$ (ensuring momentum conservation, $\boldsymbol{v}'+\boldsymbol{v}_1'=\boldsymbol{v}+\boldsymbol{v}_1$). From Eq. (4.98), $\boldsymbol{\Delta}=\frac{1}{2}(\boldsymbol{g}-g\hat{\boldsymbol{\epsilon}})$, where $\boldsymbol{g}=\boldsymbol{v}_1-\boldsymbol{v}$, $g=|\boldsymbol{g}|$, and $\hat{\boldsymbol{\epsilon}}$ is a unit vector in the scattering direction; see Fig. 4.2. Consider the volume-preserving (unity Jacobian) transformation, $\boldsymbol{v}\to\frac{1}{2}(\boldsymbol{v}+\boldsymbol{v}')$, $\boldsymbol{v}_1\to\frac{1}{2}(\boldsymbol{v}_1+\boldsymbol{v}_1')$, $\boldsymbol{v}'\to\boldsymbol{v}'-\boldsymbol{v}$, and $\boldsymbol{v}_1'\to\boldsymbol{v}_1'-\boldsymbol{v}_1$. Under this change of variables,

$$\begin{aligned} w(\boldsymbol{v},\boldsymbol{v}_1;\boldsymbol{v}',\boldsymbol{v}_1') \to w\left[\tfrac{1}{2}(\boldsymbol{v}+\boldsymbol{v}'),\tfrac{1}{2}(\boldsymbol{v}_1+\boldsymbol{v}_1');\boldsymbol{v}'-\boldsymbol{v},\boldsymbol{v}_1'-\boldsymbol{v}_1\right] &= w(\boldsymbol{v}+\tfrac{1}{2}\boldsymbol{\Delta},\boldsymbol{v}_1-\tfrac{1}{2}\boldsymbol{\Delta};\boldsymbol{\Delta},-\boldsymbol{\Delta}) \\ &\equiv w(\boldsymbol{v}+\tfrac{1}{2}\boldsymbol{\Delta},\boldsymbol{v}_1-\tfrac{1}{2}\boldsymbol{\Delta};\boldsymbol{\Delta}). \end{aligned}$$

From detailed balance, w is invariant under the interchange of incident and scattered particles, implying $w(\boldsymbol{v}+\frac{1}{2}\boldsymbol{\Delta},\boldsymbol{v}_1-\frac{1}{2}\boldsymbol{\Delta};\boldsymbol{\Delta})=w(\boldsymbol{v}+\frac{1}{2}\boldsymbol{\Delta},\boldsymbol{v}_1-\frac{1}{2}\boldsymbol{\Delta};-\boldsymbol{\Delta})$; thus, w is an even function in the variable $\boldsymbol{\Delta}$ to the right of the semicolon. Boltzmann equation's can then be written

$$\frac{\partial}{\partial t}f(\boldsymbol{v}) = \int \mathrm{d}\boldsymbol{v}_1\mathrm{d}\Omega w(\boldsymbol{v}+\tfrac{1}{2}\boldsymbol{\Delta},\boldsymbol{v}_1-\tfrac{1}{2}\boldsymbol{\Delta};\boldsymbol{\Delta})\left[f(\boldsymbol{v}_1-\boldsymbol{\Delta})f(\boldsymbol{v}+\boldsymbol{\Delta}) - f(\boldsymbol{v}_1)f(\boldsymbol{v})\right]. \tag{5.1}$$

[1]The terms comes from statistical mechanics where subsystems in thermal contact are said to be weakly coupled if their energy of interaction is small compared to subsystem energies[5, p84].

DOI: 10.1201/9781003512295-5

Boltzmann's equation in this form is well suited to the analysis of weakly coupled systems. Small interaction energies imply that the deviations $\boldsymbol{\Delta}$ are small compared to particle velocities, $|\boldsymbol{\Delta}| \ll |\boldsymbol{v}|, |\boldsymbol{v}_1|$. We can then approximate the functions in Eq. (5.1) with their Taylor expansions,

$$w(\boldsymbol{v}+\frac{1}{2}\boldsymbol{\Delta}, \boldsymbol{v}_1-\frac{1}{2}\boldsymbol{\Delta}; \boldsymbol{\Delta}) = w(\boldsymbol{v}, \boldsymbol{v}_1; \boldsymbol{\Delta}) + \frac{1}{2}\boldsymbol{\Delta}\cdot[\boldsymbol{\nabla}_{\boldsymbol{v}} - \boldsymbol{\nabla}_{\boldsymbol{v}_1}]\, w(\boldsymbol{v}, \boldsymbol{v}_1; \boldsymbol{\Delta}) + O(\Delta^2),$$

$$f(\boldsymbol{v}+\boldsymbol{\Delta}) = f(\boldsymbol{v}) + \boldsymbol{\Delta}\cdot\boldsymbol{\nabla}_{\boldsymbol{v}} f(\boldsymbol{v}) + \frac{1}{2}\boldsymbol{\Delta}\boldsymbol{\Delta}{:}\boldsymbol{\nabla}_{\boldsymbol{v}}\boldsymbol{\nabla}_{\boldsymbol{v}} f(\boldsymbol{v}) + O(\Delta^3)$$

$$f(\boldsymbol{v}_1-\boldsymbol{\Delta}) = f(\boldsymbol{v}_1) - \boldsymbol{\Delta}\cdot\boldsymbol{\nabla}_{\boldsymbol{v}_1} f(\boldsymbol{v}_1) + \frac{1}{2}\boldsymbol{\Delta}\boldsymbol{\Delta}{:}\boldsymbol{\nabla}_{\boldsymbol{v}_1}\boldsymbol{\nabla}_{\boldsymbol{v}_1} f(\boldsymbol{v}_1) + O(\Delta^3).$$

Substitute these expressions in Eq. (5.1) and keep terms through second order; we omit the algebra. The first-order terms vanish and the second-order terms simplify by integrating by parts with respect to $\boldsymbol{v}_1$. We find:

$$\frac{\partial}{\partial t}f(\boldsymbol{v}) = \frac{1}{2}\sum_{a,b}\int \mathrm{d}\boldsymbol{v}_1 \mathrm{d}\Omega \Delta_a \Delta_b \Bigg\{ w(\boldsymbol{v}, \boldsymbol{v}_1; \boldsymbol{\Delta}) \left[\frac{\partial^2 f(\boldsymbol{v})}{\partial v_a \partial v_b} f(\boldsymbol{v}_1) - \frac{\partial f(\boldsymbol{v})}{\partial v_a}\frac{\partial f(\boldsymbol{v}_1)}{\partial v_{1,b}}\right]$$

$$+ \frac{\partial w(\boldsymbol{v}, \boldsymbol{v}_1; \boldsymbol{\Delta})}{\partial v_a}\left[\frac{\partial f(\boldsymbol{v})}{\partial v_b} f(\boldsymbol{v}_1) - f(\boldsymbol{v})\frac{\partial f(\boldsymbol{v}_1)}{\partial v_{1,b}}\right]\Bigg\},$$

where $(a,b) = 1, 2, 3$. This expression can be written compactly:

$$\frac{\partial f(\boldsymbol{v})}{\partial t} = \boldsymbol{\nabla}_{\boldsymbol{v}}\cdot\int \mathrm{d}\boldsymbol{v}_1 \mathbf{G}\cdot(\boldsymbol{\nabla}_{\boldsymbol{v}} - \boldsymbol{\nabla}_{\boldsymbol{v}_1})\, f(\boldsymbol{v}) f(\boldsymbol{v}_1) \tag{5.2}$$

with the symmetric tensor

$$\mathbf{G} \equiv \frac{1}{2}\int \mathrm{d}\Omega w(\boldsymbol{v}, \boldsymbol{v}_1; \boldsymbol{\Delta})\boldsymbol{\Delta}\boldsymbol{\Delta}. \tag{5.3}$$

As we'll show, $\mathbf{G}$ can be related to the intermolecular potential.

We first develop an expression for $\boldsymbol{\Delta}$. The difference between the asymptotes $\boldsymbol{v}$ and $\boldsymbol{v}'$ can be calculated from the force exerted on the particle over its trajectory,

$$\boldsymbol{\Delta} = \boldsymbol{v}' - \boldsymbol{v} = \frac{1}{m}\int_{-\infty}^{\infty} \boldsymbol{F}\mathrm{d}t = -\frac{1}{m}\int_{-\infty}^{\infty} \boldsymbol{\nabla}\Phi\left(|\boldsymbol{r}(t) - \boldsymbol{r}_1(t)|\right)\mathrm{d}t, \tag{5.4}$$

where Φ is the energy of interaction between the particle at its current position $\boldsymbol{r}(t)$ and the other particle at position $\boldsymbol{r}_1(t)$ [instantaneous interactions assumed, with $\Phi(\boldsymbol{r}) = \Phi(|\boldsymbol{r}|)$]. By definition, $\boldsymbol{r}_1(t) - \boldsymbol{r}(t) = \boldsymbol{r}_1(0) - \boldsymbol{r}(0) + \int_0^t [\boldsymbol{v}_1(t') - \boldsymbol{v}(t')]\mathrm{d}t' \equiv \boldsymbol{a} + \int_0^t \boldsymbol{g}(t')\mathrm{d}t'$, where $\boldsymbol{a}$ is a constant and $\boldsymbol{g}(t) = \boldsymbol{v}_1(t) - \boldsymbol{v}(t)$ is the time-dependent generalization of $\boldsymbol{g} = \boldsymbol{v}_1 - \boldsymbol{v}$. Because $\boldsymbol{\Delta}$ is small (weak coupling), we approximate the argument of Φ in Eq. (5.4) as

$$\boldsymbol{r}_1(t) - \boldsymbol{r}(t) \approx \boldsymbol{a} + \boldsymbol{g}t \equiv \boldsymbol{r}_t. \tag{5.5}$$

With Eq. (5.5), we've approximated the actual, slightly curved trajectory (small deviations $\boldsymbol{\Delta}$) with one of constant velocity. We introduce the Fourier transform of the potential, $\Phi(\boldsymbol{r}) = \int \mathrm{d}\boldsymbol{k}\mathrm{e}^{\mathrm{i}\boldsymbol{k}\cdot\boldsymbol{r}}\widetilde{\Phi}_{\boldsymbol{k}}$, which places the $\boldsymbol{r}$-dependence in an exponential where it's easy to analyze. With $\Phi(\boldsymbol{r}) = \Phi(r)$ a central potential, $\widetilde{\Phi}_{\boldsymbol{k}}$ is real and a function of the magnitude of $\boldsymbol{k}$, $\widetilde{\Phi}_{\boldsymbol{k}} = \widetilde{\Phi}_k$. From Eq. (5.4),

$$\boldsymbol{\Delta} = -\frac{\mathrm{i}}{m}\int \mathrm{d}t \int \mathrm{d}\boldsymbol{k}\widetilde{\Phi}_{\boldsymbol{k}}\boldsymbol{k}\mathrm{e}^{\mathrm{i}\boldsymbol{k}\cdot\boldsymbol{r}_t}. \tag{5.6}$$

We now turn to $\mathbf{G}$ in Eq. (5.3). Referring to Fig. C.6, which shows the geometry of two-body collisions in the reference frame of an incident particle, we have $w(\boldsymbol{v}, \boldsymbol{v}_1; \boldsymbol{\Delta})\mathrm{d}\Omega \equiv g\sigma(\Omega) = 2\pi g b \mathrm{d}b$, where b is the impact parameter. Combining Eq. (5.6) with Eq. (5.3),

$$\mathbf{G} = -\frac{\pi}{m^2}\int_0^{\infty} gb\mathrm{d}b \int_{-\infty}^{\infty}\mathrm{d}t_1 \int_{-\infty}^{\infty}\mathrm{d}t_2 \int \mathrm{d}\boldsymbol{k}_1 \int \mathrm{d}\boldsymbol{k}_2 \boldsymbol{k}_1\boldsymbol{k}_2 \widetilde{\Phi}_{\boldsymbol{k}_1}\widetilde{\Phi}_{\boldsymbol{k}_2}\exp\left(\mathrm{i}[\boldsymbol{k}_1\cdot\boldsymbol{r}_{t_1} + \boldsymbol{k}_2\cdot\boldsymbol{r}_{t_2}]\right).$$

The exponential can be written $\exp(\mathrm{i}[\boldsymbol{k}_1\cdot\boldsymbol{r}_{t_1}+\boldsymbol{k}_2\cdot\boldsymbol{r}_{t_2}]) = \exp(\mathrm{i}(\boldsymbol{k}_1+\boldsymbol{k}_2)\cdot\boldsymbol{r}_{t_1})\exp(\mathrm{i}\boldsymbol{k}_2\cdot\boldsymbol{g}(t_2-t_1))$. From the scattering geometry, $2\pi\int_0^\infty b\mathrm{d}b\int_{-\infty}^{\infty}\mathrm{d}t_1 g \equiv \int \mathrm{d}\boldsymbol{r}_{t_1}$. Dropping the subscript from $\boldsymbol{r}_{t_1}$ and defining $\tau \equiv t_2 - t_1$,

$$\mathbf{G} = -\frac{1}{2m^2}\int \mathrm{d}\boldsymbol{r}\int_{-\infty}^{\infty}\mathrm{d}\tau\int \mathrm{d}\boldsymbol{k}_1\mathrm{d}\boldsymbol{k}_2\boldsymbol{k}_1\boldsymbol{k}_2\widetilde{\Phi}_{\boldsymbol{k}_1}\widetilde{\Phi}_{\boldsymbol{k}_2}\exp(\mathrm{i}[\boldsymbol{k}_1+\boldsymbol{k}_2]\cdot\boldsymbol{r})\exp(\mathrm{i}\boldsymbol{k}_2\cdot\boldsymbol{g}\tau).$$

The integrations are now straightforward. We find:

$$\mathbf{G} = \frac{8\pi^4}{m^2}\int \mathrm{d}\boldsymbol{k}\widetilde{\Phi}_k^2\delta(\boldsymbol{k}\cdot\boldsymbol{g})\boldsymbol{k}\boldsymbol{k}. \tag{5.7}$$

Thus we have expressed $\mathbf{G}$ is a function of the Fourier components of the interaction potential, $\widetilde{\Phi}_{\boldsymbol{k}}$. Note that only modes with $\boldsymbol{k}$ orthogonal to $\boldsymbol{g}$ contribute.

Using Eq. (5.7), Eq. (5.2) can be written,

$$\frac{\partial f(\boldsymbol{v}_1)}{\partial t} = \frac{8\pi^4}{m^2}\int \mathrm{d}\boldsymbol{v}_2\int \mathrm{d}\boldsymbol{k}\widetilde{\Phi}_k^2\boldsymbol{k}\cdot\boldsymbol{\nabla}_{\boldsymbol{v}_1}\delta(\boldsymbol{k}\cdot\boldsymbol{g})\boldsymbol{k}\cdot(\boldsymbol{\nabla}_{\boldsymbol{v}_1}-\boldsymbol{\nabla}_{\boldsymbol{v}_2})\,f(\boldsymbol{v}_1)f(\boldsymbol{v}_2). \tag{5.8}$$

This can be put in symmetric form by replacing $\boldsymbol{\nabla}_{\boldsymbol{v}_1}$ with $\boldsymbol{\nabla}_{\boldsymbol{v}_1} - \boldsymbol{\nabla}_{\boldsymbol{v}_2}$. For any function $F(\boldsymbol{v})$ vanishing sufficiently fast at infinity, $\int \mathrm{d}\boldsymbol{v}\boldsymbol{\nabla}_{\boldsymbol{v}}F(\boldsymbol{v}) = 0$. Thus,

$$\frac{\partial f(\boldsymbol{v}_1)}{\partial t} = \frac{8\pi^4}{m^2}\int \mathrm{d}\boldsymbol{v}_2\int \mathrm{d}\boldsymbol{k}\widetilde{\Phi}_k^2\boldsymbol{k}\cdot\boldsymbol{\nabla}_{12}\delta(\boldsymbol{k}\cdot\boldsymbol{g})\boldsymbol{k}\cdot\boldsymbol{\nabla}_{12}f(\boldsymbol{v}_1)f(\boldsymbol{v}_2), \tag{5.9}$$

where $\boldsymbol{\nabla}_{12} \equiv \boldsymbol{\nabla}_{\boldsymbol{v}_1} - \boldsymbol{\nabla}_{\boldsymbol{v}_2}$. Equation (5.9) is one form of the Landau equation[161][162][163][164]. Its solutions satisfy the H-theorem, with Maxwellians as stationary solutions (see Exercise 5.1).

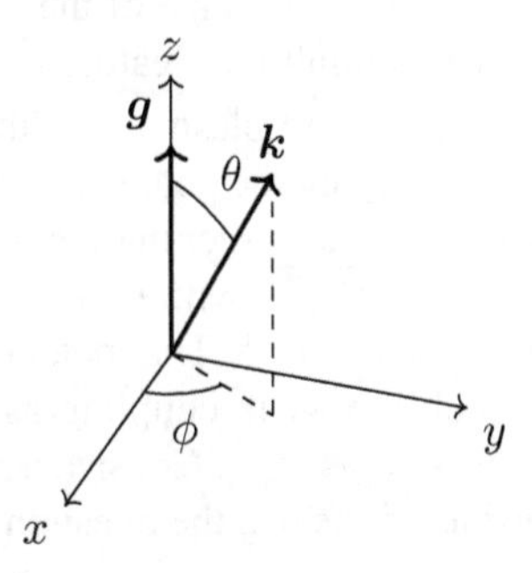

Figure 5.1 Coordinate system for evaluating the integral in Eq. (5.7) with the z-axis of $\boldsymbol{k}$-space aligned with $\boldsymbol{g}$.

The Landau equation can be put in a more useful form. First, evaluate Eq. (5.7) in a frame with the z-axis aligned with $\boldsymbol{g}$; see Fig. 5.1. In that way, the nine elements of $\mathbf{G}$ can be found as cases of:

$$G_{\left\{\begin{smallmatrix} x & x \\ y & y \\ z & z \end{smallmatrix}\right\}} = \frac{8\pi^4}{m^2}\int_0^\infty k^4\widetilde{\Phi}_k^2\mathrm{d}k\int_0^\pi \delta(kg\cos\theta)\sin\theta\mathrm{d}\theta\int_0^{2\pi}\mathrm{d}\phi\left\{\begin{matrix}\sin\theta\cos\phi\\ \sin\theta\sin\phi\\ \cos\theta\end{matrix}\right\}\times\left\{\begin{matrix}\sin\theta\cos\phi\\ \sin\theta\sin\phi\\ \cos\theta\end{matrix}\right\}.$$

In this frame, only G_{xx}, G_{yy} are nonzero,

$$\mathbf{G} = \left(\frac{B}{gm^2}\right)\begin{pmatrix}1 & 0 & 0\\ 0 & 1 & 0\\ 0 & 0 & 0\end{pmatrix}, \tag{5.10}$$

where

$$B \equiv 8\pi^5 \int_0^\infty k^3 \widetilde{\Phi}_k^2 \mathrm{d}k. \tag{5.11}$$

Note that $B \geq 0$. We can now find $\mathbf{G}$ in any frame because in this one frame it can be written as a tensor equation, $\mathbf{G} = (B/gm^2)(\mathbf{I} - \mathbf{I}_z\mathbf{I}_z)$, with $\mathbf{I}$ the unit tensor and $\mathbf{I}_z \equiv \boldsymbol{g}/g$. In an arbitrary frame, therefore,[2]

$$G_{ab} = \left(\frac{B}{gm^2}\right)\left(\delta_{ab} - \frac{g_a g_b}{g^2}\right). \tag{5.12}$$

We therefore have another expression for the Landau equation,

$$\frac{\partial f(\boldsymbol{v}_1)}{\partial t} = \frac{B}{m^2}\sum_{ab}\int \mathrm{d}\boldsymbol{v}_2 \left(\boldsymbol{\nabla}_{12}\right)_a \frac{1}{g^3}\left(g^2\delta_{ab} - g_a g_b\right)\left(\boldsymbol{\nabla}_{12}\right)_b f(\boldsymbol{v}_1)f(\boldsymbol{v}_2). \tag{5.13}$$

Equations (5.9) and (5.13) are equivalent, yet (5.13) has a property not obvious in (5.9), that details of the interaction potential appear only in B. *The mechanics of the collision operator is the same for all homogeneous weakly coupled gases.* What distinguishes one from another, B, amounts to setting the time scale. This feature is a limiting behavior associated with weak interactions. Charges influence each other, weakly, at large separations because of the long-range nature of electrical forces. The Landau kinetic equation[3] describes the evolution of $f(\boldsymbol{v}, t)$ by means of weak long-range interactions—"collisions." When applied to the Coulomb potential, however, it runs into difficulties. The Fourier transform of r^{-1}, $\widetilde{\Phi}_{\boldsymbol{k}} \sim k^{-2}$ (see Exercise 5.4), and thus from Eq. (5.11),

$$B \sim \int_0^\infty k^3 \times \frac{1}{k^4}\mathrm{d}k = \ln(\infty) - \ln(0). \tag{5.14}$$

We have a logarithmic divergence at both limits.[4] The divergence at $k \to \infty$ is due to our neglect of a hard core and that at $k \to 0$ is due to the long range of the Coulomb force, which we see is "too long." If the potential weren't purely Coulomb and featured cutoff parameters, we'd have a finite theory, $B \sim \int_{k_{\text{min}}}^{k_{\text{max}}} \mathrm{d}k/k = \ln(k_{\text{max}}/k_{\text{min}})$, and plasmas *do* involve such parameters. A system of mobile charges *screens* electric fields, implying the potential can't be that of a bare point charge. The potential is approximately Coulomb up to the screening length, the *Debye length* (appropriate for the weak coupling regime), $r_D \equiv \sqrt{\epsilon_0 kT/(ne^2)}$, with n the number density and e the magnitude of the electronic charge (SI units). Thus, $k_{\text{min}} \sim r_D^{-1}$. That potential energies be small compared with kinetic implies a distance below which the weak coupling assumption breaks down, the *Landau length*, $r_{\text{min}} \equiv e^2/(4\pi\epsilon_0 kT)$, and thus $k_{\text{max}} \sim r_{\text{min}}^{-1}$. Consistency requires the system be sufficiently dilute so that $r_D > r_{\text{min}}$. These considerations beg the question of the potential energy environment in plasmas.

5.2 CONNECTION WITH THE FOKKER-PLANCK EQUATION

In Chapter 3 we derived the Fokker-Planck equation for the probability distribution of a colloidal particle subject to small molecular impacts (Brownian motion), and, as we've now seen, the Landau equation emerges as the limiting case of the Boltzmann equation for weak interactions. Are the two related, the Landau and Fokker-Planck equations? The Landau equation can indeed be derived (in a nontrivial calculation) starting from the Fokker-Planck equation for an inverse-square force law[166]. Here we show simply that the Landau equation has the Fokker-Planck form.

[2]By the principal of covariance, a tensor equation true in one frame is true in all frames; see [6, p84].

[3]The Landau equation, like the Boltzmann equation, is a nonlinear integro-differential equation.

[4]Divergences occur frequently in modern theories of physics; the renormalization method in quantum field theory was devised to handle them. Dirac wrote that the laws of nature "... do not govern the world as it appears in our mental picture in any very direct way, but instead they control a substratum of which we cannot form a mental picture without introducing irrelevancies"[165, pvii].

From Eq. (5.2), we can write the Landau equation

$$\frac{\partial}{\partial t} f(\boldsymbol{v}, t) = \sum_i \frac{\partial}{\partial v_i}\left(- A_i + \sum_j \frac{\partial}{\partial v_j} B_{ij} \right) f(\boldsymbol{v}, t), \tag{5.15}$$

where

$$\begin{aligned} A_i &\equiv \int \mathrm{d}\boldsymbol{v}_1 \left(\sum_j G_{ij} \frac{\partial}{\partial v_{1j}} f(\boldsymbol{v}_1, t) + f(\boldsymbol{v}_1, t) \sum_j \frac{\partial}{\partial v_j} G_{ij} \right) \\ B_{ij} &\equiv \int \mathrm{d}\boldsymbol{v}_1 G_{ij} f(\boldsymbol{v}_1, t). \end{aligned} \tag{5.16}$$

Equation (5.15) has the form of the Fokker-Planck equation [see Eq. (3.37)] generalized to three dimensions with a friction vector A_i and a *diffusion tensor* B_{ij}. Note that these quantities change as $f(\boldsymbol{v}, t)$ changes, whereas they're constant in the Fokker-Planck equation. That there is a such a connection is satisfying theoretically—the Fokker-Planck equation emerges from the theory of random processes and the Boltzmann equation is an "almost" dynamical theory.[5]

5.3 INHOMOGENEOUS PLASMAS: THE VLASOV EQUATION

A plasma is an ionized gas, which we take to be electrically neutral.[6] Charges repel like charges and attract those of the opposite sign, creating clouds of opposite charge surrounding the particles of the plasma. We can use statistical mechanics to model the effect. Assume a fixed charge $Q > 0$ held in place at the origin, surrounded by a plasma. Denote the local density of negative charges $n(\boldsymbol{r})$ and that of positive charges $p(\boldsymbol{r})$. Poisson's equation connects charge density with the electrostatic potential $\varphi(\boldsymbol{r})$,

$$\nabla^2 \varphi(\boldsymbol{r}) = -\frac{e}{\epsilon_0} \left[p(\boldsymbol{r}) - n(\boldsymbol{r}) \right] - \frac{Q}{\epsilon_0} \delta(\boldsymbol{r}). \qquad (e > 0) \tag{5.17}$$

Statistical physics provides a relation between the average charge density and the potential, the barometric formula, Eq. (4.128), $p(\boldsymbol{r}) = n \exp(-e\varphi(\boldsymbol{r})/kT)$ and $n(\boldsymbol{r}) = n \exp(e\varphi(\boldsymbol{r})/kT)$, with n the density where $\varphi = 0$, which by neutrality is the same for positive and negative charges. We therefore have the *Poisson-Boltzmann equation* (which we note is nonlinear),[7]

$$\nabla^2 \varphi(\boldsymbol{r}) = \frac{2e}{\epsilon_0} n \sinh(e\varphi(\boldsymbol{r})/kT) - \frac{Q}{\epsilon_0} \delta(\boldsymbol{r}). \tag{5.18}$$

Invoking the weak coupling assumption ($\sinh x \approx x$ for $x \ll 1$), Eq. (5.18) can be linearized,

$$\nabla^2 \varphi(\boldsymbol{r}) = \frac{2}{r_D^2} \varphi(\boldsymbol{r}) - \frac{Q}{\epsilon_0} \delta(\boldsymbol{r}), \tag{5.19}$$

where[8] $r_D = \sqrt{\epsilon_0 kT/(e^2 n)}$. The spherically symmetric solution of Eq. (5.19) is (see Exercise 5.7)

$$\varphi(r) = \frac{Q}{4\pi\epsilon_0 r} \exp(-\sqrt{2} r / r_D). \tag{5.20}$$

Clearly $\varphi \to 0$ for $r \gg r_D$. The average charge surrounding Q is $-Q$ for $r \gg r_D$; Exercise 5.8.

---

[5] The molecular chaos assumption cannot be derived from the laws of mechanics.

[6] In the simplest situation, electrons and positive ions are created together in the process of forming a plasma and we have charge neutrality. Nonneutral plasmas are possible, and comprise an active field of research; see Davidson[167].

[7] Equation (5.18) is equivalent to the Debye-Hückel equation of ionic-solution theory (derived 1923); see Friedman[168].

[8] The Debye length r_D is traditionally defined without the factor of two. There are other, slightly different definitions of r_D depending on the geometry of the system.

Debye shielding, exemplified by Eq. (5.20), provides a natural interpretation of a cutoff parameter rendering the Landau theory finite. The charge Q could be from a foreign object in the plasma or it could be an electron or ion of the plasma. We can think of each electron as surrounded by positive charges, with each ion surrounded by electrons. Charges separated by more than a few Debye lengths don't influence each other because of screening. To take into account this characteristic feature of plasmas (screening), Landau's theory must be extended to inhomogeneous systems. This is done by adding a flow term to the equation and by including spatial dependence in the collision integral. Both distribution functions must be evaluated at the same location [see Eq. (5.9)], but to preserve the symmetry between particles 1 and 2 we make the replacement $f(\boldsymbol{v}_1,t)f(\boldsymbol{v}_2,t) \to \int \mathrm{d}\boldsymbol{r}_2\delta(\boldsymbol{r}_1-\boldsymbol{r}_2)f(\boldsymbol{r}_1,\boldsymbol{v}_1,t)f(\boldsymbol{r}_2,\boldsymbol{v}_2,t)$. Thus, as a first step, we have

$$\left(\frac{\partial}{\partial t}+\boldsymbol{v}_1\cdot\boldsymbol{\nabla}_{\boldsymbol{r}_1}\right)f(\boldsymbol{r}_1,\boldsymbol{v}_1,t) = \frac{8\pi^4}{m^2}\int \mathrm{d}\boldsymbol{v}_2\int \mathrm{d}\boldsymbol{r}_2\int \mathrm{d}\boldsymbol{k}\delta(\boldsymbol{r}_1-\boldsymbol{r}_2)\widetilde{\Phi}^2_{\boldsymbol{k}} \\ \times \boldsymbol{k}\cdot\boldsymbol{\nabla}_{12}\delta(\boldsymbol{k}\cdot\boldsymbol{g})\boldsymbol{k}\cdot\boldsymbol{\nabla}_{12}f(\boldsymbol{r}_1,\boldsymbol{v}_1,t)f(\boldsymbol{r}_2,\boldsymbol{v}_2,t). \tag{5.21}$$

Equation (5.21) is incomplete, however. The Boltzmann equation was derived assuming free paths between collisions, but charged particles are not free; they're subject to electromagnetic forces. Accelerated motion can be included in the Boltzmann equation for accelerations caused by known external forces; see Eq. (P4.3). For plasmas we must develop a model of *internally generated*, local forces. Noting that in homogeneous plasmas the net internal field is zero, we define a local potential energy associated with deviations from uniformity,[9]

$$U(\boldsymbol{r},t) \equiv \int \Phi(\boldsymbol{r}-\boldsymbol{r}')\mathrm{d}\boldsymbol{r}'\int \left[f(\boldsymbol{r}',\boldsymbol{v}',t)-\phi(\boldsymbol{v}',t)\right]\mathrm{d}\boldsymbol{v}', \tag{5.22}$$

where $\phi(\boldsymbol{v},t)$ is the solution of the Landau equation.[10] The negative gradient of $U(\boldsymbol{r},t)$ specifies a local force which acts as if it were produced by an external field. Adding this force to Eq. (5.21),

$$\left(\frac{\partial}{\partial t}+\boldsymbol{v}_1\cdot\boldsymbol{\nabla}_{\boldsymbol{r}_1}\right)f(\boldsymbol{r}_1,\boldsymbol{v}_1,t)-\frac{1}{m}\boldsymbol{\nabla}_{\boldsymbol{r}_1}U(\boldsymbol{r}_1,t)\cdot\boldsymbol{\nabla}_{\boldsymbol{v}_1}f(\boldsymbol{r}_1,\boldsymbol{v}_1,t) \\ = \frac{8\pi^4}{m^2}\int \mathrm{d}\boldsymbol{v}_2\int \mathrm{d}\boldsymbol{r}_2\int \mathrm{d}\boldsymbol{k}\delta(\boldsymbol{r}_1-\boldsymbol{r}_2)\widetilde{\Phi}^2_{\boldsymbol{k}}\boldsymbol{k}\cdot\boldsymbol{\nabla}_{12}\delta(\boldsymbol{k}\cdot\boldsymbol{g})\boldsymbol{k}\cdot\boldsymbol{\nabla}_{12}f(\boldsymbol{r}_1,\boldsymbol{v}_1,t)f(\boldsymbol{r}_2,\boldsymbol{v}_2,t). \tag{5.23}$$

Equation (5.23) coupled with Eq. (5.22) is the *Vlasov-Landau equation.*

The interaction potential $\Phi(\boldsymbol{r})$ occurs linearly in $U(\boldsymbol{r})$ and quadratically in the collision integral through $\widetilde{\Phi}^2_{\boldsymbol{k}}$. In the weak coupling approximation, it's meaningful to neglect collisions in comparison with the effects of long-range interactions:

$$\left(\frac{\partial}{\partial t}+\boldsymbol{v}\cdot\boldsymbol{\nabla}_{\boldsymbol{r}}\right)f(\boldsymbol{r},\boldsymbol{v},t)-\frac{1}{m}\boldsymbol{\nabla}_{\boldsymbol{r}}U(\boldsymbol{r},t)\cdot\boldsymbol{\nabla}_{\boldsymbol{v}}f(\boldsymbol{r},\boldsymbol{v},t)=0. \tag{5.24}$$

Equation (5.24), coupled with Eq. (5.22), is the *Vlasov equation* or the *collisionless Boltzmann equation.*[11] Because $\phi(\boldsymbol{v},t)$ in Eq. (5.22) (solution of the Landau equation) is independent of position, it adds a term to $U(\boldsymbol{r})$ proportional to $\widetilde{\Phi}_{\boldsymbol{k}=0}$, the gradient of which is zero. Equation (5.24) can

---

[9] "Deviations from uniformity" has been a guiding principle at several points, in deriving the off-diagonal part of the pressure tensor and in the Chapman-Enskog theory.

[10] The distribution functions $f(\boldsymbol{r},\boldsymbol{v},t)$ and $\phi(\boldsymbol{v},t)$ have the same dimensions; see Exercise 5.3.

[11] The Vlasov equation was developed in 1938[169] (see also Vlasov[170]) as a way of modeling plasma oscillations (a collective effect known since the 1920s through the work of Tonks and Langmuir[171]), a phenomenon the Boltzmann equation based on short-range interactions cannot describe. The collisionless Boltzmann equation has been applied to problems of galactic dynamics starting with the 1915 work of Jeans[172] and is a fundamental equation in that subject.

then be written with $\phi(\boldsymbol{v},t)$ omitted,[12]

$$\left(\frac{\partial}{\partial t}+\boldsymbol{v}_1\cdot\boldsymbol{\nabla}_{\boldsymbol{r}_1}\right)f(\boldsymbol{r}_1,\boldsymbol{v}_1,t)-\frac{1}{m}\left[\boldsymbol{\nabla}_{\boldsymbol{r}_1}\int \mathrm{d}\boldsymbol{r}_2\mathrm{d}\boldsymbol{v}_2\Phi(\boldsymbol{r}_1-\boldsymbol{r}_2)f(\boldsymbol{r}_2,\boldsymbol{v}_2,t)\right]\cdot\boldsymbol{\nabla}_{\boldsymbol{v}_1}f(\boldsymbol{r}_1,\boldsymbol{v}_1,t)=0. \tag{5.25}$$

For definiteness, consider a charge-neutral system where electrons are able to move through a uniform background of positive charge.[13] Assume particles interact microscopically through the Coulomb potential; the effective potential energy environment is obtained from Eq. (5.22). Thus,

$$\begin{aligned} U(\boldsymbol{r},t) &= -\frac{e^2}{4\pi\epsilon_0}\int \mathrm{d}\boldsymbol{r}'\frac{1}{|\boldsymbol{r}-\boldsymbol{r}'|}\underbrace{\int \mathrm{d}\boldsymbol{v}'\left[f(\boldsymbol{r}',\boldsymbol{v}',t)-\phi(\boldsymbol{v}',t)\right]}_{n(\boldsymbol{r}',t)-n} \\ &\equiv \frac{e}{4\pi\epsilon_0}\int \mathrm{d}\boldsymbol{r}'\frac{1}{|\boldsymbol{r}-\boldsymbol{r}'|}\rho(\boldsymbol{r}',t), \end{aligned} \tag{5.26}$$

with $n(\boldsymbol{r},t)$ the density of electrons at $(\boldsymbol{r},t)$, ne the charge density of the positive background, and $\rho(\boldsymbol{r},t)=e[n-n(\boldsymbol{r},t)]$ the net local charge density. It's straightforward to show that

$$\nabla^2 U(\boldsymbol{r},t) = -\frac{e}{\epsilon_0}\rho(\boldsymbol{r},t). \tag{5.27}$$

The Vlasov equation consists of equations (5.24) and the Poisson equation (5.27). The Vlasov-Landau equations are *self-consistent*: Charged particles respond to the same potential they generate.

5.4 LANDAU DAMPING

The Vlasov equation is time-reversal invariant [as readily seen from Eq. (5.25)] and therefore cannot describe the approach to equilibrium. Said differently, it does not have a definite direction of time. It's surprising, then, that it predicts a type of damping, *Landau damping*, derived in 1946[173] and observed in 1964[174], in which a longitudinal space charge wave is damped in *collisionless* plasmas. Plasma physics is outside the scope of this book,[14] yet an example of damping without dissipation (without entropy production) is within its purview. We outline the derivation.

Consider the simplest problem of electrons in a uniform positive-charge background of density n in an unmagnetized, neutral, collisionless plasma. The local force on electrons, $-e\boldsymbol{E}(\boldsymbol{r},t)$, is supplied by an electric field, with $e>0$. The Vlasov "equation" consists of two coupled equations:

$$\left(\frac{\partial}{\partial t}+\boldsymbol{v}\cdot\boldsymbol{\nabla}\right)f(\boldsymbol{r},\boldsymbol{v},t)-\frac{e}{m}\boldsymbol{E}(\boldsymbol{r},t)\cdot\boldsymbol{\nabla}_{\boldsymbol{v}}f(\boldsymbol{r},\boldsymbol{v},t)=0, \tag{5.28}$$

for the electron distribution function $f(\boldsymbol{r},\boldsymbol{v},t)$, where $\boldsymbol{E}(\boldsymbol{r},t)=-\boldsymbol{\nabla}\varphi(\boldsymbol{r},t)$ with

$$\nabla^2\varphi(\boldsymbol{r},t) = -\frac{e}{\epsilon_0}\left(n-\int f(\boldsymbol{r},\boldsymbol{v},t)\mathrm{d}\boldsymbol{v}\right) \tag{5.29}$$

governing the scalar potential.

[12] We can drop $\phi(\boldsymbol{v},t)$ from the dynamical equation, but not from Eq. (5.22).

[13] Because ions are so much more massive than electrons, an approximation is to ignore the motion and spatial distribution of positive ions, a model referred to as the "jellium" approximation. In semiconductors, ions—dopants attached to lattice sites—are indeed immobile, with the exception of electromigration, the gradual motion of ions under high-current conditions.

[14] Landau damping is not restricted to plasma physics. It has been applied to the collisionless motions of stars in galaxies moving through the gravitational field produced by other stars; see Lynden-Bell[175]. See also Binney and Tremaine[176].

5.4.1 The linearized Vlasov equation

The Vlasov equation is nonlinear. Even though the collision term has been dropped, the self-consistent potential renders the problem nonlinear. To make progress, linearize. Write the distribution function

$$f(\boldsymbol{r},\boldsymbol{v},t) = f_0(\boldsymbol{v}) + f_1(\boldsymbol{r},\boldsymbol{v},t), \tag{5.30}$$

where $f_1(\boldsymbol{r},\boldsymbol{v},t)$ describes small, local, time-dependent deviations from a spatially uniform time-independent distribution $f_0(\boldsymbol{v})$, with $\int f_0(\boldsymbol{v})\mathrm{d}\boldsymbol{v} = n$, and $|f_1| \ll f_0$. With Eq. (5.30), Eqs. (5.28) and (5.29) become, to first order in small quantities,

$$\left(\frac{\partial}{\partial t} + \boldsymbol{v}\cdot\boldsymbol{\nabla}\right) f_1(\boldsymbol{r},\boldsymbol{v},t) + \frac{e}{m}\boldsymbol{\nabla}\varphi(\boldsymbol{r},t)\cdot\boldsymbol{\nabla}_{\boldsymbol{v}} f_0(\boldsymbol{v}) = 0 \tag{5.31}$$

$$\nabla^2\varphi(\boldsymbol{r},t) = \frac{e}{\epsilon_0}\int f_1(\boldsymbol{r},\boldsymbol{v},t)\mathrm{d}\boldsymbol{v}. \tag{5.32}$$

We can, in the usual way, seek plane-wave solutions for small-amplitude waves, $f_1(\boldsymbol{r},\boldsymbol{v},t) = f_1(\boldsymbol{v})\exp[\mathrm{i}(\boldsymbol{k}\cdot\boldsymbol{r} - \omega t)]$ and $\varphi(\boldsymbol{r},t) = \varphi_0\exp[\mathrm{i}(\boldsymbol{k}\cdot\boldsymbol{r} - \omega t)]$, propagating in the $\boldsymbol{k}$-direction with phase speed $v_\phi \equiv \omega/k$. With these substitutions, Eqs. (5.31) and (5.32) become

$$(-\mathrm{i}\omega + \mathrm{i}\boldsymbol{k}\cdot\boldsymbol{v})f_1(\boldsymbol{v}) + \mathrm{i}\frac{e}{m}\varphi_0\boldsymbol{k}\cdot\boldsymbol{\nabla}_{\boldsymbol{v}} f_0(\boldsymbol{v}) = 0$$

$$-k^2\varphi_0 = \frac{e}{\epsilon_0}\int f_1(\boldsymbol{v})\mathrm{d}\boldsymbol{v}.$$

These equations can be combined,

$$\left[k^2 + \frac{e^2}{m\epsilon_0}\int\frac{\boldsymbol{k}\cdot\boldsymbol{\nabla}_{\boldsymbol{v}} f_0(\boldsymbol{v})}{\omega - \boldsymbol{k}\cdot\boldsymbol{v}}\mathrm{d}\boldsymbol{v}\right]\varphi_0 = 0.$$

The terms in square brackets must vanish, leaving us with a dispersion relation, the connection $\omega = \omega(\boldsymbol{k})$ between frequency and wavelength,

$$1 + \frac{e^2}{m\epsilon_0 k^2}\int\frac{\boldsymbol{k}\cdot\boldsymbol{\nabla}_{\boldsymbol{v}} f_0(\boldsymbol{v})}{\omega - \boldsymbol{k}\cdot\boldsymbol{v}}\mathrm{d}\boldsymbol{v} = 0. \tag{5.33}$$

We must find ω so that Eq. (5.33) is satisfied for given $\boldsymbol{k}$ and $f_0(\boldsymbol{v})$. There's an obvious problem, however—the integral isn't defined! There is some $\boldsymbol{v}$ for which $\omega = \boldsymbol{k}\cdot\boldsymbol{v}$ (and we integrate over all $\boldsymbol{v}$). Note that $\boldsymbol{v}$ is a particle velocity and $(\omega, \boldsymbol{k})$ are wave parameters. The singularity occurs when electron velocities match the phase velocity of the wave.

To circumvent this problem, Fourier analyze with respect to spatial variables only, leaving the time dependence intact, $f_1(\boldsymbol{r},\boldsymbol{v},t) = f_1(\boldsymbol{v},t)\exp(\mathrm{i}\boldsymbol{k}\cdot\boldsymbol{r})$, $\boldsymbol{E}(\boldsymbol{r},t) = \boldsymbol{E}(t)\exp(\mathrm{i}\boldsymbol{k}\cdot\boldsymbol{r})$, and $\varphi(\boldsymbol{r},t) = \varphi(t)\exp(\mathrm{i}\boldsymbol{k}\cdot\boldsymbol{r})$, in which case Eqs. (5.31) and (5.32) reduce to[15]

$$\left(\frac{\partial}{\partial t} + \mathrm{i}\boldsymbol{k}\cdot\boldsymbol{v}\right) f_1(\boldsymbol{v},t) = \mathrm{i}\frac{e^2}{m\epsilon_0 k^2}\boldsymbol{k}\cdot\boldsymbol{\nabla}_{\boldsymbol{v}} f_0(\boldsymbol{v})\int f_1(\boldsymbol{v}',t)\mathrm{d}\boldsymbol{v}' \tag{5.34}$$

$$\varphi(t) = -\frac{e}{\epsilon_0 k^2}\int f_1(\boldsymbol{v},t)\mathrm{d}\boldsymbol{v}. \tag{5.35}$$

The electric field amplitude $\boldsymbol{E}(t) = -\mathrm{i}\boldsymbol{k}\varphi(t)$. Note that $\boldsymbol{E}$ is parallel to $\boldsymbol{k}$, which specifies the longitudinal direction.[16] At this point, decompose $\boldsymbol{v}$ into vectors parallel and perpendicular to $\boldsymbol{k}$,

[15] All fields should be labeled with an index $\boldsymbol{k}$; ignored here to ease the notation.

[16] These waves are termed "electrostatic" because $\boldsymbol{E} = -\boldsymbol{\nabla}\varphi$ implies $\boldsymbol{k}\times\boldsymbol{E} = 0$. Longitudinal $\boldsymbol{E}$-fields should be contrasted with the transverse fields of electromagnetic waves in unbounded free space. One also finds longitudinal fields in waveguides.

$\boldsymbol{v} = \boldsymbol{u} + \boldsymbol{v}_\perp$ so that $\boldsymbol{k}\cdot\boldsymbol{v} = ku$. In this notation, the perturbed electron density

$$\delta n(t) \equiv \int f_1(\boldsymbol{v},t)\mathrm{d}\boldsymbol{v} = \int f_1(u,\boldsymbol{v}_\perp,t)\mathrm{d}u\mathrm{d}\boldsymbol{v}_\perp \equiv \int_{-\infty}^{\infty} g(u,t)\mathrm{d}u. \tag{5.36}$$

The quantity $g(u,t)$ is the perturbed distribution as a function of longitudinal speed (transverse velocities have been integrated out). Integrate Eq. (5.34) over $\boldsymbol{v}_\perp$, for which we find

$$\left(\frac{\partial}{\partial t} + \mathrm{i}ku\right) g(u,t) = -\mathrm{i}k\eta(u)\int_{-\infty}^{\infty} g(u',t)\mathrm{d}u', \tag{5.37}$$

with

$$\eta(u) \equiv -\frac{e^2}{m\epsilon_0 k^2}\frac{\mathrm{d}}{\mathrm{d}u}\int f_0(u,\boldsymbol{v}_\perp)\mathrm{d}\boldsymbol{v}_\perp \equiv -\frac{e^2}{m\epsilon_0 k^2}\frac{\mathrm{d}}{\mathrm{d}u}g_0(u) \equiv -\frac{e^2}{m\epsilon_0 k^2}g_0'(u). \tag{5.38}$$

Note that $\int_{-\infty}^{\infty} g_0(u)\mathrm{d}u = n$. Equation (5.37) is the linearized Vlasov equation, a linear integro-differential equation.

5.4.2 Initial value problem

Landau analyzed Eq. (5.37) as an initial value problem for which the perturbation f_1 is prescribed at time $t = 0$ and found for later times. This strategy underlies *not* Fourier transforming with respect to time; instead one works with Laplace transforms which naturally incorporate initial conditions.[17] The Laplace transform $\mathcal{L}$, an integral transform converting a function of a real variable $g(t)$ to a function $G(z)$ of a complex frequency variable z, is defined

$$G(z) = \mathcal{L}[g](z) \equiv \int_0^{\infty} \mathrm{e}^{-zt}g(t)\mathrm{d}t. \tag{5.39}$$

Note that $\mathcal{L}$ restricts t to positive values, thereby introducing a direction of time to the Vlasov equation.[18] The Laplace transform of derivatives is related to the transform of the function:

$$\mathcal{L}[g'](z) = \int_0^{\infty} \mathrm{e}^{-zt}\frac{\mathrm{d}g}{\mathrm{d}t}\mathrm{d}t = -g(t=0) + zG(z).$$

The inverse transform is defined

$$g(t) = \mathcal{L}^{-1}[G](t) \equiv \frac{1}{2\pi\mathrm{i}}\lim_{T\to\infty}\int_{\gamma-\mathrm{i}T}^{\gamma+\mathrm{i}T} \mathrm{e}^{zt}G(z)\mathrm{d}z = \sum_j (\text{residues})_j\,, \tag{5.40}$$

where the integration path is the line $\mathrm{Re}\,(z) = \gamma$ in the complex z-plane, with γ greater than the real part of all singularities of $G(z)$, closed with an infinite semicircle (the *Bromwich contour*[13, p244]), where the last equality follows from the Cauchy residue theorem.

Take the Laplace transform of Eq. (5.37); we find

$$-g(u,t=0) + (z+\mathrm{i}ku)\,G(u,z) = -\mathrm{i}k\eta(u)\int_{-\infty}^{\infty}\mathrm{d}u'G(u',z),$$

where $G(u,z) \equiv \mathcal{L}[g(u,t)]$. Thus, $G(u,z)$ satisfies the integral equation

$$G(u,z) = \frac{1}{z+\mathrm{i}ku}\left[g(u,t=0) - \mathrm{i}k\eta(u)\int_{-\infty}^{\infty}\mathrm{d}u'G(u',z)\right]. \tag{5.41}$$

[17] Fourier transforming with respect to time assumes the future is the same as the past, which should apply to a time-reversal invariant theory; that's what makes Landau damping subtle, a transfer of energy between wave and particle.

[18] Case[177] presents a more comprehensive analysis based on the double-sided Laplace transform.

Define

$$\Psi(z) \equiv \int_{-\infty}^{\infty} G(u,z)\mathrm{d}u. \tag{5.42}$$

Integrate Eq. (5.41) over all u, in which case we find

$$\Psi(z) = \left[1 + \mathrm{i}k\int_{-\infty}^{\infty}\mathrm{d}u\frac{\eta(u)}{z+\mathrm{i}ku}\right]^{-1}\int_{-\infty}^{\infty}\mathrm{d}u\frac{g(u,t=0)}{z+\mathrm{i}ku} \equiv \frac{1}{\varepsilon(k,z)}\int_{-\infty}^{\infty}\mathrm{d}u\frac{g(u,t=0)}{z+\mathrm{i}ku}, \tag{5.43}$$

where $\varepsilon(k,z)$ is the *plasma dielectric function*. The quantity Ψ is the Laplace transform of the perturbed electron density, $\mathcal{L}[\delta n] = \Psi(z)$, and thus from Eq. (5.35), $\mathcal{L}[\varphi] = -(e/(\epsilon_0 k^2))\Psi(z)$, so that

$$\varphi(t) = -\frac{e}{\epsilon_0 k^2}\mathcal{L}^{-1}[\Psi] = -\frac{e}{\epsilon_0 k^2}\frac{1}{2\pi\mathrm{i}}\int_C \mathrm{e}^{zt}\left[\frac{\displaystyle\int\frac{g(u,t=0)}{z+\mathrm{i}ku}\mathrm{d}u}{\displaystyle 1+\mathrm{i}k\int\mathrm{d}u\frac{\eta(u)}{z+\mathrm{i}ku}}\right]\mathrm{d}z, \tag{5.44}$$

where C denotes the Bromwich contour. By substituting $\Psi(z)$ back into Eq. (5.41), we have

$$G(u,z) = \frac{1}{z+\mathrm{i}ku}\left[g(u,t=0) - \mathrm{i}k\eta(u)\Psi(z)\right]. \tag{5.45}$$

In principle we've solved the problem: We have the Laplace transforms of $g(u,t)$ and $\varphi(t)$. What remains is to find the inverse transforms. That turns out to be a fairly involved task, which for our purposes we can forgo. At the end of the day, we'll have an expression for φ in the form

$$\varphi(t) = -\frac{e}{\epsilon_0 k^2}\sum_j \mathrm{e}^{z_j t}\,\mathrm{Res}\,(\Psi)_{z_j}, \tag{5.46}$$

where z_j denotes locations of the poles of $\Psi(z)$ with Res the residue. The potential has oscillations determined by the imaginary parts of z_j, with damping or growth controlled by the real parts.[19] Referring to Eq. (5.44), poles of Ψ are either poles of the numerator $[\int \mathrm{d}u g(u,t=0)/(z+\mathrm{i}ku)]$ or zeros of the denominator, $\varepsilon(k,z)$. The numerator has no poles for physical choices of initial perturbation $g(u,t=0)$. That leaves the task of finding the zeros of $\varepsilon(k,z)$, which are independent of initial conditions. Allowed dynamical modes are determined from solutions of the dispersion relation,

$$\varepsilon(k,z) = 1 - \mathrm{i}\frac{e^2}{m\epsilon_0 k}\int_{-\infty}^{\infty}\frac{(\mathrm{d}/\mathrm{d}u)g_0(u)}{z_j+\mathrm{i}ku}\mathrm{d}u = 0. \tag{5.47}$$

Equation (5.47) reduces to Eq. (5.33) under the substitution $z_j \to -\mathrm{i}\omega$.

Our goal is not to find the most general solution, but rather its long-time form governed by the smallest damping rate Γ (so that decaying fields vary as $\mathrm{e}^{-\Gamma t}$). That rate is given by the expression[20]

$$\Gamma = -\frac{\pi\omega_p^3}{2nk^2}g_0'(u)\bigg|_{u=v_\phi}, \tag{5.48}$$

where $\omega_p \equiv \sqrt{ne^2/(m\epsilon_0)}$ is the *plasma frequency*. The criterion for damped oscillations is that the slope of $g_0(u)$ evaluated at the wave speed is negative, $g_0'(v_\phi) < 0$, implying there are more slower particles than faster particles for speeds near the phase speed. Charged particles interact strongly with longitudinal electrostatic waves when the speeds of particles moving in the direction

[19] The reader may note that the Laplace transform in Eq. (5.39) requires $\mathrm{Re}\,(z) > 0$ and yet damped modes require $\mathrm{Re}\,(z_j) < 0$. Analytic continuation is used liberally in the derivation of Eq. (5.46). An exposition is given in Wu[178].

[20] See for example Schmidt[179, p210]. This expression holds for wavelengths larger than the Debye length, $kr_D \ll 1$. One can't have well-developed oscillations in plasmas with wavelengths smaller than the Debye length; see Exercise 5.13.

of wave propagation match the phase speed of the wave. Particles moving somewhat faster than the wave lose energy to the field (and contribute to a growing field mode); particles moving slower gain energy from the field and contribute to Landau damping.[21] Which mechanism prevails depends on the relative number of slower and faster particles for $u \approx v_\phi$.

We therefore have an example of *irreversibility not associated with dissipation.* For solutions of the Vlasov equation, Boltzmann's H-function remains unchanged, $\mathrm{d}H/\mathrm{d}t = 0$; see Exercise 5.11. Moreover, stationary solutions of the Vlasov equation are not unique; see Exercise 5.12. Processes such as Landau damping are an example of *phase mixing*[176, p269]. To avoid misunderstandings, in this book the term *irreversible* refers solely to dissipative processes.

SUMMARY

Kinetic theory can be applied to plasmas as well as ordinary gases. Plasma kinetic theory is an active field of research that we do not attempt to review (see Swanson[181]). We established its basic principles—the Landau-Vlasov equations—based on material already developed in this book.

- Starting from the Boltzmann equation, Landau derived the kinetic equation for uniform, weakly coupled gases in which the intermolecular potential is small compared to kT. Plasmas (ionized gases) approximate weakly coupled gases. Charged particles influence each other at large separations because of the long-range nature of Coulomb forces, yet such interactions are weak. For dilute plasmas, close encounters are rare; weak, long-range "collisions" are described by the Landau equation. The Landau equation can be written in the form of the Fokker-Planck equation; both equations describe systems governed by weak collisions. For all the approximations made in its derivation, the Landau equation can be derived from the BBGKY hierarchy in the limit of a weak potential; see Bogoliubov[94, p97]. Solutions of the Landau equation satisfy the Boltzmann H-theorem $\mathrm{d}H/\mathrm{d}t \leq 0$, with the Maxwell-Boltzmann distribution the stationary solution.

- The Landau equation encounters difficulties when applied to the bare Coulomb potential, not so much because of its neglect of close collisions, but because of the "too-long" nature of the Coulomb force. These issues can be overcome by parameterizing the potential with long and short-distance cutoffs. Mobile charges of opposite kinds screen electric fields, implying the intermolecular potential can be approximately Coulomb only up to the screening length, in the simplest case the Debye length.

- To take into account screening—a characteristic feature of plasmas—Landau's equation must be extended to spatially nonuniform systems. The Vlasov-Landau "equation" [two coupled equations, (5.22) and (5.23)] takes into account inhomogeneities through the development of a self-consistent potential field, wherein charges respond to the same potential they generate. A simpler version, valid within the weak-coupling approximation, known as the Vlasov or the collisionless Boltzmann equation, ignores collisions and models the system as particles streaming in the self-consistent potential, Eqs. (5.22) and (5.24).

- Landau damping is a damping mechanism of longitudinal space charge waves in collisionless plasmas not involving dissipation. It occurs as a result of energy exchange between the wave and charged particles moving with the speed of the wave. The Vlasov equation is time-reversal invariant and cannot describe the approach to equilibrium; for its solutions $\mathrm{d}H/\mathrm{d}t = 0$, yet it leads to an experimentally confirmed prediction of damping. Besides its intrinsic interest, we've presented Landau damping as an example of irreversibility without dissipation, an exception to the theme of this book that entropy production is associated with inhomogeneities.

[21] Chen[180, Section 7.5] has a nice discussion of the physics of Landau damping.

EXERCISES

5.1 Show that solutions of the Landau equation satisfy the H-theorem. Define the H-function, not as in Eq. (4.109) with a spatial integration, but only with an integration over velocity, $H(t) \equiv \int \mathrm{d}\boldsymbol{v} f(\boldsymbol{v}, t) \ln[f(\boldsymbol{v}, t)]$ (Landau's equation applies to homogeneous systems).

a. Using Landau's equation in the form of Eq. (5.2), show that

$$\frac{\mathrm{d}}{\mathrm{d}t} H(t) = \int \mathrm{d}\boldsymbol{v}_1 \left[\ln f(\boldsymbol{v}_1, t) + 1\right] \frac{\partial f(\boldsymbol{v}_1, t)}{\partial t} \tag{P5.1}$$
$$= \int \mathrm{d}\boldsymbol{v}_1 \left[\ln f(\boldsymbol{v}_1, t) + 1\right] \boldsymbol{\nabla}_{\boldsymbol{v}_1} \cdot \int \mathrm{d}\boldsymbol{v}_2 \mathbf{G} \cdot \boldsymbol{\nabla}_{12} f(\boldsymbol{v}_1, t) f(\boldsymbol{v}_2, t).$$

b. Integrate by parts on $\boldsymbol{v}_1$ to show that

$$\frac{\mathrm{d}}{\mathrm{d}t} H(t) = -\int\int \mathrm{d}\boldsymbol{v}_1 \mathrm{d}\boldsymbol{v}_2 \boldsymbol{\nabla}_{\boldsymbol{v}_1} \ln f(\boldsymbol{v}_1, t) \cdot \mathbf{G} \cdot \boldsymbol{\nabla}_{12} f(\boldsymbol{v}_1, t) f(\boldsymbol{v}_2, t). \tag{P5.2}$$

c. By interchanging the dummy variables $\boldsymbol{v}_1, \boldsymbol{v}_2$ in the integrand of Eq. (P5.2), show that

$$\frac{\mathrm{d}}{\mathrm{d}t} H(t) = -\frac{1}{2} \int\int \mathrm{d}\boldsymbol{v}_1 \mathrm{d}\boldsymbol{v}_2 \left[\boldsymbol{\nabla}_{\boldsymbol{v}_1} \ln f(\boldsymbol{v}_1, t) - \boldsymbol{\nabla}_{\boldsymbol{v}_2} \ln f(\boldsymbol{v}_2, t)\right] \cdot \mathbf{G} \cdot \boldsymbol{\nabla}_{12} f(\boldsymbol{v}_1, t) f(\boldsymbol{v}_2, t).$$

d. Define $\boldsymbol{F}(\boldsymbol{v}_1, \boldsymbol{v}_2) \equiv \boldsymbol{\nabla}_{\boldsymbol{v}_1} \ln[f(\boldsymbol{v}_1, t)] - \boldsymbol{\nabla}_{\boldsymbol{v}_2} \ln[f(\boldsymbol{v}_2, t)]$. Show that

$$\frac{\mathrm{d}}{\mathrm{d}t} H(t) = -\frac{1}{2} \int\int \mathrm{d}\boldsymbol{v}_1 \mathrm{d}\boldsymbol{v}_2 f(\boldsymbol{v}_1, t) f(\boldsymbol{v}_2, t) \boldsymbol{F}(\boldsymbol{v}_1, \boldsymbol{v}_2) \cdot \mathbf{G} \cdot \boldsymbol{F}(\boldsymbol{v}_1, \boldsymbol{v}_2).$$

The H-theorem follows if we can show the integrand is never negative.

e. Show that

$$\boldsymbol{F} \cdot \mathbf{G} \cdot \boldsymbol{F} = \left(\frac{B}{m^2 g^3}\right) \left[g^2 F^2 - (\boldsymbol{g} \cdot \boldsymbol{F})^2\right].$$

Use the Schwartz inequality to conclude that $\boldsymbol{F} \cdot \mathbf{G} \cdot \boldsymbol{F} \geq 0$, proving $\mathrm{d}H/\mathrm{d}t \leq 0$. The H-theorem had *better* be satisfied by solutions of the Landau equation which is a special case of the Boltzmann equation. Still, it's a reassuring result; there are a limited number of kinetic equations for which the H-theorem can be explicitly proven.

f. The time-invariant state $\mathrm{d}H/\mathrm{d}t = 0$ is achieved if and only if $\boldsymbol{F} \cdot \mathbf{G} \cdot \boldsymbol{F} = 0$ implying $\mathbf{G} \cdot \boldsymbol{F} = 0$ ($\mathbf{G}$ is symmetric). As one can show from Eq. (5.12), $\mathbf{G} \cdot \boldsymbol{g} = 0$. Stationary solutions are therefore such that $\mathbf{G} \cdot (\boldsymbol{F} + \beta \boldsymbol{g}) = 0$ where β is a constant.

i. Show, because $\mathbf{G}$ is arbitrary, stationary solutions are such that

$$\boldsymbol{\nabla}_{\boldsymbol{v}_1} \ln f(\boldsymbol{v}_1) - \boldsymbol{\nabla}_{\boldsymbol{v}_2} \ln f(\boldsymbol{v}_2) = -\beta(\boldsymbol{v}_1 - \boldsymbol{v}_2).$$

Note that a constant vector $\boldsymbol{u}$ could be added and subtracted on the right side.

ii. Show that stationary solutions are Maxwellians $f(\boldsymbol{v}) = K \exp\left[-(\beta/2)(\boldsymbol{v} - \boldsymbol{u})^2\right]$.

5.2 Derive Eq. (5.10).

5.3 Show that Eq. (5.13) is dimensionally correct if f has dimension $(\text{length})^{-3} \times (\text{velocity})^{-3}$.

5.4 What is the Fourier transform $\widetilde{\Phi}_{\boldsymbol{k}}$ [defined near Eq. (5.5)] of $\Phi(\boldsymbol{r}) = r^{-1}$? A straightforward attack on the problem might not prove satisfactory. One approach is to evaluate the integral (do this),

$$\int \frac{\exp(\mathrm{i}\boldsymbol{k}\cdot\boldsymbol{r})}{k^2+a^2}\mathrm{d}\boldsymbol{k} = \frac{2\pi^2}{r}\mathrm{e}^{-ar}. \qquad (a>0)$$

In the limit $a \to 0$, $\widetilde{\Phi}_{\boldsymbol{k}} = (2\pi^2 k^2)^{-1}$. (Another way to arrive at this result is to take the Fourier transform of $\nabla^2(1/r) = -4\pi\delta(\boldsymbol{r})$.)

5.5 Show from Eq. (5.11) that B diverges for power-law potentials $\Phi(r) \propto r^{-n}$ for $n > 1$.

5.6 Show that the Landau length $r_{\text{min}} = 1.67 \times 10^{-5}/T$ m, with T in Kelvin.

5.7 To solve Eq. (5.19) it's best to keep $r > 0$ and determine the solution through boundary conditions. At small distances we expect $\varphi(r)$ to approach the form of the potential of a point charge, $\varphi(r) \overset{r\to 0}{\sim} Q/(4\pi\epsilon_0 r)$. Let $\varphi(r) = Q/(4\pi\epsilon_0 r)f(r)$. Develop a spherically symmetric solution for f subject to the boundary conditions $\lim_{r\to 0} f(r) = 1$ and $\lim_{r\to\infty} f(r) = 0$.

5.8 Calculate the charge q_{ind} induced by the test charge Q at $r = 0$ in a plasma.

a. From Eq. (5.19) define (for $r > 0$) the induced charge density $\rho_{\text{ind}}(r) \equiv -(2\epsilon_0/r_D^2)\varphi(r)$. Show that

$$\rho_{\text{ind}}(r) = -\frac{Q}{2\pi r_D^2 r}\mathrm{e}^{-\sqrt{2}r/r_D}.$$

b. Calculate the induced charge in a ball of radius R, $q_{\text{ind}}(R) \equiv \int_0^R 4\pi r^2 \rho_{\text{ind}}(r)\mathrm{d}r$. Show that

$$q_{\text{ind}}(R) = -Q\left[1 - (1+\sqrt{2}R/r_D)\mathrm{e}^{-\sqrt{2}R/r_D}\right].$$

Clearly $q_{\text{ind}} \to -Q$ for $R \gg r_D$ and $q_{\text{ind}} \to 0$ as $R \to 0$.

5.9 Show that the Vlasov equation in the form of Eq. (5.25) follows from the first equation of the BBGKY hierarchy using the product approximation $F_2(\boldsymbol{r}_1, \boldsymbol{v}_1, \boldsymbol{r}_2, \boldsymbol{v}_2, t) = F_1(\boldsymbol{r}_1, \boldsymbol{v}_1, t)F_1(\boldsymbol{r}_2, \boldsymbol{v}_2, t)$; see Eq. (4.29). The Vlasov equation is therefore *formally* the Liouville equation of a single particle in an external field, yet that field is self-consistently produced by the particles themselves. The Vlasov equation is nonlinear because of self-consistency.

5.10 Derive Eq. (5.27).

5.11 a. Show, using the H-function for inhomogeneous systems, Eq. (4.109), that $\mathrm{d}H/\mathrm{d}t = 0$ for solutions of the Vlasov equation (5.24). See Exercise 4.28.

b. Show directly that the Vlasov equation is time-reversal invariant. It does not have a definite direction of time.

5.12 Consider stationary solutions of the Vlasov equation, solutions of

$$\boldsymbol{v}\cdot\boldsymbol{\nabla}f(\boldsymbol{r},\boldsymbol{v}) - \frac{1}{m}\boldsymbol{\nabla}U\cdot\boldsymbol{\nabla}_{\boldsymbol{v}}f(\boldsymbol{r},\boldsymbol{v}) = 0.$$

a. Show this partial differential equation is solved by any function of the form

$$f(\boldsymbol{r},\boldsymbol{v}) = W\left(\frac{1}{2}mv^2 + U(\boldsymbol{r})\right).$$

Stationary solutions are therefore not unique; any function W will do. This is reminiscent of the wave equation, the solutions of which in one dimension are superpositions of any functions $f(x-vt)$ and $g(x+vt)$. The parameter v is specified in the wave equation; the only parameter entering the Vlasov equation is the mass (and not temperature).

b. Try separating variables. Let $f(\boldsymbol{r},\boldsymbol{v}) = \psi(\boldsymbol{r})g(mv^2/2)$. Show you're led to the intermediate result

$$\boldsymbol{v}\cdot[\boldsymbol{\nabla}\ln\psi - (g'/g)\boldsymbol{\nabla}U] = 0.$$

This simplifies if $g'/g = \theta$, a constant, implying $g(x) = \mathrm{e}^{\theta x}$ and $\psi(\boldsymbol{r}) = \mathrm{e}^{\theta U(\boldsymbol{r})}$. One could set $\theta = -1/(kT)$, but nothing forces that conclusion.

5.13 Suppose the time-invariant, spatially uniform velocity distribution $f_0(\boldsymbol{v})$ is the Maxwell-Boltzmann distribution (see Eq. (4.121)), $f_0(\boldsymbol{v}) = n(m\beta/(2\pi))^{3/2}\exp(-m\beta v^2/2)$.

a. Find $g_0(u)$, the time-invariant, uniform distribution as a function of the longitudinal speed u, with $v^2 = u^2 + v_\perp^2$. A: $g_0(u) = n\sqrt{m\beta/(2\pi)}\exp(-m\beta u^2/2)$.

b. Verify that $\int_{-\infty}^{\infty} g_0(u)\mathrm{d}u = n$.

c. Show that the Landau damping rate in this case is given by

$$\Gamma = \sqrt{\frac{\pi}{8}}\frac{\omega_p^4}{k^3}(m\beta)^{3/2}\exp[-m\beta\omega_p^2/(2k^2)].$$

Set $u = \omega_p/k$, where ω_p is the plasma frequency.

d. Show that this expression can be rewritten in terms of the Debye length r_D with

$$\Gamma = \sqrt{\pi}\frac{\omega_p}{(2kr_D)^{3/2}}\exp\left[-\frac{1}{2(kr_D)^2}\right]$$

For the damping rate to be small compared to the oscillation frequency $\Gamma/\omega_p \ll 1$ (and thus to have well-developed oscillations), we require $kr_D \ll 1$.

CHAPTER 6

Dissipation, fluctuations, and correlations

DISSIPATION[1] is associated with fluctuations. We learned in Chapter 1 that entropy sources involve fluxes and thermodynamic forces, quantities that are linked through fluctuations; see Eqs. (1.42) and (1.43). Dissipation, associated with entropy creation, is therefore associated with fluctuations. We see this in Nyquist's theorem Eq. (2.45), $S(\omega) = 4RkT$, where the spectral density of voltage fluctuations is related to the circuit resistance R (associated with dissipating electrical energy). We find the same using the Langevin equation [see Eq. (3.7)], that R is related to temporal correlations of voltage fluctuations. In the Einstein relation Eq. (3.20), $D = kT/\alpha$, the diffusion coefficient associated with Brownian motion (fluctuating position of the particle) is inversely related to the friction parameter (associated with dissipating a colloidal particle's kinetic energy by the drag force). In this chapter, we study the connection between dissipation, fluctuations, and correlations.

6.1 FLUCTUATIONS AND THEIR CORRELATIONS

Although we've used correlation functions in this book, we need to distinguish different kinds of correlations and thus we start with a review. Consider a dynamical function, $Y(\mathbf{\Gamma})$ (see Section 4.1). Its time dependence is implicit through $\mathbf{\Gamma}$, $Y(t) \equiv Y(\mathbf{\Gamma}(t))$, where $\mathbf{\Gamma}(t)$ is the image of $\mathbf{\Gamma}(t=0)$ under the canonical motion.[2] To indicate time dependence explicitly, we write $Y(\mathbf{\Gamma}(t)) = \mathrm{e}^{\mathrm{i}\Lambda t}Y(\mathbf{\Gamma}(0))$ with Λ the Liouville operator; see Exercise 4.2. In classical mechanics, time evolution is deterministic: to every $Y(t=0)$ there is a unique $Y(t \neq 0)$.[3] We have no control over microscopic initial conditions, however, and thus, in statistical mechanics, $\mathbf{\Gamma}$ is treated as a random variable associated with a large collection of macroscopically identical systems (the ensemble). In thermal equilibrium—a macroscopic steady state—the ensemble average of Y, $\langle Y \rangle$, is independent of time,

$$\begin{aligned}\langle Y \rangle &= \int \mathrm{d}\mathbf{\Gamma} P_{\text{eq}}(\mathbf{\Gamma}) Y(\mathbf{\Gamma}) = \int \mathrm{d}\mathbf{\Gamma}(0) P_{\text{eq}}(\mathbf{\Gamma}) \mathrm{e}^{\mathrm{i}\Lambda t} Y(\mathbf{\Gamma}(0)) \\ &= \int \mathrm{d}\mathbf{\Gamma}(0) \left[\mathrm{e}^{-\mathrm{i}\Lambda t} P_{\text{eq}}(\mathbf{\Gamma})\right] Y(\mathbf{\Gamma}(0)) = \int \mathrm{d}\mathbf{\Gamma}(0) P_{\text{eq}}(\mathbf{\Gamma}) Y(\mathbf{\Gamma}(0)),\end{aligned} \tag{6.1}$$

[1]Dissipated energy is energy that, as a result of entropy creation, is diverted into a form not available for work[2, p59]. Energy isn't lost, it's channeled into microscopic degrees of freedom, the form of energy known as heat[2, p10].

[2]Every point of Γ-space evolves in time to a unique point under the action of Hamilton's equations of motion, a one-to-one mapping of Γ-space onto itself, the natural motion of phase space. See [5, p33].

[3]Quantum time evolution is deterministic in that $|\psi\rangle_{t=0}$ uniquely implies $|\psi\rangle_{t\neq 0}$. It's the act of measurement that isn't deterministic and which adds another layer of uncertainty treated in quantum statistical mechanics; see Appendix A.

DOI: 10.1201/9781003512295-6

where $\mathrm{d}\boldsymbol{\Gamma} = \mathrm{d}\boldsymbol{\Gamma}(0)$ (unity Jacobian of canonical transformations[5, p332]), the adjoint of the time evolution operator $\mathrm{e}^{\mathrm{i}\Lambda t}$ is $\mathrm{e}^{-\mathrm{i}\Lambda t}$ (see Exercise 6.1), and we've used stationarity $\Lambda P_{\mathrm{eq}}(\boldsymbol{\Gamma}) = 0$ (see [5, p47]). A measurement of Y on a randomly selected member of the ensemble will in general return a value different from $\langle Y \rangle$. Let $y(t)$ denote the fluctuation,

$$y(t) \equiv Y(t) - \langle Y \rangle. \tag{6.2}$$

Equation (6.2) underscores the stochastic nature of fluctuations: $Y(t)$ doesn't have a definite value—it's a random variable in the ensemble framework of statistical mechanics.

The only accessible information about stochastic variables is in their moments. The first moment vanishes identically, $\langle y(t) \rangle = 0$. For higher moments, start with the average of the square of $y(t)$:

$$C^0_{yy} \equiv \langle y(t)y(t) \rangle^0, \tag{6.3}$$

where $\langle\rangle^0$ denotes equilibrium average. Moments of this type are related to measurable quantities; see [5, Chapter 4]. For example, the isothermal magnetic susceptibility $\chi_T = \langle (\Delta M)^2 \rangle^0_T/(kT)$, the constant volume heat capacity $C_V = \langle (\Delta U)^2 \rangle^0_V/(kT^2)$, and the isothermal compressibility $\beta_T = (V/N^2 kT)\langle (\Delta N)^2 \rangle^0_T$ measure fluctuations in magnetization, internal energy, and particle number.[4] We can generalize Eq. (6.3) to include cross correlations,

$$C^0_{yz} \equiv \langle y(t)z(t) \rangle^0. \tag{6.4}$$

We found using fluctuation theory that $\langle \Delta U \Delta V \rangle^0 = k\beta_T V T^2 (\alpha/\beta_T - P/T)$ in a system where U and V can fluctuate [see Eq. (2.11); α is the thermal expansivity]. Time does not play a role in correlations of this type. Equilibrium fluctuations are stationary and so are their correlations; see Exercise 6.1. Another generalization is to the correlation of *local* fluctuations at different points $\boldsymbol{x}$ of physical space, a *correlation function*,[5] which in equilibrium is a function only of the spatial separation $\boldsymbol{r}$,

$$C^0_{yz}(\boldsymbol{r}) \equiv \langle y(\boldsymbol{x},t)z(\boldsymbol{x}+\boldsymbol{r},t) \rangle^0. \tag{6.5}$$

Such correlations are independent of time (equilibrium) and position $\boldsymbol{x}$ (homogeneity); we can therefore set $(\boldsymbol{x}, t) = 0$ in Eq. (6.5). An example is the correlation of spatially separated spins on a lattice, $\langle \sigma_0 \sigma_{\boldsymbol{r}} \rangle^0$, where the location of the origin is immaterial.[6] Correlations of this type contribute to the elastic-scattering structure factor associated with local fluctuations $S(\boldsymbol{q}) \equiv \int \mathrm{d}\boldsymbol{r}\mathrm{e}^{\mathrm{i}\boldsymbol{q}\cdot\boldsymbol{r}} \langle y(0)y(\boldsymbol{r}) \rangle^0$, the scattering efficiency in the direction of the wavevector transfer $\boldsymbol{q} \equiv \boldsymbol{k}_f - \boldsymbol{k}_i$; see [5, Section 6.6]. Equation (6.5) reduces to Eq. (6.4) as $\boldsymbol{r} \to 0$; $C^0_{yz}(\boldsymbol{r} = 0) = C^0_{yz}$.

We now consider the nonequilibrium counterpart,

$$C_{yz}(t) \equiv \langle y(t)z(t) \rangle, \tag{6.6}$$

where $\langle\rangle$ denotes nonequilibrium average. This quantity is time dependent because the probability distribution is not stationary. We can see that as follows (using the result of Exercise 6.1):

$$\begin{aligned} C_{yz}(t) &= \int \mathrm{d}\boldsymbol{\Gamma} y(\boldsymbol{\Gamma})z(\boldsymbol{\Gamma})P(\boldsymbol{\Gamma},t) = \int \mathrm{d}\boldsymbol{\Gamma} y(\boldsymbol{\Gamma})z(\boldsymbol{\Gamma})\mathrm{e}^{-\mathrm{i}\Lambda t}P(\boldsymbol{\Gamma},0) \\ &= \int \mathrm{d}\boldsymbol{\Gamma}\mathrm{e}^{\mathrm{i}\Lambda t}\left[y(\boldsymbol{\Gamma})z(\boldsymbol{\Gamma})\right]P(\boldsymbol{\Gamma},0) = \int \mathrm{d}\boldsymbol{\Gamma} y(\boldsymbol{\Gamma}(t))z(\boldsymbol{\Gamma}(t))P(\boldsymbol{\Gamma},0). \end{aligned} \tag{6.7}$$

Thus there are two ways of expressing $C_{yz}(t)$: one in which averages are taken with respect to a time-dependent probability distribution $P(\boldsymbol{\Gamma}, t)$ or one in which averages are taken with respect

[4] The quantities χ_T and C_V are obtained in the canonical ensemble, with β_T in the grand canonical ensemble.

[5] The set of correlations C^0_{yz} are elements of a *correlation matrix* of fluctuations in quantities (y, z).

[6] The origin can't be "too close" to boundaries where surface effects introduce inhomogeneities, a problem avoided by considering the lattice infinite in extent.

to the probability of initial conditions, $P(\mathbf{\Gamma}, 0)$. These equivalent expressions mirror the passage in quantum mechanics between the Schrödinger and Heisenberg representations (see Appendix E), where, in the former, the state of the system[7] is time dependent but observables are not, and in the latter, the state of the system is fixed but observables carry time dependence. The dynamics of fluctuations presented here differs from that introduced in Chapter 3. The stochastic nature of $\mathbf{\Gamma}(0)$, inherent in the ensemble framework of statistical mechanics, *induces* the stochastic character of $\mathbf{\Gamma}(t)$ through a deterministic dynamics; in the Fokker-Planck equation, dynamics is prescribed in terms of transition probabilities. The two ways of expressing $C_{yz}(t)$ in Eq. (6.7) show we can view its time dependence as due either to the time evolution of $P(\mathbf{\Gamma}, t)$ or to the propagation of $y(t)z(t)$ in time for fixed $P(\mathbf{\Gamma}, 0)$.

Equation (6.6) can be generalized to local fluctuations,

$$C_{yz}(t; \boldsymbol{r}, \boldsymbol{x}) \equiv \langle y(t; \boldsymbol{x}) z(t; \boldsymbol{x} + \boldsymbol{r}) \rangle. \tag{6.8}$$

Equation (6.8) specifies the *equal-time correlation function*. It depends on $\boldsymbol{x}$ (and t and $\boldsymbol{r}$); nonequilibrium states are in general nonhomogeneous. Equation (6.8) suggests the next generalization, the correlation between fluctuation y at point $\boldsymbol{x}$ at time t and fluctuation z at point $\boldsymbol{x} + \boldsymbol{r}$ at time $t + \tau$,

$$C_{yz}(\tau, t; \boldsymbol{r}, \boldsymbol{x}) \equiv \langle y(t; \boldsymbol{x}) z(t + \tau; \boldsymbol{x} + \boldsymbol{r}) \rangle. \tag{6.9}$$

Equation (6.9) defines the *two-time correlation function*, a function of two time variables and two spatial variables. For $\tau = 0$, Eq. (6.9) reduces to Eq. (6.8), $C_{yz}(\tau = 0, t; \boldsymbol{r}, \boldsymbol{x}) = C_{yz}(t; \boldsymbol{r}, \boldsymbol{x})$. Two-time correlations cannot be expressed in the Schrödinger representation. Consider, starting from the Heisenberg picture,

$$\begin{aligned}
C_{yz}(\tau, t; \boldsymbol{r}, \boldsymbol{x}) &= \int \mathrm{d}\mathbf{\Gamma} \left[\mathrm{e}^{\mathrm{i}\Lambda t} y(\mathbf{\Gamma}; \boldsymbol{x})\right] \left[\mathrm{e}^{\mathrm{i}\Lambda(t+\tau)} z(\mathbf{\Gamma}; \boldsymbol{x} + \boldsymbol{r})\right] P(\mathbf{\Gamma}, 0) \\
&= \int \mathrm{d}\mathbf{\Gamma} \mathrm{e}^{\mathrm{i}\Lambda t} \left\{y(\mathbf{\Gamma}; \boldsymbol{x}) \mathrm{e}^{\mathrm{i}\Lambda\tau} z(\mathbf{\Gamma}; \boldsymbol{x} + \boldsymbol{r})\right\} P(\mathbf{\Gamma}, 0) \\
&= \int \mathrm{d}\mathbf{\Gamma} y(\mathbf{\Gamma}; \boldsymbol{x}) \left[\mathrm{e}^{\mathrm{i}\Lambda\tau} z(\mathbf{\Gamma}; \boldsymbol{x} + \boldsymbol{r})\right] \mathrm{e}^{-\mathrm{i}\Lambda t} P(\mathbf{\Gamma}, 0) \\
&= \int \mathrm{d}\mathbf{\Gamma} y(\mathbf{\Gamma}; \boldsymbol{x}) \left[\mathrm{e}^{\mathrm{i}\Lambda\tau} z(\mathbf{\Gamma}; \boldsymbol{x} + \boldsymbol{r})\right] P(\mathbf{\Gamma}, t).
\end{aligned} \tag{6.10}$$

Two-time correlations can either be expressed in Heisenberg form or in a generalized Schrödinger representation involving the average of a τ-dependent dynamical quantity with respect to a t-dependent distribution. The Heisenberg representation is used preferentially.

If the two-time correlation function were evaluated with respect to an equilibrium distribution, the resulting correlation function would depend on τ and $\boldsymbol{r}$ but not on t and $\boldsymbol{x}$ (from stationarity and homogeneity of the equilibrium state). Using Eq. (6.10),

$$\begin{aligned}
C^0_{yz}(\tau; \boldsymbol{r}) \equiv \langle y(t; \boldsymbol{x}) z(t + \tau; \boldsymbol{x} + \boldsymbol{r}) \rangle^0 &= \int \mathrm{d}\mathbf{\Gamma} y(\mathbf{\Gamma}; 0) \left[\mathrm{e}^{\mathrm{i}\Lambda\tau} z(\mathbf{\Gamma}; \boldsymbol{r})\right] P_{\mathrm{eq}}(\mathbf{\Gamma}) \\
&= \langle y(0; 0) z(\tau; \boldsymbol{r}) \rangle^0.
\end{aligned} \tag{6.11}$$

Thus we can set $(t, \boldsymbol{x}) = 0$ in Eq. (6.9) in the case of an equilibrium distribution. Equation (6.11) generalizes $C^0_{yz}(\boldsymbol{r})$ in Eq. (6.5) to the correlation of local fluctuation $y(0; 0)$ with another $z(\tau; \boldsymbol{r})$ displaced in space and time. Correlation functions of this type are called *time-dependent correlation functions* or simply *time correlation functions* and play a significant role in our subsequent analysis. For $z = y$ we have the *autocorrelation function*; see Eq. (2.24).

---

[7] The state of a statistical mechanical system is specified by a probability distribution; see Section 4.1.

The spectral content of correlation functions is related to experimentally accessible information, that of scattering structure factors. Scattering, of electromagnetic radiation or neutrons, occurs from inhomogeneities in system properties, notably density fluctuations (whether mass, charge, current, magnetization, etc.). In inelastic scattering, the *dynamic structure factor* describes the scattering intensity associated with the wavevector $\boldsymbol{q}$ and frequency ω of scattered radiation.[8] Van Hove[183] showed that the dynamic structure factor is proportional to the spectral density of time correlation functions, $\widetilde{C}^0_{yz}(\omega;\boldsymbol{q}) = \int_{-\infty}^{\infty} \mathrm{d}\tau \int \mathrm{d}\boldsymbol{r} \mathrm{e}^{\mathrm{i}\omega\tau + \mathrm{i}\boldsymbol{q}\cdot\boldsymbol{r}} C^0_{yz}(\tau;\boldsymbol{r}) = \int_{-\infty}^{\infty} \mathrm{d}\tau \int \mathrm{d}\boldsymbol{r} \mathrm{e}^{\mathrm{i}\omega\tau + \mathrm{i}\boldsymbol{q}\cdot\boldsymbol{r}} \langle y(0;0) z(\tau;\boldsymbol{r})\rangle^0$. Thus we learn about the dynamical properties of correlations through the measured dynamic structure factor. Inelastic scattering has been extensively studied on systems near their critical points. We note that, just as scaling theories enable the study of equilibrium critical phenomena (see [5, Chapter 8]), so too is the study of dynamic critical phenomena facilitated by scaling hypotheses, a topic we lack the space to cover.[9]

So far we've made minimal use of quantum mechanics. Statistical mechanics is not just the application of classical or quantum mechanics to large systems. Rather it seeks to connect macroscopic observations with the microscopics of systems governed by Hamiltonian dynamics, whether in classical or quantum formulation; the nature of the micro-dynamics is often immaterial. There are (obviously) systems, however, where quantum treatments are necessary. Fortunately, correlation functions are easily extended to meet the requirements of quantum mechanics. Ensemble averages are found from the density operator $\hat{\rho}$, $\langle A\rangle = \mathrm{Tr}\,\hat{\rho}\hat{A}$, where $\hat{A}$ is the operator associated with A; see Eq. (A.39). A distinctly quantum issue is the product of noncommuting operators and here we follow the Weyl correspondence principle (see Section A.4) in defining two-point correlation functions as averages of symmetrized products,

$$\begin{aligned} C_{yz}(\tau,t;\boldsymbol{r},\boldsymbol{x}) &= \frac{1}{2}\mathrm{Tr}\,\hat{\rho}(t)\left[\hat{y}(t;\boldsymbol{x})\hat{z}(t+\tau;\boldsymbol{x}+\boldsymbol{r}) + \hat{z}(t+\tau;\boldsymbol{x}+\boldsymbol{r})\hat{y}(t;\boldsymbol{x})\right] \\ &= \frac{1}{2}\langle \hat{y}(t;\boldsymbol{x})\hat{z}(t+\tau;\boldsymbol{x}+\boldsymbol{r}) + \hat{z}(t+\tau;\boldsymbol{x}+\boldsymbol{r})\hat{y}(t;\boldsymbol{x})\rangle. \end{aligned} \tag{6.12}$$

6.2 LINEAR RESPONSE THEORY

Consider, for a system that in the far past ($t \to -\infty$) was in thermal equilibrium and governed by Hamiltonian[10] H^0, the problem of turning on a time-dependent external interaction $F(t)$ ($\lim_{t\to-\infty} F(t) = 0$) that couples to system quantity A with energy $H'(t) \equiv -AF(t)$ so that the Hamiltonian $H = H^0 + H'(t)$. In classical mechanics, A is a Γ-space function; in quantum mechanics A is represented by a Hermitian operator. Table 6.1 lists examples. We seek the response of

Table 6.1 Examples of time-dependent interactions

External interaction $F(t)$	System quantity A
Magnetic field $\boldsymbol{B}$	Magnetization $\boldsymbol{M}$
Electric field $\boldsymbol{E}$	Electric polarization $\boldsymbol{P}$
Sound waves	Mass density

the system to first order in the interaction, the *linear response*, the accuracy of which requires the interaction to be small.[11] The response is measured through the change $\Delta B(t)$ that occurs in observable quantity B due to the interaction $F(t)$. We develop a general formalism—*linear response theory*—applicable to any interaction in the form $H'(t) \equiv -AF(t)$, classical or quantum.

[8] We omit the derivation of the dynamic structure factor. See Stanley[182, Chapter 13], for example.
[9] See Hohenberg and Halperin[184].
[10] The Hamiltonian H^0 includes internal interactions between particles, such as in Eq. (4.18).
[11] Thus we're developing a time-dependent perturbation theory in nonequilibrium statistical mechanics.

6.2.1 Response functions: General derivations, classical and quantum

6.2.1.1 Classical

The Γ-space distribution function of classical statistical mechanics $\rho(p,q)$ satisfies Liouville's equation, (4.5), $(\partial/\partial t)\rho = -\mathrm{i}\Lambda\rho \equiv L\rho = [H,\rho]_{\mathrm{P}}$ ($[*,*]_{\mathrm{P}}$ denotes Poisson bracket). Before turning on the interaction, the system was in equilibrium with $\rho = \rho^0$ such that $[H^0,\rho^0]_{\mathrm{P}} = 0$. Write the perturbed distribution $\rho(t) = \rho^0 + \Delta\rho(t)$ with $\Delta\rho(-\infty) = 0$. Then, from Liouville's equation,

$$\frac{\partial}{\partial t}\left(\rho^0 + \Delta\rho\right) = \left[H^0 + H', \rho^0 + \Delta\rho\right]_{\mathrm{P}} = \left[H^0,\rho^0\right]_{\mathrm{P}} + \left[H^0,\Delta\rho\right]_{\mathrm{P}} + \left[H',\rho^0\right]_{\mathrm{P}} + \left[H',\Delta\rho\right]_{\mathrm{P}},$$

we have, ignoring $[H',\Delta\rho]_{\mathrm{P}}$, the equation of motion at lowest order,[12]

$$\frac{\partial}{\partial t}\Delta\rho(t) = \left[H'(t),\rho^0\right]_{\mathrm{P}} + \left[H^0,\Delta\rho(t)\right]_{\mathrm{P}} \equiv \left[H'(t),\rho^0\right]_{\mathrm{P}} + L^0\Delta\rho(t), \tag{6.13}$$

where L^0 is the Liouville operator associated with H^0. The solution of Eq. (6.13) (an inhomogeneous Liouville equation, see Exercise 6.3) is, with $\Delta\rho(-\infty) = 0$,

$$\Delta\rho(t) = \int_{-\infty}^{t} \mathrm{d}\tau \mathrm{e}^{L^0(t-\tau)} \left[H'(\tau),\rho^0\right]_{\mathrm{P}}. \tag{6.14}$$

The modification $\Delta\rho(t)$ at time t can be interpreted as a linear superposition of "sources" $\left[H'(\tau),\rho^0\right]_{\mathrm{P}}$ (disturbances to the quiescent system) propagated in time by $\mathrm{e}^{L^0(t-\tau)}$, $-\infty < \tau < t$.

The nonequilibrium ensemble average of a dynamical variable is found from

$$\langle B\rangle_t = \int \mathrm{d}\mathbf{\Gamma} B(\mathbf{\Gamma})\rho(t) = \int \mathrm{d}\mathbf{\Gamma} B\left(\mathbf{\Gamma}\right)(\rho^0 + \Delta\rho(t)) \equiv \langle B\rangle^0 + \underbrace{\int \mathrm{d}\mathbf{\Gamma} B(\mathbf{\Gamma})\Delta\rho(t)}_{\Delta B(t)}. \tag{6.15}$$

The change $\Delta B(t) \equiv \langle B\rangle_t - \langle B\rangle^0$ at lowest order, the linear response, is, using Eq. (6.14),

$$\begin{aligned}\Delta B(t) &= \int \mathrm{d}\mathbf{\Gamma} B(\mathbf{\Gamma},-\infty) \int_{-\infty}^{t} \mathrm{d}\tau \mathrm{e}^{L^0(t-\tau)} \left[H'(\tau),\rho^0\right]_{\mathrm{P}} \\ &= \int_{-\infty}^{t} \mathrm{d}\tau \int \mathrm{d}\mathbf{\Gamma} \left\{\mathrm{e}^{-L^0(t-\tau)} B(\mathbf{\Gamma},-\infty)\right\} \left[H'(\tau),\rho^0\right]_{\mathrm{P}} \\ &= \int_{-\infty}^{t} \mathrm{d}\tau \int \mathrm{d}\mathbf{\Gamma} B(\mathbf{\Gamma},t-\tau) \left[H'(\tau),\rho^0\right]_{\mathrm{P}},\end{aligned} \tag{6.16}$$

where e^{-Lt} is the adjoint of e^{Lt}. Now substitute $H'(t) = -A(p,q)F(t)$ in Eq. (6.16),

$$\Delta B(t) = -\int_{-\infty}^{t} \mathrm{d}\tau F(\tau) \int \mathrm{d}\mathbf{\Gamma} B(\mathbf{\Gamma},t-\tau) \left[A,\rho^0\right]_{\mathrm{P}}. \tag{6.17}$$

Define the *response function*,

$$\Phi_{BA}(t) \equiv \int \mathrm{d}\mathbf{\Gamma} \left[\rho^0, A\right]_{\mathrm{P}} B(\mathbf{\Gamma},t) = \int \mathrm{d}\mathbf{\Gamma} \rho^0 \left[A, B(t)\right]_{\mathrm{P}}, \tag{6.18}$$

so that

$$\Delta B(t) = \int_{-\infty}^{t} \mathrm{d}\tau \Phi_{BA}(t-\tau)F(\tau) = \int_0^{\infty} \Phi_{BA}(\theta)F(t-\theta)\mathrm{d}\theta, \tag{6.19}$$

[12] Kubo[185] shows, formally, how to include higher-order approximations, ignored in the linear theory.

where the second equality in Eq. (6.18) is established in Exercise 6.4. The response function Φ_{BA} is also known as the *after-effect function*. By the principle of causality, the response (the *effect*) cannot precede the interaction that elicits it (the *cause*), implying $\Phi_{BA}(t-\tau)=0$ for $\tau>t$ or $\Phi_{BA}(\theta<0)=0$. Any function $f(t)$ having the property $f(t<0)=0$ is known as a *causal function* (see Appendix F).

6.2.1.2 Quantum

In the quantum version of this problem, the distribution function $\rho(p,q)$ is replaced with the density operator $\hat{\rho}$, and H^0 and $H'(t)$ are replaced with operators $\hat{H}^0$ and $\hat{H}'(t)=-\hat{A}F(t)$. The state at $t=-\infty$ (thermal equilibrium) is characterized by density operator $\hat{\rho}^0$ such that $[\hat{H}^0,\hat{\rho}^0]=0$ (unadorned square brackets $[*,*]$ denote commutator). With $\hat{\rho}(t)=\hat{\rho}^0+\Delta\hat{\rho}(t)$ in Eq. (A.41), we have at lowest order[13]

$$\mathrm{i}\hbar\frac{\partial}{\partial t}\Delta\hat{\rho}(t)=\left[\hat{H}'(t),\hat{\rho}^0\right]+\left[\hat{H}^0,\Delta\hat{\rho}(t)\right]. \tag{6.20}$$

The solution to Eq. (6.20) is (as can be verified),

$$\Delta\hat{\rho}(t)=\frac{1}{\mathrm{i}\hbar}\int_{-\infty}^{t}\mathrm{d}\tau U_0(t,\tau)\left[\hat{H}'(\tau),\hat{\rho}^0\right]U_0^{\dagger}(t,\tau) \tag{6.21}$$

where [see Eq. (E.6)]

$$\mathrm{i}\hbar\frac{\partial}{\partial t}U_0(t,t')\equiv\hat{H}^0U_0(t,t'). \tag{6.22}$$

For time-independent $\hat{H}^0$ (as is the case here[14]), $U_0(t,t')=\exp[-\mathrm{i}\hat{H}^0(t-t')/\hbar]$.

The response $\Delta B(t)$ is, using Eq. (6.21),

$$\begin{aligned}\Delta B(t)=\operatorname{Tr}\Delta\hat{\rho}(t)\hat{B}&=\frac{1}{\mathrm{i}\hbar}\int_{-\infty}^{t}\mathrm{d}\tau\operatorname{Tr}\left\{U_0(t,\tau)\left[\hat{H}'(\tau),\hat{\rho}^0\right]U_0^{\dagger}(t,\tau)\hat{B}\right\}\\&=\frac{1}{\mathrm{i}\hbar}\int_{-\infty}^{t}\mathrm{d}\tau F(\tau)\operatorname{Tr}\left\{U_0(t,\tau)\left[\hat{\rho}^0,\hat{A}\right]U_0^{\dagger}(t,\tau)\hat{B}\right\}\\&=\frac{1}{\mathrm{i}\hbar}\int_{-\infty}^{t}\mathrm{d}\tau F(\tau)\operatorname{Tr}\left\{\hat{\rho}^0\left[\hat{A},U_0^{\dagger}(t,\tau)\hat{B}U_0(t,\tau)\right]\right\}=\frac{1}{\mathrm{i}\hbar}\int_{-\infty}^{t}\mathrm{d}\tau F(\tau)\langle\left[\hat{A},\hat{B}_I(t-\tau)\right]\rangle^0\\&=\frac{1}{\mathrm{i}\hbar}\int_{-\infty}^{t}\mathrm{d}\tau F(\tau)\langle\left[\hat{A}_I(\tau),\hat{B}_I(t)\right]\rangle^0\equiv\int_{-\infty}^{t}\mathrm{d}\tau\Phi_{BA}(t-\tau)F(\tau)\\&=\int_0^{\infty}\Phi_{BA}(\theta)F(t-\theta)\mathrm{d}\theta,\end{aligned} \tag{6.23}$$

where we've used $\hat{H}'(\tau)=-\hat{A}F(\tau)$, the cyclic invariance of the trace, and the interaction representation[15] (see Exercise 6.6). The response function is given by the expressions:

$$\mathrm{i}\hbar\Phi_{BA}(t-\tau)=\begin{cases}\langle\left[\hat{A},\hat{B}_I(t-\tau)\right]\rangle^0=\langle\left[\hat{A}_I(\tau),\hat{B}_I(t)\right]\rangle^0 & -\infty<\tau<t\\ 0. & \tau\geq t\end{cases} \tag{6.24}$$

The response function is causal, $\Phi_{BA}(\theta\leq 0)=0$. Equation (6.24) is known as *Kubo's formula* (for the response function[16]). Despite the appearance of i in Eq. (6.24) (unit imaginary number),

[13] Note the similarity between Eqs. (6.20) and (6.13). These equations exemplify the general correspondence between commutators and Poisson brackets, that $[\hat{A},\hat{B}]/(\mathrm{i}\hbar)\overset{\hbar\to 0}{=}[A,B]_{\mathrm{P}}$; see Dirac[165, Section 21].

[14] For time-dependent H^0, Eq. (6.22) still holds but the solution is more complicated; see Fetter and Walecka[186, p57].

[15] The interaction picture is for systems, as in linear response theory, with Hamiltonians of the form $H=H^0+H'$ where H' can be time dependent; see Appendix E. The Heisenberg representation applies to systems where H in the Schrödinger picture is time independent. The time-dependent operators in Eq. (6.23) can be considered in Heisenberg form. These are just names, however. We'll generally drop the subscript I or H and simply signify that we have time-dependent operators.

[16] The are many Kubo formulae.

Φ is real—a real response ΔB to a real interaction H' implies Φ is real; see Exercise 6.7. Moreover, Φ is independent of F, a requirement of linearity, and it depends on equilibrium quantities; the nonequilibrium response (for small interactions) is expressed in terms of time correlation functions.[17] Whereas in statistical mechanics the partition function Z is the fundamental quantity, in nonequilibrium statistical mechanics time correlation functions are fundamental. And, unlike the partition function, time correlation functions can be measured.

We can express the response function in a more explicit form for systems in the canonical ensemble with density operator $\hat{\rho}^0 = Z^{-1}(\beta)\exp(-\beta\hat{H}^0)$, where $\beta \equiv (kT)^{-1}$ and $Z(\beta) = \text{Tr}\exp(-\beta\hat{H}^0)$. To do so, we develop an identity due to Kubo[185] (see Exercise 6.8),

$$\left[\hat{A}, \text{e}^{-\beta\hat{H}}\right] = \text{e}^{-\beta\hat{H}} \int_0^\beta \text{e}^{\alpha\hat{H}}\left[\hat{H}, \hat{A}\right]\text{e}^{-\alpha\hat{H}}\text{d}\alpha = \text{e}^{-\beta\hat{H}} \int_0^\beta \left[\hat{H}, \text{e}^{\alpha\hat{H}}\hat{A}\text{e}^{-\alpha\hat{H}}\right]\text{d}\alpha, \tag{6.25}$$

where α is a dummy integration variable. In the second equality, we have a construct resembling the Heisenberg representation, $\hat{A}_H(t) \equiv \text{e}^{\text{i}\hat{H}t/\hbar}\hat{A}\text{e}^{-\text{i}\hat{H}t/\hbar}$; see Eq. (E.9). It would in fact *be* the Heisenberg operator if we were to adopt an *imaginary time*, $t \equiv -\text{i}\hbar\alpha$, a step we take:[18] $\hat{A}_H(-\text{i}\hbar\alpha) = \text{e}^{\alpha\hat{H}}\hat{A}\text{e}^{-\alpha\hat{H}}$. In the Heisenberg picture, the equation of motion is generated by the commutator: $\left[\hat{H}, \hat{A}_H(-\text{i}\hbar\alpha)\right] = -\text{i}\hbar\dot{\hat{A}}_H(-\text{i}\hbar\alpha)$; see Eq. (E.12). We can then complete the identity:[19]

$$[\text{e}^{-\beta\hat{H}}, \hat{A}] = \text{i}\hbar\text{e}^{-\beta\hat{H}} \int_0^\beta \text{d}\alpha\dot{\hat{A}}_H(-\text{i}\hbar\alpha). \tag{6.26}$$

Combine Eq. (6.26) with Eq. (6.24) and drop subscripts on time-dependent operators,

$$\begin{aligned} \Phi_{BA}(t) &= \frac{1}{\text{i}\hbar}\text{Tr}\,\hat{\rho}^0[\hat{A}, \hat{B}(t)] = \frac{1}{\text{i}\hbar}\text{Tr}[\hat{\rho}^0, \hat{A}]\hat{B}(t) \\ &= \int_0^\beta \text{d}\alpha\,\text{Tr}\,\hat{\rho}^0\dot{\hat{A}}(-\text{i}\hbar\alpha)\hat{B}(t) = -\int_0^\beta \text{d}\alpha\,\text{Tr}\,\hat{\rho}^0\hat{A}(-\text{i}\hbar\alpha)\dot{\hat{B}}(t), \end{aligned} \tag{6.27}$$

where we've used the cyclic invariance of the trace and the Heisenberg equation of motion; see Exercise 6.10.

6.2.2 The generalized susceptibility $\chi(\omega)$ and its analytic properties

Consider a monochromatic interaction,[20]

$$F(t) = E_0\text{e}^{-\text{i}\omega t} = \lim_{\epsilon\to 0+} E_0\text{e}^{-\text{i}\omega t+\epsilon t}, \tag{6.28}$$

where adding a small imaginary part $\text{i}\epsilon$ ($\epsilon > 0$) to the frequency ω ensures convergence of integrals (we require $\lim_{t\to-\infty} F(t) = 0$, "adiabatic switching"). By combining Eq. (6.28) with Eq. (6.23) we have the response

$$\Delta B(t) = E_0 \lim_{\epsilon\to 0+} \int_{-\infty}^t \text{d}\tau\Phi_{BA}(t-\tau)\text{e}^{-\text{i}(\omega+\text{i}\epsilon)\tau} = E_0 \lim_{\epsilon\to 0+} \int_0^\infty \text{d}\theta\Phi_{BA}(\theta)\text{e}^{-\text{i}(\omega+\text{i}\epsilon)(t-\theta)}.$$

Thus,

$$\Delta B(t) = \chi_{BA}(\omega)E_0\text{e}^{-\text{i}\omega t}, \tag{6.29}$$

[17] Analogous to quantum mechanics, where first-order shifts in eigenvalues depend on zeroth-order wave functions.

[18] There is a well-developed formalism in nonequilibrium statistical mechanics involving imaginary time. See Kadanoff and Baym[187], Abrikosov, Gorkov, and Dzyaloshinski[188], and Langreth[189]. Rammer[190] is a modern reference.

[19] See Eq. (3.6) of Kubo[185].

[20] The response to arbitrary (non-monochromatic) interactions can be synthesized through Fourier transformation. At some point, we have to take the real part of Eq. (6.28).

with

$$\chi_{BA}(\omega) \equiv \lim_{\epsilon \to 0+} \int_0^\infty d\theta e^{i(\omega + i\epsilon)\theta} \Phi_{BA}(\theta) \tag{6.30}$$

the *generalized susceptibility*, the Fourier transform of the response function, and the proportionality factor between response and interaction.[21] Note that the response occurs with the same frequency as the interaction, a characteristic of linear response; frequency mixing occurs in the nonlinear regime (as in nonlinear optics; see Boyd[191]).

We note some general properties of $\chi_{BA}(\omega)$. Besides causality, $\Phi(\theta \le 0) = 0$ (we drop BA), we require that a constant finite interaction ($F(t-\theta) = F_0$ for $0 \le \theta < \infty$) give rise to a constant finite response,[22] which from Eq. (6.23) implies

$$\int_0^\infty \Phi(\theta) d\theta < \infty. \tag{6.31}$$

As a consequence, $\chi(\omega = 0)$ exists [see Eq. (6.30)], implying $\chi(\omega)$ does not have a pole at $\omega = 0$. It's assumed that $\chi(\omega)$ has no poles on the real-ω line.[23] To discuss the analytic properties of $\chi(\omega)$, let's momentarily generalize the definition in Eq. (6.30),

$$\chi(\omega) = \int_0^\infty d\theta \Phi(\theta) e^{(i\omega - \nu)\theta}, \tag{6.32}$$

where $\nu > 0$ is not meant to be infinitesimal. The integral exists for all ω and vanishes for $\nu \to \infty$. *The generalized susceptibility has no poles in the closed upper half of the complex ω-plane.*[24] Such a function is said to be *holomorphic*.[25] Poles of χ lie in the lower half of the complex ω-plane.[26]

A bounded function in the upper half complex ω-plane has, as a boundary value on the real-ω line, a complex-valued function $\chi(\omega) : \mathbb{R} \to \mathbb{C}$ with real and imaginary parts satisfying the Kramers-Kronig relations (Titchmarsh's theorem; see Appendix F). In a standard notation, $\chi(\omega) = \chi'(\omega) + i\chi''(\omega) \equiv \mathrm{Re}\left[\chi(\omega)\right] + i\,\mathrm{Im}\left[\chi(\omega)\right]$, with

$$\begin{aligned} \chi'(\omega) &= \frac{1}{\pi} P \int_{-\infty}^{\infty} \frac{\chi''(\omega')}{\omega' - \omega} d\omega' \equiv \mathcal{H}(\chi'') \\ \chi''(\omega) &= -\frac{1}{\pi} P \int_{-\infty}^{\infty} \frac{\chi'(\omega')}{\omega' - \omega} d\omega' \equiv \mathcal{H}^{-1}(\chi'), \end{aligned} \tag{6.33}$$

where P denotes the Cauchy principal value integral and $\mathcal{H}$ the Hilbert transform (see Appendix F). *The real and imaginary parts of $\chi(\omega)$ are not independent*, they're Hilbert-transform pairs ($\mathcal{H}^{-1} = -\mathcal{H}$); *$\chi(\omega)$ can be reconstructed from either part.*

6.2.3 Identification of $\chi''(\omega)$ with dissipation

That the Fourier transform of the response function has real and imaginary parts χ' and χ'' satisfying the Kramers-Kronig relations is a direct consequence of its causal nature.[27] We now show that χ'' is associated with energy dissipation, a result of considerable importance.

---

[21] In statistical mechanics, we have the magnetic susceptibility $\boldsymbol{M} = \chi \boldsymbol{H}$ and the electric susceptibility $\boldsymbol{P} = \epsilon_0 \chi_e \boldsymbol{E}$, quantities having no time dependence. Here we're allowing the susceptibility to have a frequency dependence, $\chi(\omega)$. The susceptibility is also known as the *admittance* (see Kubo[185]) or the *compliance*, the "displacement/force."

[22] This is equivalent to the requirement of no energy dissipation at frequency $\omega = 0$; see Eq. (6.39).

[23] If $\chi(\omega)$ were to have poles on the real-ω line, they would represent non-dissipative, reversible contributions to the response of the system. We ignore this possibility in a theory of irreversible phenomena.

[24] The closed upper half plane is the upper half plane together with its closure, the real line.

[25] See [13, p198].

[26] By Liouville's theorem [from complex analysis, not Eq. (4.2)] a bounded entire function is a constant; $\chi(\omega)$ must have singularities for complex ω. Complex analysis is presumed familiar to the reader; see [13, Chapter 8].

[27] Analyticity and causality are closely linked; one implies the other, see Appendix F.

First, another general property, that because Φ is real, $\chi'(\omega)$ is even and $\chi''(\omega)$ is odd (see Exercise 6.11): $\chi'(-\omega) = \chi'(\omega)$ and $\chi''(-\omega) = -\chi''(\omega)$. Moreover, $\chi^*(\omega) = \chi(-\omega)$. Thus,

$$\begin{aligned}\chi'(\omega) &= \frac{1}{2}\left[\chi(\omega) + \chi(-\omega)\right] \\ \chi''(\omega) &= \frac{1}{2\mathrm{i}}\left[\chi(\omega) - \chi(-\omega)\right].\end{aligned} \tag{6.34}$$

The real part χ' is time-reversal symmetric ($\omega \leftrightarrow -\omega$), but the imaginary part χ'' is not. *Irreversibility is associated with* χ''; it knows about the arrow of time. The real part is the *reactive* part of the response function with the imaginary part the *dissipative* or the *absorptive* part.

How does the system's energy change in time? With $W \equiv \langle \hat{H} \rangle = \text{Tr}\, \hat{\rho}\hat{H}$ the average energy,

$$\frac{\mathrm{d}W}{\mathrm{d}t} = \frac{\mathrm{d}}{\mathrm{d}t}\text{Tr}\,\hat{\rho}\hat{H} = \cancelto{0}{\text{Tr}(\dot{\rho}H)} + \text{Tr}(\rho\dot{H}) = \text{Tr}(\rho\dot{H}') = -\dot{F}\,\text{Tr}(\rho A) = -\dot{F}\left(\langle A\rangle^0 + \Delta A(t)\right), \tag{6.35}$$

where we've dropped the "hat" on operators and $\text{Tr}\,\dot{\rho}H = 0$ (cyclic invariance of trace). For a monochromatic interaction of frequency ω, which we write as

$$F(t) = E_0\cos(\omega t) = \frac{1}{2}\left[E_0\mathrm{e}^{-\mathrm{i}\omega t} + E_0^*\mathrm{e}^{\mathrm{i}\omega t}\right], \tag{6.36}$$

we time average over a cycle of period $T = 2\pi/\omega$. Under $\dot{W} \to (1/T)\int_0^T (\mathrm{d}W/\mathrm{d}t)\mathrm{d}t$ in Eq. (6.35),

$$\begin{aligned}\dot{W} &= -\frac{1}{T}\int_0^T \mathrm{d}t\dot{F}\left(\langle A\rangle^0 + \Delta A(t)\right) = -\frac{1}{T}\int_0^T \mathrm{d}t\dot{F}\Delta A(t) = -\frac{1}{T}\int_0^\infty \mathrm{d}\theta\Phi(\theta)\int_0^T \mathrm{d}t\dot{F}F(t-\theta) \\ &= \frac{\mathrm{i}\omega}{4T}\int_0^\infty \mathrm{d}\theta\Phi(\theta)\int_0^T \mathrm{d}t\left(E_0\mathrm{e}^{-\mathrm{i}\omega t} - E_0^*\mathrm{e}^{\mathrm{i}\omega t}\right)\left(E_0\mathrm{e}^{-\mathrm{i}\omega(t-\theta)} + E_0^*\mathrm{e}^{\mathrm{i}\omega(t-\theta)}\right),\end{aligned}$$

where we've used Eq. (6.36) to find $\dot{F}$. Performing the integrations on t, we find the average rate at which energy changes (through the action of a driving force of frequency ω),

$$\begin{aligned}\dot{W}(\omega) &= \frac{\mathrm{i}\omega}{4}|E_0|^2\int_0^\infty \mathrm{d}\theta\Phi(\theta)\left(\mathrm{e}^{-\mathrm{i}\omega\theta} - \mathrm{e}^{\mathrm{i}\omega\theta}\right) = \frac{\mathrm{i}\omega}{4}|E_0|^2\left(\chi(\omega) - \chi(-\omega)\right) \\ &= -\frac{\omega}{2}|E_0|^2\chi''(\omega),\end{aligned} \tag{6.37}$$

where we've used Eq. (6.34). Dissipation is indicated if $\dot{W}$ is negative.

From thermodynamics (see [2, p25]), the work done *by* a system operating in a cycle, $-W_{\text{cycle}}$, is equal to the net heat absorbed in a cycle, $-W_{\text{cycle}} = Q_{\text{cycle}}$ (in our sign convention, positive values of the symbols Q, W represent energy transfers *to* the system in the form of heat and work; see [2, p9]).[28,29] Thus, from Eq. (6.37), the average rate of heat absorption is proportional to χ'',

$$\dot{Q}(\omega) = \frac{1}{2}|E_0|^2\omega\chi''(\omega). \tag{6.38}$$

Unless the system is an amplifier, we require dissipation at each frequency,

$$\omega\chi''(\omega) \geq 0. \tag{6.39}$$

Equation (6.39) is the *positivity condition*. By this condition, χ'' is positive for positive frequency and negative for negative frequency, consistent with the requirement that it be odd from the reality of the response function. There is no dissipation at $\omega = 0$, consistent with Eq. (6.31) that a constant external force gives rise to a constant finite response.

---

[28]Internal energy is a state variable and thus in a cycle $\Delta U_{\text{cycle}} = 0 = Q_{\text{cycle}} + W_{\text{cycle}}$. Internal energy is a storehouse of the adiabatic work done on systems; see [2, p8]. Adiabatically isolated systems interact with their surroundings solely through mechanical means, implying that the adiabatic work stored in a system is the average value of the Hamiltonian $\langle H\rangle$. Thus, $W \equiv \langle H\rangle$ introduced near Eq. (6.35) can be identified with the work W appearing in thermodynamics. See [5, p92].

[29]Nothing limits the conversion of work into heat. The second law places limitations on converting heat into work.

6.2.4 Explicit formulae for the quantum response

We've been able to infer important properties of the response function and its Fourier transform without exhibiting an explicit expression. We do that now starting from Eq. (6.24) in the form $\mathrm{i}\hbar\Phi_{BA}(t-\tau) = \mathrm{Tr}\,\rho^0[\hat{A}, \hat{B}(t-\tau)]$ where we work in the canonical ensemble using a basis in which the internal Hamiltonian H^0 is diagonal. This form of Eq. (6.24) is convenient because the time dependence is in one place. We work with density operator $\rho^0 = Z^{-1}(\beta)e^{-\beta H^0}$ [see Eq. (A.45)] and we assume a complete orthonormal basis $\{|n\rangle\}$ such that $\sum_n |n\rangle\langle n| = I$, $\langle n|m\rangle = \delta_{n,m}$, with $H^0|n\rangle = E_n|n\rangle$. Then, for $\tau < t$,

$$\begin{aligned}\mathrm{i}\hbar\Phi_{BA}(t-\tau) &= \sum_n \langle n|\rho^0[A, B(t-\tau)]|n\rangle = \sum_n\sum_m \langle n|\rho^0|m\rangle\langle m|[A, B(t-\tau)]|n\rangle \\ &= \frac{1}{Z(\beta)}\sum_n \mathrm{e}^{-\beta E_n}\langle n|[A, B(t-\tau)]|n\rangle. \end{aligned} \tag{6.40}$$

Let's work on $\langle n|[A, B(t-\tau)|n\rangle$:

$$\langle n|[A, B(t-\tau)]|n\rangle = \langle n|AU_0^\dagger(t-\tau)BU_0(t-\tau) - U_0^\dagger(t-\tau)BU_0(t-\tau)A|n\rangle.$$

Make judicious use of the completeness relation,

$$\begin{aligned}\langle n|[A, B(t-\tau]|n\rangle &= \sum_m \langle n|AU_0^\dagger(t-\tau)|m\rangle\langle m|BU_0(t-\tau)|n\rangle \\ &\quad - \sum_m \langle n|U_0^\dagger(t-\tau)BU_0(t-\tau)|m\rangle\langle m|A|n\rangle \\ &= \sum_m \left[\mathrm{e}^{\mathrm{i}(E_m-E_n)(t-\tau)/\hbar}A_{nm}B_{mn} - \mathrm{e}^{\mathrm{i}(E_n-E_m)(t-\tau)/\hbar}B_{nm}A_{mn}\right], \end{aligned} \tag{6.41}$$

where the matrix elements $A_{nm} \equiv \langle n|A|m\rangle$, etc. Defining $\hbar\omega_{nm} \equiv E_n - E_m$ and combining Eq. (6.41) with Eq. (6.40), we have, playing with indices,

$$\Phi_{BA}(t-\tau) = \frac{1}{\mathrm{i}\hbar Z(\beta)}\sum_n\sum_m \mathrm{e}^{-\beta E_n}\left(1 - \mathrm{e}^{\beta\hbar\omega_{nm}}\right)\mathrm{e}^{-\mathrm{i}\omega_{nm}(t-\tau)}A_{nm}B_{mn}. \tag{6.42}$$

It can be shown (as a check, see Exercise 6.13) that Φ_{BA} in Eq. (6.42) is real. The generalized susceptibility then follows using Eq. (6.30). We note first that

$$\lim_{\epsilon\to 0+}\int_0^\infty \mathrm{e}^{\mathrm{i}[\omega-\omega_{nm}+\mathrm{i}\epsilon]\theta}\mathrm{d}\theta = \lim_{\epsilon\to 0+}\frac{\mathrm{i}}{\omega-\omega_{nm}+\mathrm{i}\epsilon} = \mathrm{i}P\left(\frac{1}{\omega-\omega_{nm}}\right) + \pi\delta(\omega-\omega_{nm}), \tag{6.43}$$

where we've used the Plemelj relation, Eq. (F.20). By combining Eq. (6.42) with Eq. (6.30) and making use of Eq. (6.43), we have the real and imaginary parts of $\chi(\omega)$,

$$\begin{aligned}\chi'_{BA}(\omega) &= \frac{1}{\hbar Z(\beta)}\sum_n\sum_m \mathrm{e}^{-\beta E_n}\left(1 - \mathrm{e}^{\beta\hbar\omega_{nm}}\right)A_{nm}B_{mn}P\left(\frac{1}{\omega-\omega_{nm}}\right) \\ \chi''_{BA}(\omega) &= \frac{\pi}{\hbar Z(\beta)}\sum_n\sum_m \mathrm{e}^{-\beta E_n}\left(\mathrm{e}^{\beta\hbar\omega} - 1\right)A_{nm}B_{mn}\delta(\omega-\omega_{nm}). \end{aligned} \tag{6.44}$$

The absorptive part χ'' consists of a sequence of δ-functions at the frequencies ω_{nm} associated with differences between allowed energies—a basic idea of quantum mechanics. The reactive part χ' has a sequence of zeroes at ω_{nm}.

6.2.5 Brownian motion of a harmonically bound classical particle

We illustrate these concepts with an example from classical physics, that of a harmonic oscillator in thermal contact with its environment—the Brownian motion of a harmonically bound particle.[30] Assume a particle of mass m attached to a Hookean spring of force constant k interacting with a heat bath and subject to an external force $F(t)$. We consider a single degree of freedom (i.e., we treat this as a one-dimensional problem). From Newton's law,

$$m\ddot{x}(t) = -m\omega_0^2 x(t) + F_{\text{int}}(t) + F(t), \tag{6.45}$$

where $\omega_0^2 \equiv k/m$ and $F_{\text{int}}(t)$ denotes the instantaneous internal force of interaction of the oscillator with its surroundings.[31] Our knowledge of the internal force is limited to its statistical properties (see our treatment of the Langevin equation, Chapter 3). Thus we take the average of Eq. (6.45),

$$m\langle\ddot{x}\rangle_t + m\omega_0^2\langle x\rangle_t - \langle F_{\text{int}}\rangle_t = F(t) \tag{6.46}$$

where $\langle\rangle_t$ denotes a nonequilibrium ensemble average of systems subject to the force $F(t)$. We know phenomenologically that friction acts in a direction opposing motion with a strength proportional to the speed; we take

$$\langle F_{\text{int}}\rangle_t = -m\gamma\langle\dot{x}\rangle_t, \tag{6.47}$$

where $\gamma > 0$ is the friction coefficient. We therefore have the equation of motion

$$m\frac{\mathrm{d}^2}{\mathrm{d}t^2}\langle x\rangle_t + m\gamma\frac{\mathrm{d}}{\mathrm{d}t}\langle x\rangle_t + m\omega_0^2\langle x\rangle_t = F(t), \tag{6.48}$$

where we've equated averages of derivatives with derivatives of averages; see Eq. (2.21).

We seek the response $\langle x\rangle_t$ to the external force $F(t)$ as specified by the relation

$$\langle x\rangle_t = \int_{-\infty}^{\infty} \Phi(t-t')F(t')\mathrm{d}t'. \tag{6.49}$$

Combine Eq. (6.49) with Eq. (6.48):

$$\int_{-\infty}^{\infty}\left(m\frac{\mathrm{d}^2}{\mathrm{d}t^2} + m\gamma\frac{\mathrm{d}}{\mathrm{d}t} + m\omega_0^2\right)\Phi(t-t')F(t')\mathrm{d}t' = F(t). \tag{6.50}$$

Equation (6.50) implies

$$\left(m\frac{\mathrm{d}^2}{\mathrm{d}t^2} + m\gamma\frac{\mathrm{d}}{\mathrm{d}t} + m\omega_0^2\right)\Phi(t-t') = \delta(t-t'). \tag{6.51}$$

The response function is thus the Green function associated with the inhomogeneous differential equation describing the nonequilibrium behavior.[32] We're not using statistical mechanics to find the response function in this case; this is not a microscopic approach.

There's no one way to solve differential equations. Let's find the Fourier transform of Φ. From Eq. (6.30),[33]

$$\chi(\omega) = \int_{-\infty}^{\infty} \mathrm{d}\theta \mathrm{e}^{\mathrm{i}\omega\theta}\Phi(\theta), \tag{6.52}$$

[30] The harmonic oscillator is among a handful of exactly solvable problems in physics; in essence this is the classic classical physics example.

[31] See [2, p7] for a discussion of the distinction between system and surroundings.

[32] The method of Green functions is presumed known to the reader; see [13, Chapter 9].

[33] We don't have to put in the "causality cutoff"; it will emerge from the analysis.

where $\theta \equiv t - t'$, with the inverse relation

$$\Phi(\theta) = \frac{1}{2\pi}\int_{-\infty}^{\infty} \mathrm{d}\omega \mathrm{e}^{-\mathrm{i}\omega\theta}\chi(\omega). \tag{6.53}$$

Combine Eq. (6.53) with Eq. (6.51) and use the integral representation of the delta function, $\delta(x) = \int_{-\infty}^{\infty} \mathrm{e}^{-\mathrm{i}kx}\mathrm{d}k/(2\pi)$. We find

$$\chi(\omega) = \frac{1}{m}\frac{1}{\omega_0^2 - \omega^2 - \mathrm{i}\gamma\omega}. \tag{6.54}$$

Much depends on the location of the poles of $\chi(\omega)$ (for ω complex). It's straightforward to show that $\chi(\omega)$ has simple poles at $\omega_\pm \equiv -\mathrm{i}\gamma/2 \pm \sqrt{\omega_0^2 - (\gamma/2)^2}$. *In all cases these occur in the lower half plane*. By Titchmarsh's theorem, therefore (see Section F.2), Φ is a causal function, $\Phi(\theta < 0) = 0$. Combining Eqs. (6.54) and (6.53),

$$\Phi(\theta) = -\frac{1}{2\pi m}\int_{-\infty}^{\infty} \mathrm{d}\omega \frac{\mathrm{e}^{-\mathrm{i}\omega\theta}}{(\omega - \omega_+)(\omega - \omega_-)}. \tag{6.55}$$

It's an exercise in the residue theorem (see Exercise 6.15) to show from Eq. (6.55) that[34]

$$\Phi(\theta) = \begin{cases} \exp(-\gamma\theta/2)\dfrac{\sin\left(\sqrt{\omega_0^2 - (\gamma/2)^2}\,\theta\right)}{m\sqrt{\omega_0^2 - (\gamma/2)^2}} & \theta \geq 0 \\ 0. & \theta < 0 \end{cases} \tag{6.56}$$

For $\omega_0 > \gamma/2$, Φ is a damped sine wave; it decays without oscillation for $\omega_0 < \gamma/2$. For the real and imaginary parts of $\chi(\omega)$ (for ω real), we have

$$\begin{aligned} \chi'(\omega) &= \frac{1}{m}\frac{\omega_0^2 - \omega^2}{\left(\omega_0^2 - \omega^2\right)^2 + (\gamma\omega)^2} \\ \chi''(\omega) &= \frac{1}{m}\frac{\gamma\omega}{\left(\omega_0^2 - \omega^2\right)^2 + (\gamma\omega)^2}. \end{aligned} \tag{6.57}$$

Absorption is maximized at $\omega = \omega_0$, *resonance absorption*, and the reactive part vanishes at $\omega = \omega_0$. We also have the positivity condition, $\omega\chi''(\omega) \geq 0$.

6.2.6 The relaxation function

Consider a system subjected to a constant force $F = F_0$ for $-\infty < t < 0$ which is then switched off at $t = 0$. At that point the response $\Delta B(t)$ will begin to relax to zero, a process described by

$$\Delta B(t > 0) = F_0 \int_{-\infty}^{0} \mathrm{d}\tau \Phi_{BA}(t - \tau) = F_0 \int_t^{\infty} \Phi_{BA}(\theta)\mathrm{d}\theta, \tag{6.58}$$

where $\theta \equiv t - \tau$. The function

$$R_{BA}(t) \equiv \int_t^{\infty} \Phi_{BA}(\theta)\mathrm{d}\theta \equiv \lim_{\epsilon \to 0+} \int_t^{\infty} \Phi_{BA}(\theta)\mathrm{e}^{-\epsilon\theta}\mathrm{d}\theta \tag{6.59}$$

is the *relaxation function*, which can be defined with a convergence factor if need be. We stipulated, however, in Eq. (6.31) that $\Phi(\theta)$ is integrable, implying that

$$\lim_{t \to \infty} \Phi(t) = 0. \tag{6.60}$$

[34] Equation (6.56) is the particular solution of Eq. (6.51). What about the complementary solution; what about initial conditions? The general solution to Eq. (6.51) is given in Uhlenbeck and Ornstein[64, p834], reprinted in Wax[44, p93]. Equation (6.56) is the solution for initial conditions $x(0) = \dot{x}(0) = 0$.

6.3 FLUCTUATION-DISSIPATION THEOREM

We now establish a fundamental result—the *fluctuation-dissipation theorem*—a relation between dissipation (of the energy of external perturbations) and spontaneously occurring fluctuations in equilibrium. Fortunately we've done most of the work in making this connection. Consider the time correlation function of observables A and B [see Eq. (6.12)]

$$C_{AB}^{0}(t) = \frac{1}{2}\,\text{Tr}\,\hat{\rho}^{0}\left[\hat{A}\hat{B}(t) + \hat{B}(t)\hat{A}\right]. \tag{6.61}$$

Repeating the analysis of Eqs. (6.40) and (6.41) step by step, we have

$$\begin{aligned} C_{AB}^{0}(t) &= \frac{1}{2Z(\beta)}\sum_{n}\mathrm{e}^{-\beta E_n}\langle n|\hat{A}\hat{B}(t) + \hat{B}(t)\hat{A}|n\rangle \\ &= \frac{1}{2Z(\beta)}\sum_{n}\sum_{m}\mathrm{e}^{-\beta E_n}\left[\langle n|\hat{A}U_0^{\dagger}(t)|m\rangle\langle m|\hat{B}U_0(t)|n\rangle + \langle n|U_0^{\dagger}(t)\hat{B}U_0(t)|m\rangle\langle m|\hat{A}|n\rangle\right] \\ &= \frac{1}{2Z(\beta)}\sum_{n}\sum_{m}\mathrm{e}^{-\beta E_n}\left[\mathrm{e}^{\mathrm{i}\omega_{mn}t}A_{nm}B_{mn} + \mathrm{e}^{\mathrm{i}\omega_{nm}t}B_{nm}A_{mn}\right]. \end{aligned}$$

This can be condensed by interchanging $n \leftrightarrow m$ in the second term,

$$C_{AB}^{0}(t) = \frac{1}{2Z(\beta)}\sum_{n}\sum_{m}\mathrm{e}^{-\beta E_n}\left(\mathrm{e}^{\beta\hbar\omega_{nm}} + 1\right)\mathrm{e}^{-\mathrm{i}\omega_{nm}t}A_{nm}B_{mn}. \tag{6.62}$$

We require the Fourier transform,

$$\widetilde{C}_{AB}^{0}(\omega) \equiv \int_{-\infty}^{\infty} C_{AB}^{0}(t)\mathrm{e}^{\mathrm{i}\omega t}\mathrm{d}t. \tag{6.63}$$

Combining Eq. (6.62) with Eq. (6.63), we have

$$\widetilde{C}_{AB}^{0}(\omega) = \frac{\pi}{Z(\beta)}\sum_{n}\sum_{m}\mathrm{e}^{-\beta E_n}\left(\mathrm{e}^{\beta\hbar\omega} + 1\right)A_{nm}B_{mn}\delta(\omega - \omega_{nm}). \tag{6.64}$$

Note that the definition of $\chi(\omega)$ in Eq. (6.30) involving a one-sided Fourier transform leads to the use of the Plemelj formula in Eq. (6.43) and the subsequent division of χ into real and imaginary parts in Eq. (6.44). Equation (6.63), however, the definition of $\widetilde{C}^{0}(\omega)$, involves an integration over all times and leads to the delta function in Eq. (6.64), a purely real expression (see Exercise 6.16).

Equation (6.64) [for $\widetilde{C}^{0}(\omega)$] is almost identical to Eq. (6.44) [for $\chi''(\omega)$]; the two differ by a multiplicative constant and a plus sign. One can show that

$$\chi_{AB}''(\omega) = \frac{1}{\hbar}\frac{\mathrm{e}^{\beta\hbar\omega} - 1}{\mathrm{e}^{\beta\hbar\omega} + 1}\widetilde{C}_{AB}^{0}(\omega) = \frac{1}{\hbar}\tanh(\tfrac{1}{2}\beta\hbar\omega)\widetilde{C}_{AB}^{0}(\omega). \tag{6.65}$$

Equation (6.65) is one form of the celebrated fluctuation-dissipation theorem, which has many faces—it's been derived in different guises by different researchers. Credit is usually attributed to Callen and Welton[192], but antecedents can be seen in Nyquist's theorem, the Einstein relation, and Onsager reciprocity. Kubo[193] recounts the history of the theorem; Chester[194] and Zwanzig[195] give extensive literature reviews. Case[196] is an excellent review article. *A fluctuation-dissipation theorem is a relation between $\chi''(\omega)$ and the Fourier components of the time correlation function $\widetilde{C}^{0}(\omega)$*. It's often given as the inverse of Eq. (6.65),

$$\widetilde{C}_{AB}^{0}(\omega) = \hbar\coth(\tfrac{1}{2}\beta\hbar\omega)\chi_{AB}''(\omega) + C\delta(\omega), \tag{6.66}$$

where C is a system-dependent constant. One has to treat the $\omega = 0$ limit carefully in inverting Eq. (6.65). The classical version is found from the formal limit $\hbar \to 0$ in Eq. (6.65),

$$\chi''_{AB}(\omega) = \frac{1}{2}\beta\omega\widetilde{C}^0_{AB}(\omega). \qquad \text{(classical)} \tag{6.67}$$

Note the dimension of $\beta\omega$, (energy-time)$^{-1}$, a classical proxy for $\hbar^{-1}$ in Eq. (6.65).

Fluctuation-dissipation theorems relate two distinct physical quantities, each accessible to measurement. Fluctuations occur spontaneously, in the absence of external interactions, from the ceaseless motions of microscopic constituents of macroscopic systems.[35] Correlations of fluctuations, in space and time, are revealed through scattering structure factors. Dissipation is the irreversible absorption of the work done by external forces into microscopic degrees of freedom, and there are different ways of measuring absorption. One can regard the theorem in the form of Eq. (6.65) as specifying the efficacy of fluctuations (of frequency ω) in absorbing energy of frequency ω. The converse, Eq. (6.66), specifies the strength of fluctuations at frequency ω occurring in response to the absorption of energy of frequency ω. Both quantities are mechanical in origin—absorption of work done by external forces and fluctuations from particle motions. Note the characteristic comparison of two energies: $\hbar\omega$ and kT. The theorem explains the success of the Onsager regression hypothesis which equates the rate of change of fluctuations with dissipative fluxes; see Eq. (1.42).[36] Basically, a system doesn't "know" whether it's been brought into a nonequilibrium state by the action of an external force or as the result of a random fluctuation.

6.4 GREEN-KUBO THEORY OF TRANSPORT COEFFICIENTS

Transport coefficients, introduced phenomenologically in irreversible thermodynamics, can be expressed as integrals of time correlation functions, formulae known as *Green-Kubo relations*, from Green[197] and Kubo[185]. Transport coefficients are classified as *mechanical*, characterizing the response to external interactions, electrical conductivity for example, or *thermal*, those associated with inhomogeneities—diffusion, viscosity, or heat conduction, the type treated in Chapman-Enskog theory, Section 4.9.

6.4.1 Mechanical transport processes

As a representative example, consider a system of charges $\{e_k\}$ in an oscillating spatially uniform electric field $\boldsymbol{E}(t) = \boldsymbol{E}\mathrm{e}^{-\mathrm{i}\omega t} \equiv \boldsymbol{E}\lim_{\epsilon\to 0^+}\mathrm{e}^{-\mathrm{i}(\omega+\mathrm{i}\epsilon)t}$, where strictly speaking a convergence factor should be included. The energy of interaction[37]

$$H'(t) = -\sum_k e_k\boldsymbol{r}_k \cdot \boldsymbol{E}(t) \equiv -\mathrm{e}^{-\mathrm{i}\omega t}\boldsymbol{P}\cdot\boldsymbol{E}, \tag{6.68}$$

where e_k is the charge of the kth-particle with $\boldsymbol{r}_k$ its position vector, a dynamical variable. The total current density $\boldsymbol{J} = (1/V)\sum_k e_k\dot{\boldsymbol{r}}_k$, where V is the system volume. We write its μth-component

$$J_\mu \equiv \frac{1}{V}\sum_k e_k\dot{r}_{k,\mu}. \tag{6.69}$$

We use Greek letters $\mu \equiv 1, 2, 3$ to label vector components and Roman letters k to label particles. In many cases we can let $V = 1$; a system of unit volume.

---

[35] Consider Eq. (4.125) relating the width of the velocity distribution (at every point of physical space) to the temperature.

[36] The regression hypothesis is integral to the proof of Onsager's (experimentally confirmed) reciprocity relations.

[37] Equation (6.68) is the standard expression $U = -\boldsymbol{P}\cdot\boldsymbol{E}$ with $\boldsymbol{P}$ the electrical polarization. Note, however, that $\boldsymbol{P}$ is extensive; it scales with the size of the system. In electromagnetic theory, $\boldsymbol{P}$ is defined as a polarization density.

The change $\Delta\rho(t)$ occurring in response to the perturbation $H'(t)$ [see Eq. (6.21)] induces a nonequilibrium current (in equilibrium $\langle J_\mu \rangle^0 = \text{Tr}\,\rho^0 J_\mu = 0$),

$$\begin{aligned}\langle J_\mu \rangle_t = \text{Tr}\,\Delta\rho(t) J_\mu &= \frac{1}{\text{i}\hbar}\int_{-\infty}^{t} \text{d}\tau \text{e}^{-\text{i}\omega\tau}\, \text{Tr}\{\rho^0\,[\boldsymbol{P}, J_\mu(t-\tau)]\} \cdot \boldsymbol{E} \\ &= \frac{1}{\text{i}\hbar}\int_{-\infty}^{t} \text{d}\tau \text{e}^{-\text{i}\omega\tau}\, \text{Tr}\{[\rho^0, \boldsymbol{P}] J_\mu(t-\tau)\} \cdot \boldsymbol{E},\end{aligned} \tag{6.70}$$

from Eq. (6.23) and our old friend, cyclic invariance of the trace. No confusion should arise over the symbols in Eq. (6.70): $J_\mu(t-\tau)$ is a microscopic operator, Eq. (6.69), and $\langle J_\mu \rangle_t$ is the nonequilibrium average. Equation (6.70) implies a generalization of Ohm's law to include anisotropy,

$$\langle J_\mu \rangle_t = \sum_\nu \sigma_{\mu\nu}(\omega) E_\nu \text{e}^{-\text{i}\omega t} \tag{6.71}$$

[compare with Eq. (6.29)], where the *conductivity tensor* is the generalized susceptibility for this problem,

$$\sigma_{\mu\nu}(\omega) \equiv \frac{1}{\text{i}\hbar}\int_0^\infty \text{d}\theta \text{e}^{\text{i}\omega\theta}\, \text{Tr}\{[\rho^0, P_\nu] J_\mu(\theta)\}. \tag{6.72}$$

This can be simplified with the Kubo identity [see Eq. (6.26)],

$$[\rho^0, P_\nu] = \text{i}\hbar\rho^0 \int_0^\beta \text{d}\alpha \dot{P}_\nu(-\text{i}\hbar\alpha). \tag{6.73}$$

Combining Eqs. (6.73) and (6.72),

$$\sigma_{\mu\nu}(\omega) = \int_0^\beta \text{d}\alpha \int_0^\infty \text{d}\theta \text{e}^{\text{i}\omega\theta}\, \text{Tr}\{\rho^0 \dot{P}_\nu(-\text{i}\hbar\alpha) J_\mu(\theta)\}. \tag{6.74}$$

But $\dot{P}_\nu$ is related to the current density [see Eq. (6.69)],

$$\dot{P}_\nu = \sum_k e_k \dot{r}_{k,\nu} = V J_\nu, \tag{6.75}$$

implying the conductivity

$$\sigma_{\mu\nu}(\omega) = V\int_0^\beta \text{d}\alpha \int_0^\infty \text{d}\theta \text{e}^{\text{i}\omega\theta}\, \text{Tr}\{\rho^0 J_\nu(-\text{i}\hbar\alpha) J_\mu(\theta)\} \equiv V\int_0^\beta \text{d}\alpha \int_0^\infty \text{d}\theta \text{e}^{\text{i}\omega\theta} \langle J_\nu(-\text{i}\hbar\alpha) J_\mu(\theta)\rangle^0. \tag{6.76}$$

Equation (6.76) (the Kubo formula for the conductivity) specifies that conductivity is proportional to the time integral of the current autocorrelation function. The longer the current components stay correlated, the greater is the conductivity. The classical conductivity is found from the limit $\hbar \to 0$,

$$\sigma_{\mu\nu}(\omega) = \beta V \int_0^\infty \text{d}\theta \text{e}^{\text{i}\omega\theta} \langle J_\nu(0) J_\mu(\theta)\rangle^0. \qquad \text{(classical)} \tag{6.77}$$

If J_ν, J_μ are uncorrelated, $\sigma_{\mu\nu}$ is diagonal and the conductivity is a scalar, $\sigma = \frac{1}{3}\sum_\mu \sigma_{\mu\mu}$.

We can rewrite Eq. (6.76) by invoking the stationarity of equilibrium averages—no unique origin in time. For fixed α, the products $J_\nu(-\text{i}\hbar\alpha) J_\mu(\theta)$ generated for $0 \le \theta < \infty$ are the same as those generated by $J_\nu(\theta - \text{i}\hbar\alpha) J_\mu(0)$. Thus,

$$\sigma_{\mu\nu}(\omega) = V\int_0^\beta \text{d}\alpha \int_0^\infty \text{d}\theta \text{e}^{\text{i}\omega\theta} \langle J_\nu(\theta - \text{i}\hbar\alpha) J_\mu(0)\rangle^0. \tag{6.78}$$

Now let $\theta = \theta' + \mathrm{i}\hbar\alpha$,

$$\begin{aligned}\sigma_{\mu\nu}(\omega) &= V\int_0^\beta \mathrm{d}\alpha \int_{-i\hbar\alpha}^{\infty-i\hbar\alpha} \mathrm{d}\theta' \mathrm{e}^{\mathrm{i}\omega\theta'}\mathrm{e}^{-\hbar\omega\alpha}\langle J_\nu(\theta')J_\mu(0)\rangle^0 \\ &= V\int_0^\beta \mathrm{d}\alpha \mathrm{e}^{-\hbar\omega\alpha}\int_0^\infty \mathrm{d}\theta \mathrm{e}^{\mathrm{i}\omega\theta}\langle J_\nu(\theta)J_\mu(0)\rangle^0, \end{aligned} \tag{6.79}$$

where it's assumed that functions of the complex variable $\theta - \mathrm{i}\hbar\alpha$ are sufficiently analytic that the integration path can be shifted to the real axis.[38] Integrating over α,

$$\sigma_{\mu\nu}(\omega) = \frac{V}{\hbar\omega}\left(1 - \mathrm{e}^{-\beta\hbar\omega}\right)\int_0^\infty \mathrm{d}\theta \mathrm{e}^{\mathrm{i}\omega\theta}\langle J_\nu(\theta)J_\mu(0)\rangle^0. \tag{6.80}$$

We see from the $\omega \to 0$ limit of Eq. (6.80) that $\sigma''_{\mu\nu}(0) = 0$; no dissipation at $\omega = 0$.

6.4.2 Thermal transport processes

Linear response theory is well suited to the analysis of mechanical transport processes; just add the appropriate energy of interaction term to the Hamiltonian. There are transport processes, however, not associated with perturbations of the Hamiltonian. Diffusion, viscosity, heat conduction—processes associated with inhomogeneities and involving transport of matter, momentum, and energy—do not occur in response to forces represented by terms in the microscopic Hamiltonian.[39] Yet, as we show, Green-Kubo relations can be developed for thermal transport coefficients.

We follow the approach of Mori[199], the crux of which is the use of local thermodynamic equilibrium.[40] We introduced that concept in Chapter 1 to extend thermodynamics to systems not rigorously in thermodynamic equilibrium, but in which variations in state variables are negligible over small regions of space. Nothing in that theory, however, quantifies the size of such regions. Kinetic theory provides a characteristic length, the mean free path l. Local equilibrium holds when spatial variations over a mean free path are small (see Section 4.8 on normal solutions),

$$l\frac{|\nabla f|}{f} \ll 1, \tag{6.81}$$

where f is the single-particle distribution function. This criterion is equivalent to the collision frequency far exceeding the rate of flow processes. In collision-dominated regimes, collisions quickly establish a state that's closely related to the equilibrium state, but only locally. Solutions of the Boltzmann equation in this regime—normal solutions [for example, the local Maxwellian Eq. (4.129)]—are such that spatiotemporal variations occur through a functional dependence on the hydrodynamic fields $\rho(\boldsymbol{r},t)$, $\boldsymbol{u}(\boldsymbol{r},t)$, and $T(\boldsymbol{r},t)$.

Mori extended the concept of local equilibrium to the N-body distribution—a bold step—yet its predictions are in agreement with other approaches; see Green[200][201]. We *define* the N-body distribution function associated with local equilibrium to be a generalization of the local Maxwellian,

$$F^{\mathrm{LE}}(\boldsymbol{\Gamma}) \equiv Z^{-1}\exp\left\{-\int \mathrm{d}\boldsymbol{r}\hat{\mathcal{E}}(\boldsymbol{\Gamma},\boldsymbol{r})\Big/[kT(\boldsymbol{r})]\right\}, \tag{6.82}$$

where $\boldsymbol{r}$ is an arbitrary position in the system (and not a canonical variable), $T(\boldsymbol{r})$ is the local temperature, and $\hat{\mathcal{E}}(\boldsymbol{\Gamma},\boldsymbol{r})$ is a microscopic internal energy density (thermal kinetic energy and potential

[38] As an example, $\int_{-\infty+\mathrm{i}\alpha}^{\infty+\mathrm{i}\alpha}\exp(-x^2)\mathrm{d}x = \int_{-\infty}^{\infty}\exp(-x^2)\mathrm{d}x$ for real constant α; see Exercise 2.47.

[39] Molecules respond to external fields the same in homogeneous or inhomogeneous systems. Luttinger[198] showed that thermal transport coefficients can be calculated in a Hamiltonian formalism when suitable external potentials are posited, such as time-varying, inhomogeneous gravitational fields (which don't exist on laboratory length scales).

[40] Mori used quantum mechanics and the grand canonical ensemble, unimportant differences with our analysis.

energy of intermolecular forces),

$$\hat{\mathcal{E}}(\mathbf{\Gamma},\boldsymbol{r}) \equiv \sum_i \left[\frac{1}{2}m[\boldsymbol{v}_i - \boldsymbol{u}(\boldsymbol{r})]^2 + \frac{1}{2}\sum_{j\neq i}\Phi(\boldsymbol{r}_i,\boldsymbol{r}_j)\right]\delta(\boldsymbol{r}-\boldsymbol{r}_i)$$
$$= \sum_i \left\{\frac{1}{2}m\left[v_i^2 - 2\boldsymbol{u}\cdot\boldsymbol{v}_i + u^2\right] + \frac{1}{2}\sum_{j\neq i}\Phi(\boldsymbol{r}_i,\boldsymbol{r}_j)\right\}\delta(\boldsymbol{r}-\boldsymbol{r}_i)$$
$$\equiv \hat{\varepsilon}(\mathbf{\Gamma},\boldsymbol{r}) - \boldsymbol{u}(\boldsymbol{r})\cdot\hat{\boldsymbol{g}}(\mathbf{\Gamma},\boldsymbol{r}) + \frac{1}{2}u^2(\boldsymbol{r})\hat{\rho}(\mathbf{\Gamma},\boldsymbol{r}). \tag{6.83}$$

In this equation, $\boldsymbol{r}_i$ is the canonical position variable of the ith-particle with $m\boldsymbol{v}_i$ its momentum, $\Phi(\boldsymbol{r}_i,\boldsymbol{r}_j)$ is the two-body potential energy function [see Eq. (4.18)], the quantities $\hat{\rho}(\mathbf{\Gamma},\boldsymbol{r})$, $\hat{\boldsymbol{g}}(\mathbf{\Gamma},\boldsymbol{r})$, and $\hat{\varepsilon}(\mathbf{\Gamma},\boldsymbol{r})$ are local microscopic densities of mass, momentum, and total energy[41] [defined in Chapter 4, see Eqs. (4.30), (4.32), and (4.35)], and $\boldsymbol{u}(\boldsymbol{r})$ is the mean local velocity [see Eq. (4.33)],

$$\boldsymbol{u}(\boldsymbol{r}) \equiv \frac{\langle\hat{\boldsymbol{g}}(\mathbf{\Gamma},\boldsymbol{r})\rangle}{\langle\hat{\rho}(\mathbf{\Gamma},\boldsymbol{r})\rangle}. \tag{6.84}$$

Equation (6.83) specifies a microscopic *internal* energy density, a quantity not defined in Section 4.4.1. The average of Eq. (6.83), combined with Eq. (6.84), reproduces Eq. (4.59) for the internal energy, $\langle\hat{\mathcal{E}}\rangle = \langle\hat{\varepsilon}\rangle - \frac{1}{2}\langle\hat{\rho}\rangle u^2$. A term could be added to Eq. (6.83) having zero average.

We denote the N-body distribution[42] $F(t)$. From Liouville's equation (4.5), $\partial F(t)/\partial t = LF(t)$ [$L \equiv -\mathrm{i}\Lambda$; see Eq. (4.4)], implying $F(t) = \mathrm{e}^{Lt}F(0)$. We can then write

$$F(t) = F(0) + [F(t) - F(0)] = F(0) + \int_0^t \mathrm{d}\tau \frac{\partial}{\partial\tau}F(\tau)$$
$$= F(0) + \int_0^t \mathrm{d}\tau LF(\tau) = F(0) + \int_0^t \mathrm{d}\tau \mathrm{e}^{L\tau}LF(0).$$

Assume the system at $t = 0$ is in the local equilibrium state, $F(0) = F^{\text{LE}}$. In that case,

$$F(t) = F^{\text{LE}} + \int_0^t \mathrm{d}\tau \mathrm{e}^{L\tau}LF^{\text{LE}}. \tag{6.85}$$

Clearly we need to find the action of L on F^{LE} to have an expression for the nonequilibrium distribution $F(t)$.[43] It's straightforward to show that (see Exercise 6.17),

$$LF^{\text{LE}} = -F^{\text{LE}}\int \mathrm{d}\boldsymbol{r}\frac{1}{kT(\boldsymbol{r})}L\hat{\mathcal{E}}(\mathbf{\Gamma},\boldsymbol{r}). \tag{6.86}$$

The problem therefore reduces to finding the action of L on $\hat{\mathcal{E}}(\mathbf{\Gamma},\boldsymbol{r})$. Fortunately we've already done much of the work.

Noting that $\hat{\rho},\hat{\boldsymbol{g}},\hat{\varepsilon}$ represent observables, we have from Liouville's equation (see Exercise 4.2),

$$L\begin{Bmatrix}\hat{\rho}\\ \hat{\boldsymbol{g}}\\ \hat{\varepsilon}\end{Bmatrix} = -\frac{\partial}{\partial t}\begin{Bmatrix}\hat{\rho}\\ \hat{\boldsymbol{g}}\\ \hat{\varepsilon}\end{Bmatrix} = \begin{Bmatrix}\boldsymbol{\nabla}\cdot\hat{\boldsymbol{g}}\\ \boldsymbol{\nabla}\cdot\hat{\mathbf{P}}\\ \boldsymbol{\nabla}\cdot\hat{\boldsymbol{J}}_{\varepsilon}\end{Bmatrix}, \tag{6.87}$$

[41] In Section 4.4 we denoted microscopic densities with "hats," $\hat{\rho}(\boldsymbol{r}), \hat{\boldsymbol{g}}(\boldsymbol{r})$, and $\hat{\varepsilon}(\boldsymbol{r})$. We include the notation $\mathbf{\Gamma}$ in Eq. (6.83) to distinguish microscopic Γ-space functions [such as $\hat{\rho}(\mathbf{\Gamma},\boldsymbol{r})$] from the average quantity $\boldsymbol{u}(\boldsymbol{r})$.

[42] We haven't had much occasion to use the N-body distribution, which we denoted $\rho(\mathbf{\Gamma},t)$ in Section 4.1. We used f_s for s-body distributions in Section 4.2, with F_s the s-tuple distribution. The Boltzmann equation is for the one-particle distribution function f_1, where we've generally dropped the subscript.

[43] If the local equilibrium approximation F^{LE} were truly the equilibrium distribution, we'd have $LF^{\text{LE}} = 0$.

where we've used Eqs. (4.65), (4.69), and (4.71), the microscopic balance equations for mass, momentum, and energy (in the absence of external forces), where $\hat{\mathbf{P}} = \hat{\mathbf{P}}^K + \hat{\mathbf{P}}^\Phi$ is the microscopic momentum flux tensor [see Eqs. (4.66) and (4.68)] and $\hat{\boldsymbol{J}}_\varepsilon$ is the microscopic total energy flux; see Eq. (4.72). From Eqs. (6.83) and (6.87),

$$L\hat{\mathcal{E}}(\boldsymbol{\Gamma}, \boldsymbol{r}) = \boldsymbol{\nabla} \cdot \hat{\boldsymbol{J}}_\varepsilon(\boldsymbol{\Gamma}, \boldsymbol{r}) - \boldsymbol{u}(\boldsymbol{r}) \cdot \boldsymbol{\nabla} \cdot \hat{\mathbf{P}}(\boldsymbol{\Gamma}, \boldsymbol{r}) + \frac{1}{2} u^2(\boldsymbol{r}) \boldsymbol{\nabla} \cdot \hat{\boldsymbol{g}}(\boldsymbol{\Gamma}, \boldsymbol{r}). \tag{6.88}$$

Note the three spatial derivatives; we're going to integrate by parts.[44] From Eq. (6.86) [dropping the dependence on $(\boldsymbol{\Gamma}, \boldsymbol{r})$ in the terms for $\hat{\mathcal{E}}(\boldsymbol{\Gamma}, \boldsymbol{r})$],

$$\begin{aligned}
\int \mathrm{d}\boldsymbol{r} \frac{L\hat{\mathcal{E}}(\boldsymbol{r})}{kT(\boldsymbol{r})} &= \int \mathrm{d}\boldsymbol{r} \frac{1}{kT(\boldsymbol{r})} \left\{ \sum_\alpha \nabla_\alpha \hat{J}_{\varepsilon,\alpha} - \sum_{\alpha,\beta} u_\beta \nabla_\alpha \hat{P}_{\beta\alpha} + \frac{1}{2} u^2 \sum_\alpha \nabla_\alpha \hat{g}_\alpha \right\} \\
&= \int \mathrm{d}\boldsymbol{r} \sum_\alpha \left\{ -\hat{J}_{\varepsilon,\alpha} + \sum_\beta u_\beta \hat{P}_{\beta\alpha} - \frac{1}{2} u^2 \hat{g}_\alpha \right\} \nabla_\alpha \left(\frac{1}{kT(\boldsymbol{r})} \right) \\
&+ \int \mathrm{d}\boldsymbol{r} \frac{1}{kT(\boldsymbol{r})} \sum_{\alpha,\beta} \left(\hat{P}_{\beta\alpha} - u_\beta \hat{g}_\alpha \right) \nabla_\alpha u_\beta,
\end{aligned} \tag{6.89}$$

where integrated parts vanish. We see that temperature and velocity gradients emerge naturally in this approach. The next logical step would be to combine Eq. (6.89) with Eq. (6.86) and substitute back into Eq. (6.85). Equation (6.89) is quite complicated, however; time for approximations.[45]

We start by noting there is no local dissipation in the local equilibrium state. In this state, the average kinetic part of the heat flux vanishes, as can be seen using the local Maxwellian Eq. (4.129),[46]

$$\boldsymbol{J}_Q^{\text{LE}}(\boldsymbol{r}, t) \approx \frac{m}{2} \int (\boldsymbol{v} - \boldsymbol{u})(\boldsymbol{v} - \boldsymbol{u})^2 f_{\text{LM}}(\boldsymbol{r}, \boldsymbol{v}, t) \mathrm{d}\boldsymbol{v} = 0 \tag{6.90}$$

[integrand is odd in $(\boldsymbol{v} - \boldsymbol{u})$], and the average kinetic part of the pressure tensor is diagonal with the local pressure of an ideal gas on the diagonal,[47]

$$P_{\alpha\beta}^{\text{LE}}(\boldsymbol{r}, t) \approx m \int \mathrm{d}\boldsymbol{v} (\boldsymbol{v} - \boldsymbol{u})_\alpha (\boldsymbol{v} - \boldsymbol{u})_\beta f_{\text{LM}}(\boldsymbol{r}, \boldsymbol{v}, t) = n(\boldsymbol{r}, t) kT(\boldsymbol{r}, t) \delta_{\alpha,\beta} \equiv P(\boldsymbol{r}, t) \delta_{\alpha,\beta}. \tag{6.91}$$

Thus, *spatial inhomogeneities alone are insufficient to drive dissipative transport processes*; the distribution function must have deviations from local-equilibrium form, as we see in Eq. (6.85) and what we found in Chapman-Enskog theory, see Eq. (4.180).

Under the substitutions $\boldsymbol{v}_i \to \boldsymbol{u} + (\boldsymbol{v}_i - \boldsymbol{u})$ in $\hat{\mathbf{P}}^K$ [see Eq. (4.66)], we have for the full pressure tensor $\hat{\mathbf{P}}$,

$$\begin{aligned}
\hat{P}_{\alpha\beta}(\boldsymbol{\Gamma}, \boldsymbol{r}) =& \hat{\rho}(\boldsymbol{\Gamma}, \boldsymbol{r}) u_\alpha u_\beta + [\hat{g}_\alpha(\boldsymbol{\Gamma}, \boldsymbol{r}) - \hat{\rho}(\boldsymbol{\Gamma}, \boldsymbol{r}) u_\alpha] u_\beta + [\hat{g}_\beta(\boldsymbol{\Gamma}, \boldsymbol{r}) - \hat{\rho}(\boldsymbol{\Gamma}, \boldsymbol{r}) u_\beta] u_\alpha \\
&+ \hat{P}(\boldsymbol{\Gamma}, \boldsymbol{r}) \delta_{\alpha,\beta} + \hat{P}_{\alpha\beta}^{\text{diss}}(\boldsymbol{\Gamma}, \boldsymbol{r}),
\end{aligned} \tag{6.92}$$

where $\hat{P}(\boldsymbol{\Gamma}, \boldsymbol{r}) = \frac{1}{3} \sum_\alpha \hat{P}_{\alpha\alpha}(\boldsymbol{\Gamma}, \boldsymbol{r})$ is the local microscopic pressure [see Eq. (4.50)] and $\hat{P}_{\alpha\beta}^{\text{diss}}$ is the remainder of $\hat{\mathbf{P}}$ having the important property that it vanishes in local equilibrium,

$$\langle \hat{P}_{\alpha\beta}^{\text{diss}} \rangle^{\text{LE}} = 0. \tag{6.93}$$

[44] The quantity $L\hat{\mathcal{E}}$ is the time derivative of $-\hat{\mathcal{E}}$. Thus, in Eq. (6.88) we have a time derivative expressed in terms of spatial derivatives, similar to that in the Chapman-Enskog theory; see Eq. (4.160).

[45] The starting point Eq. (6.82) is itself an approximation. Theoretical physics is the art of approximation: It's better to be approximately right than precisely wrong.

[46] The average heat flux $J_Q = \widetilde{J}^K + \widetilde{J}^\Phi$ [see Eq. (4.60)], where $\widetilde{J}^K, \widetilde{J}^\Phi$ are average fluxes of kinetic and potential energy associated with thermal motions; see Eq. (4.58). Equation (6.90) assumes negligible potential energy contributions.

[47] The average pressure tensor $\mathbf{P} = \widetilde{\mathbf{P}}^K + \mathbf{P}^\Phi$, with $\mathbf{P}^\Phi$ the average momentum flux associated with potential energy [see Eq. (4.45)] and $\widetilde{\mathbf{P}}^K$ that due to thermal motions [see Eq. (4.48)]. Equation (6.91) ignores potential energy contributions.

Thus, we've separated momentum flux into convective and dissipative parts. Applying the same substitution to the energy flux $\hat{\boldsymbol{J}}_\varepsilon$, we have a separation into convective and dissipative parts,

$$\hat{J}_{\varepsilon,\alpha}(\boldsymbol{\Gamma},\boldsymbol{r}) = u_\alpha\hat{\varepsilon}(\boldsymbol{\Gamma},\boldsymbol{r}) + \frac{1}{2}u^2\left[\hat{g}_\alpha(\boldsymbol{\Gamma},\boldsymbol{r}) - u_\alpha\hat{\rho}(\boldsymbol{\Gamma},\boldsymbol{r})\right] + u_\alpha\hat{P}(\boldsymbol{\Gamma},\boldsymbol{r}) + \sum_\beta u_\beta\hat{P}^{\text{diss}}_{\alpha\beta}(\boldsymbol{\Gamma},\boldsymbol{r}) + \hat{J}^{\text{diss}}_{\varepsilon,\alpha}(\boldsymbol{\Gamma},\boldsymbol{r}), \tag{6.94}$$

where the remainder has the property

$$\langle\hat{J}^{\text{diss}}_{\varepsilon,\alpha}\rangle^{\text{LE}} = 0. \tag{6.95}$$

The flows represented by $\hat{P}^{\text{diss}}_{\alpha\beta}$ and $\hat{J}^{\text{diss}}_{\varepsilon,\alpha}$ (being zero in local equilibrium) are measures of the response of the system to the presence of gradients and play the same role as the response $\langle\Delta B(t)\rangle$ in linear response theory; see Section 6.2. Let's calculate the average dissipative energy flow,

$$\begin{aligned}
\langle\hat{J}^{\text{diss}}_{\varepsilon,\alpha}(\boldsymbol{r})\rangle_t &= \int \mathrm{d}\boldsymbol{\Gamma}\,\hat{J}^{\text{diss}}_{\varepsilon,\alpha}(\boldsymbol{\Gamma},\boldsymbol{r})F(\boldsymbol{\Gamma},t) = \int \mathrm{d}\boldsymbol{\Gamma}\,\hat{J}^{\text{diss}}_{\varepsilon,\alpha}(\boldsymbol{\Gamma},\boldsymbol{r})\int_0^t \mathrm{d}\tau\mathrm{e}^{L\tau}LF^{\text{LE}}(\boldsymbol{\Gamma}) \\
&= \int \mathrm{d}\boldsymbol{\Gamma}\,\hat{J}^{\text{diss}}_{\varepsilon,\alpha}(\boldsymbol{\Gamma},\boldsymbol{r})\int_0^t \mathrm{d}\tau\mathrm{e}^{L\tau}F^{\text{LE}}(\boldsymbol{\Gamma})\int \mathrm{d}\boldsymbol{r}'\frac{1}{kT(\boldsymbol{r}')}\left(-L\hat{\mathcal{E}}(\boldsymbol{\Gamma},\boldsymbol{r}')\right) \\
&= \int \mathrm{d}\boldsymbol{\Gamma}\,\hat{J}^{\text{diss}}_{\varepsilon,\alpha}(\boldsymbol{\Gamma},\boldsymbol{r})\int_0^t \mathrm{d}\tau\mathrm{e}^{L\tau}F^{\text{LE}}(\boldsymbol{\Gamma})\int \mathrm{d}\boldsymbol{r}'\times \\
&\quad\left\{\sum_\beta\left(\hat{J}_{\varepsilon,\beta}(\boldsymbol{\Gamma},\boldsymbol{r}') - \sum_\gamma u_\gamma(\boldsymbol{r}')\hat{P}_{\gamma\beta}(\boldsymbol{\Gamma},\boldsymbol{r}') + \frac{1}{2}u^2(\boldsymbol{r}')\hat{g}_\beta(\boldsymbol{\Gamma},\boldsymbol{r}')\right)\nabla_\beta\left(\frac{1}{kT(\boldsymbol{r}')}\right)\right. \\
&\quad\left.-\frac{1}{kT(\boldsymbol{r}')}\sum_{\beta,\gamma}\left[\hat{P}_{\gamma\beta}(\boldsymbol{\Gamma},\boldsymbol{r}') - u_\gamma(\boldsymbol{r}')\hat{g}_\beta(\boldsymbol{\Gamma},\boldsymbol{r}')\right]\nabla_\beta u_\gamma(\boldsymbol{r}')\right\},
\end{aligned} \tag{6.96}$$

where in the second equality we've used Eq. (6.85), in the third, Eq. (6.86), and in the fourth, Eq. (6.89). Equation (6.96) is quite complicated. It simplifies, however, when we recognize that we're only interested in the linear response. We can therefore evaluate the quantities multiplying gradients at lowest order, the state of $\boldsymbol{u} = 0$, implying the local equilibrium distribution function reduces to that of the canonical ensemble, $F^{\text{LE}}(\boldsymbol{\Gamma}) \to f_0(\boldsymbol{\Gamma}) \equiv Z^{-1}\exp(-\beta H(p,q))$. In this approximation, Eq. (6.96) reduces to

$$\langle\hat{J}^{\text{diss}}_{\varepsilon,\alpha}(\boldsymbol{r})\rangle_t = \sum_\beta\left[\int \mathrm{d}\boldsymbol{\Gamma}\,\hat{J}^{\text{diss}}_{\varepsilon,\alpha}(\boldsymbol{\Gamma},\boldsymbol{r})\int_0^t \mathrm{d}\tau\mathrm{e}^{L\tau}f_0(\boldsymbol{\Gamma})\int \mathrm{d}\boldsymbol{r}'\,\hat{J}^{\text{diss}}_{\varepsilon,\beta}(\boldsymbol{\Gamma},\boldsymbol{r}')\right]\nabla_\beta\left(\frac{1}{kT(\boldsymbol{r})}\right). \tag{6.97}$$

In arriving at Eq. (6.97), we've: 1) used that the vector flow $\hat{J}^{\text{diss}}_{\varepsilon,\alpha}$ cannot couple to the tensor flow $\hat{P}^{\text{diss}}_{\gamma\beta}$ (Curie's theorem,[48] Section 1.5); 2) evaluated the gradient locally at position $\boldsymbol{r}$; and 3) used that $\hat{J}_{\varepsilon,\alpha}$ in Eq. (6.94) reduces to $\hat{J}^{\text{diss}}_{\varepsilon,\alpha}$ in this approximation.

The terms in square brackets in Eq. (6.97) are basically the thermal conductivity tensor. This can be brought out with a few steps. Define a spatially averaged flux,

$$\mathcal{J}_\alpha(\boldsymbol{\Gamma}) \equiv \frac{1}{V}\int \mathrm{d}\boldsymbol{r}\,\hat{J}^{\text{diss}}_{\varepsilon,\alpha}(\boldsymbol{\Gamma},\boldsymbol{r}) \tag{6.98}$$

[48] Our treatment therefore relies on the phenomenological theory of Chapter 1. We note that Luttinger's nominally microscopic treatment[198] also requires the validity of the phenomenological theory.

where V is the system volume. Then, from Eq. (6.97),

$$\begin{aligned}\langle \mathcal{J}_\alpha \rangle &= V \sum_\beta \left[\int \mathrm{d}\boldsymbol{\Gamma} \mathcal{J}_\alpha(\boldsymbol{\Gamma}) \int_0^t \mathrm{d}\tau \mathrm{e}^{L\tau} f_0(\boldsymbol{\Gamma}) \mathcal{J}_\beta(\boldsymbol{\Gamma}) \right] \nabla_\beta \left(\frac{1}{kT(\boldsymbol{r})} \right) \\ &= -\frac{V}{kT^2} \sum_\beta \left[\int_0^t \mathrm{d}\tau \int \mathrm{d}\boldsymbol{\Gamma} f_0(\boldsymbol{\Gamma}) \mathcal{J}_\beta(\boldsymbol{\Gamma}) \left(\mathrm{e}^{-L\tau} \mathcal{J}_\alpha(\boldsymbol{\Gamma}) \right) \right] \nabla_\beta T. \end{aligned} \tag{6.99}$$

Assume the time integral reaches a plateau value over a time on the order of the relaxation time. The limit of integration can then be extended to infinity, leaving us with an expression for the *thermal conductivity tensor*,

$$\kappa_{\alpha\beta} = \frac{V}{kT^2} \int_0^\infty \mathrm{d}\tau \langle \mathcal{J}_\alpha(\tau) \mathcal{J}_\beta(0) \rangle^0, \tag{6.100}$$

a result in the Green-Kubo form, proportional to an integral of the heat flow autocorrelation function. An analogous calculation for the viscosity of an isotropic system yields

$$\eta = \frac{V}{kT} \int_0^\infty \mathrm{d}\tau \langle \mathcal{P}_{xy}(\tau) \mathcal{P}_{xy}(0) \rangle^0, \tag{6.101}$$

where $\mathcal{P}_{\alpha\beta} \equiv V^{-1} \int \mathrm{d}\boldsymbol{r} \hat{P}^{\text{diss}}_{\alpha\beta}(\boldsymbol{r})$. Viscosity is proportional to the time integral of the momentum flux autocorrelation function.

6.4.3 Einstein relation

In Section 3.1 we found a fundamental connection between transport coefficients, Eq. (3.23) (reproduced here),

$$D = \frac{kT}{q} \mu, \tag{3.23}$$

where D is the diffusion coefficient, μ is the electrical mobility, and q is the charge of the particle. Mobility is a mechanical transport coefficient, the proportionality factor between drift velocity and applied electric field, $\langle \boldsymbol{v} \rangle_{\text{d}} = \mu \boldsymbol{E}$, and D is a thermal transport coefficient, the proportionality between particle current and density gradient, $\boldsymbol{J}_n = -D \boldsymbol{\nabla} n$. Is the Einstein relation "contained" in a Green-Kubo formula? The answer is yes.

Start with the zero-frequency conductivity, from Eq. (6.80),

$$\sigma_{\mu\nu} \equiv \sigma_{\mu\nu}(\omega = 0) = \frac{V}{kT} \int_0^\infty \mathrm{d}t \langle J_\mu(0) J_\nu(t) \rangle^0, \tag{6.102}$$

where $J_\mu = nqv_\mu$ is the electrical current density with v_μ the μth-velocity component. Thus,

$$\sigma_{\mu\nu} = \frac{V}{kT} n^2 q^2 \int_0^\infty \mathrm{d}t \langle v_\mu(0) v_\nu(t) \rangle^0. \tag{6.103}$$

Allowing that at long times $\langle v_\mu(0) v_\nu(t) \rangle^0 \to 0$ as $t \to \infty$, we can take

$$\lim_{t \to \infty} \langle v_\mu(0) v_\nu(t) \rangle^0 = \langle v_\mu \rangle^0 \langle v_\nu \rangle^0 = 0. \tag{6.104}$$

Then, we can write the integral in Eq. (6.103), using Eq. (6.104),

$$\begin{aligned}\int_0^\infty \mathrm{d}t\langle v_\mu(0)v_\nu(t)\rangle^0 &= \lim_{T\to\infty}\frac{1}{2T}\int_0^T\int_0^T \langle v_\mu(t)v_\nu(t')\rangle^0 \mathrm{d}t\mathrm{d}t' \\ &= \lim_{T\to\infty}\frac{1}{2T}\int_0^T\int_0^T \left\langle \frac{\mathrm{d}x_\mu}{\mathrm{d}t}\frac{\mathrm{d}x_\nu}{\mathrm{d}t'}\right\rangle^0 \mathrm{d}t\mathrm{d}t' \\ &= \lim_{T\to\infty}\frac{1}{2T}\langle (x_\mu(T)-x_\mu(0))\left(x_\nu(T)-x_\nu(0)\right)\rangle^0 \\ &\equiv D_{\mu\nu}, \end{aligned} \tag{6.105}$$

where the final limit defines a diffusion tensor[49] as a generalization of Eq. (2.62). The factor of 2 is to prevent over counting; from the stationarity of equilibrium averages, $\langle v_\mu(t)v_\nu(t')\rangle^0 = f(|t-t'|)$. Combining Eq. (6.105) with Eq. (6.103),

$$\sigma_{\mu\nu} = \frac{V}{kT}n^2q^2D_{\mu\nu} = N\frac{nq^2}{kT}D_{\mu\nu}, \tag{6.106}$$

a generalized Einstein relation. A connection between conductivity and diffusivity is inevitable—charge is attached to particles and particles undergo diffusive motion.[50]

6.4.4 Comments

Transport coefficients can be expressed as integrals of autocorrelation functions of microscopic flows, formulae that are remarkably compact. Green-Kubo relations, however, may not be the best starting point for practical calculations. Evaluating autocorrelation functions is a difficult undertaking; one is faced with the full N-body problem, and the use of local equilibrium is a strong assumption difficult to justify in many-body theory. We saw in Chapter 4 that thermal transport coefficients can be calculated in Chapman-Enskog theory—also not an easy task. Resibois[202] showed the equivalence of thermal transport coefficients obtained from kinetic equations and correlation functions.

6.5 GENERALIZED LANGEVIN EQUATION

In Chapter 3, we introduced the Langevin equation featuring a random force as a way to model Brownian motion, colloidal particles incessantly in motion as a result of collisions with molecules of the suspension medium. The Langevin equation could be considered an "inspired guess" as a generalization of Newton's second law with the random force a new kind of force. In nonequilibrium statistical physics we seek the slowest processes in the approach to equilibrium. Is there a systematic way to develop Langevin-like equations for arbitrary slow modes? There is, and such equations are known as generalized Langevin equations. We follow Mori[203].

6.5.1 Derivation

Suppose we have a set of dynamical variables $\{A_1,\ldots,A_f\}$ that we deem "relevant" (most likely because they represent macroscopic degrees of freedom). These variables comprise a subspace (the $\boldsymbol{A}$-space) of the infinite-dimensional space of *all* dynamical variables, for which a scalar product

[49] One way to define tensors of a given rank is through products of lower-rank tensors; in this case a second-rank tensor ($D_{\mu\nu}$ or $\sigma_{\mu\nu}$) is defined by the outer product of vectors (first-rank tensors), $x_\mu x_\nu$ or $v_\mu v_\nu$. See [6, p81].

[50] Conduction occurs in charge-neutral systems, the motion of light charge carriers through a system of more massive ions. We can treat charge carriers as ostensibly independent particles. Note that $\mu = \sigma/(nq)$

must be specified before it can be called Hilbert space.[51] We indicate scalar products with the notation,

$$(U,V) \equiv \begin{cases} \int \mathrm{d}\boldsymbol{\Gamma} f_{\text{eq}}(\boldsymbol{\Gamma}) U^*(\boldsymbol{\Gamma}) V(\boldsymbol{\Gamma}) & \text{(classical)} \\ \mathrm{Tr}\, \rho_{\text{eq}} U^\dagger V. & \text{(quantum)} \end{cases} \tag{6.107}$$

Note that, by definition,[52] $(U,V)^* = (V,U)$ and that we're using correlation functions to define the scalar product, $C_{AB}(t) = (A, B(t))$. The quantities $A_1, \ldots, A_f$ comprise a basis for $\boldsymbol{A}$-space.[53] We can define an operator $\mathcal{P}$ that projects from an arbitrary dynamical variable X its components along the $\boldsymbol{A}$-variables,

$$\mathcal{P}X \equiv \sum_{k=1}^{f}\sum_{l=1}^{f} A_k G_{kl}^{-1} \left(A_l, X\right), \tag{6.108}$$

where G_{kl}^{-1} is the matrix inverse of $G_{ij} \equiv (A_i, A_j)$. $\mathcal{P}$ satisfies the defining property of projection operators, $\mathcal{P} \circ \mathcal{P}X = \mathcal{P}X$ for all X; see Exercise 6.18.[54]

We have the identity,

$$X = \mathcal{P}X + \left[\mathcal{I} - \mathcal{P}\right] X, \tag{6.109}$$

where $\mathcal{I}$ is the identity operator. We can consider $\mathcal{P}X$ the macroscopic part of X and $[\mathcal{I} - \mathcal{P}]X$ its microscopic part, an interpretation based on the orthogonality of $[\mathcal{I} - \mathcal{P}]\, X$ and $\mathcal{P}X$ (see Exercise 6.18),

$$\left(\mathcal{P}X, \left[\mathcal{I} - \mathcal{P}\right] X\right) = 0. \tag{6.110}$$

Thus, $[\mathcal{I} - \mathcal{P}]\, X$ has no component along the macroscopic part $\mathcal{P}X$ and what is not macroscopic is microscopic.[55] We encountered that idea in our treatment of the Langevin equation where the random force is independent of B-particle velocity; see Eq. (3.10). It's traditional to denote the projection operator $\mathcal{I} - \mathcal{P} \equiv \mathcal{Q}$. One can show that $\mathcal{Q} \circ \mathcal{Q} = \mathcal{Q}$. The Liouville operator can be split into orthogonal parts[56] by applying the identity $\mathcal{I} = \mathcal{P} + \mathcal{Q}$,

$$L = \mathcal{I}L = \mathcal{P}L + \mathcal{Q}L. \tag{6.111}$$

Here $\mathcal{P}L$ represents the time rate of change of the "relevant" components[57] of X with $\mathcal{Q}L$ the rate of change of the components of X orthogonal to $\mathcal{P}X$.

We seek the decomposition of the time evolution operator. We state the result and then discuss how it's found,

$$\mathrm{e}^{Lt} = \mathrm{e}^{\mathcal{Q}Lt} + \int_0^t \mathrm{d}\tau \mathrm{e}^{L(t-\tau)} \mathcal{P}L \mathrm{e}^{\mathcal{Q}L\tau}. \tag{6.112}$$

This identity can be verified directly through differentiation (see Exercise 6.19), but let's derive it. The Laplace transform of e^{Lt} generates the resolvent (see Appendix D) of L,[204, p341]

$$\int_0^\infty \mathrm{e}^{-zt} \mathrm{e}^{Lt} \mathrm{d}t = \frac{1}{z - L}, \tag{6.113}$$

---

[51] Hilbert space is a complete inner-product space[13, p45]. It's understood that Hilbert space is infinite dimensional; finite-dimensional spaces are automatically complete. The term "finite-dimensional Hilbert space" is a misnomer.

[52] Scalar products, like many quantities in mathematics, are defined by axioms; see [13, p10]. The property $(U,V)^* = (V,U)$ is an axiom for the inner product used in many areas of physics.

[53] Any N linearly independent vectors form a basis for an N-dimensional space. It won't necessarily be an orthonormal basis, but such bases can always be constructed with the Gram-Schmidt method; see [13, p14].

[54] A projection operator $\mathcal{P}$ on Hilbert space $\mathcal{H}$ is a linear mapping $\mathcal{P}\colon \mathcal{H} \to \mathcal{H}$ such that $\mathcal{P} \circ \mathcal{P} = \mathcal{P}$.

[55] This is the same logic used in Chapter 1 to derive the entropy balance equation, the division of internal energy into heat and work; see discussion near Eq. (1.34). In [2, p10] we noted that, "The division of internal energy into work and heat lines up with the distinction between macroscopic and non-macrosopic, i.e., microscopic. ... Work is associated with changes in observable macroscopic quantities. What is not work is heat, energy transferred to microscopic degrees of freedom."

[56] For any X, $(\mathcal{P}LX, \mathcal{Q}LX) = 0$. Note that $\mathcal{P}\mathcal{Q} \equiv 0$.

[57] From $LX(\boldsymbol{\Gamma}) = -\partial X/\partial t$, $\mathcal{P}LX = -\mathcal{P}\partial X/\partial t = -(\partial/\partial t)\mathcal{P}X$, because $\mathcal{P}$ has no explicit time dependence.

for $|z| > ||L||$, the operator norm of L (a condition we blithely assume). Using the identity in Eq. (D.21),[58]

$$\frac{1}{z-A} = \frac{1}{z-B} - \frac{1}{z-A}(B-A)\frac{1}{z-B}, \tag{6.114}$$

we have, with $B \equiv \mathcal{Q}L$ and $A \equiv L = (\mathcal{P}+\mathcal{Q})L$,

$$\frac{1}{z-L} = \frac{1}{z-\mathcal{Q}L} + \frac{1}{z-L}\mathcal{P}L\frac{1}{z-\mathcal{Q}L}. \tag{6.115}$$

The inverse Laplace transform of Eq. (6.115) (with the convolution theorem) is Eq. (6.112).

Now for some algebraic steps. First, using Eq. (6.108),

$$\mathrm{e}^{Lt}\mathcal{P}LA_j = \sum_k A_k(t)\sum_l G^{-1}_{kl}(A_l, LA_j) \equiv \sum_k A_k(t)\Omega_{kj}, \tag{6.116}$$

where the Ω_{kj} comprise the elements of a matrix of frequencies

$$\Omega_{kj} = \sum_l G^{-1}_{kl}(A_l, LA_j) = \sum_l G^{-1}_{kl}\left(A_l, \frac{\mathrm{d}}{\mathrm{d}t}A_j\Big|_{t=0}\right) \equiv \sum_l G^{-1}_{kl}\left(A_l, \dot{A}_j(0)\right). \tag{6.117}$$

In Eq. (6.116), $\mathrm{e}^{Lt}\mathcal{P}LA_j$, the time development of the projection of the initial slope $\dot{A}_j(0)$ onto $\boldsymbol{A}$-space, is represented by a linear combination of the $A_k(t)$ weighted by Ω_{kj}. If all $\boldsymbol{A}$-variables have the same signature under time reversal, then $\Omega_{kj} = 0$ [because in that case $\left(A_l(0), \dot{A}_j(0)\right) = 0$].[59] Then, with

$$\frac{\mathrm{d}}{\mathrm{d}t}\left(\mathrm{e}^{Lt}A_j\right) = \mathrm{e}^{Lt}LA_j = \mathrm{e}^{Lt}\mathcal{P}LA_j + \mathrm{e}^{Lt}\mathcal{Q}LA_j,$$

we have, using Eq. (6.116),

$$\frac{\mathrm{d}}{\mathrm{d}t}A_j(t) - \sum_k A_k(t)\Omega_{kj} = \underbrace{\mathrm{e}^{\mathcal{Q}Lt}\mathcal{Q}LA_j}_{f_j(t)} + \int_0^t \mathrm{d}\tau \mathrm{e}^{L(t-\tau)}\mathcal{P}L\underbrace{\mathrm{e}^{\mathcal{Q}L\tau}\mathcal{Q}LA_j}_{f_j(\tau)}, \tag{6.118}$$

where we've right-multiplied Eq. (6.112) by $\mathcal{Q}LA_j$. We denote by $f_j(t)$ the term $\mathrm{e}^{\mathcal{Q}Lt}\mathcal{Q}LA_j$, the time development (under $\mathrm{e}^{\mathcal{Q}Lt}$) of the projection of the initial slope $\dot{A}_j(0)$ onto the orthogonal complement of $\boldsymbol{A}$-space. It plays the role of a driving force in the equation of motion for $A_j(t)$ and is called the "random force," yet dimensionally $[f_j] = [A_j]/\text{time}$. There is a separate force term for each $\boldsymbol{A}$-variable, each having no projection onto $\boldsymbol{A}$-space, $\mathcal{P}f_j(t) \equiv 0$. The f_j are therefore uncorrelated with $\boldsymbol{A}$-variables, $(A_i, f_j(t)) = (\mathcal{P}A_i, f_j(t)) = (A_i, \mathcal{P}f_j(t)) = 0$, where we've used $\mathcal{P}A_j = A_j$. In that regard, $f_j(t)$ behaves like the random force postulated in the Langevin equation; see Eq. (3.10). The lack of correlation between $f_j(t)$ and initial values $A_i(0)$ lies at the foundation of the Onsager regression hypothesis.

The integrand of Eq. (6.118) involves $\mathcal{P}Lf_j$, the projection of the initial slope $Lf_j \equiv \dot{f}_j(0)$ onto $\boldsymbol{A}$-space,

$$\begin{aligned}\mathcal{P}Lf_j(\tau) &= \sum_k\sum_l A_k G^{-1}_{kl}\left(A_l, Lf_j(\tau)\right) = -\sum_k\sum_l A_k G^{-1}_{kl}\left(LA_l, f_j(\tau)\right) \\ &= -\sum_k\sum_l A_k G^{-1}_{kl}\left((\mathcal{P}+\mathcal{Q})LA_l, f_j(\tau)\right) = -\sum_k\sum_l A_k G^{-1}_{kl}\left(\mathcal{Q}LA_l, f_j(\tau)\right) \\ &= -\sum_k\sum_l A_k G^{-1}_{kl}\left(f_l(0), f_j(\tau)\right) \equiv -\sum_k A_k M_{kj}(\tau),\end{aligned} \tag{6.119}$$

[58] Equation (6.113) is the negative of the resolvent defined in Appendix D; let $R \to -R$ in Eq. (D.21).

[59] Said differently, the frequencies Ω_{kj} vanish unless A_l and A_j have different time-reversal symmetries.

where we've used that[60] $(X, LY) = -(LX, Y)$ for any X, Y and we've introduced *memory functions*,

$$M_{kj}(t) \equiv \sum_l G^{-1}_{kl} \left(f_l(0), f_j(t)\right). \tag{6.120}$$

Memory functions are related to time correlation functions of random forces, which Kubo[193] termed the *second fluctuation-dissipation theorem.*[61] We note that the force-force correlation function is stationary,

$$(f_i(t), f_j(t')) = \left(e^{QLt}QLA_i, e^{QLt'}QLA_j\right) = \left(\dot{A}_i(0), Qe^{QL(t'-t)}Q\dot{A}_j(0)\right). \tag{6.121}$$

With these definitions, Eq. (6.118) can be written[62]

$$\frac{d}{dt}A_j(t) - \sum_k A_k(t)\Omega_{kj} + \sum_k \int_0^t d\tau A_k(t-\tau)M_{kj}(\tau) = f_j(t). \tag{6.122}$$

Equation (6.122), the *generalized Langevin equation*, is an exact rewrite of the microscopic equations of motion. It was derived by Mori[203]; see discussions in Forster[205] and Zwanzig[206]. It separates the dynamics into three effects associated with $(\Omega_{kl}, M_{kj}(t), f_j(t))$: 1) collective oscillations[63] caused by correlations among $\boldsymbol{A}$-space variables; 2) time-lagged memory effects, mediated by force-force time correlations, where past values of $\boldsymbol{A}$-variables control their present rates of change; and 3) driving forces $f_j(t)$ uncorrelated with $\boldsymbol{A}$-variables. The generalized Langevin equation is not easier to solve than the Liouville equation; it's a platform for introducing approximations.

6.5.2 Recovering the Langevin equation for Brownian motion

Let's verify that the generalized Langevin equation reduces to the Langevin equation for Brownian motion established in Chapter 3. The relevant variables are the velocity components of the B-particle of mass M, $A_j = v_j$, $j = 1, 2, 3$. The frequencies $\Omega_{kj} = 0$ because the v_j all have the same signature under time reversal. The correlation matrix for these variables is diagonal, $G_{ij} = (v_i, v_j) = \delta_{ij} v^2_{\text{th}}$, where $v_{\text{th}} \equiv \sqrt{kT/M}$, implying $G^{-1}_{ij} = \delta_{ij}/v^2_{\text{th}}$. We assume the driving forces vary significantly faster than the particle velocity and can be treated as delta-correlated,

$$(f_l(0), f_j(t)) = \Gamma \delta_{lj} \delta(t), \tag{6.123}$$

where Γ is to be determined. With Eq. (6.123), we have from Eq. (6.120),

$$M_{kj}(t) = \frac{\Gamma}{v^2_{\text{th}}} \delta_{kj} \delta(t). \tag{6.124}$$

There are no memory effects for delta-correlated forces. We therefore have from Eq. (6.122),

$$\frac{d}{dt}v_j + \frac{\Gamma}{2v^2_{\text{th}}} v_j(t) = f_j(t), \tag{6.125}$$

where the factor of two is from integrating over "half" a delta function. Equation (6.125) is identical to the Langevin equation (3.9) when we set $\Gamma = 2kT\alpha/(M^2)$, with α the drag-force friction coefficient. (This choice differs from Eq. (3.14) by the factor of M^2 because f_j in the present case is an acceleration, whereas in the Brownian motion problem the random force is actually a force.)

[60] The Liouville operator Λ as defined in Eq. (4.4) is Hermitian, but we're using the abbreviation $L = i\Lambda$. The operator L is *skew-Hermitian*.

[61] The first fluctuation-dissipation theorem, Eq. (6.65), relates the imaginary part of the generalized susceptibility to the spectral properties of time correlation functions. The second, Eq. (6.120), relates memory functions (and their spectral properties) to time correlation functions of the random force (and their spectral properties).

[62] The time-lagged memory integral can equally well be written $\int_0^t d\tau A_k(\tau) M_{kj}(t-\tau)$.

[63] When we substitute $L = i\Lambda$, these terms represent oscillatory motions.

6.6 MEMORY FUNCTIONS

Let's turn to the larger theme of this chapter, correlation functions. Consider the set of correlation functions among the "relevant" variables introduced in the last section,

$$C_{ij}(t) \equiv (A_i(0), A_j(t)), \tag{6.126}$$

where it's convenient to continue with the inner product notation of Eq. (6.107). Take the time derivative of $C_{ij}(t)$ and use the generalized Langevin equation. We find

$$\frac{\mathrm{d}}{\mathrm{d}t}C_{ij}(t) = \sum_k C_{ik}(t)\Omega_{kj} - \sum_k \int_0^t \mathrm{d}\tau C_{ik}(t-\tau)M_{kj}(\tau). \tag{6.127}$$

The driving force does not appear because $(A_i, f_j) = 0$; it's implicit in the memory function $M_{kj}(\tau)$. Equation (6.127) is the *memory function equation.* It's an exact consequence of the microscopic equations of motion; it's a powerful tool for calculating time correlation functions when particular forms of the memory function are assumed; see Boon and Yip[207].

Noting the convolution form, introduce the Laplace transform

$$\widetilde{C}_{ij}(z) \equiv \int_0^\infty \mathrm{e}^{-zt} C_{ij}(t)\mathrm{d}t. \tag{6.128}$$

Equation (6.127) can then be written in a form that's useful if the Laplace transforms $\widetilde{M}_{kj}(z) \equiv \int_0^\infty \mathrm{e}^{-zt} M_{kj}(t)\mathrm{d}t$ are simpler objects than the $\widetilde{C}_{ik}(z)$,

$$\sum_k \left[z\delta_{kj} - \Omega_{kj} + \widetilde{M}_{kj}(z)\right]\widetilde{C}_{ik}(z) = C_{ij}(t=0). \tag{6.129}$$

One solves Eq. (6.129) for the $\widetilde{C}_{ik}(z)$ and then inverse transforms to infer $C_{ik}(t)$.

SUMMARY

We began this chapter with a review of correlation functions, of which there are different kinds, equilibrium and nonequilibrium. Of particular interest are time correlation functions, $C_{AB}(t) = \langle A(0)B(t)\rangle^0$, where the superscript indicates an average with respect to an equilibrium probability distribution. Time correlation functions comprise the third leg of the stool supporting nonequilibrium statistical mechanics; the other two are stochastic dynamics and kinetic theory.

- Measurements on macroscopic systems are described in terms of correlation functions—a fundamental reason to study them. Spectral properties of correlation functions are directly measured through scattering structure factors, elastic and inelastic. In addition, the response of macroscopic systems to perturbations is expressed in terms of correlation functions. For a time-dependent interaction, the energy of which is $H'(t) = -AF(t)$, where A is a Γ-space function (classical) or a Hermitian operator (quantum) and $F(t)$ is the time history of turning on the interaction with $\lim_{t\to-\infty} F(t) = 0$, the nonequilibrium response of system quantity B to interaction A is specified by the response function $\Phi_{BA}(t)$ with $\Delta B(t) = \int_{-\infty}^t \mathrm{d}\tau \Phi_{BA}(t-\tau)F(\tau)$. The response function is defined with correlation functions,

$$\Phi_{BA}(t) = \begin{cases} \int \mathrm{d}\mathbf{\Gamma} f(\mathbf{\Gamma})\,[A, B(t)]_{\mathrm{P}} & \text{(classical)} \\ 1/(\mathrm{i}\hbar)\,\mathrm{Tr}\,\rho\,[A, B(t)]\,, & \text{(quantum)} \end{cases}$$

  where $f(\mathbf{\Gamma})$ is the phase-space distribution function, $[*,*]_{\mathrm{P}}$ denotes Poisson bracket, ρ is the density matrix, and $[*,*]$ denotes commutator. The response function has the property $\Phi_{BA}(t<0) = 0$. Any function $f(t)$ such that $f(t<0) = 0$ is known as a causal function.

- For a monochromatic interaction with $F(t) \equiv \lim_{\epsilon \to 0^+} E_0 e^{-i\omega t + \epsilon t}$, the response can be written $\Delta B(t) = \chi_{BA}(\omega) E_0 e^{-i\omega t}$, where $\chi_{BA}(\omega)$, the Fourier transform of $\Phi_{BA}(t)$, is known as the generalized susceptibility, $\chi_{BA}(\omega) \equiv \lim_{\epsilon \to 0^+} \int_0^\infty d\theta e^{i(\omega + i\epsilon)\theta} \Phi_{BA}(\theta)$. The quantity $\chi(\omega)$ (we drop BA) has no poles on the real-ω line. Moreover, it has no poles in the upper half of the complex-ω plane. Any poles of $\chi(\omega)$ lie in the lower half of the complex-ω plane. Mathematically (Titchmarsh's theorem), such a function has as a boundary value on the real-ω line a complex-valued function $\chi(\omega): \mathbb{R} \to \mathbb{C}$ with real and imaginary parts satisfying the Kramers-Kronig relations. The real and imaginary parts of $\chi(\omega)$ are therefore not independent, they're Hilbert transform pairs; $\chi(\omega)$ can be reconstructed from knowledge of either part. In a standard notation for complex-valued functions, $\chi(\omega) = \chi'(\omega) + i\chi''(\omega)$. The imaginary part $\chi''(\omega)$ is associated with energy dissipation and satisfies the positivity condition $\omega\chi''(\omega) \geq 0$ for $-\infty < \omega < \infty$. The real part $\chi'(\omega)$ is termed the reactive part, with the imaginary part the dissipative or absorptive part.

- The fluctuation-dissipation theorem is a fundamental relation between $\chi''(\omega)$ and the Fourier transform of the time correlation function, $\widetilde{C}(\omega)$,

$$\chi''_{AB}(\omega) = \frac{1}{\hbar} \tanh(\tfrac{1}{2}\beta\hbar\omega) \widetilde{C}_{AB}(\omega),$$

  where $\beta \equiv (kT)^{-1}$. Any connection between $\chi''(\omega)$ and $\widetilde{C}(\omega)$ is known as a fluctuation-dissipation theorem. Fluctuation-dissipation theorems connect two distinct physical quantities, each accessible to measurement. Fluctuations occur spontaneously, in the absence of external interactions, from the ceaseless motions of microscopic constituents of macroscopic systems. Correlations of fluctuations, in space and time, are revealed through scattering structure factors. Dissipation is the irreversible absorption of the work done by external forces into microscopic degrees of freedom, and there are different ways of measuring absorption. Both quantities are mechanical in origin, the absorption of work done by external forces and fluctuations from particle motions. Note the characteristic comparison of two energies: $\hbar\omega$ and kT. The theorem explains the success of the Onsager regression hypothesis which equates the rate of change of fluctuations with dissipative fluxes. Basically, a system doesn't "know" whether it's been brought into a nonequilibrium state by the action of an external force or as the result of a random fluctuation.

- Transport coefficients can be calculated from Green-Kubo relations, integrals of time correlation functions of microscopic flows. Transport coefficients are classified as mechanical, characterizing the response to external perturbations, electrical conductivity for example, and thermal, those associated with system inhomogeneities, thermal conductivity for example. Both types can be expressed by Green-Kubo relations.

- Using projection operators onto a space of physically relevant dynamical variables $A_j(t)$ (usually those with a clear separation in time scales between slow and fast variables), we derived the generalized Langevin equation, Eq. (6.122), an exact rewrite of the microscopic equations of motion having the mathematical form of Langevin equations. This analysis introduces three quantites: a set of frequencies Ω_{ij} describing collective oscillations of the relevant variables controlled by correlations among them; a driving force $f_j(t)$ having no projection onto the space of relevant variables; and a set of memory functions, $M_{jk}(t)$, controlled by time correlation functions of the random force. The oscillation frequencies vanish for dynamical modes of definite time-reversal symmetry. Perhaps the most important finding of this analysis is that the random forces $f_j(t)$ are uncorrelated with the relevant variables $A_j(t)$.

EXERCISES

6.1 We've used, several times and without proof, properties of the time evolution operator $U(t) \equiv e^{-i\Lambda t}$, with Λ the Liouville operator. Let's establish some of these properties.

a. First, $U(t)$ is unitary. This is evident because $P(\mathbf{\Gamma}, t)$ is normalized for all times, $\int P(\mathbf{\Gamma}, t)d\mathbf{\Gamma} = 1$, with $P(\mathbf{\Gamma}, t) = e^{-i\Lambda t}P(\mathbf{\Gamma}, 0)$; see Eq. (4.6). A deeper reason is *Stone's theorem*: A self-adjoint operator H on Hilbert space $\mathcal{H}$ generates a one-parameter continuous family $\{U(\alpha)\}$ (or *Lie group*) of unitary operators on $\mathcal{H}$ with $U(\alpha) = e^{i\alpha H}$, where $-\infty < \alpha < \infty$,[64] having the properties $U(0) = I$ (identity operator) and $U(\alpha)U(\beta) = U(\alpha + \beta)$, and thus is an abelian group. Each $U(\alpha)$ has an inverse, $U^{-1}(\alpha) = U(-\alpha)$, which, because it's unitary, has adjoint $U^\dagger = U^{-1}$; see [13, p34]. Group elements are continuous: $U(\alpha) \to U(\beta)$ as $\alpha \to \beta$. Another way of stating the theorem is that an operator $e^{i\alpha H}$, with α a real continuous parameter, is unitary if and only if H is self-adjoint. The operator H is called the generator of the group. Thus, inasmuch as Λ is self-adjoint (see [5, p335]), $U(t) = e^{-it\Lambda}$ is unitary.

b. An equivalent form of Stone's theorem is that, for a given self-adjoint operator H, there is a continuous one-parameter group of unitary transformations $\{U(\alpha)\}$ with $iH = \lim_{\alpha\to 0}(1/\alpha)\left[U(\alpha) - I\right]$ as its generating infinitesimal transformation. The idea is that a transformation $U(\alpha)$ for finite α can be realized from a succession of infinitesimal transformations (Euler definition of exponential),

$$U(\alpha) = e^{i\alpha H} = \lim_{N\to\infty}\left[I + i\frac{\alpha}{N}H\right]^N .$$

The entire group $\{U(\alpha)\}$ is generated by iH. It's usually easier to prove results on infinitesimal transformations than for finite, knowing that finite transformations can be built up from the compound effect of many infinitesimal transformations.[65]

For $\mathcal{D}_H \subseteq \mathcal{H}$ the domain of H, and for $\phi \in \mathcal{D}_H$, suppose that $U(\alpha)\phi \in \mathcal{D}_H$. Then, as $\epsilon \to 0$, $(1/i\epsilon)\left[U(\epsilon) - I\right]U(\alpha)\phi \to HU(\alpha)\phi$, implying that $(1/i\epsilon)\left[U(\alpha+\epsilon) - U(\alpha)\right]\phi \to HU(\alpha)\phi$ as $\epsilon \to 0$. Define

$$\lim_{\epsilon\to 0}\frac{1}{\epsilon}\left[U(\alpha+\epsilon) - U(\alpha)\right] \equiv \frac{d}{d\alpha}U(\alpha),$$

so that

$$\frac{d}{d\alpha}U(\alpha)\phi = iHU(\alpha)\phi.$$

Introduce the notation $\phi(0) = U(0)\phi = I\phi = \phi$ and $\phi(\alpha) = U(\alpha)\phi$. Then we have

$$\frac{d}{d\alpha}\phi(\alpha) = iH\phi(\alpha),$$

an equation of motion for vectors in $\mathcal{D}_H$ as they change under $U(\alpha)$.

c. Let (ϕ, χ, ψ) be vectors in $\mathcal{D}_H$ and assume that χ is obtained from the product $\chi = \phi\psi$. The transformation $U(\alpha) = e^{i\alpha H}$ associates with χ another vector, $\chi(\alpha) = U(\alpha)\chi$. The

[64] See Stone[208] and Riesz and Sz.-Nagy[209, Section 137]. The theorem can be stated in the converse: For $\{U(\alpha)\}$ a continuous one-parameter unitary group on $\mathcal{H}$, there is a unique self-adjoint operator H such that $U(\alpha) = e^{i\alpha H}$ for each α. Note that H is uniquely associated with a *family* of operators, $\{U(\alpha)\}$. For a single operator e^{iH}, the choice of generator is not unique—H can be replaced with $B = H + 2\pi nI$ where n is an integer so that $e^{iB} = e^{iH}$.

[65] This idea is used frequently in theoretical physics, e.g., rotations and Lorentz transformations; see [6, Chapter 6].

vectors ϕ and ψ are also transformed, $\phi(\alpha) = U(\alpha)\phi$ and $\psi(\alpha) = U(\alpha)\psi$. To preserve algebraic relations,[66] we require $\chi(\alpha) = \phi(\alpha)\psi(\alpha)$ for each α, or that

$$U(\alpha)[\phi\psi] = U(\alpha)\phi U(\alpha)\psi.$$

This result is highly useful in calculations. Prove this relation by showing it for infinitesimal transformations, assuming that H satisfies the product rule, $H(\phi\psi) = \phi H\psi + \psi H\phi$. A:

$$\begin{aligned} U(\alpha)\phi U(\alpha)\psi &= (\phi + \mathrm{i}\alpha H\phi)(\psi + \mathrm{i}\alpha H\psi) + O(\alpha^2) = \phi\psi + \mathrm{i}\alpha\phi H\psi + \mathrm{i}\alpha\psi H\phi + O(\alpha^2) \\ &= \phi\psi + \mathrm{i}\alpha H(\phi\psi) + O(\alpha^2) = U(\alpha)[\phi\psi]. \end{aligned}$$

It's shown in Exercise 4.1 that the Liouville operator satisfies the product rule.

d. Show, following similar steps to those in Eq. (6.1), that the equilibrium correlation function defined in Eq. (6.4) is independent of time. Hint: Use the relation just derived. A:

$$\begin{aligned} C_{yz}^0 &= \int \mathrm{d}\boldsymbol{\Gamma} y(t)z(t)P_{\text{eq}}(\boldsymbol{\Gamma}) = \int \mathrm{d}\boldsymbol{\Gamma} \left[\mathrm{e}^{\mathrm{i}\Lambda t}(yz)\right] P_{\text{eq}}(\boldsymbol{\Gamma}) \\ &= \int \mathrm{d}\boldsymbol{\Gamma} y(\boldsymbol{\Gamma})z(\boldsymbol{\Gamma})\mathrm{e}^{-\mathrm{i}\Lambda t}P_{\text{eq}}(\boldsymbol{\Gamma}) = \int \mathrm{d}\boldsymbol{\Gamma} y(\boldsymbol{\Gamma})z(\boldsymbol{\Gamma})P_{\text{eq}}(\boldsymbol{\Gamma}). \end{aligned}$$

6.2 Derive the following relations (refer to Section 6.1 for their definitions):

$$\int C_{yz}^0(\omega; \boldsymbol{q})\mathrm{d}\boldsymbol{q} = 2\pi C_{yz}^0(\omega);$$

$$\int_{-\infty}^{\infty} C_{yz}^0(\omega)\mathrm{d}\omega = 2\pi C_{yz}^0;$$

$$\int_{-\infty}^{\infty} C_{yz}^0(\omega; \boldsymbol{q})\mathrm{d}\omega = 2\pi \int \mathrm{d}\boldsymbol{r}\mathrm{e}^{\mathrm{i}\boldsymbol{q}\cdot\boldsymbol{r}} C_{yz}^0(0; \boldsymbol{r}).$$

6.3 In this exercise we find the solution to the inhomogeneous Liouville equation:

$$\frac{\partial}{\partial t} f(t) = L f(t) + g(t). \tag{P6.1}$$

a. First find the complementary solution. Show that the solution of $(\partial/\partial t)f(t) = Lf(t)$ has the form $f(t) = U(t)f(0)$, where U satisfies $(\partial/\partial t)U(t) = LU(t)$ with $U(0) = I$. Clearly, $U(t) = \mathrm{e}^{Lt}$.

b. Show that the solution of Eq. (P6.1) can be written

$$f(t) = U(t)f(0) + \int_0^t \mathrm{d}\tau U(t-\tau)g(\tau).$$

Use the Leibniz integral rule.

6.4 Poisson brackets have a property analogous to the cyclic invariance of the trace [for operators $\hat{A}, \hat{B}, \hat{C}$, $\text{Tr}\, \hat{A}\hat{B}\hat{C} = \text{Tr}\, \hat{B}\hat{C}\hat{A} = \text{Tr}\, \hat{C}\hat{A}\hat{B}$], that for dynamical variables A, B, C,

$$\int \mathrm{d}\boldsymbol{\Gamma} A\, [B, C]_{\text{P}} = \int \mathrm{d}\boldsymbol{\Gamma} B\, [C, A]_{\text{P}} = \int \mathrm{d}\boldsymbol{\Gamma} C\, [A, B]_{\text{P}}. \tag{P6.2}$$

Derive this result. Integrate by parts *selectively*, where for example to "pull" B out of $A[B, C]_{\text{P}}$, integrate by parts only those derivatives involving B. Assume integrated parts vanish, a boundary condition that would have to be checked before using Eq. (P6.2).

[66] A transformation preserving algebraic structures is called an *automorphism*.

6.5 Verify that Eq. (6.22) solves Eq. (6.20).

6.6 Show for time-independent operators $\hat{A}$, $\hat{B}$ that (used in deriving Eq. (6.23))

$$\mathrm{Tr}\left\{\hat{\rho}^0\left[\hat{A}, U_0^\dagger(t,\tau)\hat{B}U_0(t,\tau)\right]\right\} = \mathrm{Tr}\left\{\hat{\rho}^0\left[\hat{A}_I(\tau), \hat{B}_I(t)\right]\right\} \tag{P6.3}$$

where $\hat{\rho}^0 \equiv Z^{-1}(\beta)\mathrm{e}^{-\beta\hat{H}^0}$ is the canonical density operator, $U_0(t,\tau) = \exp[-i\hat{H}^0(t-\tau)]$, H^0 is time independent, and we're working in the interaction representation (see Appendix E). Use cyclic invariance of the trace and that $[H^0, \hat{\rho}^0] = 0$.

6.7 We stated on general grounds that the response function Φ in Eq. (6.24) must be real (a real response to a real interaction implies Φ is real). This can be demonstrated explicitly, however. Show for Hermitian operators $\hat{A}$, $\hat{B}$ that $\mathrm{i}[\hat{A},\hat{B}]$ is Hermitian.

6.8 Derive Eq. (6.25).

a. For operators $\hat{A}$ and $\hat{H}$, show that

$$\frac{\mathrm{d}}{\mathrm{d}\alpha}\left(\mathrm{e}^{\alpha\hat{H}}\hat{A}\mathrm{e}^{-\alpha\hat{H}}\right) = \mathrm{e}^{\alpha\hat{H}}\left[\hat{H},\hat{A}\right]\mathrm{e}^{-\alpha\hat{H}}. \tag{P6.4}$$

b. Integrate Eq. (P6.4) between 0 and β and multiply by $\mathrm{e}^{-\beta\hat{H}}$ from the left to arrive at the middle expression in Eq. (6.25).

c. Show that $\mathrm{e}^{\alpha\hat{H}}\left[\hat{H},\hat{A}\right]\mathrm{e}^{-\alpha\hat{H}} = \left[\hat{H},\mathrm{e}^{\alpha\hat{H}}\hat{A}\mathrm{e}^{-\alpha\hat{H}}\right]$.

6.9 What is the dimension of the parameter α in Eq. (6.25)? What is the dimension of $\hbar\alpha$?

6.10 Derive the equality in Eq. (6.27),

$$\int_0^\beta \mathrm{d}\alpha\, \mathrm{Tr}\,\hat{\rho}^0\dot{\hat{A}}(-\mathrm{i}\hbar\alpha)\hat{B}(t) = -\int_0^\beta \mathrm{d}\alpha\, \mathrm{Tr}\,\hat{\rho}^0\hat{A}(-\mathrm{i}\hbar\alpha)\dot{\hat{B}}(t).$$

This result follows from the Heisenberg equation of motion, Eq. (E.12) and the cyclic invariance of the trace.

6.11 Show for a real-valued function $f(t)$ that

a. Its Fourier transform $g(\omega)$ is such that $g(\omega) = g^*(-\omega)$.

b. Its real part is an even function of ω and its imaginary part is odd.

6.12 Show that $\mathrm{Tr}\,\dot{\rho}H = 0$. Hint: Eq. (A.41).

6.13 We've said on general grounds that the response function is real. Show that the result in Eq. (6.42) is real.

a. Show from Eq. (6.42) that

$$\Phi_{BA}(t) - \Phi^*_{BA}(t) = -\frac{1}{\hbar Z}\sum_{n,m}\Big\{\left(\mathrm{e}^{-\beta E_n} - \mathrm{e}^{-\beta E_m}\right)\big[\mathrm{i}\cos(\omega_{nm}t)\left(A_{nm}B_{mn} + B_{nm}A_{mn}\right) + \sin(\omega_{nm}t)\left(A_{nm}B_{mn} - B_{nm}A_{nm}\right)\big]\Big\}.$$

The matrices $[A_{nm}]$ and $[B_{nm}]$ are Hermitian.

b. Show that the terms in curly braces are odd under $n \leftrightarrow m$ and thus Φ is real. A symmetric sum over an antisymmetric function vanishes, $\sum_{n,m} S_{nm} = 0$ if $S_{nm} = -S_{mn}$.

6.14 Show that the poles of $\chi(\omega)$ in Eq. (6.54) (with ω a complex variable) occur in the negative half plane for all ω_0.

6.15 Show that Eq. (6.56) follows from Eq. (6.55), a task best done using the residue theorem. Show that the contour must be closed in the lower half plane for $\theta > 0$ and in upper half plane for $\theta < 0$. Beware minus signs! Closing in the lower half plane is not standard.

6.16 Show that the expression for $\widetilde{C}^0_{AB}(\omega)$ in Eq. (6.64) is real. A, B are Hermitian matrices.

6.17 Derive Eq. (6.86). Hint: The Liouville operator acts on canonical variables only (not on $\boldsymbol{r}$).

6.18 For the projection operator $\mathcal{P}$ defined in Eq. (6.108) and inner product (X, Y) defined in Eq. (6.107), show that:

a. $\mathcal{P} \circ \mathcal{P}X = \mathcal{P}X$ for any X;

b. $\mathcal{P}A_j = A_j$ for A_j an element of the space of relevant dynamical variables;

c. $G_{ij} \equiv (A_i, A_j)$ is a Hermitian matrix;

d. $\mathcal{P}$ is self-adjoint, i.e., $(\mathcal{P}X, Y) = (X, \mathcal{P}Y)$ for any X, Y. Hint: The inverse of a Hermitian matrix is Hermitian;

e. The eigenvalues of $\mathcal{P}$ are 0 or 1;

f. $(\mathcal{P}X, [\mathcal{I} - \mathcal{P}]X) = 0$, where $\mathcal{I}$ is the identity operator. Thus for any X, $[\mathcal{I} - \mathcal{P}]X$ is orthogonal to $\mathcal{P}X$;

g. The operator $\mathcal{Q} \equiv \mathcal{I} - \mathcal{P}$ is a projection operator, $\mathcal{Q} \circ \mathcal{Q}X = \mathcal{Q}X$.

6.19 Verify Eq. (6.112) directly. Define a function

$$F(t) \equiv \mathrm{e}^{\mathcal{Q}Lt} + \int_0^t \mathrm{d}\tau \mathrm{e}^{L(t-\tau)} \mathcal{P} L \mathrm{e}^{\mathcal{Q}L\tau} - \mathrm{e}^{Lt}.$$

Note that $F(0) = 0$. Show that

$$\frac{\mathrm{d}F}{\mathrm{d}t} = LF(t).$$

Use the Leibniz integral rule. Conclude that $F(t) = 0$ for all t.

Statistical mechanics

STATISTICAL mechanics must handle two sources of indeterminacy: 1) fluctuations—variations in measurements on *macroscopically identical* systems that force us to interpret measured values as the mean of a large number of measurements, and 2) quantum, that even on well-prepared microscopic systems, repeated measurements do not produce the same outcomes. We review the foundations of statistical mechanics required in this book, classical and quantum.

A.1 CLASSICAL ENSEMBLES: PROBABILITY DENSITY FUNCTIONS

A.1.1 Classical dynamics of many particle systems; Γ-space

The state of N interacting point particles in three spatial dimensions is specified at an instant of time by a point (the *system point*) in the $6N$-dimensional space (Γ-space) spanned by the canonical variables[1] $(p_1, \ldots, p_{3N}; q_1, \ldots, q_{3N})$. The time evolution of the system point is governed by $6N$ coupled first-order differential equations, Hamilton's equations of motion,[2]

$$\dot{p}_k = -\frac{\partial H}{\partial q_k} \qquad \dot{q}_k = \frac{\partial H}{\partial p_k}, \qquad k = 1, \ldots, 3N \tag{A.1}$$

where $H = H(p_1, \ldots, p_{3N}; q_1, \ldots, q_{3N})$. The time history of the system is a trajectory in Γ-space. Consider a phase-space function, $A(p, q, t) \equiv A(p_1(t), \ldots, p_n(t), q_1(t), \ldots, q_n(t), t)$. Form its total time derivative:

$$\begin{aligned}\frac{\mathrm{d}A}{\mathrm{d}t} &= \frac{\partial A}{\partial t} + \sum_k \left(\frac{\partial A}{\partial q_k}\dot{q}_k + \frac{\partial A}{\partial p_k}\dot{p}_k \right) = \frac{\partial A}{\partial t} + \sum_k \left(\frac{\partial A}{\partial q_k}\frac{\partial H}{\partial p_k} - \frac{\partial A}{\partial p_k}\frac{\partial H}{\partial q_k} \right) \\ &\equiv \frac{\partial A}{\partial t} + [A, H]_{\mathrm{P}}, \end{aligned} \tag{A.2}$$

where the second line specifies the *Poisson bracket*[3] of A with H. The Poisson bracket between phase-space functions $u(p, q)$ and $v(p, q)$ is defined as

$$[u, v]_{\mathrm{P}} \equiv \sum_k \left(\frac{\partial u}{\partial q_k}\frac{\partial v}{\partial p_k} - \frac{\partial u}{\partial p_k}\frac{\partial v}{\partial q_k} \right). \tag{A.3}$$

[1] Canonical variables are by definition those that satisfy Hamilton's equations of motion.

[2] Canonical variables at time t evolve under the action of Hamilton's equations to new canonical variables at time $t + \Delta t$ (a canonical transformation, see texts on classical mechanics or [5, p334]). Canonical coordinates thus stay canonical in their evolution, with the Jacobian of canonical transformations equal to unity, an essential part of the proof of Liouville's theorem; see [5, p47]. The application of Liouville's theorem to statistical mechanics is taken up in Chapter 4.

[3] We denote Poisson brackets with subscript P to avoid confusion with the commutator bracket of quantum mechanics.

DOI: 10.1201/9781003512295-A

Poisson brackets have the algebraic property $[A, B]_{\text{P}} = -[B, A]_{\text{P}}$, implying that $[A, A]_{\text{P}} = 0$ for any A. An immediate consequence is, if $\partial H/\partial t = 0$ (appropriate to statistical mechanics), then

$$\frac{\mathrm{d}}{\mathrm{d}t}H(p, q) = [H, H]_{\text{P}} = 0.$$

Thus the Hamiltonian is a *constant of the motion* (if $\partial H/\partial t = 0$): The value of $H(p, q)$ stays constant in time when its arguments are replaced by solutions of Hamilton's equations of motion.

A.1.2 The transition to statistical mechanics

Measurements on macroscopic systems represent time averages of fluctuating quantities. To calculate time averages from first principles would require us to know the detailed, microscopic time histories of interacting particles, a task beyond our computational reach for the practical reason of limited computing resources and for the fundamental reason that we're unable to control initial conditions. The equilibrium state is one of maximum entropy.[4] *Numerous* microstates are consistent with the same macrostate. With each microstate specified by $6N$ initial conditions (and hence a point in Γ-space), consider a collection of distinct microstates in Γ-space at the same time,[5] (indicated schematically in Fig. A.1). Because we lack the ability to control initial conditions, our

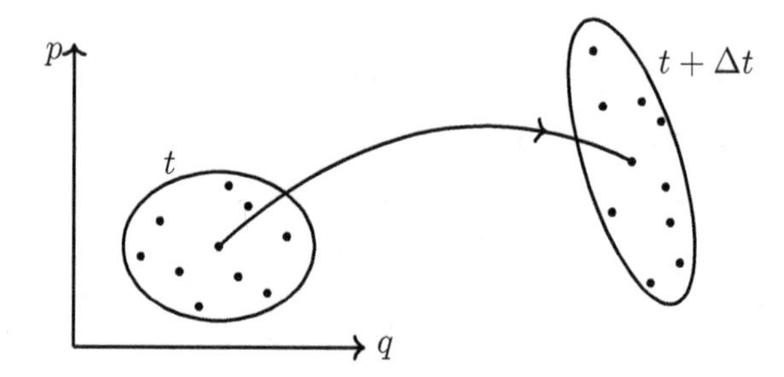

Figure A.1 A collection of points in Γ-space at time t flows under Hamiltonian dynamics to the collection at time $t + \Delta t$ such that the density of points remains fixed (Liouville's theorem). Axes p, q represent the $3N$ axes associated with $p_i, q_i, i = 1, \ldots, 3N$.

knowledge of the dynamics is consistent with the motion of a swarm of equivalent system points in Γ-space, equivalent in the sense of distinct microstates representing the same macrostate. By Liouville's theorem,[6] *any collection of points in Γ-space moves as an incompressible fluid* (whether representing equilibrium systems or not)[5, p55].

An ensemble is an abstract collection of macroscopically identical systems.[7] Even though prepared identically in meeting constraints, their system points in Γ-space are not identical. We can ask for the probability that the system point of a randomly selected member of the ensemble lies in the range (p, q) to $(p + \mathrm{d}p, q + \mathrm{d}q)$ (where p and q denote $(p_1, \ldots, p_{3N})$ and $(q_1, \ldots, q_{3N})$, with $\mathrm{d}p \equiv \mathrm{d}p_1 \cdots \mathrm{d}p_{3N}$ and $\mathrm{d}q \equiv \mathrm{d}q_1 \cdots \mathrm{d}q_{3N}$). Such a probability is proportional to $\mathrm{d}p\mathrm{d}q$, which we can write

$$\rho(p, q)\frac{\mathrm{d}p\mathrm{d}q}{h^{3N}} \equiv \rho(p, q)\mathrm{d}\Gamma, \tag{A.4}$$

[4]That is, the maximum value entropy *can* have subject to prescribed values of state variables such as V and T[2, p37].

[5]There is a fundamental restriction on *how* distinct two microstates must be so as not to be counted as the same. A point in Γ-space cannot be specified with greater precision than to lie within a volume h^{3N}; see [2, p110].

[6]Derived in 1838. Its significance for statistical mechanics was first recognized by Gibbs; see Gibbs[131, Chapter 1].

[7]The ensemble concept is due to Gibbs[131, p5]: "Let us imagine a great number of independent systems, identical in nature, but differing in phase, that is, in their condition with respect to configuration and velocity."

where h^{3N} ensures that the phase-space probability density function[8] $\rho(p,q)$ is dimensionless, which we require to be normalized to unity,

$$\int_{\Gamma} \rho(p,q)\mathrm{d}\Gamma = 1. \tag{A.5}$$

If we knew $\rho(p,q)$, then for a phase-space function $A(p,q)$, we could evaluate its ensemble average

$$\langle A \rangle \equiv \int_{\Gamma} \rho(p,q)A(p,q)\mathrm{d}\Gamma. \tag{A.6}$$

The working assumption of statistical mechanics is that experimentally measurable quantities correspond to appropriate ensemble averages[5, p46].

The task of finding $\rho(p,q)$ is enabled by Liouville's theorem, another form of which[9] is that $\rho(p,q)$ is a constant of the motion[5, p47]:

$$\frac{\mathrm{d}}{\mathrm{d}t}\rho(p,q) = 0. \tag{A.7}$$

The function $\rho(p,q)$ is certainly a phase-space function, and thus, combining Eqs. (A.2) and (A.7),

$$\frac{\partial \rho}{\partial t} + [\rho, H]_{\mathrm{P}} = 0. \tag{A.8}$$

Equation (A.8) is *Liouville's equation*, often written in the form of Eq. (4.5). Statistical mechanics is the study of the solutions of Liouville's equation.

Any constant of the motion solves Eq. (A.7); conversely any solution of Eq. (A.7) is a constant of the motion. The most general solution of Liouville's equation is therefore a function of the constants of the motion.[10] In thermal equilibrium $\partial\rho/\partial t = 0$, and thus we seek solutions of

$$[\rho, H]_{\mathrm{P}} = 0. \tag{A.9}$$

Standard statistical mechanics proceeds on the assumption that the Hamiltonian is the only relevant constant of the motion, a procedure justified *a posteriori* through comparison of the predictions of theory with experimental results. Thus, we assume that ρ is a function of the Hamiltonian:

$$\rho(p,q) = F(H(p,q)). \tag{A.10}$$

It remains to determine F. All microscopic information about a system's components and their interactions enters through the Hamiltonian. The nature of F depends on the types of interactions the systems comprising the ensemble allow with the environment[5, Chapter 4].

A.1.3 Canonical ensemble, closed systems

For systems of fixed NVT allowing energy exchanges with the environment (closed systems)—the canonical ensemble—the equilibrium phase-space probability density function is found to be[11]

$$\rho(p,q) = \frac{1}{Z(\beta)}\mathrm{e}^{-\beta H(p,q)} = \frac{1}{N!Z_{\mathrm{can}}(\beta)}\mathrm{e}^{-\beta H(p,q)}, \tag{A.11}$$

[8]The difference between probabilities and probability densities is reviewed in Appendix B.

[9]If the flow of ensemble points in Γ-space is incompressible, then the probability density is constant along streamlines.

[10]See the example on the ideal gas in Section 4.1.

[11]Deriving Eq. (A.11) is a major task in the development of statistical mechanics; see [5, Chapter 4]. Feynman[3, p1] stated: "This fundamental law is the summit of statistical mechanics, and the entire subject is either the slide-down from this summit, as the principle is applied to various cases, or the climb-up to where the fundamental law is derived ...." It's straightforward to show that $[\mathrm{e}^{-\beta H(p,q)}, H]_{\mathrm{P}} = 0$ for any β.

where $\beta \equiv (kT)^{-1}$ and the canonical partition function

$$Z_{\text{can}}(N,V,T) \equiv \frac{1}{N!}\int \mathrm{d}\Gamma \mathrm{e}^{-\beta H(p,q)}. \tag{A.12}$$

The probability density can be written in an equivalent way as a function of energy,

$$\rho(E) = \frac{1}{N! Z_{\text{can}}} \mathrm{e}^{-\beta E}\Omega(E), \tag{A.13}$$

where $\Omega(E)$ (density of states function) is such that $\Omega(E)\mathrm{d}E$ is the number of energy states in $[E, E+\mathrm{d}E]$ and where

$$Z_{\text{can}}(N,V,T) = \frac{1}{N!}\int \mathrm{e}^{-\beta E}\Omega(E)\mathrm{d}E. \tag{A.14}$$

The Boltzmann factor $\mathrm{e}^{-\beta E}$ is a measure of the extent to which the energy states of the system are populated at temperature T, with $Z = \int_0^\infty \mathrm{e}^{-\beta E}\Omega(E)\mathrm{d}E$ the total number of possible states accessible to the system at temperature T. The probability $P(E)\mathrm{d}E = \mathrm{e}^{-\beta E}\Omega(E)\mathrm{d}E/Z$ is the ratio of two numbers: the number of states available to the system at energy E when it has temperature T, to the total number of states available to the system at temperature T. The connection between statistical mechanics and thermodynamics[12] is contained in the relation

$$Z_{\text{can}}(N,V,T) = \mathrm{e}^{-\beta F(N,V,T)}, \tag{A.15}$$

where $F(N,V,T) = U - TS$ is the Helmholtz free energy function[5, p96].

Once one has obtained $Z_{\text{can}}(N,V,T)$, the standard quantities of thermodynamics can be obtained through partial differentiation, e.g.,

$$\mu = -kT\frac{\partial}{\partial N}\ln Z_{\text{can}}\bigg|_{T,V} \qquad P = kT\frac{\partial}{\partial V}\ln Z_{\text{can}}\bigg|_{T,N} \qquad S = k\frac{\partial}{\partial T}(T\ln Z_{\text{can}})\bigg|_{V,N}. \tag{A.16}$$

Fluctuations in the canonical ensemble follow from second partial derivatives of Z_{can}, e.g.,

$$\frac{\partial^2 \ln Z_{\text{can}}}{\partial \beta^2} = \langle (H - \langle H \rangle)^2 \rangle = kT^2 C_V. \tag{A.17}$$

A.1.4 Grand canonical ensemble, open systems

The grand canonical ensemble is based on systems of fixed μVT allowing particle exchanges with the environment as well as energy (open systems), where the chemical potential μ is the energy required to add another particle to the system at constant S, V or at constant T, P[2, p39]. The ensemble average of a phase-space function $A(p,q,N)$ is[5, p100]

$$\langle A \rangle = \frac{1}{Z_G}\sum_{N=0}^{\infty}\frac{\mathrm{e}^{\beta N\mu}}{N!}\int_{\Gamma_N} A(p,q,N)\mathrm{e}^{-\beta H_N(p,q)}\mathrm{d}\Gamma_N = \frac{1}{Z_G}\sum_{N=0}^{\infty}\mathrm{e}^{\beta\mu N} Z_{\text{can},N}\langle A \rangle_N, \tag{A.18}$$

where the grand partition function

$$Z_G(\mu,V,T) = \sum_{N=0}^{\infty}\frac{1}{N!}\mathrm{e}^{\beta\mu N}\int_{\Gamma_N}\mathrm{e}^{-\beta H_N(p,q)}\mathrm{d}\Gamma_N = \sum_{N=0}^{\infty}\mathrm{e}^{\beta\mu N} Z_{\text{can}}(N,V,T), \tag{A.19}$$

[12] The laws of thermodynamics are reproduced by statistical mechanics; see [5, pp92–96].

with $H_N(p,q)$ the Hamiltonian for an N-particle system. If these infinite summations worry your inner mathematician, μ is usually a negative quantity.[13] The grand partition function is related to a thermodynamic potential, the grand potential $\Phi(T,V,\mu) \equiv F - N\mu$[5, p99],

$$Z_G(T,V,\mu) = \mathrm{e}^{-\beta\Phi(T,V,\mu)}. \tag{A.20}$$

Once one has obtained $Z_G(\mu,V,T)$, the other variables follow from partial differentiation:

$$N = kT\frac{\partial \ln Z_G}{\partial \mu}\bigg|_{T,V} \qquad P = kT\frac{\partial \ln Z_G}{\partial V}\bigg|_{T,\mu} \qquad S = k\frac{\partial}{\partial T}\left(T\ln Z_G\right)\bigg|_{V,\mu}. \tag{A.21}$$

Fluctuations in the grand canonical ensemble are obtained from second derivatives of Z_G[5, p101].

A.2 QUANTUM ENSEMBLES: PROBABILITY DENSITY OPERATORS

A.2.1 Quantum indeterminacy

The quantum nature of matter adds another source of indeterminacy. Even in well-defined quantum systems, measurements produce a range of values and we must seek the mean of a number of measurements.[14] The state of a quantum system at time t is represented by a *state vector*, $|\psi(t)\rangle$, an element of an abstract Hilbert space, with every observable (measurable) quantity A represented by a Hermitian operator, $\hat{A}$. The expectation value $\langle A\rangle_\psi$ on systems in state $|\psi\rangle$ is found from the expression (when $|\psi\rangle$ is normalized, $\langle\psi|\psi\rangle = 1$),

$$\langle A\rangle_\psi = \int \mathrm{d}x\psi^*(x)\hat{A}\psi(x) \equiv \langle\psi|\hat{A}\psi\rangle. \tag{A.22}$$

The act of measurement introduces an uncontrollable interaction with the system that modifies what we seek to measure. This can be seen using the eigenfunctions of $\hat{A}$, a set of functions $\{\phi_n\}_{n=0}^{\infty}$ such that $\hat{A}\phi_n = \lambda_n\phi_n$, where (because $\hat{A}$ is Hermitian) the λ_n are real numbers and the eigenfunctions are orthonormal, $\langle\phi_n|\phi_m\rangle = \delta_{nm}$. Eigenfunctions of Hermitian operators are *complete*:[15] An arbitrary square-integrable function can be represented by an infinite linear combination,

$$\psi(x,t) = \sum_{n=0}^{\infty} a_n(t)\phi_n(x), \tag{A.23}$$

where the expansion coefficients $a_n(t) = \langle\phi_n|\psi(t)\rangle$ carry the time dependence. That $|\psi(t)\rangle$ remains normalized in time[16] implies

$$\sum_{n=0}^{\infty} |a_n(t)|^2 = 1. \tag{A.24}$$

As one can show,

$$\langle\psi|\hat{A}\psi\rangle = \sum_{n=0}^{\infty} |a_n(t)|^2\lambda_n. \tag{A.25}$$

The expectation value of A for systems in state $|\psi\rangle$ is therefore an average of the eigenvalues of $\hat{A}$ over the probability distribution $|a_n(t)|^2$. We can only measure the eigenvalues of $\hat{A}$. To what extent, however, can quantum states be controlled? We must introduce a *second* kind of average, an ensemble average over a system's quantum states.

[13] A classical argument for $\mu < 0$ is given in [2, p39]. At the quantum level, it's required that $\mu \leq 0$ for bosons, whereas for fermions there is no restriction on μ[5, p141]. Cases where $\mu \gtrsim 0$ for fermions occur in high-density systems where the effects of the Pauli exclusion principle restrict occupation numbers to $N = 0, 1$, obviating concerns about convergence.

[14] Classically, the trajectories of particles subject to the same forces with the same initial conditions are reproducibly the same; in quantum mechanics the same measurements on identically prepared systems do not produce the same values.

[15] See for example [13, Chapter 2].

[16] The evolution of $|\psi(t)\rangle$ is unitary with $\langle\psi(t)|\psi(t)\rangle = \langle\psi(0)|\psi(0)\rangle$. Probability is locally conserved: The Schrödinger equation implies a continuity equation, $\partial|\psi(\boldsymbol{r},t)|^2/\partial t + \boldsymbol{\nabla}\cdot\boldsymbol{J} = 0$, with $\boldsymbol{J} = \hbar/(2m\mathrm{i})\left(\psi^*\boldsymbol{\nabla}\psi - \psi\boldsymbol{\nabla}\psi^*\right)$.

A.2.2 Many-particle wave functions

Statistical mechanics utilizes wave functions of many particle systems, and here we run into a new piece of physics, the *indistinguishability of identical particles.* Wave functions of an assembly of identical particles are either symmetric or antisymmetric under interchange of particles (see Appendix D of [5]). Under the interchange of particles at positions $\boldsymbol{r}_j$ and $\boldsymbol{r}_k$ (leaving all other particles unchanged), the basic principles of quantum mechanics require that (for identical particles)

$$\psi(\ldots, \boldsymbol{r}_j, \ldots, \boldsymbol{r}_k, \ldots, t) = \theta\psi(\ldots, \boldsymbol{r}_k, \ldots, \boldsymbol{r}_j, \ldots, t), \tag{A.26}$$

where

$$\theta = \begin{cases} +1 & \text{for bosons} \\ -1. & \text{for fermions} \end{cases}$$

In either case, $|\psi(\ldots, \boldsymbol{r}_k, \ldots, \boldsymbol{r}_l, \ldots, t)|^2 = |\psi(\ldots, \boldsymbol{r}_l, \ldots, \boldsymbol{r}_k, \ldots, t)|^2$, implying the fundamental result that identical particles can't be labeled.[17] To construct an N-particle wave function, we use as basis functions, products of N single-particle wave functions,[18]

$$\psi(\boldsymbol{r}_1, \ldots, \boldsymbol{r}_N, t) = \sum_{m_1} \cdots \sum_{m_N} c(m_1, \ldots, m_N, t)\phi_{m_1}(\boldsymbol{r}_1)\phi_{m_2}(\boldsymbol{r}_2)\cdots\phi_{m_N}(\boldsymbol{r}_N), \tag{A.27}$$

where $\{\phi_k\}_{k=0}^{\infty}$ is some complete set (whatever is convenient for the problem at hand), and the $c(m_1, \ldots, m_N, t)$ are expansion coefficients. Equation (A.27) is to many-particle wave functions what Eq. (A.23) is to single-particle wave functions. The expansion coefficients $c(m_1, \ldots, m_N, t)$ "do the work" of providing exchange symmetries (because the basis functions do not), i.e.,

$$c(\ldots, m_k, \ldots, m_l, \ldots, t) = \theta c(\ldots, m_l, \ldots, m_k, \ldots, t). \tag{A.28}$$

If $m_l = m_k = m$, antisymmetry requires that $c(\ldots, m, \ldots, m, \ldots, t) = 0$, the usual form of the Pauli principle for fermions.

To avoid cumbersome notation, let's write Eq. (A.27) (many-particle wave function) in the form

$$\psi(x, t) = \sum_r c_r(t)\phi_r(x), \tag{A.29}$$

so that r is a multi-index symbol and x denotes the collection of spatial coordinates (which drops out of the theoretical description). The expectation value of $\hat{A}$ is (using two copies of Eq. (A.29))

$$\langle\psi(t)|\hat{A}\psi(t)\rangle = \sum_r\sum_s c_r^*(t)c_s(t)\langle\phi_r|\hat{A}\phi_s\rangle \equiv \sum_r\sum_s c_r^*(t)c_s(t)A_{rs}, \tag{A.30}$$

where $A_{rs} = \langle\phi_r|\hat{A}\phi_s\rangle$ are the matrix elements of $\hat{A}$ in the basis set used in Eq. (A.27).

A.2.3 The density operator

A quantum system known to be in state $|\psi\rangle$ is said to be in a *pure state*. It's almost always the case that we have incomplete knowledge of the microscopic state of the system (a *mixed state*). And what do we do in the face of incomplete knowledge? I hope you're saying: "Resort to probability." Let p_i denote the probability that a randomly selected member of the ensemble is in state $|\psi^{(i)}\rangle$, such that

$$\sum_i p_i = 1. \tag{A.31}$$

[17] In that case, why base a theory on ill-defined quantities (wave functions involving particle positions)? The occupation number formalism (reviewed in Appendix D of [5]) is devoid of coordinate labels.

[18] The functions $\{\phi_n(\boldsymbol{r})\}$ form a complete set on the space of position coordinates for a single particle. *Products* of such functions form a complete set in the union of the domains of the spatial coordinates for each particle[72, p56].

The wave function $\psi^{(i)}(x)$ of each possible state can be expressed as in Eq. (A.29):[19]

$$\psi^{(i)}(x) = \sum_r c_r^{(i)} \phi_r(x). \tag{A.32}$$

The quantum expectation value of $\hat{A}$ in state $|\psi^{(i)}\rangle$ is, from Eq. (A.30) (introducing new notation),

$$\overline{A}^{(i)} \equiv \langle \psi^{(i)} | \hat{A} \psi^{(i)} \rangle = \sum_r \sum_s c_r^{(i)*} c_s^{(i)} A_{rs}.$$

The ensemble average is therefore

$$\langle A \rangle \equiv \sum_i p_i \overline{A}^{(i)} = \sum_i p_i \sum_r \sum_s c_r^{(i)*} c_s^{(i)} A_{rs} = \sum_r \sum_s \left(\sum_i p_i c_r^{(i)*} c_s^{(i)} \right) A_{rs}. \tag{A.33}$$

There are *two* averages: the quantum expectation value $\overline{A}^{(i)}$ of $\hat{A}$ in state $|\psi^{(i)}\rangle$ and an average over the states $|\psi^{(i)}\rangle$ which occur with probabilities p_i.

Define the *density matrix* ρ, with matrix elements (note the order of the indices)

$$\rho_{sr} \equiv \sum_i p_i c_r^{(i)*} c_s^{(i)}. \tag{A.34}$$

Combining Eqs. (A.34) and (A.33), we have an expression for the ensemble average,

$$\langle A \rangle = \sum_r \sum_s \rho_{sr} A_{rs} = \sum_s (\rho A)_{ss} = \text{Tr}\, \rho A, \tag{A.35}$$

where the trace is the sum of diagonal elements, $\text{Tr}\, M \equiv \sum_s M_{ss}$. From Eq. (A.34), the trace of the density matrix is unity,

$$\text{Tr}\, \rho \equiv \sum_s \rho_{ss} = \sum_s \sum_i p_i |c_s^{(i)}|^2 = \sum_i p_i \sum_s |c_s^{(i)}|^2 = \sum_i p_i = 1, \tag{A.36}$$

where we've used Eqs. (A.24) and (A.31).

New features appear in quantum statistical mechanics having no classical counterpart. Write Eq. (A.35) separating the diagonal from the off-diagonal matrix elements,

$$\langle A \rangle = \sum_s \rho_{ss} A_{ss} + \sum_s \sum_{r \neq s} \rho_{sr} A_{rs}. \tag{A.37}$$

From Eq. (A.34), $\rho_{ss} = \sum_i p_i |c_s^{(i)}|^2 \geq 0$, and from Eq. (A.36), $\sum_s \rho_{ss} = 1$. The diagonal element ρ_{ss} is the probability of finding the system in state s. The off-diagonal elements have no definite sign and cannot be interpreted as probabilities; they are without classical counterpart.[20] If in a given basis the density matrix is diagonal, the definition of $\langle A \rangle$ in Eq. (A.35) is the same as in the classical theory. The diagonal character of a matrix is not basis independent, however.

Define the *density operator* (or the *statistical operator*)

$$\hat{\rho} \equiv \sum_i p_i |\psi^{(i)}\rangle \langle \psi^{(i)}|, \tag{A.38}$$

so that its matrix elements are those of the density matrix ρ_{sr} (in the same basis), i.e., $\langle \phi_s | \hat{\rho} | \phi_r \rangle = \sum_i p_i \langle \phi_s | \psi^{(i)} \rangle \langle \psi^{(i)} | \phi_r \rangle = \sum_i p_i c_s^{(i)} c_r^{(i)*} = \rho_{sr}$. Using Eq. (A.35), we have

$$\langle A \rangle = \sum_{sr} \rho_{sr} A_{rs} = \sum_{sr} \langle \phi_s | \hat{\rho} | \phi_r \rangle \langle \phi_r | \hat{A} | \phi_s \rangle = \sum_s \langle \phi_s | \hat{\rho} \hat{A} | \phi_s \rangle = \text{Tr}\, \hat{\rho} \hat{A} = \text{Tr}\, \hat{A} \hat{\rho}, \tag{A.39}$$

[19] The basis functions are a complete set; they can be used to represent arbitrary wave functions.

[20] The density matrix is positive definite, however; see [5, p123].

where $\sum_r |\phi_r\rangle\langle\phi_r| = I$ (completeness relation in Dirac notation). The trace is *basis independent*[13, p34] and thus we can associate the trace *with the operator itself.* The last equality reflects the *cyclic invariance of the trace*, $\text{Tr}\, AB = \text{Tr}\, BA$[13, p36]. Note that

$$\text{Tr}\,\hat{\rho} = \sum_s \langle\phi_s|\hat{\rho}|\phi_s\rangle = \sum_s\sum_i p_i\langle\phi_s|\psi^{(i)}\rangle\langle\psi^{(i)}|\phi_s\rangle = \sum_s\sum_i p_i c_s^{(i)} c_s^{(i)*} = 1, \tag{A.40}$$

where we've used Eq. (A.36).[21] Equations (A.39) and (A.40) are the quantum analogs of Eqs. (A.6) and (A.5). Thus, we've replaced the classical ensemble average of the phase-space function $A(p, q)$, $\langle A\rangle = \int_\Gamma \rho(p, q)A(p, q)\text{d}\Gamma$, involving the probability density *function* $\rho(p, q)$, with the trace of the density *operator* $\hat{\rho}$ acting on the operator $\hat{A}$ representing the observable, $\langle A\rangle = \text{Tr}\,\hat{\rho}\hat{A}$.

Generalizing $\hat{\rho}(t)$ to include time, it satisfies the *von Neumann equation,*

$$\text{i}\hbar\frac{\partial}{\partial t}\hat{\rho}(t) = \left[\hat{H}, \hat{\rho}(t)\right], \tag{A.41}$$

where the square brackets denote the commutator $\left[\hat{A}, \hat{B}\right] \equiv \hat{A}\hat{B} - \hat{B}\hat{A}$, with $\hat{H}$ the Hamiltonian operator. The von Neumann equation is the quantum analog of the Liouville equation, Eq. (A.8); quantum statistical mechanics is the study of the solutions of Eq. (A.41). Deriving Eq. (A.41) is straightforward. Take the partial time derivative of Eq. (A.38) (with the p_i time independent) and use the time-dependent Schrödinger equation (which generates the time dependence of state vectors): $\text{i}\hbar(\partial/\partial t)|\psi^{(i)}\rangle = \hat{H}|\psi^{(i)}\rangle$ and $-\text{i}\hbar(\partial/\partial t)\langle\psi^{(i)}| = \langle\psi^{(i)}|\hat{H}$.

In thermal equilibrium $\partial\hat{\rho}/\partial t = 0$, implying that $\hat{\rho}$ commutes with $\hat{H}$ in equilibrium. Commuting operators have a common set of eigenfunctions. Using the eigenfunctions of $\hat{H}$ as a basis, $\hat{H}\phi_n = E_n\phi_n$ (the *energy representation*), the density matrix is diagonal,

$$\rho_{rs} = \rho_r\delta_{rs}. \tag{A.42}$$

A.2.4 Canonical ensemble

Consider the TVN ensemble. In the energy representation the density matrix is diagonal, implying the diagonal elements are the probability the system has energy E_n,

$$\rho_n = \frac{1}{Z(\beta)}g_n\text{e}^{-\beta E_n}, \tag{A.43}$$

where, for discrete energies,

$$Z(\beta) = \sum_n g_n\text{e}^{-\beta E_n}, \tag{A.44}$$

with g_n the degeneracy of the nth energy state, the number of linearly independent eigenfunctions each having eigenvalue E_n. The density operator may be evaluated using Eq. (A.38),

$$\hat{\rho} = \sum_n \rho_n|\phi_n\rangle\langle\phi_n| = \frac{1}{Z}\sum_n g_n\text{e}^{-\beta E_n}|\phi_n\rangle\langle\phi_n| = \frac{1}{Z}\text{e}^{-\beta\hat{H}}\sum_n|\phi_n\rangle\langle\phi_n| = \frac{1}{Z(\beta)}\text{e}^{-\beta\hat{H}}, \tag{A.45}$$

where we've used $\text{e}^{-\beta\hat{H}}\phi_n = \text{e}^{-\beta E_n}\phi_n$ and completeness. The ensemble average of an observable quantity A is therefore

$$\langle A\rangle = \text{Tr}\left(\hat{\rho}\hat{A}\right) = \frac{1}{Z}\text{Tr}\,\text{e}^{-\beta\hat{H}}\hat{A} = \frac{1}{Z(\beta)}\sum_n g_n\text{e}^{-\beta E_n}\langle\phi_n|\hat{A}\phi_n\rangle. \tag{A.46}$$

Equation (A.46) is one of the more useful equations in quantum statistical mechanics.[22]

[21] The density operator is self-adjoint, $\hat{\rho}^\dagger = \hat{\rho}$, positive semi-definite, $\langle\psi|\hat{\rho}|\psi\rangle \geq 0$, and $\text{Tr}\,\hat{\rho} = 1$.

[22] It appears on page 1 of Feynman[3].

A.2.5 Grand canonical ensemble

For $\hat{\rho}$ to be diagonal in the grand canonical ensemble (systems specified by $TV\mu$), it must have a representation in which $\hat{H}$ and the number operator $\hat{N}$ (eigenvalues $0, 1, 2, \ldots$; see Appendix D of [5]) are diagonal, implying that $\hat{\rho}$ must commute with $\hat{H}$ and $\hat{N}$. For this system,

$$\hat{\rho} = \frac{1}{Z_G} \mathrm{e}^{-\beta(\hat{H} - \mu\hat{N})}, \tag{A.47}$$

where

$$Z_G = \operatorname{Tr} \mathrm{e}^{-\beta(\hat{H} - \mu\hat{N})} = \sum_{N=0}^{\infty} \mathrm{e}^{\beta\mu N} Z_N(\beta) \tag{A.48}$$

is the grand partition function, with Z_N the canonical partition function for an N-particle system. Ensemble averages are then given by

$$\langle A \rangle = \frac{1}{Z_G} \sum_{N=0}^{\infty} \mathrm{e}^{\beta\mu N} Z_N(\beta) \langle A \rangle_N, \tag{A.49}$$

where $\langle A \rangle_N$ denotes the canonical ensemble average for an N-particle system.

A.3 ENTROPY

Entropy, a state variable discovered by Clausius was an unexpected development (as with anything new) and providing a microscopic interpretation of this quantity became a task for statistical physics. For isolated systems (microcanonical ensemble) we have the Boltzmann entropy formula $S = k \ln W$, with W the number of ways that macrostates are realized from microstates. Gibbs developed another expression for closed systems (canonical ensemble), the *Gibbs entropy* S_G,

$$S_G \equiv -k \int \mathrm{d}\Gamma \rho(p, q) \ln \rho(p, q) \equiv -k \langle \ln \rho \rangle, \tag{A.50}$$

with $\rho(p, q)$ the Γ-space probability density. The Gibbs formula reduces to the Boltzmann formula in the microcanonical ensemble (see Section 4.6.3). Von Neumann generalized the Gibbs entropy for quantum systems,[23]

$$S_V \equiv -k \operatorname{Tr} \rho \ln \rho, \tag{A.51}$$

where now ρ is the density matrix.[24] The trace of a function of a matrix A, $\operatorname{Tr} f(A) = \sum_j f(\lambda_j)$, where $\{\lambda_j\}$ are the eigenvalues of A.[25] Thus, for λ_j the eigenvalues of ρ,

$$S_V = -k \sum_j \lambda_j \ln \lambda_j. \tag{A.52}$$

The von Neumann entropy vanishes for pure-state systems, e.g., $\lambda_1 = 1$ and $\lambda_{j \neq 1} = 0$.

[23] Von Neumann's 1927 work on entropy (in German) is referenced in [210]. The topic is developed in von Neumann's book on quantum mechanics[211, Chapter 5]. Today there are many books on quantum entropy; see for example [212].

[24] The logarithm of a matrix is defined by a Taylor series. The expansion of the function $\ln(1 + x)$ around $x = 0$,

$$\ln(1 + x) = x - \frac{1}{2}x^2 + \frac{1}{3}x^3 - \frac{1}{4}x^4 + \cdots = \sum_{n=1}^{\infty} \frac{(-1)^{n+1}}{n} x^n,$$

is extended to a matrix A,

$$\ln A = \ln(A - I + I) \equiv \ln(I + B) = \sum_{n=1}^{\infty} \frac{(-1)^{n+1}}{n} B^n = \sum_{n=1}^{\infty} \frac{(-1)^{n+1}}{n} (A - I)^n,$$

where I is the identity matrix. Note that $\ln I = 0$, the zero matrix.

[25] The formula $\operatorname{Tr} f(A) = \sum_j f(\lambda_j)$ can be derived 1) by assuming f is an analytic function (possesses a power series) and 2) by evaluating the trace in a basis of the normalized eigenfunctions of A, knowing that the trace is basis independent.

The von Neumann entropy $S_V = S_V(\lambda_1, \lambda_2, \ldots)$ is a continuous function of the eigenvalues of the density matrix. It therefore doesn't have to be associated with systems in thermal equilibrium; it applies to any density matrix, not necessarily that for equilibrium systems. A generalization of entropy applicable to *any* distribution of probabilities $\{p_i\}_{i=1}^N$ was introduced by Shannon (mentioned in Section 4.6.3; see [2, Chapter 12]) having the form of Eq. (A.52), $-K\sum_{i=1}^N p_i \ln p_i$ (K is a positive constant), known as *information* or the *Shannon entropy*. The von Neumann entropy is therefore the information-theoretic Shannon entropy of the spectrum of the density matrix. There are other types of entropy associated with von Neumann entropy (relative entropy, conditional entropy, the list is not small) in the framework of quantum information theory to characterize the entropy of quantum entanglement (see for example Nielsen and Chuang[213]). The subject of quantum information is rapidly developing.

A.4 THE WEYL CORRESPONDENCE PRINCIPLE

Statistical mechanics applies to systems governed by Hamiltonian dynamics whether in its classical or quantum formulation, with the roles played by classical and quantum mechanics appearing similar. What the classical and quantum theories share is a common algebraic structure of Poisson brackets and commutators; both are Lie algebras. In quantum mechanics, the canonical coordinates q, p are associated with abstract linear operators $\hat{q}, \hat{p}$ acting on an abstract Hilbert space. These are Hermitian and noncommuting, $[\hat{q}, \hat{p}] = \mathrm{i}\hbar I$, with I the identity operator. The state of a quantum system is represented by an element of the Hilbert space rather than as a point in Γ-space (which is forbidden by the Heisenberg uncertainty principle). Dynamical functions are represented as operators on the Hilbert space instead of functions on Γ-space. Dirac[165, Section 21] found a connection between the commutator of operators associated with physical observables $[\hat{A}, \hat{B}]$ and the Poisson brackets of their classical representations $[A, B]_{\mathrm{P}}$, $[\hat{A}, \hat{B}] \leftrightarrow \mathrm{i}\hbar\,[A, B]_{\mathrm{P}}\, I + O(\hbar^2)$. This leaves open the question of how we find the operator $\hat{A}$ associated with a given classical dynamical function $A(p, q)$.

Weyl developed a general rule for associating self-adjoint operators with classical dynamical functions.[26] The first step is to represent Γ-space functions as Fourier integrals,

$$A(p, q) = \int\!\!\int a(\alpha, \beta) \mathrm{e}^{\mathrm{i}(\alpha p + \beta x)/\hbar} \mathrm{d}\alpha \mathrm{d}\beta, \tag{A.53}$$

where α, β are real parameters with α having the dimension of length and β the dimension of momentum.[27] The *Weyl correspondence* defines $\hat{A}(\hat{p}, \hat{q})$ through the extension of Eq. (A.53) under $p \to \hat{p}$, $q \to \hat{q}$,

$$\hat{A}(\hat{p}, \hat{q}) \equiv \int\!\!\int a(\alpha, \beta) \mathrm{e}^{\mathrm{i}(\alpha \hat{p} + \beta \hat{q})/\hbar} \mathrm{d}\alpha \mathrm{d}\beta. \tag{A.54}$$

The reality of $A(p, q)$ implies the self-adjointness of $\hat{A}(\hat{p}, \hat{q})$.[28] Keep in mind that the Weyl prescription is a postulate, the veracity of which is determined by comparison with experiment.

We therefore have constructed a self-adjoint operator $\hat{A}(\hat{p}, \hat{q})$ from a weighted superposition of unitary operators $\hat{E}(\alpha, \beta) \equiv \mathrm{e}^{\mathrm{i}(\alpha\hat{p} + \beta\hat{q})/\hbar}$, where the weighting factor $a(\alpha, \beta)$ is the Fourier transform of the classical function $A(p, q)$. It behooves us therefore to understand the properties of $\hat{E}(\alpha, \beta)$. It's readily seen to be unitary, $\hat{E}^\dagger(\alpha, \beta) = \hat{E}(-\alpha, -\beta) = \hat{E}^{-1}(\alpha, \beta)$. It can be shown

[26] See Weyl[214, p275] and also McCoy[215]. Curtright, Fairlie, and Zachos[216] is a modern reference.

[27] Planck's constant has the dimension of action. For simplicity we work in one dimension. In actuality, α is an abbreviation for a set of transform variables $(\alpha_1, \ldots, \alpha_N)$, $\mathrm{d}\alpha \equiv \mathrm{d}\alpha_1 \ldots \mathrm{d}\alpha_N$, and $\alpha p = \sum_i \alpha_i p_i$; the same for β.

[28] The reality of $A(p, q)$ implies $a^*(\alpha, \beta) = a(-\alpha, -\beta)$ where α, β each extend from $-\infty$ to ∞.

using the Baker-Campbell-Hausdorff equation[29] that $\hat{E}$ can be written in two ways:

$$\hat{E}(\alpha,\beta) = \begin{cases} e^{i\alpha\beta/(2\hbar)} \cdot e^{i\beta\hat{q}/\hbar} e^{i\alpha\hat{p}/\hbar} & \text{(qp-ordered form)} \\ e^{-i\alpha\beta/(2\hbar)} \cdot e^{i\alpha\hat{p}/\hbar} e^{i\beta\hat{q}/\hbar}. & \text{(pq-ordered form)} \end{cases} \tag{A.55}$$

Thus, $e^{i\beta\hat{q}/\hbar}e^{i\alpha\hat{p}/\hbar} = e^{-i\alpha\beta/\hbar}e^{i\alpha\hat{p}/\hbar}e^{i\beta\hat{q}/\hbar}$. Another way of writing $\hat{E}$ (in two versions) is

$$\hat{E}(\alpha,\beta) = e^{i\beta\hat{q}/(2\hbar)}e^{i\alpha\hat{p}/\hbar}e^{i\beta\hat{q}/(2\hbar)} = e^{i\alpha\hat{p}/(2\hbar)}e^{i\beta\hat{q}/\hbar}e^{i\alpha\hat{p}/(2\hbar)}. \tag{A.56}$$

The trace of $\hat{E}$ has the form

$$\text{Tr}\,\hat{E}(\alpha,\beta) = 2\pi\hbar\delta(\alpha)\delta(\beta). \tag{A.57}$$

To show this, rely momentarily on the coordinate representation, knowing that the trace is basis independent. With $\hat{p} = -i\hbar d/dx$, $\exp[i\alpha\hat{p}/(2\hbar)] = \exp[(\alpha/2)d/dx]$, so that, for any function $f(x)$, $\exp[(\alpha/2)d/dx]f(x) = f(x+\alpha/2)$; $\exp[(\alpha/2)d/dx]$ acting on an analytic function generates its Taylor series. Assume a complete set of functions such that $\sum_n \psi_n^*(x)\psi_n(y) = \delta(x-y)$. Then, using the second equality in Eq. (A.56),

$$\begin{aligned} \text{Tr}\,\hat{E} &= \sum_n \langle n|\hat{E}|n\rangle = \sum_n \int dx \psi_n^*(x) \exp[(\alpha/2)d/dx] e^{i\beta x/\hbar} \exp[(\alpha/2)d/dx]\psi_n(x) \\ &= \sum_n \int dx \psi_n^*(x) e^{i\beta(x+\alpha/2)/\hbar}\psi_n(x+\alpha) = \delta(\alpha)\int dx e^{i\beta(x+\alpha/2)/\hbar} = 2\pi\hbar\delta(\alpha)\delta(\beta). \end{aligned}$$

We also have the identities

$$\begin{aligned} \hat{E}(\alpha',\beta')\hat{E}(\alpha'',\beta'') &= e^{i(\alpha'\beta''-\beta'\alpha'')/(2\hbar)}\hat{E}(\alpha'+\alpha'',\beta'+\beta'') \\ &= e^{i(\alpha'\beta''-\beta'\alpha'')/\hbar}\hat{E}(\alpha'',\beta'')\hat{E}(\alpha',\beta'). \end{aligned} \tag{A.58}$$

As a consequence of Eqs. (A.57) and (A.58), we have another trace property

$$\text{Tr}\,\hat{E}(\alpha',\beta')\hat{E}^\dagger(\alpha,\beta) = 2\pi\hbar\delta(\alpha'-\alpha)\delta(\beta'-\beta). \tag{A.59}$$

Equation (A.59) can be used to show that, starting from Eq. (A.54),

$$a(\alpha,\beta) = \frac{1}{2\pi\hbar}\text{Tr}\,\hat{A}(\hat{p},\hat{q})\hat{E}^\dagger(\alpha,\beta). \tag{A.60}$$

Thus we have a way of inverting Eq. (A.54). Fourier analysis on functions $A(p,q)$ implies Fourier analysis on operators $\hat{A}(\hat{p},\hat{q})$. By combining Eq. (A.60) with Eq. (A.54), we have the identity,

$$\hat{A} = \frac{1}{2\pi\hbar}\int\int d\alpha d\beta\,\text{Tr}\left(\hat{A}\hat{E}^\dagger(\alpha,\beta)\right)\hat{E}(\alpha,\beta). \tag{A.61}$$

Let's construct the operator associated with the function $A(p,q) = qp$. The first task is to find $a(\alpha,\beta)$. Starting from the inverse of Eq. (A.53), $a(\alpha,\beta) = \left(1/(2\pi\hbar)^2\right)\int\int dp dq e^{-i(\alpha p+\beta q)/\hbar} A(p,q)$, we must evaluate

$$a(\alpha,\beta) = \frac{1}{(2\pi\hbar)^2}\int\int pq e^{-i(\alpha p+\beta q)/\hbar} dp dq.$$

[29]The Baker-Campbell-Hausdorff equation is often presented in quantum texts as a specialized version of a more general formula, usually without derivation. The theorem states, for operators $\hat{A}$, $\hat{B}$, that $e^{\hat{A}}e^{\hat{B}} = e^{C(\hat{A},\hat{B})}$ with $C(\hat{A},\hat{B}) = \hat{A}+\hat{B}+\frac{1}{2}[\hat{A},\hat{B}]+\frac{1}{12}[\hat{A},[\hat{A},\hat{B}]]+\frac{1}{12}[[\hat{A},\hat{B}],\hat{B}]+\cdots$. See for example Veltman[217, Appendix A.4]. Only when $\hat{A}$, $\hat{B}$ commute with their commutator do we have the closed-form expression $e^{\hat{A}}e^{\hat{B}} = e^{\hat{A}+\hat{B}+\frac{1}{2}[\hat{A},\hat{B}]}$.

Noting that $pe^{-i\alpha p/\hbar} = i\hbar(\partial/\partial\alpha)e^{-i\alpha p/\hbar}$ and $qe^{-i\beta q/\hbar} = i\hbar(\partial/\partial\beta)e^{-i\beta q/\hbar}$, we have

$$
\begin{aligned}
a(\alpha,\beta) &= \frac{1}{(2\pi\hbar)^2}(i\hbar)^2 \int\int dpdq \frac{\partial}{\partial\alpha}\frac{\partial}{\partial\beta} e^{-i(\alpha p+\beta q)/\hbar} = -\frac{1}{4\pi^2}\frac{\partial}{\partial\alpha}\frac{\partial}{\partial\beta}\int\int dpdq e^{-i(\alpha p+\beta q)/\hbar} \\
&= -\hbar^2\left(\frac{\partial}{\partial\alpha}\delta(\alpha)\right)\left(\frac{\partial}{\partial\beta}\delta(\beta)\right) \equiv -\hbar^2\delta'(\alpha)\delta'(\beta),
\end{aligned}
$$

where the derivative of the Dirac delta function,[30] $\delta'(\alpha) \equiv d\delta(\alpha)/d\alpha$. We see the pattern now. For $A(p,q) = p^n q^m$, (n,m) integers, we have with $p^n \leftrightarrow (i\hbar\partial/\partial\alpha)^n$ and $q^m \leftrightarrow (i\hbar\partial/\partial\beta)^m$ that $a(\alpha,\beta) = (i\hbar)^{n+m}\delta^{(n)}(\alpha)\delta^{(m)}(\beta)$ where $\delta^{(n)}(\alpha) \equiv d^n\delta(\alpha)/d\alpha^n$. The quantities $a(\alpha,\beta)$ are therefore highly singular. The operator $\hat{A}$ associated with $A(p,q) = qp$ is, from Eq. (A.54),

$$
\begin{aligned}
\hat{A} &= -\hbar^2 \int\int d\alpha d\beta \delta'(\beta)\delta'(\alpha)\hat{E}(\alpha,\beta) = -\hbar^2\int\int d\alpha d\beta\delta(\beta)\delta(\alpha)\frac{\partial^2}{\partial\beta\partial\alpha}\hat{E}(\alpha,\beta) \\
&= -\hbar^2\left(\frac{\partial^2}{\partial\beta\partial\alpha}\hat{E}(\alpha,\beta)\right)\Bigg|_{\alpha=0,\beta=0},
\end{aligned}
\tag{A.62}
$$

where we've integrated by parts on α and β. Using the qp-ordered form of $\hat{E}$ in Eq. (A.55) we find $\hat{A} = \hat{q}\hat{p} - i(\hbar/2)I$, whereas with the pq-ordered form $\hat{A} = \hat{p}\hat{q} + i(\hbar/2)I$. The Weyl rule therefore leads us to

$$
A(p,q) = qp \implies \hat{A} = \begin{cases} \hat{q}\hat{p} - \frac{i}{2}\hbar I & \text{(qp-ordered form)} \\ \hat{p}\hat{q} + \frac{i}{2}\hbar I. & \text{(pq-ordered form)} \end{cases}
\tag{A.63}
$$

The two are equivalent by the commutation relation $[\hat{q},\hat{p}] = i\hbar I$. Under the Weyl correspondence principle,

$$
A(p,q) = qp \underset{\text{Weyl}}{\implies} \hat{A} = \frac{1}{2}\left(\hat{q}\hat{p} + \hat{p}\hat{q}\right).
\tag{A.64}
$$

The Weyl procedure is used to construct Hermitian operators associated with products of non-commuting operators. One might guess for example that the operator associated with the function q^2p is $\frac{1}{2}(\hat{q}^2\hat{p} + \hat{p}\hat{q}^2)$ or $\hat{q}\hat{p}\hat{q}$ (both are Hermitian). The Weyl rule leads to $\frac{1}{3}\left(\hat{q}^2\hat{p} + \hat{q}\hat{p}\hat{q} + \hat{p}\hat{q}^2\right)$.

[30] Derivatives of the Dirac delta function of all orders are defined inside integrals; see Lighthill[99, p19].

APPENDIX B

Probability theory

A review of basic probability theory is given, one that presumes prior exposure to the subject. More details can be found in Feller[218] or Parzen[38] or [5, Chapter 3].

B.1 EVENTS, SAMPLE SPACE, AND PROBABILITY

Event—a primitive concept in probability theory—denotes the outcome of an experiment that can be repeated many times. The set of all possible outcomes of an experiment is known as *sample space*, denoted Ω. *Elementary events* are represented by single points (elements) of Ω. *Compound events* are represented by subsets of Ω having two or more elementary events. For example, the sample space for the number of dots showing on the toss of a six-sided die is $\Omega = (1,2,3,4,5,6)$. These are elementary events. The event consisting of an even number of dots, i.e., the subset $(2,4,6)$, is a compound event. The entire sample space Ω is referred to as the *certain event*; the *impossible event* (no elementary event) is denoted $\varnothing$. Many results in elementary probability theory follow from set theory. For $A \subset \Omega$ and $B \subset \Omega$, one can form their union $A \cup B$ (events belonging to A or B) and their intersection $A \cap B$ (events belonging to A and B). Events A and B are *mutually exclusive*[1] if $A \cap B = \varnothing$.

For sample spaces with denumerably many elementary events E_i, $\Omega \equiv (E_1, E_2, \ldots, E_n, \ldots)$, a *probability function* P is defined as any function[2] satisfying the *Kolmogorov axioms*,

1. $P(E_i) \geq 0$ for every event E_i,
2. $P(\Omega) = 1$ for the certain event Ω,
3. $P(A \cup B) = P(A) + P(B)$ if $A \cap B = \varnothing$.

The edifice of probability theory follows as a logical consequence of these axioms. It can be shown (from the axioms) that $P(\varnothing) = 0$ and that $0 \leq P(E) \leq 1$ for any event E. Other familiar properties are consequences of these axioms, such as the probability of the union of two events, a generalization of the third axiom,

$$P(A \cup B) = P(A) + P(B) - P(A \cap B). \tag{B.1}$$

The probability of A and B occurring, $P(A \cap B)$, is calculated using the *conditional probability* $P(B|A)$, the probability that B occurs *given* the occurrence of A. The quantity $P(A \cap B)$ is defined

$$P(A \cap B) = P(A)P(B|A) = P(B)P(A|B). \tag{B.2}$$

Events A and B are said to be *statistically independent* if $P(A \cap B) = P(A)P(B)$. Independent events are such that $P(B|A) = P(B)$ and $P(A|B) = P(A)$.

[1] Elementary events are mutually exclusive.

[2] Functions assign numbers to each element of its domain. The domain of a probability function is the sample space.

DOI: 10.1201/9781003512295-B

Although any function P satisfying the axioms constitutes a probability function, we have yet to specify one. It's often the case that elementary events occur with equal likelihood (as in the throw of a fair die). Elementary events $(E_1, \ldots, E_N)$ occurring with equal likelihood all have the same probability,

$$P(E_1) = P(E_2) = \cdots = P(E_N) = \frac{1}{N}. \tag{B.3}$$

For an event $A \subset \Omega$ consisting of N_A elementary events, $P(A)$ is the ratio of the number of sample points associated with the occurrence of A to the number of points in the sample space, N_Ω:

$$P(A) = \frac{N_A}{N_\Omega}. \tag{B.4}$$

Equation (B.4) is the *frequency interpretation* of probability, formulated by Laplace in 1812. It reflects the fact that in experiments with well-controlled conditions, the property under observation does not always occur with the same value—so no deterministic regularity—but over the course of many trials there is an empirically observed *statistical regularity* that A occurs a fraction of the time given by $P(A)$.[3] Probabilities computed using Eq. (B.4) satisfy the axioms, and thus they satisfy any theorems that follow logically from the axioms. Equation (B.4) requires us to have the ability to compute the *size* of sets. Counting denumerable collections of objects[4] is a familiar activity, yet there are specialized techniques for large sets. *Combinatorics* is a branch of mathematics devoted to just that. The reader is presumed to have a proficiency with basic combinatorics: permutations, combinations, binomial coefficients, and Stirling's approximation.

B.2 RANDOM VARIABLES AND PROBABILITY DISTRIBUTIONS

One of the most useful concepts in probability theory is that of *random variable*. The sample space for the throw of two coins could be described $\{HH, HT, TH, TT\}$ (H for heads, T for tails). Or it could be represented by points in the x-y plane, $\{(1,1), (1,0), (0,1), (0,0)\}$, where we assign H the number 1 and T the number 0. How we display the sample space is immaterial, yet some ways are better than others. We'd like to *parameterize* the sample space in a way that's aligned with the business of calculating averages.

Definition. *A random variable X assigns a real number (a random number[5]) to each point of sample space.*[6]

As an example, let a random variable X represent the number of heads showing in the toss of two coins with $X = 0$ for no heads (TT), 1 for one head (TH or HT), and 2 for two heads, HH. Under such an assignment, the probabilities associated with the sample points are $1/4$ for $X = 0$, $1/2$ for $X = 1$, and $1/4$ for $X = 2$. As another example, let X denote the sum of the dots showing in the roll of two dice, with $f(X)$ the probability associated with the value of X. Enumeration of the possibilities shows that $f(2) = f(12) = 1/36$, $f(3) = f(11) = 2/36$, $f(4) = f(10) = 3/36$, $f(5) = f(9) = 4/36$, $f(6) = f(8) = 5/36$ and $f(7) = 6/36$.

[3] A *random* phenomenon is empirically characterized by the property that its observation under controlled conditions does not always result in the same value but rather different outcomes occur with statistical regularity. Randomness reflects our inability to control microscopic initial conditions.

[4] Naive extensions of probability to nondenumerable sets lead to logical difficulties that are resolved only with more advanced mathematics. The size of nondenumerable sets is the province of *measure theory* (beyond the level of this book). See Loève[39, Chapters 1–8] and Cramér[31, Chapters 4–9] for introductions to measure theory.

[5] The word *random* is used because the elements of the sample space are associated with physical experiments in which the outcome of any one experiment is unpredictable and associated with randomness.

[6] To be more mathematically precise, random variables are real-valued *functions* on sample space, $X : \Omega \to S$, where S is the *state space*, the space of measured values of X, a subset of $\mathbb{R}$. Probabilities are also functions on Ω, but the mapping is onto $[0, 1]$. The values of random variables inherit the ordering of points on the real line.

B.2.1 Probability distributions on discrete sample spaces; joint distributions

The set of probabilities associated with the range of a random variable is a *probability distribution*.[7] For each value x_j of a random variable X, the aggregate of sample points associated with x_j form the event having probability $P(X = x_j)$.

Definition. *A function $f(x)$ such that for each x_j, $f(x_j) = P(X = x_j)$, is the probability distribution of X.*

For the range of X, $f(x_j) \geq 0$ and $\sum_j f(x_j) = 1$.

One can have more than one random variable on the same sample space. Consider random variables X and Y having the values $x_1, x_2, \ldots$ and $y_1, y_2, \ldots$, and let the corresponding probability distributions be $f(x_j)$ and $g(y_k)$. The aggregate of events for which $X = x_j$ and $Y = y_k$ form the event having probability $P(X = x_j, Y = y_k)$.

Definition. *A function $p(x, y)$ such that, for each x_j and y_k, $p(x_j, y_k) = P(X = x_j, Y = y_k)$, is the joint probability distribution of X and Y.*

Clearly, $p(x_j, y_k) \geq 0$ and $\sum_{jk} p(x_j, y_k) = 1$. Moreover, for fixed x_j

$$\sum_k p(x_j, y_k) = f(x_j), \tag{B.5}$$

while, for fixed y_k,

$$\sum_j p(x_j, y_k) = g(y_k). \tag{B.6}$$

Adding the probabilities for all events y_k for fixed x_j produces the probability distribution for x_j and adding the probabilities for all events x_j produces the probability distribution for y_k. This idea generalizes to n random variables $X_1, X_2, \ldots X_n$, such that $p(x_{j1}, x_{j2}, \ldots, x_{jn}) = P(X_1 = x_{j1}, X_2 = x_{j2}, \ldots, X_n = x_{jn})$.

Definition. *If the joint probability distribution $f(x_1, x_2, \ldots, x_n)$ can be factored in the form $f(x_1, x_2, \ldots, x_n) = f(x_1)f(x_2)\cdots f(x_n)$, where $f(x_i)$ is the probability distribution of X_i, then the random variables $X_1, X_2, \ldots, X_n$ are statistically independent.*

B.2.2 Probability densities on continuous sample spaces; joint densities

A wide class of experiments involve continuous sample spaces (those not characterized by isolated sample points), on which *continuous* random variables are defined. The probability distribution of a continuous random variable is called a *probability density* $f(x)$, of which $f(x)\mathrm{d}x$ represents the probability that its value lies between x and $x + \mathrm{d}x$. No measurement of a continuous quantity is ever perfectly precise;[8] one can only specify a probability that the value of a continuous random variable lies within a *window* $[x, x + \mathrm{d}x]$.

Definition. *A probability density is a function $f(x)$ such that:*

$$f(x) \geq 0 \qquad \int_{-\infty}^{\infty} f(x)\mathrm{d}x = 1 \qquad \int_a^b f(x)\mathrm{d}x = P(a < x < b). \tag{B.7}$$

[7] The term probability distribution can be ambiguous. We take it to be a function $f(x)$ defined on the range of X such that $f(x_j) = P(X = x_j)$. It can also mean the *cumulative* probability up to and including the value x, $F(x) \equiv \sum_{x_i \leq x} f(x_i)$. We use the term exclusively in the first sense.

[8] How tall are you, *exactly*? Whatever your answer, it can only lie within experimental uncertainties. You can only specify a continuous quantity within so many decimal places of accuracy.

A probability density for several continuous random variables is a straightforward generalization.

Definition. *A joint probability density is a function $f(x_1, \ldots, x_n)$ having the properties:*

$$
\begin{aligned}
& f(x_1, \ldots, x_n) \geq 0 \\
& \int_{-\infty}^{\infty} \cdots \int_{-\infty}^{\infty} f(x_1, \ldots, x_n) \mathrm{d}x_1 \cdots \mathrm{d}x_n = 1 \\
& \int_{a_n}^{b_n} \cdots \int_{a_1}^{b_1} f(x_1, \ldots, x_n) \mathrm{d}x_1 \cdots \mathrm{d}x_n = P(a_1 < x_1 < b_1, \ldots, a_n < x_n < b_n). \qquad \text{(B.8)}
\end{aligned}
$$

B.2.3 Moments of distributions

Definition. *The kth moment about the origin of a probability distribution $f(x)$ is found from*

$$\mu'_k \equiv \sum_j (x_j)^k f(x), \qquad (k = 0, 1, 2, \ldots) \qquad \text{(B.9)}$$

where the prime indicates that the moment is defined about the origin of the range of the random variable X.

Moments characterize the *shape* of probability distributions.[9] It can happen that the sum in Eq. (B.9) fails to exist for some value $k = r$. When the rth moment exists, all moments for $k \leq r$ exist. The moment $\mu'_0 = 1$ is the normalization of the distribution.

Definition. *The moment associated with $k = 1$ is the average, or the mean, or the expectation value, indicated with a variety of notations,*

$$\mu'_1 = \overline{x} = \langle x \rangle \equiv \sum_j x_j f(x_j). \qquad \text{(B.10)}$$

Equation (B.10) generalizes to functions of random variables, because a function $\phi(X)$ of a random variable X is a new random variable having values $\phi(X = x_j) = \phi(x_j)$. Thus,

$$\overline{\phi} = \langle \phi \rangle = \sum_j \phi(x_j) f(x_j). \qquad \text{(B.11)}$$

Definition. *The kth moment about the mean is defined*

$$\mu_k \equiv \sum_j (x_j - \overline{x})^k f(x_j). \qquad \text{(B.12)}$$

Clearly $\mu_1 = 0$. The moments μ'_1 and $\sqrt{\mu_2}$ are given special symbols: $\mu \equiv \mu'_1$ and $\sigma \equiv \sqrt{\mu_2}$. The quantity μ_2 is known as the *variance* of the distribution, with σ the *standard deviation*. Equations (B.10) and (B.11) generalize to continuous random variables where sums are replaced by integrals,

$$\mu'_k = \int x^k f(x) \mathrm{d}x \qquad \overline{\phi} = \langle \phi \rangle = \int \phi(x) f(x) \mathrm{d}x. \qquad \text{(B.13)}$$

This concludes our review of elementary probability theory (which largely suffices for statistical mechanics). Additional tools from probability theory required in nonequilibrium statistical physics are developed in Chapter 2.

[9] We have defined moments as moments of the probability distribution; moments are also referred to as moments of random variables.

Elastic collisions

THE theory of two-body elastic collisions (see for example Landau and Lifshitz[219]), essential in formulating the Boltzmann equation (see Chapter 4), is reviewed in this appendix. Figure C.1 shows two masses m_1, m_2, located by position vectors $\boldsymbol{r}_1, \boldsymbol{r}_2$ relative to origin O,

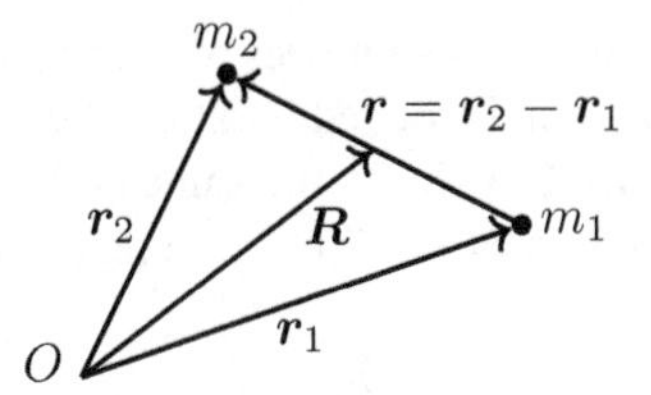

Figure C.1 Particles $1, 2$ are located by position vectors $\boldsymbol{r}_1, \boldsymbol{r}_2$. Center of mass is located by $\boldsymbol{R}$.

the laboratory reference frame. The vector $\boldsymbol{R} \equiv (m_1\boldsymbol{r}_1 + m_2\boldsymbol{r}_2)/(m_1 + m_2)$ locates a point (the center of mass) on the line joining the masses, the separation vector $\boldsymbol{r} \equiv \boldsymbol{r}_2 - \boldsymbol{r}_1$. As one can show, $\boldsymbol{r}_1 = \boldsymbol{R} - [m_2/(m_1 + m_2)]\boldsymbol{r}$ and $\boldsymbol{r}_2 = \boldsymbol{R} + [m_1/(m_1 + m_2)]\boldsymbol{r}$; particles can be located either with $\boldsymbol{r}_1, \boldsymbol{r}_2$ or $\boldsymbol{R}, \boldsymbol{r}$. Now let them have velocities $\boldsymbol{v}_1, \boldsymbol{v}_2$ in the lab frame. The particles have canonical coordinates $(\boldsymbol{p}_1, \boldsymbol{r}_1), (\boldsymbol{p}_2, \boldsymbol{r}_2)$, where $\boldsymbol{p}_1 = m_1\boldsymbol{v}_1$, $\boldsymbol{p}_2 = m_2\boldsymbol{v}_2$. By differentiating the formula for $\boldsymbol{R}$, we have $M\dot{\boldsymbol{R}} = m_1\dot{\boldsymbol{r}}_1 + m_2\dot{\boldsymbol{r}}_2 = \boldsymbol{p}_1 + \boldsymbol{p}_2$, where $M \equiv m_1 + m_2$. We can therefore associate with the center of mass a momentum $\boldsymbol{P} \equiv M\dot{\boldsymbol{R}}$ equal to the total momentum, $\boldsymbol{P} = \boldsymbol{p}_1 + \boldsymbol{p}_2$. With the *reduced mass*[1] $\mu \equiv m_1 m_2/M$ we can associate a momentum with the relative separation, $\boldsymbol{p} \equiv \mu\dot{\boldsymbol{r}} = (m_1\boldsymbol{p}_2 - m_2\boldsymbol{p}_1)/M$. With these definitions, $\boldsymbol{p}_1 = (m_1/M)\boldsymbol{P} - \boldsymbol{p}$ and $\boldsymbol{p}_2 = (m_2/M)\boldsymbol{P} + \boldsymbol{p}$; momenta can be described either with $\boldsymbol{p}_1, \boldsymbol{p}_2$ or $\boldsymbol{P}, \boldsymbol{p}$. The transformation from $(\boldsymbol{p}_1, \boldsymbol{r}_1, \boldsymbol{p}_2, \boldsymbol{r}_2)$ to $(\boldsymbol{P}, \boldsymbol{R}, \boldsymbol{p}, \boldsymbol{r})$ is measure preserving (has unit Jacobian).

Assume particles interact through a central potential energy function $V(|\boldsymbol{r}_2 - \boldsymbol{r}_1|) = V(|\boldsymbol{r}|) \equiv V(r)$ that has a repulsive core.[2,3] The Hamiltonian is, in either set of variables,

$$H(\boldsymbol{p}_1, \boldsymbol{r}_1, \boldsymbol{p}_2, \boldsymbol{r}_2) = \frac{1}{2m_1}p_1^2 + \frac{1}{2m_2}p_2^2 + V(r) = \frac{1}{2M}P^2 + \frac{1}{2\mu}p^2 + V(r) = H(\boldsymbol{P}, \boldsymbol{R}, \boldsymbol{p}, \boldsymbol{r}). \quad (C.1)$$

The kinetic energy is thus the sum of the kinetic energy of the center of mass and that of a particle of mass μ, and, because $\boldsymbol{R}$ is a cyclic coordinate, $\boldsymbol{P}$ is conserved.[4] Thus, $P^2/(2M)$ is a constant

[1] Note that $\mu = (m_2/M)m_1 = (m_1/M)m_2$; μ is smaller than either of m_1, m_2. If $m_1 = m_2 = m$, $\mu = \frac{1}{2}m$.
[2] The Lennard-Jones potential for example has a repulsive core. See Fig. 6.1 in [5].
[3] We consider *elastic* collisions which don't involve any change in the internal state of particles.
[4] These are instances of general theorems in classical mechanics; see Goldstein[220, Chapter 1].

DOI: 10.1201/9781003512295-C

and can be dropped from the Hamiltonian. The motion of two interacting particles in a central force field reduces to the dynamics of a single particle of mass μ subject to the same force.

Figure C.2 depicts the motion of two interacting particles in the lab frame. The dashed line

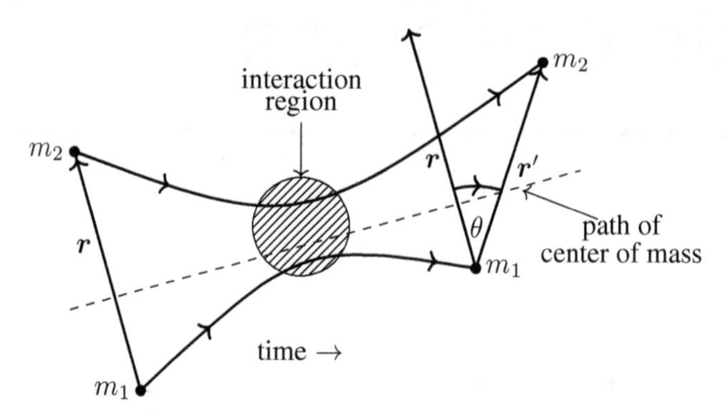

Figure C.2 Motion of interacting particles in the lab frame (not drawn to scale).

indicates the path of the center of mass, which moves with constant velocity. Note that $\boldsymbol{r}$, the relative separation vector before the interaction, is rotated into that after the interaction, $\boldsymbol{r}'$. Primed variables denote after-collision values. Figure C.3 shows the same collision in the center-of-mass reference frame ($\boldsymbol{R} = 0$). The rotation angle θ in Fig. C.3 is the same as that in Fig. C.2.

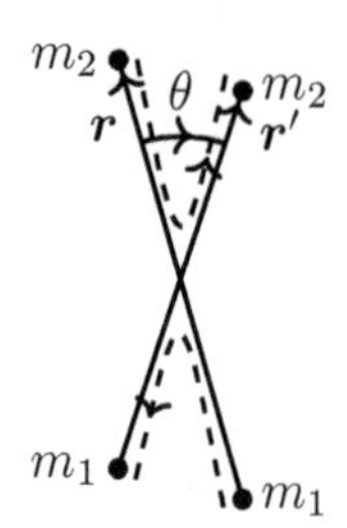

Figure C.3 Motion of interacting particles in the center-of-mass frame.

There are three regimes associated with scattering: before, during, and after the interaction. *Before* and *after* are provided meaning by potentials having a finite range r_0 so that $V(r > r_0) = 0$, indicated in Fig. C.2 with the hatched circle, the *region of interaction*. Particles outside the interaction region are free and move with constant velocities. We will use Fig. C.4 to depict two-body

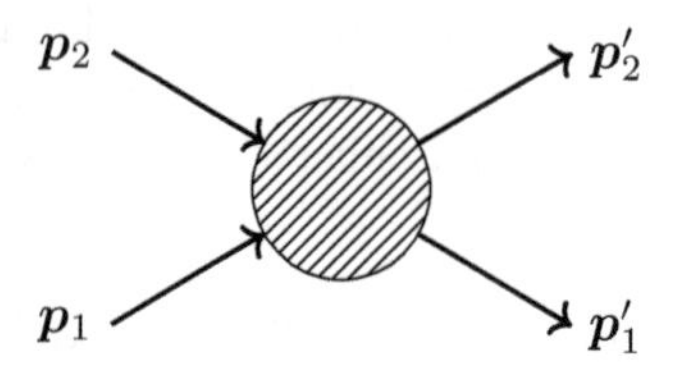

Figure C.4 Schematic illustration of a two-body collision in the lab frame.

collisions in the lab frame when the detail shown in Fig. C.2 is not required. The before and after-collision variables pertain to free particles, implying the energy and momentum conservation laws

$$\boldsymbol{p}_1 + \boldsymbol{p}_2 = \boldsymbol{p}_1' + \boldsymbol{p}_2' \qquad \frac{1}{2m_1}p_1^2 + \frac{1}{2m_2}p_2^2 = \frac{1}{2m_1}p_1'^2 + \frac{1}{2m_2}p_2'^2. \tag{C.2}$$

Angular momentum is conserved (central forces). For given $\boldsymbol{p}_1, \boldsymbol{p}_2$, a range of possible momenta $\boldsymbol{p}'_1, \boldsymbol{p}'_2$ exist satisfying the conservation laws (which are kinematic and apply regardless of the interaction potential). That is, conservation laws don't uniquely determine the after-collision momenta;[5] knowledge of the force law is required. One can't for example predict the scattering angle on the basis of conservation laws alone. Let's show that.

As we've seen, two-body motion in a central force field reduces to the dynamics of a fictitious particle of mass μ located by the vector $\boldsymbol{r}$ with $H = p^2/(2\mu) + V(r)$. Central-force motion is planar,[6] implying two generalized coordinates are needed to specify $\boldsymbol{r}$. Choose polar coordinates (r, ϕ) where the angle ϕ is relative to any line in the plane. The Hamiltonian is

$$H = \frac{1}{2\mu}p_r^2 + \frac{1}{2\mu r^2}p_\phi^2 + V(r).$$

Angular momentum is conserved, set $p_\phi = L$; H is a constant of the motion, set $H = E$. Thus, with $p_r = \mu\dot{r}$,

$$E = \frac{\mu}{2}\left(\frac{\mathrm{d}r}{\mathrm{d}t}\right)^2 + \frac{L^2}{2\mu r^2} + V(r) \implies \frac{\mathrm{d}r}{\mathrm{d}t} = \pm\sqrt{\frac{2}{\mu}[E - L^2/(2\mu r^2) - V(r)]}.$$

The particle μ moves in the *effective potential* $V_{\text{eff}}(r) \equiv V(r) + L^2/(2\mu r^2)$. The differential relation $\mathrm{d}\phi/\mathrm{d}r = \dot{\phi}/\dot{r} = L/(\mu r^2\dot{r})$ can be integrated to provide the orbit equation,

$$\phi(r) = \frac{L}{\sqrt{2\mu}}\int\frac{\mathrm{d}r}{r^2\sqrt{E - V(r) - L^2/(2\mu r^2)}} + \text{constant}. \tag{C.3}$$

Thus, to calculate ϕ one must know $V(r)$, E, and L. Note that our analysis applies to scattering or unbound orbits. Assuming $\lim_{r\to\infty} V(r) = 0$, unbound trajectories occur for $E > 0$.

Figure C.5 shows scattering in the frame of the separation vector $\boldsymbol{r}$, i.e., in the reference frame

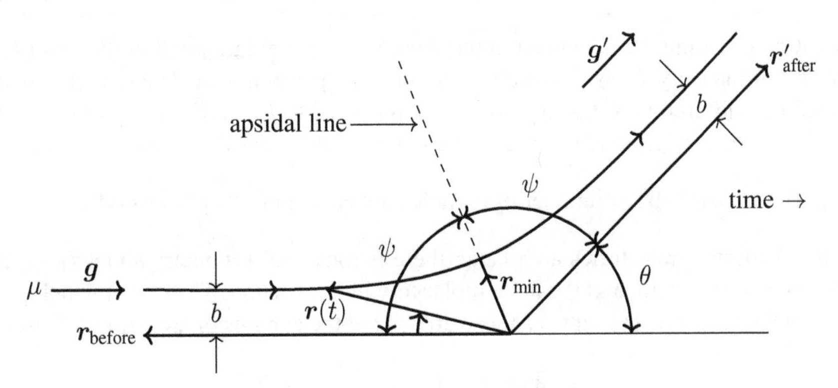

Figure C.5 Scattering in the frame of the separation vector $\boldsymbol{r}$ (particle 1 in Fig. C.1).

of particle 1. For $t \to -\infty$, $\boldsymbol{r}(t) \to \boldsymbol{r}_{\text{before}}$ and as $t \to \infty$, $\boldsymbol{r}(t) \to \boldsymbol{r}'_{\text{after}}$. The particle of mass μ approaches the scattering center with velocity $\boldsymbol{g} \equiv \dot{\boldsymbol{r}} = \boldsymbol{v}_2 - \boldsymbol{v}_1$, displaced from $\boldsymbol{r}_{\text{before}}$ by the *impact parameter* b. The angular momentum $\boldsymbol{L} = \boldsymbol{r} \times \boldsymbol{p}$ has magnitude $L = \mu bg$ (b is the lever arm associated with the cross product). By energy conservation, $E = \mu g^2/2 = \mu g'^2/2$, and hence $g' = g$. The relative speed is not a constant of the motion; g' takes the value g after the collision. Angular momentum conservation implies $b' = b$. The impact parameter b and the speed g are

[5]For given $\boldsymbol{p}_1, \boldsymbol{p}_2$, the unknowns $\boldsymbol{p}'_1, \boldsymbol{p}'_2$ in Eq. (C.2) imply six scalar quantities constrained by four equations. In d dimensions, $2d$ scalar quantities are constrained by $d + 1$ equations. Only for $d = 1$ (head-on collisions) does $d + 1 = 2d$.

[6]The vectors $\boldsymbol{r}$ and $\boldsymbol{p}$ are each orthogonal to $\boldsymbol{L} = \boldsymbol{r} \times \boldsymbol{p}$ and define a plane. $\boldsymbol{L}$ is a constant vector under central forces.

therefore intrinsic properties of a collision. At the point of closest approach, the *apsis* of the orbit, $\dot{r} = 0$; r_{min} is found from the solution to

$$V(r_{\text{min}}) + L^2/(2\mu r_{\text{min}}^2) = E. \tag{C.4}$$

The trajectory is symmetric about the apsidal line (a consequence of angular momentum conservation). The apsidal angle ψ can be calculated using Eq. (C.3) (for $\dot{r} > 0$),

$$\psi = \frac{L}{\sqrt{2\mu}} \int_{r_{\text{min}}}^{\infty} \frac{\mathrm{d}r}{r^2\sqrt{E - V(r) - L^2/(2\mu r^2)}}. \tag{C.5}$$

The scattering angle θ (between incoming and outgoing velocities) is $\theta = \pi - 2\psi$.

The scattering angle is thus determined by g, b, and $V(r)$. In experiments, scatterers are exposed to a *beam* of particles of the same speed but of varying impact parameters. Figure C.6 illustrates

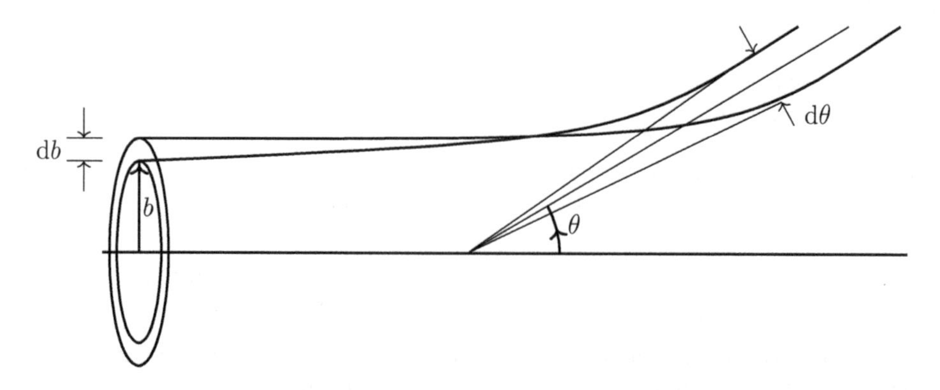

Figure C.6 Variation of impact parameter $\mathrm{d}b$ results in variation of scattering angle $\mathrm{d}\theta$.

how a variation $\mathrm{d}b$ in impact parameter about b leads to an angular variation $\mathrm{d}\theta$ about θ. A steady particle beam of intensity I has I particles per unit time per unit area. The rate at which particles are scattered into differential solid angle $\mathrm{d}\Omega$ is proportional to I and $\mathrm{d}\Omega$. The proportionality factor is termed the *differential scattering cross section*, $\sigma(\Omega)$, which has the dimension of area. Thus,

$$I\sigma(\Omega)\mathrm{d}\Omega = \text{number of particles deflected into } \mathrm{d}\Omega \text{ per second.}$$

Because the scattering trajectories are planar, there is rotational symmetry about $\boldsymbol{r}_{\text{before}}$ and $\mathrm{d}\Omega \equiv \sin\theta \mathrm{d}\theta \mathrm{d}\phi$ (ϕ is the azimuth angle) can be replaced with $2\pi \sin\theta \mathrm{d}\theta$. The rate at which particles are scattered into $2\pi \sin\theta \mathrm{d}\theta$ is the same as those arriving at the annulus of area $2\pi b \mathrm{d}b$. Thus

$$I \cdot 2\pi b \mathrm{d}b = I \cdot \sigma(\theta) 2\pi \sin\theta \mathrm{d}\theta,$$

implying

$$\sigma(\theta) = \frac{b(\theta)}{\sin\theta} \left| \frac{\mathrm{d}b}{\mathrm{d}\theta} \right|. \tag{C.6}$$

The functional relation between b and θ is established by $\theta = \pi - 2\psi$, with ψ determined from Eq. (C.5) with $L = \mu b g$ and $E = \mu g^2/2$. Once $b = b(\theta)$ is known for a given potential, the differential scattering cross section is found from Eq. (C.6). The *total scattering cross section*

$$\sigma_t \equiv 2\pi \int_0^{\pi} \sigma(\theta) \sin\theta \mathrm{d}\theta \tag{C.7}$$

is the effective area of the target particle for producing a scattering event.

Integral equations and resolvents

A knowledge of integral equations is crucial to finding solutions of the Boltzmann equation (see Section 4.8). Resolvents, a topic pertaining to linear operators and used in nonequilibrium statistical mechanics (see Section 6.5), have a connection with integral equations. We review both.

Linear integral equations have the general form (see for example [13, Chapter 10]),

$$\beta y(x) - \lambda \int_a^b K(x,z)y(z)\mathrm{d}z = f(x), \tag{D.1}$$

where $y(x)$ and $f(x)$ are continuous functions, λ and β are constants, and $K(x,z)$ is the *kernel* of the integral equation, a function continuous in both variables. Equation (D.1) specifies a transformation of a continuous function $y(x)$ into another continuous function $f(x)$. The problem is to find $y(x)$ for given $f(x)$ and $K(x,z)$.[1]

Integral equations are classified in several ways.

- In *Fredholm equations* the limits of integration in Eq. (D.1) are constants. In *Volterra equations* the limits are functions of x.
- If $\beta = 0$, Eq. (D.1) is an integral equation of the *first kind*; if $\beta = 1$, it's an integral equation of the *second kind.*
- If $f(x) = 0$, Eq. (D.1) is *homogeneous*; otherwise it's *inhomogeneous.*

Values of λ for which homogeneous integral equations $y = \lambda \int_a^b Ky$ have nontrivial solutions are *eigenvalues* of K with the solutions $y(x)$ eigenfunctions. Every eigenvalue has finite multiplicity r, the number of linearly independent eigenfunctions[72, p113].

The *Fredholm theorems* are the basic theorems of integral equations; see [72, Chapter 3] or Smithies[221]. In the following, set $\beta = 1$ in Eq. (D.1).

1. Either the inhomogeneous equation (D.1) has a unique solution $y(x)$ for any function $f(x)$ (when λ is a *regular value* of K, *not* an eigenvalue), or the associated homogeneous equation

$$\psi(x) = \lambda \int_a^b K(x,z)\psi(z)\mathrm{d}z \tag{D.2}$$

has $r \geq 1$ linearly independent solutions $\psi_1, \ldots, \psi_r$, i.e., λ *is* an eigenvalue.

[1] Equation (D.1) is a linear transformation in that corresponding to the superposition $c_1y_1(x)+c_2y_2(x)$ is the analogous combination $c_1f_1(x) + c_2f_2(x)$.

DOI: 10.1201/9781003512295-D

2. If λ is not an eigenvalue associated with Eq. (D.1) (the first alternative—this is known as the *Fredholm alternative theorem*), λ is also not an eigenvalue of the "transposed" equation (swap x, z in K),

$$g(x) = \phi(x) - \lambda \int_a^b K(z,x)\phi(z)\mathrm{d}z, \tag{D.3}$$

which has a unique solution ϕ for every g. If λ is an eigenvalue (the second alternative) then it's also an eigenvalue of the transposed homogeneous equation

$$\chi(x) = \lambda \int_a^b K(z,x)\chi(z)\mathrm{d}z \tag{D.4}$$

with $r \geq 1$ linearly independent solutions $\chi_1, \ldots, \chi_r$. The latter is the analog of a matrix and its transpose having the same eigenvalues.

3. If $\lambda = \lambda_i$ is an eigenvalue of K, the inhomogeneous equation (D.1) has a solution if and only if $f(x)$ is orthogonal to the eigenfunctions of the transposed kernel,

$$\int_a^b f(x)\chi_i(x)\mathrm{d}x = 0. \qquad (i = 1, \ldots, r) \tag{D.5}$$

This follows if we multiply Eq. (D.1) by $\chi_i(x)$ and integrate over x. In this case the solution of Eq. (D.1) is determined only up to linear combinations of eigenfunctions, $\sum_{i=1}^r c_i\psi_i$. The constants are uniquely determined by the additional requirements $\int_a^b y(x)\psi_i(x)\mathrm{d}x = 0$, $i = 1, \ldots, r$. The Fredholm theorems mirror similar theorems in linear algebra[72, p6].

These theorems are used in finding solutions of the Boltzmann equation.

A kernel is symmetric if $K(x,z) = K(z,x)$; Hermitian if $K(x,z) = K^*(z,x)$. The same theorems hold for Hermitian kernels as for operators: Real eigenvalues and orthogonal eigenfunctions associated with distinct eigenvalues; see [13, p299]. For degenerate eigenvalues, the associated eigenfunctions can be orthogonalized with the Gram-Schmidt method.

A key result is that functions $g(x)$ generated by the *integral transform* of $y(x)$ with Hermitian kernel,

$$g(x) = \int_a^b K(x,z)y(z)\mathrm{d}z, \tag{D.6}$$

can be expanded in the eigenfunctions $\psi_n(x)$ of K [$\psi_n(x) = \lambda_n \int_a^b K(x,z)\psi_n(z)\mathrm{d}z$], $g(x) = \sum_{n=1}^\infty a_n\psi_n(x)$ with a_n expansion coefficients[72, p136]. Inhomogeneous integral equations can always be written in the form of integral transforms,

$$g(x) \equiv y(x) - f(x) = \lambda \int_a^b K(x,z)y(z)\mathrm{d}z. \tag{D.7}$$

The solution of Eq. (D.7) can then be expressed

$$y(x) = f(x) + \lambda \sum_{i=1}^\infty \frac{\langle\psi_i|f\rangle}{\lambda_i - \lambda}\psi_i(x), \tag{D.8}$$

where $\langle y_i|f\rangle \equiv \int_a^b y_i^*(x)f(x)\mathrm{d}x$. The steps leading to Eq. (D.8) are shown in [13, p300].

For inhomogeneous integral equations of the second kind, which we write in the form

$$y(x) = f(x) + \lambda \int_a^b K(x,z)y(z)\mathrm{d}z, \tag{D.9}$$

an iterative solution strategy presents itself:

$$y(x) = f(x) + \lambda \int_a^b K(x,z)\left(f(z) + \lambda \int_a^b K(z,z')y(z')\mathrm{d}z'\right)\mathrm{d}z$$
$$= f(x) + \lambda \int_a^b K(x,z)f(z)\mathrm{d}z + \lambda^2 \int_a^b \int_a^b K(x,z)K(z,z')y(z')\mathrm{d}z'\mathrm{d}z. \quad \text{(D.10)}$$

Equation (D.10) is exact, but it suggests an approximation scheme, the *Neumann series*, with $y_n(x)$ the nth-approximant,

$$y_n(x) = f(x) + \sum_{p=1}^{n} \lambda^p \int_a^b K_p(x,z)f(z)\mathrm{d}z, \quad \text{(D.11)}$$

where K_p is defined recursively ($K_0 \equiv 1$),

$$K_i(x,z) \equiv \int_a^b K(x,z')K_{i-1}(z',z)\mathrm{d}z'. \qquad (i = 1,2,\ldots) \quad \text{(D.12)}$$

As $n \to \infty$, Eq. (D.11) becomes an infinite series, which if convergent is the solution of second-kind integral equations, $y(x) = \lim_{n\to\infty} y_n(x)$. The Neumann series converges if[72, p141]

$$(b-a)\max|K(x,y)| < \frac{1}{|\lambda|}. \quad \text{(D.13)}$$

The solution to Eq. (D.9) can be written (for sufficiently small λ, summing the series),

$$y(x) = f(x) + \lambda \int_a^b R(x,z;\lambda)f(z)\mathrm{d}z, \quad \text{(D.14)}$$

where

$$R(x,z;\lambda) \equiv \sum_{m=0}^{\infty} \lambda^m K_{m+1}(x,z) \quad \text{(D.15)}$$

is known as the *resolvent kernel*. Using Eq. (D.8), the resolvent associated with a Hermitian kernel can be written

$$R(x,z;\lambda) = \sum_{i=1}^{\infty} \frac{\psi_i(x)\psi_i^*(z)}{\lambda_i - \lambda}. \quad \text{(D.16)}$$

Equation (D.16) formally resembles the eigenfunction expansion of the Green function associated with self-adjoint inhomogeneous differential equations.[2] Equation (D.16), although derived for sufficiently small $|\lambda|$, provides for an analytic continuation into the complex λ-plane with the eigenvalues λ_i appearing as simple poles. The resolvent is a meromorphic function of λ.

Equation (D.13) can be replaced by a less restrictive condition,[209, p147]

$$\int_a^b \int_a^b |K(x,y)|^2 \mathrm{d}x\mathrm{d}y < \frac{1}{\lambda^2}. \quad \text{(D.17)}$$

Kernels are therefore square integrable and belong to a Hilbert space associated with the domain of integration $a \le x \le b$, $a \le y \le b$. We could come up with a special notation for L^2 defined on a two-dimensional integration domain, but we'd soon dispense with it. The integral $\int_a^b |K(x,y)|^2\mathrm{d}y$ exists for almost all x (*Fubini's theorem*[209, p84]) and thus $k(x) \equiv [\int_a^b |K(x,y)|^2\mathrm{d}y]^{1/2}$ is an

[2] See [13, p258]. The resemblance shouldn't be taken literally, however. Eigenvalues in the theory of integral equations correspond to the inverse of eigenvalues in linear algebra and Sturm-Liouville theory. There is a connection between Green functions and kernels of integral equations[13, p287].

element of L^2 defined in the usual way as a space of square-integrable functions of a single variable x for $a \leq x \leq b$. Let $h(x)$ be a member of L^2. The integral $\int_a^b K(x,y)h(y)\mathrm{d}y$ has meaning for all x where $k(x)$ is well defined and defines a function belonging to L^2. Thus, kernels generate linear transformations on Hilbert space, $K : L^2 \to L^2$, $Kh \to g$, an abbreviation for $Kh(x) = g(x) \equiv \int_a^b K(x,y)h(y)\mathrm{d}y$. The kernel is an *integral operator* ([13, p17] or Stone[222, p99]) and can be treated on the same footing as differential operators or matrices.[3]

Consider the inhomogeneous equation

$$Ty - ly = g, \tag{D.18}$$

where T is a linear operator, l is a given complex number, and g is a given element of Hilbert space; the task is to find y. Equation (D.18) becomes Eq. (D.9) under the substitutions $g = -f/\lambda, l = \lambda^{-1}$, and $T = K$. For Eq. (D.18) to have a solution y for every g, l must not equal an eigenvalue of T. In that case, the inverse $(T - lI)^{-1}$ exists and $y = (T - lI)^{-1}g$. The operator-valued function

$$R(l;T) \equiv (T - lI)^{-1} \tag{D.19}$$

is called the *resolvent*[4] of T. The resolvent provides a rigorous way to characterize the spectrum of T through the analytic structure of $R(l;T)$. The set of complex numbers different from any eigenvalue of T is called the *resolvent set* of T, denoted $\rho(T)$; the resolvent is well-defined for $l \in \rho(T)$. The set of points in the complex l-plane such that $T - lI$ has no inverse is called the *point spectrum* of T.[5] Kato[223] is a useful reference for the properties of resolvents.

We note two identities, the *resolvent equations*.

1. For comparing the resolvents of the same operator T, for $l_1, l_2 \in \rho(T)$,

$$R(l_1;T) - R(l_2;T) = (l_1 - l_2)R(l_1;T)R(l_2;T). \tag{D.20}$$

Thus, $R(l_1;T), R(l_2;T)$ commute. Equation (D.20) follows from a one-line proof:

$$R(l_1) - R(l_2) = R(l_1)R^{-1}(l_2)R(l_2) - R(l_1)R^{-1}(l_1)R(l_2).$$

2. For comparing the resolvents of two distinct operators A and B, for $l \in \rho(A) \cap \rho(B)$,

$$R(l;A) - R(l;B) = R(l;A)(B - A)R(l;B). \tag{D.21}$$

Equation (D.21) follows in one line:

$$R(l;A) - R(l;B) = R(l;A)\left[R^{-1}(l;B) - R^{-1}(l;A)\right]R(l;B).$$

Equation (D.21) is used in Chapter 6.

---

[3]Matrix operations in Hilbert space in principle require infinite matrices.

[4]In the theory of integral equations, $\lambda K(I - \lambda K)^{-1}$ is also called the resolvent of K; the two definitions have the same analytic structure. The term *resolvent* was coined by Hilbert.

[5]There are two other classifications of spectrum based on the properties of the resolvent that need not concern us here: the continuous spectrum of T and the residual spectrum of T; there are no other possibilities (besides the resolvent set and the point spectrum)[222, p129].

APPENDIX E

Dynamical representations in quantum mechanics

IT'S usually taught in one's first exposure to the subject that quantum systems evolve in time through the time dependence of state vectors $|\psi\rangle$, elements of a Hilbert space $\mathcal{H}$, with observables represented by time-independent Hermitian operators[1] A defined on $\mathcal{H}$. Expectation values of A for systems in state $|\psi\rangle$ are treated as inner products on $\mathcal{H}$, $\langle\psi|A\psi\rangle$. Inner products, however, are invariant under unitary transformations[13, p33], $\langle\psi|\phi\rangle = \langle U\psi|U\phi\rangle$ with $U^\dagger U = I$. One has leeway therefore in introducing unitary transformations on vectors and operators so long as expectation values are preserved. Such transformations are referred to as *representations* in quantum mechanics, the subject of this appendix.[2]

E.1 SCHRÖDINGER REPRESENTATION

We start with the *Schrödinger representation*, where the system state at time t is represented by the vector $|\psi_S(t)\rangle$, the time dependence of which is governed by the Schrödinger equation

$$\mathrm{i}\hbar\frac{\partial}{\partial t}|\psi_S(t)\rangle = H_S|\psi_S(t)\rangle, \tag{E.1}$$

where H_S is the Hamiltonian operator, with the subscript S indicating the Schrödinger form of these quantities. The expectation value at time t of an observable represented by operator A_S for a system in state $|\psi_S(t)\rangle$ is given by

$$\langle A\rangle_t = \langle\psi_S(t)|A_S|\psi_S(t)\rangle. \tag{E.2}$$

The time rate of change of $\langle A\rangle_t$ can be found from:

$$\begin{aligned}\mathrm{i}\hbar\frac{\partial}{\partial t}\langle A\rangle_t &= \left(\mathrm{i}\hbar\frac{\partial}{\partial t}\langle\psi_S(t)|\right)A_S|\psi_S(t)\rangle + \langle\psi_S(t)|A_S\,\mathrm{i}\hbar\frac{\partial}{\partial t}|\psi_S(t)\rangle \\ &= -\langle\psi_S(t)|H_SA_S|\psi_S(t)\rangle + \langle\psi_S(t)|A_SH_S|\psi_S(t)\rangle = \langle\psi_S(t)|[A_S,H_S]|\psi_S(t)\rangle,\end{aligned} \tag{E.3}$$

where we've used the rules of Hermitian conjugation on Eq. (E.1), $-\mathrm{i}\hbar(\partial/\partial t)\langle\psi_S(t)| = \langle\psi_S(t)|H_S^\dagger = \langle\psi(t)|H_S$ (self-adjoint), and the commutator $[A,B] \equiv AB - BA$.

[1] We dispense with the "hat" notation for operators.

[2] Representation is an unfortunate choice of word, as it also connotes the representation of vectors or operators in a particular basis or representations of abstract groups. The term *dynamical picture* is often used instead[224, p312–314]. Messiah[225, p312] refers to dynamical representations in quotation marks: "representation."

DOI: 10.1201/9781003512295-E

A natural way to represent the dynamics of state vectors is with a linear operator, the *time evolution operator*, $U(t,t_0)$ that maps the state at time t_0 to that at time t, $|\psi_S(t_0)\rangle \rightarrow |\psi_S(t)\rangle$:

$$|\psi_S(t)\rangle \equiv U(t,t_0)|\psi_S(t_0)\rangle. \tag{E.4}$$

To construct $U(t,t_0)$, we note that it must preserve normalization,[3]

$$\langle\psi_S(t)|\psi_S(t)\rangle = \langle\psi_S(t_0)|U^\dagger(t,t_0)U(t,t_0)|\psi_S(t_0)\rangle = \langle\psi_S(t_0)|\psi_S(t_0)\rangle, \tag{E.5}$$

i.e., it must be unitary, $U^\dagger U = I$. This requirement implies $U(t_0,t_0) = I$ and that it has the composition property $U(t,t_0) = U(t,t_1)U(t_1,t_0)$: The evolution from t_0 to t may be viewed as a two-step process,[4] the evolution from t_0 to t_1 followed by that from t_1 to t. Most importantly, it must be consistent with the dynamics of $|\psi_S(t)\rangle$ as obtained from the Schrödinger equation,

$$\mathrm{i}\hbar\frac{\partial}{\partial t}|\psi_S(t)\rangle = \mathrm{i}\hbar\frac{\partial}{\partial t}U(t,t_0)|\psi_S(t_0)\rangle = H_S|\psi_S(t)\rangle = H_S U(t,t_0)|\psi_S(t_0)\rangle, \tag{E.6}$$

implying it satisfies the first-order differential equation with initial condition $U(t_0,t_0) = I$:

$$\mathrm{i}\hbar\frac{\partial}{\partial t}U(t,t_0) = H_S U(t,t_0). \tag{E.7}$$

If the Hamiltonian H_S is time independent, Eq. (E.7) has the formal solution

$$U(t,t_0) = \exp(-\mathrm{i}H_S(t-t_0)/\hbar), \tag{E.8}$$

where the exponential of an operator is defined by its series expansion, $\mathrm{e}^A = I + A + \frac{1}{2}A^2 + \cdots$. If H_S is time dependent, however, there is not a closed-form expression for $U(t,t_0)$ and one must adopt more advanced methods.[5]

E.2 HEISENBERG REPRESENTATION

Another way to treat time dependence is to work with time-dependent operators instead of time-dependent state vectors.[6] In the *Heisenberg representation*, operators in the Schrödinger picture acquire a time dependence specified by (where t_0 is an arbitrary reference time),

$$A_H(t) \equiv U^\dagger(t,t_0)A_S U(t,t_0), \tag{E.9}$$

with the state vector propagated backward in time to t_0,

$$|\psi_H\rangle \equiv U^\dagger(t,t_0)|\psi_S(t)\rangle = U^\dagger(t,t_0)U(t,t_0)|\psi_S(t_0)\rangle = |\psi_S(t_0)\rangle. \tag{E.10}$$

One can show that expectation values are invariant under these transformations:

$$\langle\psi_H|A_H(t)|\psi_H\rangle = \langle\psi_S(t)|A_S|\psi_S(t)\rangle. \tag{E.11}$$

The time-dependent operator $A_H(t)$ satisfies an equation of motion. One can show that

$$\mathrm{i}\hbar\frac{\partial}{\partial t}A_H(t) = [A_H(t), H_S], \tag{E.12}$$

where we've used that H_S commutes with $U(t)$. Equation (E.12) is the *Heisenberg equation of motion*. It's basically the same as calculating the rate of change of the density operator, Eq. (A.41).

[3] Probability is locally conserved in the Schrödinger equation; there is a continuity equation associated with probability.

[4] Note the similarity with the SCK equation of Markov processes, Eq. (2.29).

[5] See for example Fetter and Walecka[186, p58].

[6] The two ways of viewing quantum dynamics as due either to the time dependence of operators or state vectors is analogous to the difference between passive and active transformations. Consider the mapping of a vector $\boldsymbol{r} \rightarrow \boldsymbol{r}'$, with $\boldsymbol{r}' = R\boldsymbol{r}$. In the active transformation the vector is actively changed relative to a fixed coordinate system, and in the passive the vector is left unchanged but the coordinate system is changed. See [6, p58].

E.3 INTERACTION REPRESENTATION

The *interaction representation* (also called the *Dirac picture*) is intermediate between the Schrödinger and Heisenberg pictures. Assume a system Hamiltonian of the general form

$$H = H_0 + H', \tag{E.13}$$

where H_0, the unperturbed part, contains the effects of internal interactions among system components (as well as kinetic energies) and H', the interaction term, contains the effects of external interactions; H_0 and H' are noncommuting operators.[7] Whereas in the Schrödinger and Heisenberg pictures the state vector *or* the operators carry time dependence, in the interaction picture both carry part of the time dependence of expectation values.

Consider the problem of finding the time evolution operator for systems described by Eq. (E.13),

$$\mathrm{i}\hbar\frac{\partial}{\partial t}U(t,t_0) = (H_0 + H')\,U(t,t_0). \tag{E.14}$$

Even if H_0, H' were time independent, we would have little use for a propagator in the form $U(t,t_0) = \exp[-\mathrm{i}\,(H_0 + H')\,(t - t_0)\,/\hbar]$ because H_0 and H' don't commute; in particular[8] $U(t,t_0) \neq \exp[-\mathrm{i}H_0(t-t_0)/\hbar]\exp[-\mathrm{i}H'(t-t_0)/\hbar]$. Nevertheless, we're going to *force* $U(t,t_0)$ to have the factored form

$$U(t,t_0) \equiv U_0(t,t_0)U'(t,t_0), \tag{E.15}$$

where, by definition,

$$\mathrm{i}\hbar\frac{\partial}{\partial t}U_0(t,t_0) \equiv H_0 U_0(t,t_0). \tag{E.16}$$

For H_0 time independent (assumed here), $U_0(t,t_0) = \exp(-\mathrm{i}H_0(t-t_0)/\hbar)$. Combine Eqs. (E.15) and (E.14) and make use of Eq. (E.16) to show that U' must satisfy

$$\mathrm{i}\hbar\frac{\partial}{\partial t}U'(t,t_0) = \left[U_0^\dagger(t,t_0)H'U_0(t,t_0)\right]U'(t,t_0) \equiv H_I'(t)U'(t,t_0), \tag{E.17}$$

which defines the interaction Hamiltonian in the interaction picture, $H_I'(t) \equiv U_0^\dagger(t,t_0)H'U_0(t,t_0)$. Equation (E.17) is exact; no linearizations are involved.

We define, for any operator A_S, its counterpart in the interaction picture,

$$A_I(t) \equiv U_0^\dagger(t,t_0)A_S U_0(t,t_0). \tag{E.18}$$

Equation (E.18) is the same as the Heisenberg form of the operator, except for the Hamiltonian H_0. For expectation values,

$$\begin{aligned}
\langle\psi_S(t)|A_S|\psi_S(t)\rangle &= \langle\psi_S(t_0)|U^\dagger(t,t_0)A_S U(t,t_0)|\psi_S(t_0)\rangle \\
&= \langle\psi_S(t_0)|U'^\dagger(t,t_0)U_0^\dagger(t,t_0)A_S U_0(t,t_0)U'(t,t_0)|\psi_S(t_0)\rangle \\
&= \langle U'(t,t_0)\psi_S(t_0)|U_0^\dagger(t,t_0)A_S U_0(t,t_0)|U'(t,t_0)\psi_S(t_0)\rangle \\
&\equiv \langle\psi_I(t)|A_I(t)|\psi_I(t)\rangle,
\end{aligned} \tag{E.19}$$

where, to preserve inner products, we define the state vector in the interaction representation

$$|\psi_I(t)\rangle \equiv U'(t,t_0)|\psi_S(t_0)\rangle = U_0^\dagger(t,t_0)|\psi_S(t)\rangle. \tag{E.20}$$

---

[7] If H_0, H' were commuting operators, we'd be adding a constant of the motion to the Hamiltonian; H' can't commute with H_0 and constitute a dynamical interaction effecting observable changes in the system.

[8] See discussion of the Baker-Campbell-Hausdorff equation in Section A.4.

Thus we have the identities $\langle\psi_I(t)|A_I(t)|\psi_I(t)\rangle = \langle\psi_S(t)|A_S|\psi_S(t)\rangle = \langle\psi_H|A_H(t)|\psi_H\rangle$ which show the role of time in the representations. One can show that $|\psi_I(t)\rangle$ satisfies a Schrödinger-like equation in which its evolution is controlled by the interaction Hamiltonian,

$$\mathrm{i}\hbar\frac{\partial}{\partial t}|\psi_I(t)\rangle = H_I'(t)|\psi_I(t)\rangle. \tag{E.21}$$

The interaction form of the operator satisfies the equation of motion,

$$\mathrm{i}\hbar\frac{\partial}{\partial t}A_I(t) = [A_I(t), H_0]. \tag{E.22}$$

The interaction and Schrödinger pictures coincide for H_0, with $H_{0,I}(t) = H_{0,S} \equiv H_0$. One should compare Eqs. (E.3), (E.12), and (E.22).

E.4 DENSITY MATRIX

The density matrix is another representation of the state of a quantum system; see Appendix A. It's used in quantum statistical mechanics to calculate ensemble averages; see Eq. (A.39). Here we note it's form in the three pictures. In the Schrödinger picture,

$$\rho_S(t) = \sum_n p_n|\psi_{n,S}(t)\rangle\langle\psi_{n,S}(t)| = U(t,t_0)\rho_S(t_0)U^\dagger(t,t_0), \tag{E.23}$$

where p_n denotes the probability that a randomly selected member of the ensemble is in state $|\psi_n\rangle$. In the Heisenberg picture,

$$\rho_H = \sum_n p_n|\psi_{n,H}\rangle\langle\psi_{n,H}| = \rho_S(t_0). \tag{E.24}$$

In the interaction picture,

$$\rho_I(t) = \sum_n p_n|\psi_{n,I}(t)\rangle\langle\psi_{n,I}(t)| = U^\dagger(t,t_0)\rho_S(t)U(t,t_0). \tag{E.25}$$

Table E.1 summarizes the results of this appendix.

Table E.1 State vector $|\psi\rangle$, operator A, expectation value $\langle\psi|A|\psi\rangle$, and density matrix ρ in the Schrödinger, Heisenberg, and interaction representations. H_S is assumed time independent and $t_0 = 0$.

	Schrödinger	Heisenberg	Interaction
$\lvert\psi(t)\rangle$	$\mathrm{e}^{-\mathrm{i}H_S t/\hbar}\lvert\psi_S(0)\rangle$	$\lvert\psi_H\rangle = \text{constant}$	$\lvert\psi_I(t)\rangle = \mathrm{e}^{\mathrm{i}H_0 t/\hbar}\lvert\psi_S(t)\rangle$
A	$A_S = \text{constant}$	$A_H(t) = \mathrm{e}^{\mathrm{i}H_S t/\hbar}A_S\mathrm{e}^{-\mathrm{i}H_S t/\hbar}$	$A_I(t) = \mathrm{e}^{\mathrm{i}H_0 t/\hbar}A_S\mathrm{e}^{-\mathrm{i}H_0 t/\hbar}$
$\langle\psi\lvert A\rvert\psi\rangle$	$\langle\psi_S(t)\lvert A_S\rvert\psi_S(t)\rangle$	$\langle\psi_H\lvert A_H(t)\rvert\psi_H\rangle$	$\langle\psi_I(t)\lvert A_I(t)\rvert\psi_I(t)\rangle$
$\rho(t)$	$\mathrm{e}^{-\mathrm{i}H_S t/\hbar}\rho_S(0)\mathrm{e}^{\mathrm{i}H_S t/\hbar}$	$\rho_H = \text{constant}$	$\mathrm{e}^{\mathrm{i}H_0 t/\hbar}\rho_S(t)\mathrm{e}^{-\mathrm{i}H_0 t/\hbar}$

APPENDIX F

Causality and analyticity

CAUSALITY (in physics) means that effects cannot precede causes.[1] The principle is used in Chapter 6 to define the response function $\Phi_{BA}(\theta)$, the time-dependent response of quantity B to interaction A, so that $\Phi_{BA}(\theta < 0) = 0$. Any function having the property $f(t < 0) = 0$ is known as a *causal function*. We review the mathematics of causality: Kramers-Kronig relations, Titchmarsh's theorem, and Fourier transforms of causal functions.

F.1 KRAMERS-KRONIG RELATIONS

Let F be a function analytic in the upper half of the complex plane $\mathbb{C}$, with all poles having negative imaginary parts.[2] By the Cauchy integral formula[13, p210], we have, for integration path C along the real axis and closed in the upper half plane in a counterclockwise semicircle of infinite radius,

$$\frac{1}{2\pi\mathrm{i}}\int_C \frac{F(\zeta)}{\zeta - z}\mathrm{d}\zeta = \begin{cases} F(z) & \text{if } \mathrm{Im}\,(z) > 0 \\ 0 & \text{if } \mathrm{Im}\,(z) < 0. \end{cases} \tag{F.1}$$

Let $z = x + \mathrm{i}\epsilon$ with x real and $\epsilon > 0$ and assuming F vanishes on the infinite semicircle,

$$F(x + \mathrm{i}\epsilon) = \frac{1}{2\pi\mathrm{i}}\int_{-\infty}^{\infty} \frac{F(x')}{x' - x - \mathrm{i}\epsilon}\mathrm{d}x'. \tag{F.2}$$

By letting $\epsilon \to 0^+$ in Eq. (F.2), we have, by continuity, and in the usual way with a pole on the integration path,

$$F(x) = \frac{1}{2\pi\mathrm{i}}P\int_{-\infty}^{\infty} \frac{F(x')}{x' - x}\mathrm{d}x' + \frac{1}{2}F(x)$$

where P denotes Cauchy principal value integral[13, p227]. Under these assumptions, F is constrained to satisfy

$$F(x) = \frac{1}{\pi\mathrm{i}}P\int_{-\infty}^{\infty} \frac{F(x')}{x' - x}\mathrm{d}x'. \tag{F.3}$$

In Eq. (F.3), F is restricted to the real line, $F : \mathbb{R} \to \mathbb{C}$. Separating F into real and imaginary parts,

$$\begin{aligned} \mathrm{Re}\,[F(x)] &= \frac{1}{\pi}P\int_{-\infty}^{\infty} \frac{\mathrm{Im}\,[F(x')]}{x' - x}\mathrm{d}x' \\ \mathrm{Im}\,[F(x)] &= -\frac{1}{\pi}P\int_{-\infty}^{\infty} \frac{\mathrm{Re}\,[F(x')]}{x' - x}\mathrm{d}x'. \end{aligned} \tag{F.4}$$

[1]See Bunge[226]. The idea is refined in the theory of relativity, where the temporal order of events can be reference-frame dependent; causal influences lie within or on the past lightcone of effects; see [6, p67]. The term *causality* has different meanings in philosophy and in statistics.

[2]In math speak, $F : \mathbb{C} \to \mathbb{C}$ is meromorphic in the lower half plane and holomorphic in the upper half plane.

DOI: 10.1201/9781003512295-F

The real and imaginary parts are therefore interrelated.[3] Said differently, the real and imaginary parts of $F(x)$ are not independent; the function can be reconstructed from just one of its parts. These formulae are known as the *Kramers-Kronig relations*.[4]

F.2 TITCHMARSH'S THEOREM

The integrals in Eq. (F.4) are *Hilbert transforms*[13, p238], which for a real function $f(t)$ is defined

$$\hat{f}(t) = (\mathcal{H}f)(t) \equiv \frac{1}{\pi} P \int_{-\infty}^{\infty} \frac{f(\tau)}{\tau - t} \mathrm{d}\tau \tag{F.5}$$

and we've denoted the transform as the result of an operator, $\mathcal{H}$. Note that it's linear. For complex-valued functions $F : \mathbb{R} \to \mathbb{C}$, with $F(x) = g(x) + \mathrm{i}h(x)$, we have from the Kramers-Kronig relations, $g(x) = \mathcal{H}h(x)$ and $h(x) = -\mathcal{H}g(x)$. By combining these results, we have[5] $\mathcal{H}\mathcal{H}f = -f$ for any f or that the inverse Hilbert transform $\mathcal{H}^{-1} = -\mathcal{H}$. The functions in the Kramers-Kronig relations are therefore Hilbert-transform pairs, $g(x) = \mathcal{H}h(x)$, $h(x) = \mathcal{H}^{-1}g(x)$. The Hilbert transform is the convolution (in the Fourier sense) of f with $g(t) \equiv -1/(\pi t)$:

$$(\mathcal{H}f)(t) = \frac{1}{\pi} P \int_{-\infty}^{\infty} \frac{f(\tau)}{\tau - t} \mathrm{d}\tau \equiv \int_{-\infty}^{\infty} f(\tau) g(t-\tau) \mathrm{d}\tau \equiv f(t) * g(t).$$

By the convolution theorem (see [13, p113]),

$$\mathcal{F}(\mathcal{H}f)(\alpha) = \mathcal{F}[f(t) * g(t)](\alpha) = (\mathcal{F}f)(\alpha) \times (\mathcal{F}g)(\alpha) = \mathrm{i}\,\mathrm{sgn}(\alpha)(\mathcal{F}f)(\alpha), \tag{F.6}$$

where the *signum function*,

$$\mathrm{sgn}(x) \equiv \begin{cases} 1 & \text{if } x > 0 \\ 0 & \text{if } x = 0 \\ -1 & \text{if } x < 0 \end{cases} \tag{F.7}$$

and $(\mathcal{F}f)(\alpha) \equiv \int_{-\infty}^{\infty} f(x) \mathrm{e}^{-\mathrm{i}\alpha x} \mathrm{d}x$; $\mathcal{F}[-1/(\pi x)](\alpha) = \mathrm{i}\,\mathrm{sgn}(\alpha)$ (use $\int_{-\infty}^{\infty} \mathrm{d}u \sin(u)/u = \pi$). The Hilbert transform thus introduces a phase shift on, but preserves the magnitude of $\mathcal{F}f$,

The Kramers-Kronig relations are part of *Titchmarsh's theorem* (see theorem 95 of [80, p128] and [13, p240]), necessary and sufficient conditions for a complex-valued square-integrable function $F(x)$ on the real line [a member of $L^2(-\infty, \infty)$] to be the boundary value of a holomorphic function $F(z)$ in the upper half plane. Titchmarsh's theorem is that the following statements are equivalent:

- $F(x)$ is the limit $z \to x$ of an analytic function $F(z)$ bounded in the upper half plane, $\int_{-\infty}^{\infty} |F(x + \mathrm{i}y)|^2 \mathrm{d}x < K$ for some number K and $y > 0$;
- For $F(x) = f(x) - \mathrm{i}g(x)$, $g(x) = \mathcal{H}f(x)$;
- The Fourier transform of $F(x)$, $(\mathcal{F}F)(\omega) = 0$ for $\omega < 0$.

The last statement is readily shown:

$$\mathcal{F}(F) = \mathcal{F}(f) - \mathrm{i}\mathcal{F}(g) = \mathcal{F}(f) - \mathrm{i}\mathcal{F}(\mathcal{H}f) = \mathcal{F}(f) - \mathrm{i}(\mathrm{i}\,\mathrm{sgn}(\omega))\mathcal{F}(f) = (1 + \mathrm{sgn}(\omega))\,\mathcal{F}(f).$$

For $\omega < 0$, $(\mathcal{F}F)(\omega) = 0$.

[3] The real and imaginary parts of holomorphic functions are already interrelated through the Cauchy-Riemann conditions at every point of the upper half plane; the Kramers-Kronig relations are satisfied at the boundary, on the real line.

[4] See for example Landau and Lifshitz[227, Section 62] or Born and Wolf[228, Section 10.2].

[5] The result $\mathcal{H}^2 f = -f$ is one of the most important properties of the Hilbert transform and is used in signal processing. It can be derived without invoking the Kramers-Kronig relations, but we arrive at the same conclusion faster this way.

F.3 CAUSAL TRANSFORMS

We first show how to find causal functions—most functions are not causal. Any real function $f(t)$ can be expressed as a sum of even and odd functions,

$$f(t) = f_e(t) + f_o(t) \equiv \frac{1}{2}[f(t) + f(-t)] + \frac{1}{2}[f(t) - f(-t)]. \tag{F.8}$$

Clearly, $f_e(-t) = f_e(t)$ and $f_o(-t) = -f_o(t)$. The Fourier transform of f can be expressed in terms of cosine and sine transforms,

$$\mathcal{F}(f)(\omega) \equiv \int_{-\infty}^{\infty} e^{-i\omega t} f(t)dt = \int_{-\infty}^{\infty} f_e(t)\cos(\omega t)dt - i\int_{-\infty}^{\infty} f_o(t)\sin(\omega t)dt. \tag{F.9}$$

For any real function f, $\text{Re}[\mathcal{F}f](\omega)$ is even and $\text{Im}[\mathcal{F}f](\omega)$ is odd: $\text{Re}[\mathcal{F}f](-\omega) = \text{Re}[\mathcal{F}f](\omega)$ and $\text{Im}[\mathcal{F}f](-\omega) = -\text{Im}[\mathcal{F}f](\omega)$.

For a given odd function $f_o(t)$, $f_e(t) \equiv \text{sgn}(t)f_o(t)$ is even. The even function obtained this way is, for $t < 0$, the mirror image of $f_o(t)$ for $t > 0$. An analytic representation of causal functions results from the sum of $\text{sgn}(t)f_o(t)$ and f_o,

$$f_c(t) = [1 + \text{sgn}(t)]\, f_o(t), \tag{F.10}$$

where the subscript on f_c denotes causal. Clearly, $f_c(t < 0) = 0$. Thus we have a recipe for constructing a causal function f_c given an odd function f_o. Note that $f_c(0) = 0$.

Consider the Fourier transform of f_c:

$$\begin{aligned}
\mathcal{F}[f_c](\omega) &= \int_{-\infty}^{\infty} dt f_c(t)e^{-i\omega t} = \int_{-\infty}^{\infty} dt\,[1 + \text{sgn}(t)]\, f_o(t)e^{-i\omega t} \\
&= \int_{-\infty}^{\infty} \text{sgn}(t) f_o(t)\cos(\omega t)dt - i\int_{-\infty}^{\infty} f_o(t)\sin(\omega t)dt \\
&= \int_{-\infty}^{\infty} \text{sgn}(t) f_o(t)e^{-i\omega t}dt + \int_{-\infty}^{\infty} f_o(t)e^{-i\omega t}dt.
\end{aligned} \tag{F.11}$$

With $\mathcal{F}f_c = \text{Re}[\mathcal{F}f_c] - i\,\text{Im}[\mathcal{F}f_c]$, we have

$$\begin{aligned}
\text{Re}[\mathcal{F}f_c](\omega) &= \int_{-\infty}^{\infty} \text{sgn}(t) f_o(t)e^{-i\omega t}dt \\
\text{Im}[\mathcal{F}f_c](\omega) &= i\int_{-\infty}^{\infty} f_o(t)e^{-i\omega t}dt.
\end{aligned} \tag{F.12}$$

The parity properties of $(\mathcal{F}f_c)(\omega)$ are the same as those for $(\mathcal{F}f)(\omega)$; causal functions are real.

The relations in Eq. (F.12) are the Kramers-Kronig relations in disguise, as we now show. The real part of $\mathcal{F}f_c$ is the Fourier transform of the product $\text{sgn}(t)f_o(t)$. It can be written as a convolution of the Fourier transforms of $\text{sgn}(t)$ and $f_o(t)$,

$$\text{Re}[\mathcal{F}f_c](\omega) = \int_{-\infty}^{\infty} dt e^{-i\omega t}\int_{-\infty}^{\infty} d\omega'\Sigma(\omega')e^{-i\omega' t}\int_{-\infty}^{\infty} d\omega'' e^{-i\omega'' t}F_o(\omega'') \tag{F.13}$$

where

$$\begin{aligned}
\text{sgn}(t) &= \int_{-\infty}^{\infty} e^{-i\omega t}\Sigma(\omega)d\omega \Longleftrightarrow \Sigma(\omega) = \frac{1}{2\pi}\int_{-\infty}^{\infty} e^{i\omega t}\,\text{sgn}(t)dt \\
f_o(t) &= \int_{-\infty}^{\infty} e^{-i\omega t}F_o(\omega)d\omega \Longleftrightarrow F_o(\omega) = \frac{1}{2\pi}\int_{-\infty}^{\infty} e^{i\omega t} f_o(t)dt.
\end{aligned} \tag{F.14}$$

Equation (F.13) can then be written

$$\begin{aligned}\text{Re}\,[\mathcal{F}f_c]\,(\omega) &= \int_{-\infty}^{\infty} d\omega' \int_{-\infty}^{\infty} d\omega'' \Sigma(\omega') F_o(\omega'') \int_{-\infty}^{\infty} dt e^{-it(\omega+\omega'+\omega'')} \\ &= 2\pi \int_{-\infty}^{\infty} d\omega'' F_o(\omega'') \int_{-\infty}^{\infty} \Sigma(\omega')\delta(\omega+\omega'+\omega'')d\omega' \\ &= 2\pi \int_{-\infty}^{\infty} d\omega'' F_o(\omega'')\Sigma(\omega-\omega'') \\ &= 2\pi F_o(\omega) * \Sigma(\omega),\end{aligned} \tag{F.15}$$

where we've used that $\text{Re}\,[\mathcal{F}f_c](\omega)$ is even. It's straightforward to show (see [13, p238]) that the Fourier transform of $\text{sgn}(x)$ is

$$\Sigma(\omega) = \frac{i}{\pi\omega}. \tag{F.16}$$

Note also that

$$F_o(\omega) = \frac{i}{2\pi}\,\text{Im}\,[\mathcal{F}f_c]\,(\omega) \tag{F.17}$$

Combining Eqs. (F.16) and (F.17) with Eq. (F.15), we find

$$\text{Re}\,[\mathcal{F}f_c]\,(\omega) = \frac{1}{\pi} P \int_{-\infty}^{\infty} d\omega' \frac{\text{Im}\,[\mathcal{F}f_c]\,(\omega')}{\omega'-\omega} = \mathcal{H}\left(\text{Im}\,[\mathcal{F}f_c]\right)(\omega). \tag{F.18}$$

Equation (F.18) implies (by the inverse Hilbert transform) the other Kramers-Kronig relation,

$$\text{Im}\,[\mathcal{F}f_c] = \mathcal{H}^{-1}\left(\text{Re}\,[\mathcal{F}f_c]\right) = -\mathcal{H}\left(\text{Re}\,[\mathcal{F}f_c]\right). \tag{F.19}$$

Fourier transforms of causal functions, *causal transforms*, obey the Kramers-Kronig relations.

F.4 THE PLEMELJ FORMULA

The *Plemelj formula* (also known as the Sokhotski-Plemelj theorem), is a useful relation among generalized functions[6] that often occurs in applications involving causal functions,

$$\lim_{\epsilon\to 0} \frac{1}{x \pm i\epsilon} = P\frac{1}{x} \mp i\pi\delta(x). \tag{F.20}$$

To establish Eq. (F.20) we must show that

$$\lim_{\epsilon\to 0} \int_{-\infty}^{\infty} \frac{f(x)}{x \pm i\epsilon} dx = P \int_{-\infty}^{\infty} \frac{f(x)}{x} dx \mp i\pi f(0), \tag{F.21}$$

where $f(x)$ is a smooth function vanishing sufficiently fast as $x \to \pm\infty$ that integrals are convergent. Equation (F.20) is shorthand for Eq. (F.21).

We start with the identity

$$\frac{1}{x \pm i\epsilon} = \frac{x \mp i\epsilon}{x^2+\epsilon^2}. \tag{F.22}$$

Then, for a smooth function $f(x)$,

$$\int_{-\infty}^{\infty} \frac{f(x)}{x \pm i\epsilon} dx = \int_{-\infty}^{\infty} \frac{xf(x)}{x^2+\epsilon^2} dx \mp i\epsilon \int_{-\infty}^{\infty} \frac{f(x)}{x^2+\epsilon^2} dx. \tag{F.23}$$

[6]Generalized functions have meaning only inside integrals, where they are integrated together with smooth functions; see Lighthill[99] or [13, p66]. *Smooth function* is generally a code word in mathematics for "all derivatives exist." Lighthill[99, p15] defines a *good function* as one which is everywhere differentiable any number of times and is such that it and all its derivatives are $O(|x|^{-N})$ as $|x| \to \infty$ for all N; $f(x) = \exp(-x^2)$ is a good function.

We're interested in the limit as $\epsilon \to 0$. The first integral on the right of Eq. (F.23) can be written (where we'll take the limit $\delta \to 0$)

$$\int_{-\infty}^{\infty} \frac{xf(x)}{x^2+\epsilon^2}\mathrm{d}x = \int_{-\infty}^{-\delta} \frac{xf(x)}{x^2+\epsilon^2}\mathrm{d}x + \int_{\delta}^{\infty} \frac{xf(x)}{x^2+\epsilon^2}\mathrm{d}x + \int_{-\delta}^{\delta} \frac{xf(x)}{x^2+\epsilon^2}\mathrm{d}x. \tag{F.24}$$

The limit $\epsilon \to 0$ can safely be taken in the first two integrals on the right of Eq. (F.24). In the third integral, we can, for sufficiently small δ, replace $f(x) \approx f(0)$ for $|x| < \delta$. Thus,

$$\begin{aligned}\lim_{\epsilon\to 0}\int_{-\infty}^{\infty} \frac{xf(x)}{x^2+\epsilon^2}\mathrm{d}x &= \lim_{\epsilon\to 0}\lim_{\delta\to 0}\left[\int_{-\infty}^{-\delta} \frac{xf(x)}{x^2+\epsilon^2}\mathrm{d}x + \int_{\delta}^{\infty} \frac{xf(x)}{x^2+\epsilon^2}\mathrm{d}x\right] + f(0)\int_{-\delta}^{\delta}\frac{x}{x^2+\epsilon^2}\mathrm{d}x \\ &\equiv P\int_{-\infty}^{\infty}\frac{f(x)}{x}\mathrm{d}x,\end{aligned}$$

where the last integral vanishes by symmetry and the terms in square bracket define the Cauchy principal value integral. For the second integral on the right of Eq. (F.23), the only significant contribution to the integral is, for small ϵ, from the integration region $x \approx 0$, so that

$$\epsilon\int_{-\infty}^{\infty}\frac{f(x)}{x^2+\epsilon^2}\mathrm{d}x \approx \epsilon f(0)\int_{-\infty}^{\infty}\frac{\mathrm{d}x}{x^2+\epsilon^2} = \epsilon f(0)\cdot\frac{1}{\epsilon}\tan^{-1}(x/\epsilon)\Big|_{-\infty}^{\infty} = \pi f(0). \tag{F.25}$$

We've therefore established Eq. (F.21).

An alternate version of the Plemelj relation applies to one-sided Fourier transforms. Consider the integrals where we put in convergence factors,

$$\begin{aligned} I_+(\omega) &\equiv \int_0^{\infty} \mathrm{e}^{\mathrm{i}\omega t}\mathrm{d}t = \lim_{\lambda\to 0^+}\int_0^{\infty}\mathrm{e}^{\mathrm{i}\omega t}\mathrm{e}^{-\lambda t}\mathrm{d}t = \lim_{\lambda\to 0}\frac{1}{\lambda - \mathrm{i}\omega} \\ I_-(\omega) &\equiv \int_{-\infty}^{0} \mathrm{e}^{\mathrm{i}\omega t}\mathrm{d}t = \lim_{\lambda\to 0^+}\int_{-\infty}^{0}\mathrm{e}^{\mathrm{i}\omega t}\mathrm{e}^{\lambda t}\mathrm{d}t = \lim_{\lambda\to 0}\frac{1}{\lambda + \mathrm{i}\omega}. \end{aligned} \tag{F.26}$$

The sum $I_+(\omega) + I_-(\omega) = 2\pi\delta(\omega)$. Using Eq. (F.20), we have the result, useful for dealing with causal functions,

$$\int_0^{\infty}\mathrm{e}^{\pm\mathrm{i}\omega t}\mathrm{d}t = \pi\delta(\omega) \pm \mathrm{i}P\left(\frac{1}{\omega}\right). \tag{F.27}$$

Gel'Fand and Shilov[229, p172] present a rigorous derivation. Equation (F.27) is the Fourier transform of the *unit step function*

$$\theta(x) = \begin{cases} 1 & \text{if } x \geq 0 \\ 0. & \text{if } x < 0 \end{cases} \tag{F.28}$$

The step function often appears inside integrals and thus the value of $\theta(x=0)$ isn't usually important; the value of a function at a single point doesn't affect the value of the integral. Sometimes the step function is defined so that $\theta(0) = \frac{1}{2}$.

F.5 TAKE AWAY

The Fourier transform of a causal function in the time domain is, in the frequency domain, an analytic function in the closed upper half of the complex frequency plane,[7] implying it obeys the Kramer-Kronig relations on the real line: Causality implies analyticity and analyticity implies causality. As summarized by J.S. Toll[230]: *A function of integrable square is zero for all negative values of its argument if and only if its Fourier transform is a causal transform.*

[7]The closed upper half plane is the upper half plane together with its closure, the real line.

Bibliography

[1] J.W. Gibbs. *The Collected Works of J. Willard Gibbs: Volume I Thermodynamics*. Yale University Press, 1948.

[2] J.H. Luscombe. *Thermodynamics*. CRC Press, 2018.

[3] R.P. Feynman. *Statistical Mechanics*. W.A. Benjamin, 1972.

[4] D.D. Nolte. The tangled tale of phase space. *Physics Today*, 63:33–38, 2010.

[5] J.H. Luscombe. *Statistical Mechanics: From Thermodynamics to the Renormalization Group*. CRC Press, 2021.

[6] J.H. Luscombe. *Core Principles of Special and General Relativity*. CRC Press, 2019.

[7] S.R. de Groot and P. Mazur. *Non-Equilibrium Thermodynamics*. North-Holland Publishing Company, 1962.

[8] R. Clausius. *The Mechanical Theory of Heat*. John van Voorst, 1867.

[9] I. Prigogine and R. Defay. *Chemical Thermodynamics*. Longman Greens, 1954.

[10] H.B. Callen. *Thermodynanics*. John Wiley, 1960.

[11] T.L. Hill. *Thermodynamics of Small Systems*, volume 1. W.A. Benjamin, 1963.

[12] G.K. Batchelor. *An Introduction to Fluid Dynamics*. Cambridge University Press, 1967.

[13] B. Borden and J. Luscombe. *Mathematical Methods in Physics, Engineering, and Chemistry*. Wiley, 2020.

[14] P.M. Morse and H. Feshbach. *Methods of Theoretical Physics*. McGraw-Hill, 1953.

[15] L.D. Landau and E.M. Lifshitz. *Fluid Mechanics*. Pergamon Press, 1959.

[16] P.G. de Gennes and J. Prost. *The Physics of Liquid Crystals*. Oxford University Press, 1993.

[17] Th. De Donder and P. Van Rysselberghe. *Thermodynamic Theory of Affinity*. Stanford University Press, 1936.

[18] D.D. Fitts. *Nonequilibrium Thermodynamics*. McGraw-Hill, 1962.

[19] L. Onsager. Reciprocal relations in irreversible processes. I. *Phys. Rev.*, 37:405–426, 1931.

[20] L. Onsager. Reciprocal relations in irreversible processes. II. *Phys. Rev.*, 38:2265–2279, 1931.

[21] H.B.G. Casimir. On Onsager's principle of microscopic reversibility. *Reviews of Modern Physics*, 17:343–350, 1945.

[22] D.G. Miller. Thermodynamics of irreversible processes: The experimental verification of the Onsager reciprocal relations. *Chemical Review*, 60:15–37, 1960.

[23] G.S. Yablonsky, A.N. Gorban, D. Constales, V.V. Galvita, and G.B. Marin. Reciprocal relations between kinetic curves. *Europhysics Letters*, 93:20004, 2011.

[24] C. Kittel. *Elementary Statistical Physics*. John Wiley, 1958.

[25] H.J. Kreuzer. *Nonequilibrium Thermodynamics and its Statistical Foundations*. Oxford University Press, 1981.

[26] I. Prigogine. *Thermodynamics of Irreversible Processes*. Wiley, 1961.

[27] E.T. Jaynes. The minimum entropy production principle. *Annual Review of Physical Chemistry*, 31:579–601, 1980.

[28] G.D. Birkhoff. Proof of the ergodic theorem. *Proc. Natl. Acad. Sci. USA*, 17:656–660, 1931.

[29] A. Pais. *Subtle is the Lord: The Science and Life of Albert Einstein*. Oxford University Press, 1982.

[30] K. Pearson. Notes on the history of correlation. *Biometrika*, 13:25–45, 1920.

[31] H. Cramér. *Mathematical Methods of Statistics*. Princeton University Press, 1946.

[32] J.A. Shohat and J.D. Tamarkin. *The Problem of Moments*. American Mathematical Society, 1943.

[33] R.F. Greene and H.B. Callen. On the formalism of thermodynamic fluctuation theory. *Phys. Rev.*, 83:1231–1235, 1951.

[34] P.W. Anderson. More is different. *Science*, 177:393–396, 1972.

[35] N. Wiener. *Extrapolation, Interpolation, and Smoothing of Stationary Time Series*. MIT Press, 1950.

[36] J.L. Doob. *Stochastic Processes*. John Wiley, 1953.

[37] R.M. Gray. *Probability, Random Processes, and Ergodic Properties*. Springer, 1987.

[38] E. Parzen. *Modern Probability Theory and Its Applications*. John Wiley, 1960.

[39] M. Loève. *Probability Theory*. D. Van Nostrand, 1955.

[40] B. Oksendal. *Stochastic Differential Equations*. Springer, 6th edition, 2003.

[41] E. Parzen. *Stochastic Processes*. Holden-Day, 1962.

[42] M. Kac. *Probability and Related Topics in Physical Sciences*. Interscience Publishers, 1959.

[43] A. Papoulis. *Probability, Random Variables, and Stochastic Processes*. McGraw-Hill, 1991.

[44] N. Wax, editor. *Selected Papers on Noise and Stochastic Processes*. Dover Publications, 1954.

[45] H. Nyquist. Thermal agitation of electric charge in conductors. *Physical Review*, 32:110–113, 1928.

[46] J. Johnson. Thermal agitation of electricity in conductors. *Physical Review*, 32:97–109, 1928.

[47] E.B. Moullin. *Spontaneous Fluctuations of Voltage*. Oxford University Press, 1938.

[48] R. Fürth, editor. *Investigations on the Theory of the Brownian Movement by Albert Einstein*. Dover Publications, 1956.

[49] A. Nordsieck, W.E. Lamb, and G.E. Uhlenbeck. On the theory of cosmic-ray showers. *Physica*, 7:344–360, 1940.

[50] V. Lakshmikantham and M. Rama Mohana Rao. *Theory of Integro-differential Equations*. CRC Press, 1995.

[51] N.G. van Kampen. *Stochastic Processes in Physics and Chemistry*. North-Holland Publishing Company, 1992.

[52] E.P. Wigner. Derivations of Onsager's reciprocal relations. *J. Chem. Phys.*, 22:1912–1915, 1954.

[53] R. Bellman. *Introduction to Matrix Analysis*. McGraw-Hill, 1960.

[54] Y.G. Sinai. *Probability Theory*. Springer-Verlag, 1992.

[55] M.G. Kendall and A. Stuart. *The Advanced Theory of Statistics*, volume 1. Hafner, 1963.

[56] J. Marcinkiewicz. Sur une propriété de la loi de Gauss. *Mathematische Zeitschrift*, 44: 612–618, 1939.

[57] R. Kubo. Generalized cumulant expansion method. *J. Phys. Soc. Jpn.*, 17:1100–1120, 1962.

[58] J. Aczel. *Lectures on Functional Equations and Their Applications*. Academic Press, 1966.

[59] J.L. Doob. The Brownian movement and stochastic equations. *Annals of Mathematics*, 43:351–369, 1942.

[60] F. Reif. *Fundamentals of Statistical and Thermal Physics*. McGraw-Hill, 1965.

[61] G. Polya. *Mathematics and Plausible Reasoning*, volume 1. Princeton University Press, 1954.

[62] J. Perrin. *Atoms*. D. Van Nostrand, 1916.

[63] D.S. Lemon and A. Gythiel. Paul Langevin's 1908 paper "On the theory of Brownian motion". *American Journal of Physics*, 65:1079–1081, 1997.

[64] G.E. Uhlenbeck and L.S. Ornstein. On the theory of the Brownian motion. *Physical Review*, 36:823–841, 1930.

[65] R.M. Mazo. *Brownian Motion*. Oxford University Press, 2002.

[66] W.T. Coffey. Development and application of the theory of Brownian motion. In M.W. Evans, editor, *Advances in Chemical Physics*, volume 63, pages 69–252. Wiley, 1985.

[67] H. Risken. *The Fokker-Planck Equation*. Springer-Verlag, 1989.

[68] S. Kheifets, A. Simha, K. Melin, T. Li, and M.G. Raizen. Observation of Brownian motion in liquids at short times: Instantaneous velocity and memory loss. *Science*, 343:1493–1496, 2014.

[69] H. Jeffries and B. Jeffries. *Methods of Mathematical Physics*. Cambridge University Press, 1956.

[70] R. Paley and N. Wiener. *Fourier Transforms in the Complex Domain.* American Mathematical Society, 1934.

[71] S.A. Adelman. Fokker-Planck equations for simple non-Markovian systems. *Journal of Chemical Physics*, 64:124–130, 1976.

[72] R. Courant and D. Hilbert. *Methods of Mathematical Physics*, volume I. Interscience Publishers, 1953.

[73] I.N. Sneddon. *Elements of Partial Differential Equations.* McGraw-Hill, 1957.

[74] R. Courant and D. Hilbert. *Methods of Mathematical Physics*, volume II. Interscience Publishers, 1962.

[75] M.C. Wang and G.E. Uhlenbeck. On the theory of the Brownian motion II. *Reviews of Modern Physics*, 17:323–342, 1945.

[76] H.A. Kramers. Brownian motion in a field of force and the diffusion model of chemical reactions. *Physica*, 7:284–304, 1940.

[77] P. Dennery and A. Krzywicki. *Mathematics for Physicists.* Harper and Row, 1967.

[78] G. Szego. *Orthogonal Polynomials.* American Mathematical Society, 1939.

[79] E.D. Rainville. *Special Functions.* Chelsea Publishing, 1971.

[80] E.C. Titchmarsh. *Introduction to the Theory of Fourier Integrals.* Oxford University Press, 1948.

[81] J.J. Sakurai and J. Napolitano. *Modern Quantum Mechanics.* Pearson, 2011.

[82] S. Chandrasekhar. Stochastic problems in physics and astronomy. *Reviews of Modern Physics*, 15:1–89, 1943.

[83] J.E. Moyal. Stochastic processes and statistical physics. *Journal of the Royal Statistical Society. Series B (Methodological)*, 11:150–210, 1949.

[84] N.G. van Kampen. A power series expansion of the master equation. *Canadian Journal of Physics*, 39:551–567, 1961.

[85] R.F. Pawula. Generalizations and extensions of the Fokker-Planck-Kolmogorov equation. *IEEE Transactions on Information Theory*, 13:33–41, 1967.

[86] R.F. Pawula. Approximation of the linear Boltzmann equation by the Fokker-Planck equation. *Physical Review*, 162:186–188, 1967.

[87] M. Abramowitz and I.A. Stegun. *Handbook of Mathematical Functions.* US Government Printing Office, 1964.

[88] I.S. Gradshteyn and I.M. Ryzhik. *Table of Integrals, Series, and Products.* Academic Press, 1980.

[89] S.G. Brush. *The Kinetic Theory of Gases.* Imperial College Press, 2003.

[90] H. Bohr. *Almost Periodic Functions.* Chelsea Publishing, 1947.

[91] J. Yvon. La théorie statistique des fluides et l'équation d'état. *Actualités scientifiques et industrielles. Hermann et Cie., Paris*, No. 203, 1935.

[92] H.S. Green. *The Molecular Theory of Liquids*. North-Holland Publishing Company, 1952.

[93] J. Yvon. *Correlations and Entropy in Classical Statistical Mechanics*. Pergamon Press, 1969.

[94] N.N. Bogoliubov. Problems of a dynamical theory in statistical physics. In J. de Boer and G.E. Uhlenbeck, editors, *Studies in Statistical Mechanics*, volume I, pages 1–118. North-Holland Publishing Company, 1962.

[95] M. Born and H.S. Green. A general kinetic theory of liquids I. The molecular distribution functions. *Proc. Roy. Soc. A*, 188:10–18, 1946.

[96] M. Born and H.S. Green. *A General Kinetic Theory of Liquids*. Cambridge University Press, 1949.

[97] J.G. Kirkwood. The statistical mechanical theory of transport processes I. General theory. *J. Chem. Phys.*, 14:180–201, 1946.

[98] J.G. Kirkwood. The statistical mechanical theory of transport processes II. Transport in gases. *J. Chem. Phys.*, 15:72–76, 1947.

[99] M.J. Lighthill. *Introduction to Fourier Analysis and Generalised Functions*. Cambridge University Press, 1958.

[100] J.H. Irving and J.G. Kirkwood. Statistical mechanical theory of transport processes. IV. The equations of hydrodynamics. *J. Chem. Phys.*, 18:817–829, 1950.

[101] B.D. Todd, D.J. Evans, and P.J. Daivis. Pressure tensor for inhomogeneous fluids. *Physical Review E*, 52:1627–1638, 1995.

[102] D. Massignon. *Mécanique Statistique des Fluides*. Dunod, 1957.

[103] P. Resibois and M. De Leener. *Classical Kinetic Theory of Fluids*. John Wiley, 1977.

[104] J.R.T. Seddon, H.J.W. Zandvliet, and D. Lohse. Knudsen gas provides nanobubble stability. *Phys. Rev. Lett.*, 107:116101, 2011.

[105] L. Boltzmann. *Lectures on Gas Theory*. University of California Press, 1964.

[106] S. Chapman and T.G. Cowling. *The Mathematical Theory of Non-Uniform Gases*. Cambridge University Press, 1970.

[107] G.E. Uhlenbeck and G.W. Ford. *Lectures in Statistical Mechanics*. American Mathematical Society, 1963.

[108] E.C.G. Sudarshan and N. Mukunda. *Classical Dynamics: A Modern Perspective*. John Wiley, 1974.

[109] C. Cercignani. *Theory and Application of the Boltzmann Equation*. Scottish Academic Press, 1975.

[110] S. Dodelson and F. Schmidt. *Modern Cosmology*. Academic Press, 2020.

[111] R.E. Masterson. *Introduction to Nuclear Reactor Physics*. CRC Press, 2017.

[112] J. Meng and X. Li. Lattice Boltzmann model for traffic flow. *Physical Review E*, 77:036108, 2008.

[113] D.K. Ferry and S.M. Goodnick. *Transport in Nanostructures*. Cambridge University Press, 1997.

[114] J.L. Lebowitz and E.W. Montroll, editors. *Nonequilibrium Phenomena I. The Boltzmann Equation.* North-Holland Publishing Company, 1983.

[115] H. Grad. Principles of the kinetic theory of gases. In S. Flugge, editor, *Handbuch der Physik*, volume 12, pages 205–294. Springer, 1958.

[116] I. Prigogine. *Non-Equilibrium Statistical Mechanics.* Wiley, 1962.

[117] R.L. Liboff. *Kinetic Theory.* Prentice Hall, 1990.

[118] E.G.D. Cohen. The Boltzmann equation and its generalization to higher densities. In E.G.D. Cohen, editor, *Fundamental Problems in Statistical Mechanics*, pages 110–156. North-Holland Publishing Company, 1962.

[119] J.R. Dorfman and H. van Beijeren. The kinetic theory of gases. In B.J. Berne, editor, *Statistical Mechanics: Time-Dependent Processes*, pages 65–179. Plenum Press, 1977.

[120] G.F. Mazenko and S. Yip. Renormalized kinetic theory of dense fluids. In B.J. Berne, editor, *Statistical Mechanics: Time-Dependent Processes*, pages 181–231. Plenum Press, 1977.

[121] S.G. Brush. *Kinetic Theory, Volume 3: The Chapman-Enskog solution of the transport equation for moderately dense gases.* Pergamon Press, 1972.

[122] P. Ehrenfest and T. Ehrenfest. *The Conceptual Foundations of the Statistical Approach in Mechanics.* Cornell University Press, 1959.

[123] S. Chapman. Boltzmann's H-theorem. *Nature*, 139:931, 1937.

[124] R.C. Tolman. *The Principles of Statistical Mechanics.* Oxford University Press, 1938.

[125] J. Jeans. *An Introduction to the Kinetic Theory of Gases.* Cambridge University Press, 1952.

[126] A. Sommerfeld. *Thermodynamics and Statistical Mechanics.* Academic Press, 1964.

[127] S. Harris. *An Introduction to the Theory of the Boltzmann Equation.* Holt, Rinehart, and Winston, 1971.

[128] T.H. Gronwall. A functional equation in the kinetic theory of gases. *Annals of Mathematics*, 17:1–4, 1915.

[129] G.-J. Su. Modified law of corresponding states for real gases. *Industrial and Engineering Chemistry*, 38:803–806, 1946.

[130] R.P. Feynman. Simulating physics with computers. *International Journal of Theoretical Physics*, 21:467–488, 1982.

[131] J.W. Gibbs. *The Collected Works of J. Willard Gibbs: Volume II Elementary Principles in Statistical Mechanics.* Yale University Press, 1948.

[132] C. Shannon. A mathematical theory of communication. *Bell System Technical Journal*, 27:379–423, 623–656, 1948.

[133] C. Shannon and W. Weaver. *Mathematical Theory of Communication.* University of Illinois Press, 1949.

[134] E.T. Jaynes. Gibbs vs Boltzmann entropies. *American Journal of Physics*, 33:391–398, 1965.

[135] P. Zupanovic and D. Kuic. Relation between Boltzmann and Gibbs entropy and example with multinomial distribution. *J. Phys. Commun.*, 2:045002, 2018.

[136] P.W. Bridgman. *The Nature of Thermodynamics*. Harvard University Press, 1941.

[137] E.P. Culverwell. Dr. Watson's proof of Boltzmann's theorem on permanence of distributions. *Nature*, 50:617, 1894.

[138] R. Landauer. Irreversibility and heat generation in the computing process. *IBM Journal of Research and Development*, 5:183–191, 1961.

[139] R. Landauer. Information is a physical entity. *Physica A*, 263:63–67, 1999.

[140] A. Berut, A. Arakelyan, A. Petrosyan, et al. Experimental verification of Landauer's principle linking information and thermodynamics. *Nature*, 483:187–189, 2012.

[141] J. Hong, B. Lambson, S. Dhuey, et al. Experimental test of Landauer's principle in single-bit operations on nanomagnetic memory bits. *Sci. Adv.*, 2:e1501492, 2016.

[142] Y. Jun, M. Gavrilov, and J. Bechhoefer. High-precision test of Landauer's principle in a feedback trap. *Physical Review Letters*, 113:190601, 2014.

[143] J.C. Slater. Atomic radii in crystals. *J. Chem. Phys.*, 41:3199–3204, 1964.

[144] D. Hilbert. *Grundzüge einer allgemeinen Theorie der linearen Integralgleichungen*. Teubner, 1912.

[145] D. Hilbert. Begründung der kinetischen Gastheorie. *Mathematische Annalen*, 72:562–577, 1912.

[146] C. Cercignani, R. Illner, and M. Pulvirenti. *The Mathematical Theory of Dilute Gases*. Springer-Verlag, 1994.

[147] L. Saint-Raymond. *Hydrodynamic Limits of the Boltzmann Equation*. Springer, 2009.

[148] C.M. Bender and S.A. Orszag. *Advanced Mathematical Methods for Scientists and Engineers*. McGraw-Hill, 1978.

[149] M. van Dyke. *Perturbation Methods in Fluid Mechanics*. Academic Press, 1964.

[150] H. Grad. Theory of rarefied gases. In F.M. Devienne, editor, *Rarefied Gas Dynamics*, pages 100–138. Pergamon Press, 1960.

[151] H. Grad. Asymptotic theory of the Boltzmann equation. *Physics of Fluids*, 6:147–181, 1963.

[152] D. Burnett. The distribution of molecular velocities and the mean motion in a non-uniform gas. *Proceedings of the London Mathematical Society*, 40:382–435, 1936.

[153] A. Santos, J.J. Brey, and J.W. Dufty. Divergence of the Chapman-Enskog expansion. *Phys. Rev. Lett.*, 56:1571–1574, 1986.

[154] L.S. Garcia-Colin, R.M. Velasco, and F.J. Uribe. Beyond the Navier-Stokes equations: Burnett hydrodynamics. *Physics Reports*, 465:149–189, 2008.

[155] C.S. Wang Chang and G.E. Uhlenbeck. The kinetic theory of gases. In J. de Boer and G.E. Uhlenbeck, editors, *Studies in Statistical Mechanics, Volume V*, pages 1–100. North-Holland Publishing Company, 1970.

[156] Y.P. Pao. Boltmann collision operator with inverse-power intermolecular potentials, I. *Communications on Pure and Applied Mathematics*, 27:407–428, 1974.

[157] E.T. Whittaker and G.N. Watson. *A Course of Modern Analysis*. Cambridge University Press, 1927.

[158] J.O. Hirschfelder, C.F. Curtiss, and R. B. Bird. *Molecular Theory of Gases and Liquids*. John Wiley, 1954.

[159] N.W. Ashcroft and N.D. Mermin. *Solid State Physics*. Saunders College Publishing, 1976.

[160] L.D. Landau. Die kinetische Gleichung für den Fall Coulombscher Wechselwirkung. *Phys. Z. Sowjetunion*, 10:154–164, 1936.

[161] E.M. Lifshitz and L.P. Pitaevskii. *Physical Kinetics*. Pergamon Press, 1981.

[162] Y.L. Klimontovich. *The Statistical Theory of Non-Equilibrium Processes in a Plasma*. MIT Press, 1967.

[163] R. Balescu. *Statistical Mechanics of Charged Particles*. John Wiley, 1963.

[164] A. Bobylev, I. Gamba, and I. Potapenko. On some properties of the Landau kinetic equation. *Journal of Statistical Physics*, 161:1327–1338, 2015.

[165] P.A.M. Dirac. *The Principles of Quantum Mechanics*. Oxford University Press, 1958.

[166] M.N. Rosenbluth, W.M. MacDonald, and D.L. Judd. Fokker-Planck equation for an inverse-square force. *Physical Review*, 107:1–6, 1957.

[167] R.C. Davidson. *Theory of Nonneutral Plasmas*. Addison-Wesley, 1990.

[168] H.L. Friedman. *Ionic Solution Theory*. Interscience Publishers, 1962.

[169] A.A. Vlasov. On vibration properties of electron gas. *Journal of Experimental and Theoretical Physics*, 8:291–318, 1938.

[170] A.A. Vlasov. *Many Particle Theory and Its Application to Plasma*. Gordon and Breach, 1961.

[171] L. Tonks and I. Langmuir. Oscillations in ionized gases. *Physical Review*, 33:195–210, 1929.

[172] J.H. Jeans. On the theory of star-streaming and the structure of the universe. *Monthly Notices of the Royal Astronomical Society*, 76:70–84, 1915.

[173] L. Landau. On the vibrations of the electronic plasma. *JETP*, 16:574–86, 1946.

[174] J.H. Malmberg and C.B. Wharton. Collisionless damping of electrostatic plasma waves. *Physical Review Letters*, 13:184–186, 1964.

[175] D. Lynden-Bell. The stability and vibrations of a gas of stars. *Monthly Notices of the Royal Astronomical Society*, 124:279–296, 1962.

[176] J. Binney and S. Tremaine. *Galactic Dynamics*. Princeton University Press, 1987.

[177] K.M. Case. Plasma oscillations. *Annals of Physics*, 7:349–364, 1959.

[178] T.Y. Wu. *Kinetic Equations of Gases and Plasmas*. Addison-Wesley, 1966.

[179] G. Schmidt. *Physics of High Temperature Plasmas*. Academic Press, 1966.

[180] F.F. Chen. *Introduction to Plasma Physics and Controlled Fusion*. Plenum Press, 1984.

[181] D.G. Swanson. *Plasma Kinetic Theory.* CRC Press, 2008.

[182] H.E. Stanley. *Introduction to Phase Transitions and Critical Phenomena.* Oxford University Press, 1971.

[183] L. van Hove. Correlations in space and time and Born approximation scattering in systems of interacting particles. *Physical Review*, 95:249–262, 1954.

[184] P.C. Hohenberg and B.I. Halperin. Theory of dynamic critical phenomena. *Reviews of Modern Physics*, 49:435–479, 1977.

[185] R. Kubo. Statistical-mechanical theory of irreversible processes. I. *Journal of Physical Society of Japan*, 12:570–586, 1957.

[186] A.L. Fetter and J.D. Walecka. *Quantum Theory of Many-Particle Systems.* McGraw-Hill, 1971.

[187] L.P. Kadanoff and G. Baym. *Quantum Statistical Mechanics.* Benjamin-Cummings Publishing, 1962.

[188] A.A. Abrikosov, L.P. Gorkov, and I.E. Dzyaloshinski. *Methods of Quantum Field Theory in Statistical Physics.* Prentice Hall, 1963.

[189] D.C. Langreth. Linear and nonlinear response theory with applications. In J.T. Devreese and V.E. van Doren, editors, *Linear and Nonlinear Electron Transport in Solids*, pages 3–32. Plenum Press, 1976.

[190] J. Rammer. *Quantum Field Theory of Non-Equilibrium States.* Cambridge University Press, 2007.

[191] R.W. Boyd. *Nonlinear Optics.* Academic Press, 2020.

[192] H.B. Callen and T.A. Welton. Irreversibility and generalized noise. *Phys. Rev.*, 83:34–40, 1951.

[193] R. Kubo. The fluctuation-dissipation theorem. *Reports on Progress in Physics*, 29:255–284, 1966.

[194] G.V. Chester. The theory of irreversible processes. *Reports on Progress in Physics*, 26: 411–472, 1963.

[195] R. Zwanzig. Time correlation functions and transport coefficients in statistical mechanics. *Annual Review of Physical Chemistry*, 16:67–102, 1965.

[196] K.M. Case. On fluctuation-dissipation theorems. *Transport Theory and Statistical Physics*, 2:129–176, 1972.

[197] M.S. Green. Brownian motion in a gas of noninteracting molecules. *The Journal of Chemical Physics*, 19:1036–1046, 1951.

[198] J.M. Luttinger. Theory of thermal transport coefficients. *Physical Review*, 135: A1505–A1514, 1964.

[199] H. Mori. Statistical-mechanical theory of transport in fluids. *Physical Review*, 112: 1829–1842, 1958.

[200] M.S. Green. Markoff random processes and the statistical mechanics of time-dependent phenomena. II. Irreversible processes in fluids. *The Journal of Chemical Physics*, 22: 398–413, 1954.

[201] M.S. Green. Comment on a paper of Mori on time-correlation expressions for transport properties. *Physical Review*, 119:829–830, 1960.

[202] P. Resibois. On the connection between the kinetic approach and the correlation-function method for thermal transport coefficients. *The Journal of Chemical Physics*, 41:2979–2992, 1964.

[203] H. Mori. Transport, collective motion, and Brownian motion. *Progress of Theoretical Physics*, 33:423–454, 1965.

[204] E. Hille and R.S. Phillips. *Functional Analysis and Semi-groups*. American Mathematical Society, 1957.

[205] D. Forster. *Hydrodynamic Fluctuations, Broken Symmetry, and Correlation Functions*. W.A. Benjamin, 1975.

[206] R. Zwanzig. *Nonequilibrium Statistical Mechanics*. Oxford University Press, 2001.

[207] J.P. Boon and S. Yip. *Molecular Hydrodynamics*. McGraw-Hill, 1980.

[208] M.H. Stone. On one-parameter unitary groups in Hilbert space. *Annals of Mathematics*, 33:643–648, 1932.

[209] F. Riesz and B. Sz.-Nagy. *Functional Analysis*. Ungar, 1955.

[210] I. Bengtsson and K. Zyczkowski. *Geometry of Quantum States: An Introduction to Quantum Entanglement*. Cambridge University Press, 2020.

[211] J. von Neumann. *Mathematical Foundations of Quantum Mechanics*. Princeton University Press, 1955.

[212] M. Ohya and D. Petz. *Quantum Entropy and Its Use*. Springer, 2004.

[213] M.A. Nielsen and I.L. Chuang. *Quantum Computation and Quantum Information*. Cambridge University Press, 2000.

[214] H. Weyl. *The Theory of Groups and Quantum Mechanics*. Dover, 1950.

[215] N.H. McCoy. On the function in quantum mechanics which corrresponds to a given function in classical mechanics. *Proc. Natl. Acad. Sci. USA*, 18:674–676, 1932.

[216] T.L. Curtright, D.B Fairlie, and C.K. Zachos. *A Concise Treatise on Quantum Mechanics in Phase Space*. World Scientific Publishing, 2014.

[217] M. Veltman. *Diagrammatica*. Cambridge University Press, 1994.

[218] W. Feller. *An Introduction to Probability Theory and Its Applications*, volume I. John Wiley, 1957.

[219] L.D. Landau and E.M. Lifshitz. *Mechanics*. Butterworth-Heinemann, 1976.

[220] H. Goldstein. *Classical Mechanics*. Addison-Wesley, 1950.

[221] F. Smithies. *Integral Equations*. Cambridge University Press, 1958.

[222] M.H. Stone. *Linear Transformations in Hilbert Space.* American Mathematical Society, 1932.

[223] T. Kato. *Perturbation Theory for Linear Operators.* Springer-Verlag, 1966.

[224] C. Cohen-Tannoudji, B. Diu, and F. Laloe. *Quantum Mechanics*, volume I. John Wiley, 1977.

[225] A. Messiah. *Quantum Mechanics.* North-Holland Publishing Company, 1965.

[226] M. Bunge. *Causality.* Harvard University Press, 1959.

[227] L.D. Landau and E.M. Lifshitz. *Electrodynamics of Continuous Media.* Pergamon Press, 1960.

[228] M. Born and E. Wolf. *Principles of Optics.* Cambridge University Press, sixth edition, 1980.

[229] I.M. Gel'Fand and G.E. Shilov. *Generalized Functions. Vol. 1 Properties and Operations.* Academic Press, 1964.

[230] J.S. Toll. Causality and the dispersion relation: Logical foundations. *Physical Review*, 104:1760–1770, 1956.

Index